Nanophase and Nanocomposite Materials II

MATERIALS RESEARCH SOCIETY
SYMPOSIUM PROCEEDINGS VOLUME 457

Nanophase and Nanocomposite Materials II

Symposium held December 2–5, 1996, Boston, Massachusetts, U.S.A.

EDITORS:

Sridhar Komarneni
Pennsylvania State University
University Park, Pennsylvania, U.S.A.

John C. Parker
Nanophase Technologies Corporation
Burr Ridge, Illinois, U.S.A.

Heinrich J. Wollenberger
Hahn-Meitner Institute
Berlin, Germany

MRS

MATERIALS
RESEARCH
SOCIETY

PITTSBURGH, PENNSYLVANIA

CAMBRIDGE
UNIVERSITY PRESS

32 Avenue of the Americas, New York NY 10013-2473, USA

Cambridge University Press is part of the University of Cambridge.

It furthers the University's mission by disseminating knowledge in the pursuit of
education, learning and research at the highest international levels of excellence.

www.cambridge.org
Information on this title: www.cambridge.org/9781558993617

CODEN: MRSPDH

Copyright 1996 by Materials Research Society.

A catalogue record for this publication is available from the British Library

Library of Congress Cataloguing in Publication data

Nanophase and Nanocomposite Materials II : symposium held December 2–5,
    1996, Boston, Massachusetts, U.S.A. / editors, Sridhar Komarneni,
    John C. Parker, Heinrich J. Wollenberger
        p. cm—(Materials Research Society symposium proceedings ; v. 457)
        Includes bibliographical references and index.
        ISBN 1-55899-361-4
        1.   Nanostructure materials—Congresses.   2.   Composite materials—
    Congresses.   3.   Nanotechnology—Congresses.   I.   Komarneni, Sridhar
    II.   Parker, John C.   III.   Wollenberger, Heinrich J.   IV.   Series: Materials
    Research Society symposium proceedings ; v. 457.
TA418.9.N35N335   1997   97-6975
620.1′1—dc21   CIF

ISBN 978-1-558-99361-7 Hardback

CONTENTS

PART I: NANOPHASE OXIDES

*Invited Paper

*Invited Paper

*Invited Paper

*Invited Paper

ix

*Invited Paper

PART VII: NANOCOMPOSITES OF LAYERED AND MESOPOROUS MATERIALS

PART VIII: <u>LATE PAPER ACCEPTED</u>

PREFACE

This volume contains the proceedings of the symposium "Nanophase and Nanocomposite Materials II," held in Boston, MA, on December 2-5, 1996, as part of the 1996 MRS Fall Meeting. The field of nanophase and nano-composite materials has grown tremendously since the first symposium was held on this topic at the 1992 MRS Fall Meeting. The purpose of this second symposium was to provide a platform for the presentation and discussion of advances from throughout the world in this rapidly growing field. This symposium attracted scientists from various scientific and engineering disciplines and covered a multitude of topics such as synthesis, processing, properties and applications of numerous types of nanophase and nanocomposite materials. A total of 147 papers from 19 countries were presented from a large number of academic institutions, government research laboratories and private industries. The 81 papers published in this proceedings have been refereed and presented as eight chapters.

Sridhar Komarneni
John C. Parker
Heinrich J. Wollenberger

January, 1997

MATERIALS RESEARCH SOCIETY SYMPOSIUM PROCEEDINGS

MATERIALS RESEARCH SOCIETY SYMPOSIUM PROCEEDINGS

Prior Materials Research Society Symposium Proceedings available by contacting Materials Research Society

Part I

Nanophase Oxides

ENHANCED THERMAL CONDUCTIVITY THROUGH THE DEVELOPMENT OF NANOFLUIDS

J.A. EASTMAN*, U.S. CHOI**, S. LI*, L.J. THOMPSON*, AND S. LEE**
*Materials Science Division, **Energy Technology Division, Argonne National Laboratory, 9700 S. Cass Ave., Bldg. 212, Argonne, IL 60439
(JEastman@ANL.GOV)

ABSTRACT

Low thermal conductivity is a primary limitation in the development of energy-efficient heat transfer fluids required in many industrial applications. To overcome this limitation, a new class of heat transfer fluids is being developed by suspending nanocrystalline particles in liquids such as water or oil. The resulting "nanofluids" possess extremely high thermal conductivities compared to the liquids without dispersed nanocrystalline particles. For example, 5 volume % of nanocrystalline copper oxide particles suspended in water results in an improvement in thermal conductivity of almost 60% compared to water without nanoparticles. Excellent suspension properties are also observed, with no significant settling of nanocrystalline oxide particles occurring in stationary fluids over time periods longer than several days. Direct evaporation of Cu nano-particles into pump oil results in similar improvements in thermal conductivity compared to oxide-in-water systems, but importantly, requires far smaller concentrations of dispersed nanocrystalline powder.

INTRODUCTION

Despite considerable previous research and development focusing on industrial heat transfer requirements, major improvements in cooling capabilities have been held back because of a fundamental limit in the heat transfer properties of conventional fluids. It is well known that metals in solid form have orders-of-magnitude larger thermal conductivities than those of fluids. For example, the thermal conductivity of copper at room temperature is about 700 times greater than that of water and about 3000 times greater than that of engine oil, as shown in Table 1. The thermal conductivities of metallic liquids are much larger than those of nonmetallic liquids. Therefore, fluids containing suspended solid metallic particles are expected to display significantly enhanced thermal conductivities relative to conventional heat transfer fluids.

Numerous theoretical and experimental studies of the effective thermal conductivity of dispersions containing particles have been conducted since Maxwell's theoretical work was published more than 100 years ago [1]. However, all previous studies of the thermal conductivity of suspensions have been confined to those containing mm- or micron-sized particles. Maxwell's model shows that the effective thermal conductivity of suspensions containing spherical particles increases with the volume fraction of the solid particles. It is also known that the thermal conductivity of suspensions increases with the ratio of the surface area to volume of the particle. Using Hamilton and Crosser's model [2], calculations [3] have been performed by one of us (USC) that predict, for constant particle size, the thermal conductivity of

3

Table 1.   Thermal conductivities of various solids and liquids.

	Material	Thermal Conductivity (W/m-K)
Metallic Solids:	Silver	429
	Copper	401
	Aluminum	237
Nonmetallic Solids:	Silicon	148
	Alumina (Al_2O_3)	40
Metallic Liquids:	Sodium @644K	72.3
Nonmetallic Liquids:	Water	0.613
	Ethylene Glycol	0.253
	Engine Oil	0.145

a suspension containing large particles is more than doubled by decreasing the sphericity of the particles from a value of 1.0 to 0.3 (the sphericity is defined as the ratio of the surface area of a particle with a perfectly spherical shape to that of a non-spherical particle with the same volume). Since the surface area to volume ratio is 1000 times larger for particles with a 10 nm diameter than for particles with a 10 mm diameter, a much more dramatic improvement in effective thermal conductivity is expected as a result of decreasing the particle size in a solution than can obtained by altering the particle shapes of large particles.

Nanofluids are expected to have superior properties compared to conventional heat transfer fluids, as well as fluids containing micron-sized metallic particles. The much larger relative surface areas of nanophase powders, compared to those of conventional powders, should not only markedly improve heat transfer capabilities, but also should increase the stability of the suspensions. Conventional micron-sized particles cannot be used in practical heat-transfer equipment because of severe clogging problems. However, nanophase metals are believed ideally suited for applications in which fluids flow through small passages because the metallic nanoparticles are small enough to behave similarly to liquid molecules. Therefore, nanometer-sized particles will not clog flow passages, but will improve the thermal conductivity of the fluids. This will open up the possibility of using nanoparticles even in microchannels for many envisioned high-heat-load applications. This report describes the synthesis, suspension properties, and heat transfer behavior of fluids containing nanocrystalline alumina, copper oxide, and copper.

EXPERIMENTAL PROCEDURE

Two procedures were used in synthesizing nanofluids for this investigation. In the first, nanocrystalline powders were prepared by the gas condensation (IGC) process [4] and then were subsequently dispersed in deionized water. Nanocrystalline Cu and Al_2O_3 powders were produced at Argonne, while nanocrystalline CuO and additional Al_2O_3 powder were purchased [5]. No special procedures were required to form stable suspensions of commercial oxide powders in water. Difficulties in dispersing Cu and Al_2O_3 powders synthesized at Argonne will be described in the next section. Transmission electron microscopy was used to characterize particle sizes and agglomeration behavior.

To successfully produce a nanofluid with dispersed nanocrystalline Cu particles, a second preparation method was used based on the vacuum evaporation onto a running oil substrate (VEROS) technique of Yatsuya *et al.* [6]. With this technique, nanocrystalline particles are produced by direct evaporation into a low vapor pressure liquid. The system built at Argonne (Fig. 1) was based on a modification of the VEROS technique and is similar to an earlier design by Günther and co-workers at the Fraunhofer Institute for Applied Materials Research in Bremen [7]. With the system shown schematically in Fig. 1, nanocrystalline Cu was evaporated resistively into two types of pump oil [8]. These oils were chosen because they are designed for vacuum applications and thus have extremely low vapor pressures. Low vapor pressures are required with the VEROS technique to prevent vaporization of the liquid during the evaporation process.

Thermal conductivities were measured using a transient hot-wire apparatus shown schematically in Fig. 2. This system measures the electrical resistivity of the fluid systems and thermal conductivities are then calculated from the known relationship between electrical and thermal conductivities.

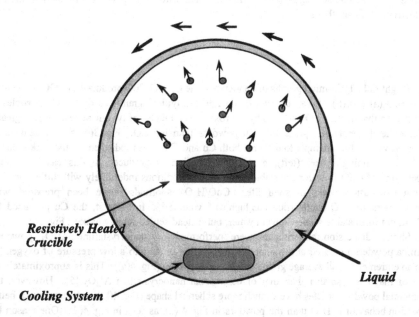

Resistively Heated Crucible

Cooling System

Liquid

Figure 1. Schematic diagram of a VEROS-type evaporation system [6,7] designed for direct evaporation of nanocrystalline particles into low vapor pressure liquids. The liquid is located in a cylinder that is rotated to continually transport a thin layer of liquid above a resistively-heated evaporation source. The liquid is cooled to prevent an undesirable increase in vapor pressure due to radiant heating during the evaporation.

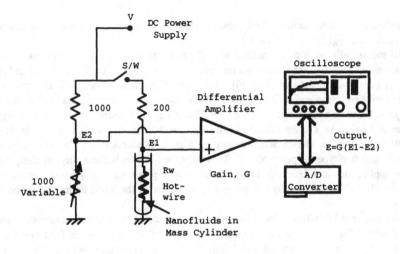

Figure 2.   Schematic diagram of a transient hot-wire apparatus for measuring thermal conductivities of nanofluids.

RESULTS

Bright field TEM micrographs of nanocrystalline Cu and CuO produced by IGC are shown in Figure 3 (a) and (b). It can be seen that while the typical grain size of the Cu particles is smaller than that of the CuO particles by approximately a factor of two, it appears that a greater degree of agglomeration is present in Cu powders than in CuO. Fig. 3(c) shows beakers of deionized water after attempts to disperse both Cu and CuO. As indicated by the dark color of the CuO-containing beaker (left), a stable suspension is produced in this case.   In fact, suspensions of CuO in water are stable under static conditions indefinitely with little apparent settling even after weeks of shelf life.   CuO/H_2O nanofluids have been prepared with concentrations of CuO nanoparticles as high as 5 volume %. In contrast, the Cu produced by IGC did not form stable suspensions in water, but instead settled rapidly as seen Fig. 3(c).

Similar dispersion experiments were performed on nanocrystalline Al_2O_3 powders. Alumina powders prepared at Argonne by evaporation of Al into a low pressure of oxygen [9] have an extremely small average grain size of less than 3 nm (Fig. 4(a)). This is approximately an order of magnitude smaller than that of commercial nanocrystalline Al_2O_3 [5]. However, the commercial powder particles have a much more spherical shape (Fig. 4(b)) and show much better dispersion behavior in H_2O than the powders in Fig. 4 (a), as seen in Fig. 4(d). One reason for the poor dispersion behavior of the smaller Al_2O_3 particles is that they appear to actually dissolve in H_2O and re-precipitate as much larger particles, as seen in Fig. 4(c).

The thermal conductivity as a function of nano-particle volume fraction is shown in Fig. 5 for nanofluids of CuO or Al_2O_3 dispersed in deionized water. As expected based on earlier calculations [3], the conductivity increases linearly with increasing particle concentration. Very significant enhancements in thermal conductivity are seen, in particular in the case of CuO where

(a)   (b)

CuO   Cu

(c)

Figure 3.   Bright-field micrographs showing the typical agglomeration size of nanocrystalline (a) Cu produced by IGC and (b) CuO [5].   While the Cu in (a) exhibits a smaller grain size of approximately 18 nm compared to the typical grain size of 36 nm for CuO in (b), the CuO agglomerate sizes are smaller than those of Cu and thus CuO forms much more stable dispersions in deionized water, as seen in (c).

increases of approximately 60% are seen for 5 volume % suspensions. Increases in conductivity of Al_2O_3 are also seen, although the effect is less significant than for CuO, which has a higher intrinsic conductivity than Al_2O_3. These results clearly demonstrate the commercial potential for significantly enhancing industrial heating and cooling capabilities through the use of nanofluids.

While a 60% improvement in thermal conductivity is very significant, comparison of the thermal conductivities of oxides and metals (Table 1) suggests that even larger effects can be obtained if metal nanoparticles such as Cu can be successfully suspended in liquids. Attempts t , suspend Cu in water by first suspending CuO and then reducing the CuO to Cu by heating the

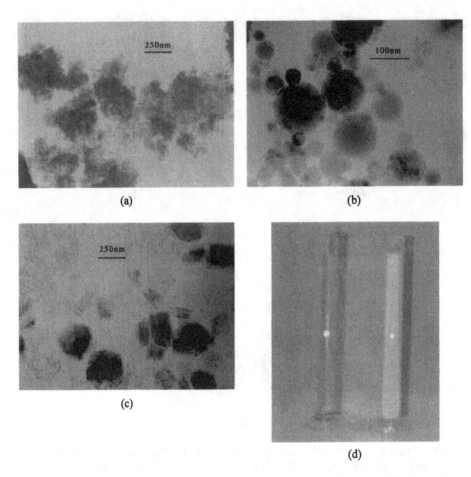

Figure 4.   Bright field TEM micrographs of nanocrystalline Al_2O_3 (a) produced by IGC at Argonne (average grain size < 3 nm) and (b) purchased from Nanophase Technologies [5]. The average grain size in (b) is 33 nm.   Both (a) and (b) show powders prior to dispersion in deionized water, while (c) shows the powder in (a) after dispersion.   Large grains are seen after dispersion. Beakers of Al_2O_3 dispersed in H_2O are seen in (d). The beaker on the right contains a stable solution of powders from (b), while the powders in the beaker on the left (corresponding to (a) and (c)) settle rapidly and are thus unsuitable for heat transfer applications.

fluid while bubbling hydrogen gas through it resulted in the precipitation of large Cu particles that displayed poor suspension behavior. Therefore, an alternative method of nanofluid preparation was undertaken based on the VEROS technique [6,7]. Using a system shown schematically in Fig. 1, Cu was evaporated into two types of low vapor pressure pump oil [8].   A TEM

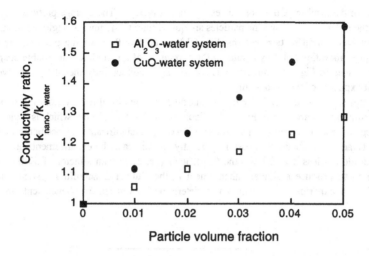

Figure 5.   Dramatic improvements in the thermal conductivity of deionized water are observed with increasing volume fraction of dispersed copper oxide or alumina nanoparticles.

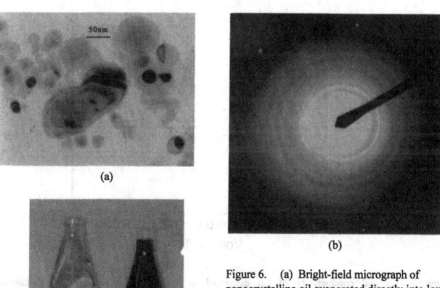

(a)

(b)

(c)

Figure 6.   (a) Bright-field micrograph of nanocrystalline oil evaporated directly into low vapor pressure pump oil The electron diffraction pattern in (b) shows rings corresponding to the fcc Cu phase. (c) Oil prior to (left) and after evaporation of Cu (right).

micrograph of the resulting Cu nanoparticles is seen in Fig. 6(a). The average particle size in this case is on the order of 35 nm and the particles are quite round with minimal agglomeration. Most importantly, electron diffraction patterns (Fig. 6(b)) reveal that the resulting particles are fcc Cu. Thus, direct evaporation not only results in the formation of a nanofluid with stable suspension properties as seen in Fig. 6(c), but also retains the higher conductivity metallic phase even after subsequent exposure of the fluid to air.

The thermal conductivity behavior of nanofluids formed by direct evaporation of Cu into two types of pump oil are seen in Fig. 7. Particularly for the HE-200 oil, which is a high purity oil designed for use in pumps such as Roots blowers [8], significant enhancements in thermal conductivity are again obtained. Most importantly, in this case the enhancements are obtained with the addition of less than 0.1 volume % particles (i.e. more than a factor of 50 less material than required to produce a similar enhancement in the CuO/H$_2$O nanofluid system shown in Fig. 5). Further experiments are required to determine if even larger enhancements in thermal

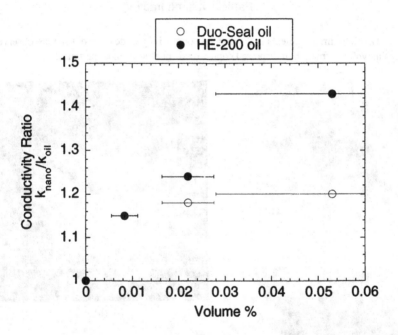

Figure 7. Thermal conductivity of Cu dispersed in oil. Note that similar substantial enhancements in thermal conductivity are seen compared to the oxide-in-water systems shown in Figure 5. However, in the case of Cu-in-oil, this improved behavior is obtained with almost two orders of magnitude less dispersed nanocrystalline powder.

conductivity can be obtained with larger concentrations of Cu nanoparticles. The large enhancement in thermal conductivity with a very small concentration of Cu particles seen in Fig. 7 is also significant because the concentration of particles is small enough that negligible changes in fluid viscosity will accompany the improved thermal behavior.

CONCLUSIONS

These preliminary results demonstrate the feasibility of significantly improving the heat transfer performance of commercial heating and cooling fluids such as water or oil by suspending nanocrystalline particles in the liquid to produce nanofluids. In the case of oxide nanoparticles suspended in water, increases in thermal conductivity of approximately 60% can be obtained with 5 volume % particles. The use of Cu nanoparticles results in even larger improvements in thermal conductivity behavior, with very small concentrations of particles producing major increases in the thermal conductivity of oil.

Further work remains to demonstrate the full potential of nanofluids. In addition to determining if additional improvements in conductivity can be obtained through the suspension of larger volume fractions of Cu particles in oil, numerous other experiments are required to determine other important properties of nanofluids. These include determining the effects of nanoparticle suspensions on the flow, corrosive, and abrasive properties of fluids.

ACKNOWLEDGMENTS

This work was supported by the U.S. Department of Energy, BES-Materials Science, under Contract W-31-109-Eng-38 and by a grant from Argonne's Coordinating Council for Science and Technology. We thank Carl Youngdahl for the micrograph of nanocrystalline Cu shown in Fig. 3(a).

REFERENCES

1. J.C. Maxwell, A Treatise on Electricity and Magnetism, 2nd Ed., 1, 435, Clarendon Press (1881).
2. R.L. Hamilton and O.K. Crosser, I and EC Fundamentals, 1, no. 3, 187 (1962).
3. U.S. Choi, in Developments and Applications of Non-Newtonian Flows, eds. D.A. Siginer and H.P. Wang, (American Society of Mechanical Engineers: New York), Vol. 231/MD-Vol. 66, 99 (1995).
4. C. G. Granqvist and R.A. Buhrman, J. Appl. Phys., 47, 2200 (1976).
5. Nanophase Technologies Corporation , Burr Ridge, IL
6. S. Yatsuya, Y. Tsukasaki, K. Mihama, and R. Uyeda, J. Cryst. Growth, 43, 490 (1978).
7. M. Wagener and B. Günther, these proceedings.
8. HE-200, produced by Leybold-Heraeus Vacuum Products Inc., Export, PA
   Duo-Seal #1407K-11, produced by Welch Vacuum Technology Inc., Skokie, IL.
9. J.A. Eastman, L.J. Thompson, and D.J. Marshall, Nanostruct. Mater. 2, 377 (1993).

Synthesis of Polyvanadates from Solutions

J. LIVAGE*, L. BOUHEDJA*, C. BONHOMME*, M. HENRY**

* Chimie de la Matière Condensée, Université P.M. Curie, Paris, France

** Chimie Moléculaire des Solides, Université Louis Pasteur, Strasbourg, France

ABSTRACT

A wide range of polyvanadates can be synthesized from aqueous solutions. Vanadium oxide gels V_2O_5,nH_2O are formed around the point of zero charge ($pH \approx 2$). They exhibit a ribbon-like structure. Weak interactions between these ribbons lead to the formation of mesophases in which vanadium oxide gels or sols behave as nematic liquid crystals. Organic species can be easily intercalated between these oxide ribbons leading to the formation of hybrid nanocomposites made of alternative layers of organic and inorganic components. Hybrid nanophases can also be formed above the point of zero charge, in the presence of large organic ions such as $[N(CH_3)_4]^+$. They often exhibit layered structures in which organic cations lie between the polyvanadate planes. Cluster shell polyvanadates have been obtained in the presence of anions such as Cl^- or I^-. They are made of negatively charged polyvanadate hollow spheres in which the anion is encapsulated. Organic cations then behave as counter ions for the formation of the crystal network.

INTRODUCTION

Nanocomposite materials are made of several components mixed together at a nanoscale. Several problems have to be solved for the synthesis of nanocomposites. The size of each nanophase and the nature of the interactions between them have to be controlled in order to avoid phase separation. This paper addresses the first point and discusses the parameters allowing a chemical control of the size and shape of polyoxovanadate nanophases synthesized from aqueous solutions.

Vanadium in its higher oxidation states (V^V, V^{IV}) gives a large number of isopolyvanadates that exhibit a wide range of structures, ranging from chain metavanadates $[VO_3^-]_n$, to layered oxides $[V_2O_5]$, compact polyanions $[V_{10}O_{28}]^{6-}$ and polyanionic hollow cages such as $[V_{15}O_{36}]^{5-}$ [1][2]. This is due to the ability of vanadium to adopt a variety of coordination geometries and various oxidation states. Many organic-based polyvanadates can be synthesized at low temperature ($<200°C$) from aqueous solutions via sol-gel chemistry or hydrothermal syntheses. Layered vanadium pentoxide gels V_2O_5,nH_2O are formed around the point of zero charge ($pH \approx 2$) via the condensation of vanadic acid. These gels are capable of intercalating a variety of organic species such as alkylamines, organic solvents, TTF, cobalticinium...[3]. The polymerization of intercalated guest molecules (polyaniline, polythiophene, polyethylene oxide...) leads to the formation of vanadium oxide nanocomposites [4][5]. Bidentate amines have been used as templates for hydrothermal syntheses leading to hybrid organic-inorganic layered nanocomposites [6-9]. Hollow polyoxovanadate spheres encapsulating anionic or molecular species have also been reported [10].

One of the main advantages of wet chemistry methods for the synthesis of nanophase materials is that weak interactions (hydrogen bonds, van der Waals, hydrophilic-hydrophobic interactions...) are not broken at low temperature. They can play an important role during the self assembling of molecular precursors. Subtle changes in the chemical conditions lead to completely different supramolecular associations and therefore to different nanostructures. The chemical control over the formation of these nanophases requires a good knowledge of the reactions that occur in the solution in order to be able to tailor the shape, size and structure of the different components at a nanoscale. This paper describes some chemical parameters that are responsible for the formation of layered or spherical polyvanadate species from aqueous solutions. The role of cations and anions during the synthesis of these polyvanadates is discussed.

EXPERIMENTS AND RESULTS

Layered nanophases formed via the condensation of vanadic acid

Vanadium pentoxide gels V_2O_5,nH_2O can be easily prepared via the acidification of aqueous solutions of sodium metavanadate $NaVO_3$ through a proton exchange resin [3]. A yellow solution containing a mixture of decavanadic acid $[H_nV_{10}O_{28}]^{(6-n)-}$ and dioxovanadium cations $[VO_2]^+$ is first obtained as shown by ^{51}V NMR experiments [11]. This clear solution polymerizes slowly within few hours. It becomes progressively more viscous, its coloration turns to dark red and depending on vanadium concentration, gels or colloidal solutions can be obtained. The sol-gel transition occurs for a vanadium concentration $[V_2O_5] \approx 0.2 mol.l^{-1}$, corresponding to a composition close to $V_2O_5,250H_2O$. Xerogels are obtained upon drying in air at room temperature. They still contain almost two water molecules per V_2O_5. All water molecules cannot be removed at room temperature, even under vacuum and the hydration-dehydration process remains reversible. Anhydrous V_2O_5 can only be obtained upon heating above 200°C, but rehydration is no longer possible [3].

Transmission electron microscopy shows that vanadium pentoxide gels are formed of flat ribbons about 0.1μm long, 100Å large and 10Å thick. The internal structure of these ribbons is close to that observed in the ab planes of orthorhombic V_2O_5 (Fig.1). A stacking of the ribbons is observed when these gels are deposited onto a flat substrate. A series of *001* peaks is observed by X-ray diffraction. The 1-D Patterson map suggests that V_2O_5 layers are made of two V_2O_5 sheets facing each other at a distance of 2.8Å [12]. The basal distance between ribbons depends on the amount of water in V_2O_5,nH_2O. It first increases by steps of about 2.8Å corresponding to the intercalation of one water layer (d=8.7Å for n= 0.5, d=11.5Å for n=1.8). A continuous swelling is then observed beyond d≈20Å [13][14].

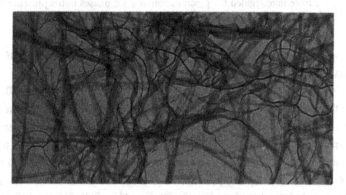

Figure 1. Transmission electron microscopy showing
the ribbon-like structure of V_2O_5,nH_2O gels

Vanadium oxide gels exhibit a large oxide/water interface and the reactions occuring at this interface become important. The adsorption-dissociation of water molecules lead to the formation of a fully hydroxylated surface. Due to the high positive charge of the V^{5+} ions, V-OH groups exhibit acid properties and acid dissociation occurs at the oxide-water interface as follows:

$$V\text{-}OH + H_2O \Rightarrow V\text{-}O^- + H_3O^+ \tag{1}$$

V_2O_5,nH_2O gels could also be described as polyvanadic acids $H_xV_2O_5,nH_2O$ (x≈0.3).

Oxide ribbons being negatively charged, they tend to repulse each other preventing the precipitation of a 3-D V_2O_5 network. Moreover, these electrostatic repulsions favor the formation of lyotropic mesophases. P. Davidson showed that when observed between crossed polarizers, vanadium oxide sols and gels display optical textures typical of liquid crystals. They actually behave like nematic solutions of semi-rigid polymers confirming that the length of the ribbons is at least one order of magnitude larger than their width [14][15]. Vanadium oxide gels offer a versatile host structure for the intercalation of a wide range of guest molecules and many layered nanocomposites have been reported, made of alternative layers of organic polymers and vanadium oxide [4][5].

Formation of polyvanadates in the presence of foreign ions.

Organic cations

Anionic precursors are formed above the point of zero charge (pH≈2) leading to negatively charged polyanions. Cations have then to be added to the solution in order to precipitate solid compounds. In order to understand their role during the formation of nanophases, we have investigated the precipitation of solid phases in the presence of large organic tetramethyl cations (TMA = N(CH$_3$)$_4$).

V$_2$O$_5$ was dissolved in an aqueous solution of TMAOH (c≈1M) within half an hour upon gentle heating at 60°C.

$$V_2O_5 + 6OH^- \Rightarrow 2[VO_4]^{3-} + 3 H_2O \qquad (2)$$

The pH of the solution prior to dissolution is close to 13. Vanadate ions [VO$_4$]$^{3-}$ are formed but according to the previous equation, the pH of the solution decreases during dissolution. Vanadate species are then protonated [H$_n$VO$_4$]$^{(3-n)-}$, allowing condensation to take place via oxolation reactions. The ^{51}V NMR spectrum of the solution after dissolution (pH≈7) exhibits two series of peaks corresponding to a mixture of metavanadates [V$_4$O$_{12}$]$^{4-}$ (δ = -583, -594 ppm) and decavanadates [V$_{10}$O$_{28}$]$^{6-}$ (δ = -423, -500, -515 ppm) (Fig.2). The intensity of metavanadate peaks decreases when the pH of the solution decreases and only decavanadates are observed below pH≈6. The nature of precipitated phases then mainly depends on the final pH of the solution.

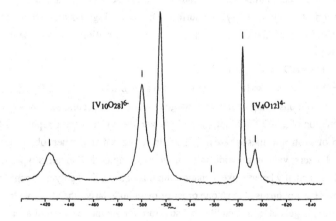

Figure 2. ^{51}V NMR of aqueous solutions formed via the dissolution of V$_2$O$_5$ in TMAOH

Tetra-alkylammonium metavanadates or decavanadates have been precipitated from such solutions at pH=8.5 and pH=7 respectively [16][17]. They contain discrete $[V_4O_{12}]^{4-}$ or $[V_{10}O_{28}]^{6-}$ anions and tetraalkylammonium cations $[NR_4]^+$ (R = Me, Et, Bu...). The polyvanadate network is formed via the electrostatic interactions between large inorganic anions and organic cations.

Layered structures are obtained via the hydrothermal treatment of a mixture of V_2O_5 and TMAOH (200°C, 48h). Black crystals of $N(CH_3)_4V_4O_{10}$ are formed. Their structure was described recently by M.S. Whittingham et al. [18]. It is close to that of orthorhombic V_2O_5. The $[V_4O_{10}]^-$ anionic layers are made up of double chains of edge sharing $[VO_5]$ tetragonal pyramids linked together by corners. Large alkylammonium cations are distributed between the oxide layers where they should interact weakly with negatively charged terminal oxygen (V=O). They behave as counter cations and the structure of the oxide network is mainly governed by the dipolar interactions between adjacent V=O bonds.

In the presence of Li^+ and NMe_4^+, a new $Li_xV_{2-\delta}O_{4-\delta}H_2O$ layered phase is formed [19]. It is made of planes of $[VO_5]$ units sharing edges and corners. Li^+ ions and water molecules are between the vanadium oxide sheets. Despite the fact that this phase cannot be obtained in the absence of NMe_4^+, there is no evidence of any incorporation of this ion. Only the smaller and more polarizing Li^+ interact with the negatively charged vanadium oxide layers.

Anionic templates

Unless the pH is very low, high-valent cations (Si^{IV}, V^V, W^{VI}...) lead to the formation of anionic or polyanionic precursors. Therefore mainly cationic species such as alkylammonium ions $[NR_4]^+$ have been used as templates for the hydrothermal synthesis of oxide materials [20]. It can be easily understood that electrostatic interactions favor the formation of the negatively charged oxide network around the cationic template. This is no more the case when anions are used as templates as was reported earlier by Müller et al. for oxopolyvanadates [10]. The synthesis described in the literature may be quite complicated but we have shown that cluster shell oxopolyvanadates can be formed simply via the hydrothermal treatment (200°C, 48h) of a mixture of V_2O_5 ($\approx5.10^{-3}$ mole), NMe_4X (X = Cl, I) ($\approx2.10^{-3}$ mole) and NMe_4OH (5.10^{-3} mole).

The solution contains both meta and decavanadate species, as shown by ^{51}V NMR. Hexagonal black crystals are formed with Cl^- (Fig.3a). They have the same structure as the $(NMe_4)_6[V_{15}O_{36}Cl]4H_2O$ crystals synthesized by A. Müller using thiovanadates $[VS_4]^{3-}$ as precursors [21]. The hollow anion $[V_{15}O_{36}]^{5-}$ is made of tetragonal $[VO_5]$ pyramids sharing edges and corners (Fig.4a). The vanadium atoms are placed at the surface of a sphere at a distance $Cl \cdots V \approx 3.43\text{Å}$ from the center of the cluster where the entrapped Cl^- anion resides. All short V=O bonds are oriented toward the outside of the sphere. Some vanadium reduction occurs during the reaction and the $[V_{15}O_{36}Cl]^{6-}$ anion is a mixed valence compound containing seven V^V and eight V^{IV}.

(a)   (b)

Figure 3. Scanning electron microscopy of cluster shell polyvanadate crystals
(a) $(NMe_4)_6[V_{15}O_{36}Cl]4H_2O$, (b) $(NMe_4)_{10}[H_3V_{18}O_{42}I]\ 3H_2O$
(size of the crystals $\approx 300\ \mu m$)

Octahedral black crystals are formed in the same conditions with I^- (Fig.3b). The crystal structure of this new phase has not yet been completely resolved, but first results show the presence of $[V_{18}O_{42}I]$ anionic spheres similar to those found for the cesium salt $Cs_9[H_4V_{18}O_{42}I]\ 12H_2O$ synthesized by A. Müller in the presence of HI and N_2H_4OH as a reducing agent [22][23]. According to thermal and chemical analyses, our compound should rather correspond to $(NMe_4)_{10}[H_3V_{18}O_{42}I]3H_2O$ and all vanadium are in the V^{IV} oxidation state. The $[V_{18}O_{42}]^{12-}$ hollow sphere is built from $[VO_5]$ square pyramids sharing edges. Vanadium atoms are at an average distance $I \cdots V \approx 3.75\text{Å}$ from the central encapsulated I^-. As previously all short V=O double bonds are opposite to the encapsulated anion (Fig.4b).

It has to be pointed out that cluster shell polyoxovanadates only form when V_2O_5, TMAX (X=Cl, I) and TMAOH are mixed together before heating. They do not precipitate when V_2O_5 is first dissolved in TMAOH at 60°C giving a solution of metavanadates $[V_4O_{12}]^{4-}$ and decavanadates $[V_{10}O_{28}]^{6-}$. Further hydrothermal heating, even in the presence of I^-, then leads to the precipitation of $NMe_4V_4O_{10}$. I^- or Cl^- remain dissolved in the solution and do not behave as templates.

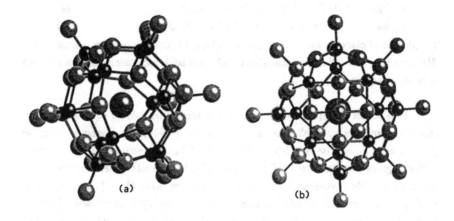

Figure 4. Molecular structure of cluster shell polyvanadates
(a) $[V_{15}O_{36}Cl]^{6-}$, (b) $[V_{18}O_{42}I]^{13-}$

DISCUSSION

Aqueous chemistry of vanadate precursors

A large variety of V^V species can be found in aqueous solutions. At room temperature they mainly depend on vanadium concentration and pH. When dissolved in water, V^{5+} ions are solvated by dipolar water molecules giving $[V(OH_2)_6]^{5+}$ species. However, due to the strong polarizing power of V^{5+} and the Lewis acid properties of H_2O, deprotonation of the coordinated water molecules occurs spontaneously as follows [23]:

$$[V(OH_2)_6]^{5+} + hH_2O \Rightarrow [V(OH)_h(OH_2)_{6-h}]^{(5-h)+} + h\ H_3O^+ \qquad (3)$$

Dioxovanadium cations $[VO_2]^+$ are formed at very low pH (h=4, pH<2) and vanadate anions $[VO_4]^{3-}$ above pH=13. More condensed species cannot be formed under these conditions as these species do not contain any V-OH group. The coordination of V^V increases with decreasing pH, being tetrahedral in alkaline conditions, five-fold in neutral and slightly acid solutions and approaching six at very low pH. This coordination change can be followed easily by the naked eye as $[VO_5]$ gives a yellow-orange color while $[VO_4]$ is colorless.

Monomeric species can only be observed in very diluted solutions ($c < 10^{-4}$ mol.l^{-1}). Condensation occurs in the pH range where V-OH groups are formed. Condensation follows two main mechanisms:

oxolation $V-OH + HO-V \Rightarrow V-O-V + H_2O$ (4)

olation $V-OH + V-OH_2 \Rightarrow V-OH-V + H_2O$ (5)

The kinetics of olation mainly depends on the lability of coordinated water molecules. For d^0 cations such as V^V, these molecules are usually very labile and olation reactions are much faster than oxolation [26].

Oxo-anions $[VO_4]^{3-}$ in which vanadium is surrounded by four equivalent oxygen atoms are formed in highly alkaline aqueous solutions (pH$\approx$14). Protonation occurs as the pH decreases giving rise to $[H_nVO_4]^{3-n}$ species. The condensation of $[HVO_4]^{2-}$ via oxolation gives pyrovanadates $[V_2O_7]^{4-}$, whereas $[H_2VO_4]^-$ leads to metavanadates in the pH range 6-9. Metavanadates form cycles $[V_4O_{12}]^{4-}$ or chains $[VO_3^-]_n$ made of corner sharing $[VO_4]^{3-}$ tetrahedra. Vanadium coordination increases up to six at lower pH (pH<6) leading to decavanadate species such as $[H_2V_{10}O_{28}]^{4-}$ made of ten edge-sharing $[VO_6]$ octahedra. Decavanadic species are strong acids and further condensation does not occur. Such polyanions cannot be precursors for more condensed species. In this pH range, the formation of a solid network is obtained when positive cations are added to the solution of anionic polyvanadates, leading to the precipitation of meta or decavanadates [1].

At very low pH, decavanadic acid $[H_2V_{10}O_{28}]^{4-}$ is dissociated into dioxovanadium cations as follows:

$[H_2V_{10}O_{28}]^{4-} + 14H_3O^+ + 18H_2O \Rightarrow 10[VO_2(OH_2)_4]^+$ (6)

Precipitation is obtained by adding anions to the $[VO_2]^+$ cation in order to form a neutral solid network.

Formation of vanadium pentoxide gels.

In the absence of foreign ions, vanadium pentoxide V_2O_5 is formed around pH$\approx$2, where decavanadic acid is the predominant species. However it has been shown that, depending on pH, decavanadic acid can be dissociated into $[VO_2]^+$ or $[VO_3^-]_n$ species [11][25]. The ^{51}V NMR of an aqueous solution formed via the acidification of vanadates through a proton exchange resin shows the presence of $[VO_2]^+$ cations (h=4) and $[V_{10}O_{28}]^{6-}$ anions (h=5.5) [11]. None of them can be considered as a precursor for the formation of an oxide network. Vanadium oxide gels should then be formed from the intermediate (h=5) neutral precursor $[V(OH)_5(OH_2)]^0$, around the point of zero charge, giving rise to the precipitation of a neutral oxide network [26]. Such a precursor has never been observed by ^{51}V NMR, presumably because condensation is very fast and the lifetime of $[V(OH)_5(OH_2)]^0$ is too short. In this precursor, vanadium ions should be surrounded by five OH groups and one water molecule. However, because of the strong polarizing power of the V^{5+} ion, some internal proton transfer occurs and a V=O double bond is formed in which a π electron transfer from oxygen to vanadium decreases the positive charge of V^V. This leads to the $[VO(OH)_3(OH_2)_2]^0$ molecular precursor in which one water molecule lies along the z axis opposite to the short V=O double bond (Fig.5).

Condensation cannot proceed along this z direction as there is no V-OH group. It can only proceed in the xy plane and layered condensed phases are expected to be obtained. Moreover, one $V-OH_2$ and three V-OH bonds are formed in this plane so that x and y directions are not equivalent. Olation reactions, along the $HO-V-OH_2$ direction proceed faster than oxolation reactions along the HO-V-OH direction giving rise to a ribbon-like structure (Fig.5). X-ray diffraction experiments show that these ribbons are formed of double chains of edge and corner sharing $[VO_5]$ pyramids. Dipolar interactions between neighboring V=O favor the formation of layered nanostructures. The V_2O_5 layer is stabilized because the V=O groups of edge-sharing pyramids are situated on different sides of the plane. The layer 'n+1' forms on layer 'n' through weak O=V···O=V interactions. Water molecules can be easily intercalated between these layers through hydrogen bonds. The acid dissociation of V-OH groups at the oxide-water interface leads to the formation of a negatively charged polyvanadic acid network. Electrostatic and Van der Waals interactions between oxide ribbons favor the formation of mesophases in vanadium pentoxide gels and sols. Once the ordered layered structure is formed, intercalation of organic species can proceed via ion exchange or redox reactions leading to the formation of hybrid layered nanocomposites.

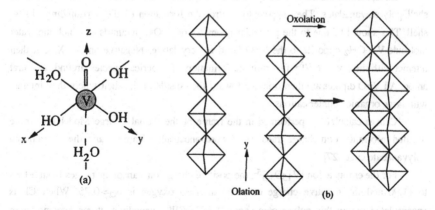

Figure 5. Formation of V_2O_5,nH_2O gels via the condensation of vanadic acid.
(a) neutral precursor, (b) olation and oxolation lead to edge and corner sharing pyramids

Template behavior of foreign ions

Cluster shell species are formed in the pH range were vanadate polyanions are observed in the solution. Therefore cations are required to form a neutral solid network. These cations interact with the negatively charged oxygen atoms. Protonation for instance occurs on the bridging oxygen, but larger cations should interact mainly with terminal V=O oxygen.

The pyramidal $[VO_5]$ unit with a short V=O double bond should plays a central role in the formation polyoxovanadates. As previously described, condensation cannot occur along the $H_2O \cdots V=O$ direction. It proceeds only in the 'xy' plane where V-OH groups are present. However the manner two adjacent $[VO_5]$ pyramids are linked together is directed by weak interactions of the foreign species X with the $V^{\delta+}=O^{\delta-}$ bond. Cationic species interact with $V=O^{\delta-}$ dipoles. The strength of this interaction depends on the polarizing power of the cation. For monovalent ions, it increases when the ionic radius decreases. In the case of the hydrothermal treatment of $V_2O_5 + NMe_4OH$, the tetramethyl ion $[N(CH_3)_4]^+$ is quite large ($r \approx 3.47 \text{Å}$) and these interactions are weak. Dipolar interactions between adjacent V=O bonds prevail and the linkage of $[VO_5]$ pyramids is close to that observed in V_2O_5 leading to the formation of $NMe_4V_4O_{10}$ made of negative $[V_4O_{10}^-]_n$ layers. However as soon as Li^+ ions are added to the solution, large $[NMe_4]^+$ are no more involved in the formation of the solid phase. They are replaced by more polarizing Li^+ cations and a $Li_xV_2O_4$ compound in which Li^+ ions are intercalated between oxide layers, is formed [19].

Encapsulated anions obviously behave as templates during the formation of cluster shell polyoxovanadates. They appear to control the formation of the surrounding cluster shell. This should be due to the particular geometry of $[VO_5]$ pyramids in which the water molecule $V \cdots OH_2$ opposite to the V=O bond is very labile. Negative anions X^- can then interact with the positive $V^{\delta+}$ favoring the formation of spherical species around a central anion. All V=O dipoles are then oriented toward the outside of the shell where they interact with large organic counter cations.

Charge calculations performed in the frame of the Partial Charge Model give more detailed information on charge transfers between encapsulated anions and the surrounding polyvanadate cage [27].

For the empty anion $[V_{15}O_{36}]^{5-}$, the positive charge on vanadium ranges from +1.83 to +1.90 and the negative charge on terminal V=O oxygen is $\delta_O \approx -0.75$. When Cl^- is encapsulated within this polyanionic cage, $[V_{15}O_{36}Cl]^{6-}$, vanadium atoms become more positive ($\delta_V \approx +1.92$ to +1.99) and terminal oxygens become more negative ($\delta_O \approx -0.82$). The negative charge of bridging μ_3-O is not modified ($\delta_O \approx -1.06$). The charge on Cl^- is close to unity $\delta_{Cl} \approx -1$, suggesting that their is no charge transfer between Cl^- and $V^{\delta+}$. Encapsulation of Cl^- leads to a polarization of the vanadate cage, favoring some charge transfer from vanadium to oxygen within the V=O double bond opposite to the $Cl\cdots V$ direction. The overall electrostatic energy E is negative. It increases from -155 eV for the empty anion to E = -498 eV when Cl lies inside the cage. Electrostatic attractions are then observed between $Cl^{\delta-}$ and $V^{\delta+}$ leading to a strong electrostatic stabilization of the cage.

A similar behavior is observed for $[V_{18}O_{42}I]^{13-}$. The charge of encapsulated I^- is close to unity ($d_I \approx -1$), but some some charge transfer occurs within V=O bonds. The average positive charge δ_V increases from +1.76 to +1.83 while the negative charge of terminal oxygen goes from -0.98 to -1.05. As before the charges of bridging oxygens do not vary ($\delta_O \approx -1.09$). The electrostatic stabilization energy increases from -74 eV without I^- to -455 eV when I^- lies in the center of the cage.

The encapsulation of anions within spherical polyvanadate cages results in a strong electrostatic stabilization of the cluster. Moreover, an electron transfer occurs toward terminal V=O oxygen. Ab initio SCF calculations performed on $[V_{18}O_{42}]^{12-}$ cages confirm that most negative potentials are situated outside of the sphere in a direction opposite to the encapsulated anion [29]. Electrostatic attractions between large anionic polyvanadates and large organic cations lead to the crystallization of hybrid 'ionic' crystals with cells parameters of the order of two nanometers.

Our experiments suggest that I^- anions cannot behave as templates when more polarizing cations are present in the aqueous solution. A mixture of V_2O_5 + NMe_4I + NMe_4OH leads to the formation of the $[V_{18}O_{42}]^{12-}$ cluster in which I^- is encapsulated. The monovalent $[NMe_4]^+$ cation is larger than the monovalent I^- anion. Therefore the manner $[VO_5]$ pyramids are linked in the 'xy' plane is directed by the $I^-...V^{\delta+}$ interactions. The cluster shell is formed around the anion giving rise to a spherical structure in which V=O dipoles are on the same side, but are tilted from the parallel position in order to minimize dipolar repulsions (Fig.4). Spherical clusters are no more formed in the presence of smaller cations such as K^+ ($r_{K^+} = 1.33\text{Å}$). Using KI instead of NMe_4I for the hydrothermal synthesis leads to the precipitation of $K_2V_3O_8$ crystals. They are formed of $[V_3O_8^{2-}]_n$ layers made of $[V^{4+}O_5]$ pyramids and $[V^{5+}_2O_7]$ corner sharing tetrahedra. K^+ cations lie between these planes where they are surrounded by oxygen atoms. This phase could be described as $K_2(VO)(V_2O_7)$ [28].

CONCLUSION

This work shows that small changes in the synthesis conditions can lead to completely different nanostructures. Several ionic species are always simultaneously present in an aqueous solution (H_3O^+, OH^-, M^{z+}...). The formation of a solid network first depends on the nature of the molecular precursor $[M(OH)_h(OH_2)_{6-h}]^{(z-h)+}$, that mainly depends on pH. The way these primary building units self-assemble results from a competition between all ionic species. It seems to be governed by the most polarizing ions. Anions don't behave as templates when small cations are present in the solution.

Three parameters seem to be important, the charge, the size and the redox potential of the foreign ions. In the case of Cl^-, a mixed valence $[V_{15}O_{36}]^{6-}$ anion is formed while a larger cluster, $[V_{18}O_{42}]^{12-}$ in which all vanadium are in the V^{IV} oxidation state is formed with I^-. These differences should be due to the larger size of I^- ($r_{I^-} = 2.16$Å, $r_{Cl^-} = 1.81$Å) and the redox potential of the I^-/I^0 couple ($E°(Cl_2/Cl^-) = 1.36$ V, $E°(V^{5+}/V^{4+}) = 1$ V, $E°(I_2/I^-) = 0.54$ V). I^- can reduce V^V into V^{IV} giving rise to the formation of I_2 that can be easily detected by its pink coloration.

Polyvanadates offer an interesting example. The formation of a large variety of nanostructures seems to be due to the versatility of the $[VO_5]$ basic unit with a strong V=O double bond and very weak $H_2O\cdots V$ interactions along the 'z' direction. The positively charged $V^{\delta+}$ ion is therefore accessible to anions allowing the formation of cluster shells. This is no more the case for tetrahedral $[VO_4]$ or octahedral $[VO_6]$ basic units in which the vanadium atom is protected by the surrounding oxygens. Anionic species can no longer interact with $V^{\delta+}$ in the precursor. Spherical nanostructures should then be obtained in the pH range where the coordination of V^V turns from four to six or with reduced V^{IV} vanadates in which the $[V^{4+}O_5]$ coordination is quite usual.

Care has to be taken for the description of the syntheses performed under hydrothermal conditions. The aqueous chemistry of cations may then be quite different from what is observed at room temperature. The viscosity and dielectric constant of water decrease when the temperature increases. At 200°C for instance the dielectric constant of water is close to 30 instead of 80 at room temperature. Higher temperatures favor the ionic dissociation of water so that the pH scale is narrower and anions such as Cl^- become more basic.

REFERENCES

1. M.T. Pope, Hetero and isopolymetallates, Springer-Verlag, Berlin (1983)

2. M.T. Pope and A. Müller, Angew. Chem. Int. Ed. Engl., 30, p.34 (1991)

3. J. Livage, Chem. Mater., 3, p.578 (1991)

4. Y.J. Liu, J.L. Schindler, D.C. DeGroot, C.R. Kannewurf, W. Hirpo, M.G. Kanatzidis, Chem. Mater., 8, p.525 (1996)

5. C.G. Wu, D.C. DeGroot, H.O. Marcy, J.L. Schindler, C.R. Kannewurf, Y.J. Liu, W. Hirpo, G. Kanatzidis, Chem. Mater., 8, p.1992 (1996)

6. D. Rioux, G. Ferey, Inorg. Chem., 34, p.6520 (1995)

7. D. Riou, G. Ferey, J. Solid State Chem., 120, p.137 (1995)

8. L.F. Nazar, B.E. Koene, J.F. Britten, Chem. Mater., 8, p.327 (1996)

9. Y. Zhang, R.C. Haushalter, A. Clearfield, Inorg. Chem., 35, p.4950 (1996)

10. A. Müller, H. Reuter and S. Dilinger, Angew. Chem. Int. Ed. Engl., 34, p.2328 (1995)

11. G.A. Pozarnsky, A.V. McCormick, Chem. Mater., 6, p.380 (1994)

12. T. Yao, Y. Oka, N. Yamamoto, Mat. Res. Bull., 27, p.669 (1992)

13. N. Baffier, P. Aldebert, J. Livage, H.W. Haesslin, J. Colloids Interface Sci., 141, p.467 (1991).

14. P. Davidson, C. Bourgaux, L. Schoutteten, P. Sergot, C. Williams, J. Livage, J. Phys. II France, 5, p.1577 (1995).

15. P. Davidson, A. Garreau, J. Livage, Liq. Cryst., 16, p.905 (1994).

16. M.T. Averbuch-Pouchot, A. Durif, Eur. J. Solid State Inorg. Chem., 31, p.567 (1994)

17. M.T. Averbuch-Pouchot, Eur. J. Solid State Inorg. Chem., 31, p.557 (1994)

18. P.Y. Zavalij, M.S. Whittingham, E.A. Boylan, Zeit. Krist., 211 (1996)

19. T. Chirayil, P. Zavalij, M.S. Whittingham, Solid State Ionics, 84, p.163 (1996)

20. M.S. Whittingham, J-D. Guo, T. Chirayil, G. Janauer, P. Zavalij, Solid State Ionics, 75, p.257 (1995)

21. A. Müller, E. Krickemeyer, M. Penk, H.J Walberg, H. Bögge, Angew. Chem. Int. Ed. Engl., 26, p.1045 (1987)

22. A. Müller, M. Penk, R. Rohlfing, E. Krickemeyer, J. Döring, Angew. Chem., Int. Ed. Engl., 29, p.926 (1990)]

23. G.K. Johnson, E.O. Schlemper, J. Amer. Chem. Soc., 100, p.3645 (1978)

24. C.F. Baess, R.E. Mesmer, Hydrolysis of cations, Wiley, New York (1976)

25. P. Comba, L. Helm, Helvetica Chim. Acta, 71, p.1406 (1988)

26. M. Henry, J.P. Jolivet, J. Livage, Structure and Bonding, 77, p.153 (1992)

27. M. Henry, Mater. Sci. Forum, 152-153, p.355 (1994)

28. J. Galy, A. Carpy, Acta Cryst., B31, p.1794 (1975)

29. M. Bénard, private communication

REFERENCES

1. J.M.F. Proc. Printig and Reprodution Hbhn. S angor Vallag. Schm (1953)
2. M.T. Rupp and A. Suline, Angew. Cren. Int. Ed. Engl. 30, 2, 3 (1991).
3. J. Livage, Chem. Mater. 3, p. 578 (1991).
4. O. Lelan, J.E. Schmidt, G.D. Deloost, C.R. Kllnswood V. Sapond, D. Brun etc. Chem. Mater. 8, p. 525 (1996).
5. G.C. Wu, E.C. DeGroot, H.O. Marcy, J.L. Schindler, C.R. Kassiwan, Y.J. Lu etc. Bhya. C. Kintwalds, Chem. Mater. 8, p. 1992 (1996).
6. D. Bhya, O. Teter, Inorg. Chem. 14, p. 3558 (1975).
7. D. Bian, C. Perry, J.Sold Stat. Chem. 130, p.157 (1995).
8. J.R. Mina, R.D. Rooke, E. Dunka, Chem. Mater. 6, p.212 (1994).
9. C. Sang, J.L. Huschhand, J. Chatfieh, Inorg. Chem. 35, p.990 (1990).
10. A. Muller, R. Rorsefal, S. Dornpan, Angew Chem. Int. Ed. Engl. 34, p.2122 (1995).
11. C.J. Brinkman, A.V. Gschneidge, Chem. Mater. 7, p.280 (1995).
12. T.V. Fan, Y. Chou, P. Goncalon, J.M. Res. Bull. 27, p.649 (1992).
13. R.E. Jorge, A.Herbert, J. Livage, H.W. Roesky, Colloid Interfaces. Sci. 163, p.651, 2000.

14. R.D. Shannon, J. Gavin, A.J. Schonmaten A. Sorger, J. William J. Collid J. Bryd. II Phase. 5, p.1879 (1995).

15. T. Derkason, A. Gianen, J. Livage, J.M. Chym. 16, p.146 (2000).
16. M.T. Pocon-Parder, A. Dordi, Bar, J.S. Sol Stae Inorg. Chem. 33 p. 581 (1996).
17. T.A. Johansmken, E.J. Sel d State Inorg. Chem. ... p. 557 (1997).
18. E. V. Loryfil, M.S. Whitangham, E.A. Hoylan, J.M. Res. p.211, 1999.
19. C.J. Fargon, Y.C. and, M.S. Whittngham, S olid State Iom. 58, p.81 (1990).
20. M.S. Whittngham, H. Zwo, T.Zhang, G. Jacobson, J. Livag, S. Plue Interfaces. 15, p. (1999).

21. M.R. K. Soms, L.L. Clng, M. Posch, H. Walter, A. Mecke, J.L. Ssabo, R. Engc. 20, p.1055 (1999).

22. A. Muller, P.J. Paul, A. Doblmg, P. F.H. Kgerner, J. Dohm, Angew. Chem. Int. Ed. Engl. 35, p.926 (1996).

23. O.R. Johnson, T.O. Sahampson J. Amer. Cram Soc. 106 p.885 (1980).
24. C.F. Dees, T.T. Moore, Phil. Stat. Mag. Nu, Whey Nevy York (1996).
25. G.H. Coulte, J. Phia, Polydor. Ana. Aca. 74, p.665 (1991).
26. W. Irons, J.F. Jonas, J. Livage, Naturn phae, phee handling 72, pp.313-319.
27. M.T. Hoy, Math. Phil Plane. 152, p19, 355 (1985).
28. J. Wes, Chem. Acta Cryst. Sci. p.172-8 (1973).

ELECTRICAL/DIELECTRIC PROPERTIES OF NANOCRYSTALLINE CERIUM OXIDE

JIN-HA HWANG*, THOMAS O. MASON*† and EDWARD J. GARBOCZI**

*Department of Materials Science and Engineering, Northwestern University, 2225 N. Campus Dr., Evanston, IL 60208-3108
** National Institute of Standards & Technology, Bldg. 226, Rm. B-350, Gaithersburg, MD 20899

† Author to whom correspondence should be addressed

ABSTRACT

Electrical/dielectric properties of nanocrystalline cerium oxide have been studied using impedance spectroscopy, thermopower, and DC 4-point conductivity. The combined techniques identified the effect of poor electroding on impedance spectra. Incomplete contact between the specimen and the electrode induces an additional arc in the impedance spectra. The additional high resistance feature results from the geometric constriction of current flow at the specimen/electrode interface and can be misinterpreted as a grain boundary response. The defect chemistry, nonstoichiometry, and transport properties were investigated in nanoscale ceria and compared with those of microcrystalline material.

INTRODUCTION

Nanocrystalline ceramics are a relatively new class of materials whose physical/chemical properties have yet to be fully exploited. The unique properties arise from a high fraction of grain boundaries or surfaces, reduced diffusion distances, and enhanced chemical reactivity. There is therefore a widespread interest in mechanical, catalytic, magnetic, and electrical properties. The high fraction of grain boundaries in nanoceramics can be employed to investigate the role of grain boundaries in electroceramics. Such efforts have been reported in ZnO and CeO_2 [1,2]. However, insufficient experimental data exist to permit a complete understanding of the chemistry and physics of grain boundaries in nanoscale materials.

Impedance spectroscopy has been extensively employed in electroceramics, due to its capabilities i) to resolve grain boundary from bulk electrical properties, ii) to calculate materials constants (conductivity and dielectric constant), and iii) to probe the electrical homogeneity. These powerful capabilities can be applied to nanocrystalline ceramics, in order to investigate the physical properties of grain boundaries in nanoscale ceramics. In addition to AC impedance characterization, simultaneous measurement of conductivity and thermopower allows determination of charge carrier concentration and mobility, and the establishment of conduction mechanism(s). Detailed information in nanoscale ceramics is lacking. Therefore, combined AC and DC characterization provides the underlying information which has not been available to date.

The current work addresses experimental techniques in nanocrystalline ceramics in order to make reliable electrical property measurements and thereby characterize the defect-related transport phenomena in nanophase ceria.

EXPERIMENTAL SECTION

Nanocrystalline cerium oxide powder (Nanophase Technologies, Inc.) was compacted, cold-isostatically pressed, and sintered at 700°C for approximately 1 hr. The sintered specimens were cut into rectangular plates and bars for AC impedance characterization and

Mat. Res. Soc. Symp. Proc. Vol. 457 ©1997 Materials Research Society conductivity and thermopower measurements, respectively. Au was sputtered as an electroding material. The experimental details are described elsewhere [3]. The oxygen partial pressure was controlled between 1 and 10^{-3} atm at 450-550°C using inert gas/oxygen mixtures.

Impedance spectra were acquired using an HP 4192A (Hewlett-Packard) impedance analyzer and the collected spectra were corrected using standard null-corrections [4]. The deconvolution of the arcs was performed using "EQUICVRT.PAS"[5]. Simultaneous measurements of thermopower and conductivity were performed following the procedure described elsewhere [3]. The electrical properties were measured as a function of temperature and oxygen partial pressure.

RESULTS

Nanocrystalline ceramics are expected to have much smaller grain boundary capacitance than ~nF range, from the conventional brick layer model. The decreased capacitance can be affected by experimental artifacts, such as apparatus contributions, as reported before [4]. Also, poor contact between the specimen/electrode leads to an additional feature (i.e., an arc) as shown in Fig. 1. Poor electroding causes current-spreading at the specimen/electrode interface.

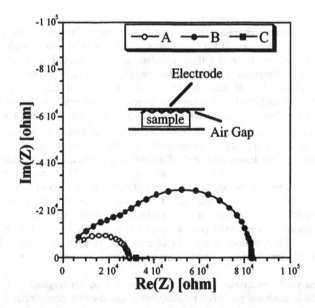

Figure 1. Impedance spectra of partially sintered ceria (~ 15 nm grain size) without(w/o) contact effect and with (w/) contact resistance effect, at 550°C A: Two sides of the specimen were polished down to 1 μm and coated with Au (0.15 μm). B: One side was polished (down to 1 μm) and coated with Au. The other side was abraded with 400 grit abrasive paper. C: True sample resistance calculated from DC 4-point conductivity.

The current is limited to narrow regions, decreasing the effective cross sectional area for current flow and increasing the corresponding resistance. Though the additional arc is closely linked to the electrode/specimen interface, the origin can be nonohmic or ohmic. The similar values of work function of ceria and Au and a linear DC 2-point I-V characteristic proved that the contact is ohmic. An equation relating spreading resistance to the effective radii of point contacts was derived by Holm[6]

$$R_C = \frac{1}{4\sigma \Sigma a} \tag{1}$$

where Σa is the sum of the radii of the individual contacts. The corresponding capacitance is given by

$$C_C = 4\varepsilon\varepsilon_o \Sigma a \tag{2}$$

The time constant of the contact effect ($R_C C_C$) should be the same as that of the bulk. A large resistance means a small capacitance. The same time constant of these processes will result in a single arc. However, the additional arc at low frequencies (large Re(Z)) (See Fig. 1) indicates that the time constant due to the contact effect is larger than that of the bulk. The increased time constant results from a parallel contribution of the air gap capacitance, as shown in the inset diagram of Fig. 1. This capacitance swamps out the small point contact capacitance (Eq. (2)).

As seen in Eq. (1), the contact resistance is a function of the conductivity (of the system under investigation) and total contact radius. The effect of contact radius can be tested by varying the mechanical contact pressure, since contact pressure changes the total contact radius. As shown in Fig. 2(a), the mechanical contact pressure controls the impedance arc at

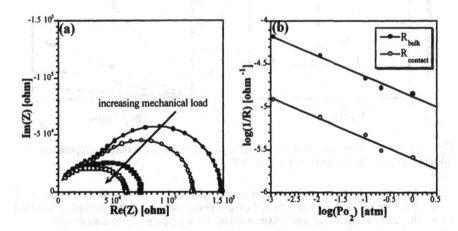

Fig. 2. (a) The variation of impedance spectra at 550°C with applied mechanical load, and (b) oxygen partial pressure dependence of the bulk and contact resistance at 500°C.

the low frequency region. However, the high frequency arc remains constant. The increased mechanical pressure increases the total contact radius, reducing the corresponding resistance. Fig. 2(b) shows the oxygen partial pressure dependence of bulk and contact resistances at 500°C. They exhibit the same trend, since both bulk and contact resistances are inversely proportional to the conductivity of the materials being measured.

A properly electroded nanophase specimen exhibits what appears to be a single arc (see Fig. 1A), where capacitance can be due to bulk ceria only, or a contribution of grain interior and grain boundary responses [7].

The electrical conductivity and thermopower were measured as a function of oxygen partial pressure. Very high values of thermopower (-900~-1000 μV/K) were observed in nanocrystalline cerium oxide. The nonideal slope of the conductivity vs. P_{O_2} (see Fig. 3(a)) and chemical analysis (SSMS) suggest that oxygen vacancies due to aliovalent impurities are the major defect species (e.g., $[V_O^{\cdot\cdot}] = 1/2[La_{Ce}^{'}]$ or $1/2[Gd_{Ce}^{'}]$) and contribute significantly to the electrical conductivity. Since both vacancies and electrons are mobile, mixed conduction results. The resultant total conductivity (σ_t) can be given by

$$\sigma_t = \sigma_{el} + \sigma_{ion} = kP_{O_2}^{-1/4} + \sigma_{ion} \qquad (3)$$

where σ_{el} is the electronic conductivity and σ_{ion} is the ionic conductivity. Fig. 3(b) shows a plot of σ_t vs. $P_{O_2}^{-1/4}$, from which σ_{ion} was calculated.

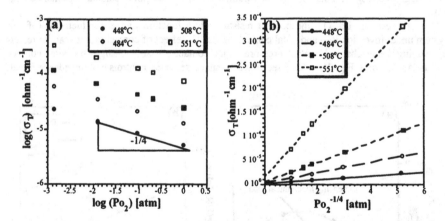

Fig 3. (a) Conductivity as a function of oxygen partial pressure at different temperatures, and (b) replot of the total conductivity (σ_t) vs. $P_{O_2}^{-1/4}$.

In this way, the individual components (ionic and electronic) could be determined. The ionic conduction increases with temperature and oxygen partial pressure, but is limited to less than 25%. The resultant electronic conductivity exhibits a 1/4 dependence, consistent with

$$O_O^x \rightarrow V_O^{\cdot\cdot} + 2e' + \frac{1}{2}O_2(g) \qquad (4)$$

if $[V_O^{\bullet\bullet}]$=constant. (Note: acceptor impurities were ~ 2000 ppm.)

Thermopower is dictated by the charge carrier concentration irrespective of the corresponding conduction mechanism, i.e., band conduction or small polaron conduction. Bulk cerium oxide is known to conduct through small polarons between Ce^{+3} and Ce^{+4} [8]. The temperature dependence of thermopower (Fig. 4(a)) was used to determine the reduction enthalpy. The calculated value of 1.84 eV is much smaller than the value of 4.67~4.98 eV in bulk cerium oxide. In combination with the temperature dependence of the electronic conductivity (Fig. 4(b)), the enthalpy of electron mobility was determined to be 0.47 eV, which is consistent with small polaron conduction. Additionally, the calculated mobility satisfied the criterion of <0.1 cm^2/Vsec [9].

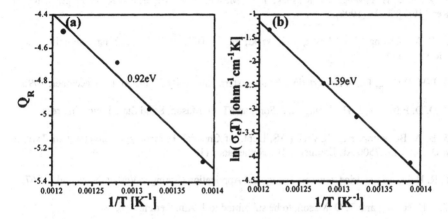

Fig. 4.(a) Plot of reduced thermopower, Q_R (-Q/2.303(k/e)) vs. inverse temperature, and (b) plot of logarithm of electronic conductivity vs. inverse temperature.

Based upon the chemical analysis through Spark Source Mass Spectrometry (SSMS) and the measured thermopower, the nonstoichiometry was estimated for undoped nanocrystalline ceria. It is three orders of magnitude larger than for the bulk ceria. This effect can be introduced by space charge layer formation which can mean a different activation enthalpy for defect formation (reduction). Nanocrystalline cerium oxide possesses an extremely large fraction of grain boundaries. The effect of a space charge can be amplified and thereby alter the conductivity. Although the grain boundaries are believed to have higher conductivity and higher dielectric constant, the resultant time constant is not significantly larger than the grain interiors (or the bulk). The resultant impedance spectra can be a single arc, as shown in this work.

CONCLUSIONS

Poor electroding between the specimen and the electrodes induced an additional arc which can be mistakenly analyzed as a grain boundary response. The feature results from the geometric constriction at point contacts on the specimen/electrode interface. Also, the air gap capacitance is connected with the contact effect, leading to an discrete arc at low frequencies.

Acceptor impurities introduce mixed conduction in nanocrystalline ceria. The ionic contribution increases with increasing temperature and oxygen partial pressure. The electrical conduction takes place through small polarons between Ce^{+3} and Ce^{+4}. However, the reduction enthalpy is much smaller than that of the microcrystalline counterpart. It is believed that this altered defect chemistry is associated with space charge layers near grain boundaries.

ACKNOWLEDGMENTS

This work was supported by the Department of Energy (Grant No. FG02-84ER45097).

REFERENCES

1. J. Lee, J.-H. Hwang, J. J. Mashek, T. O. Mason, A. E. Miller, and R. W. Siegel, J. Mater. Res., **10[9]**, 2295 (1995).

2. Y. M. Chiang, E. B. Lavik, I. Kosacki, H. L. Tuller and J. Y., Ying, submitted to J. Electroceramics.

3. J.-H. Hwang, Ph.D. Dissertation, Northwestern University Evanston, IL, December, 1996.

4. D. D. Edwards, J.-H. Hwang, S. J. Ford, and T. O. Mason, Solid State Ionics, in press.

5. B. A. Boukamp. EQUICVRT.PAS, Dept. of Chemial Technology, University of Twente, P.O. Box 217, 7500 AE Enschede, The Netherlands (1990).

6. R. Holm, in Electric Contacts: Theory and Application (Springer-Verlag, New York, 1967).

7. J.-H. Hwang and T. O. Mason, to be submitted to J. Am. Ceram. Soc.

8. H. L. Tuller and A. S. Nowick, J. Phys, Chem. Solids, **38**, 859 (1977).

9. A. J. Bosman and H. J. van Daal, Adv. Phys. **19[77]**, 1 (1970).

PREPARATION AND SINTERING OF SILICA-DOPED ZIRCONIA
BY COLLOIDAL PROCESSING

T. UCHIKOSHI, Y. SAKKA, K. OZAWA and K. HIRAGA
National Research Institute for Metals, 1-2-1, Sengen, Tsukuba, Ibaraki 305, Japan

ABSTRACT

Silica-doped (SiO_2= 0-1.0 mass%) zirconia (3 mol% Y_2O_3-doped tetragonal ZrO_2) compacts are prepared from hetero-coagulated and well-dispersed suspensions by colloidal processing. The suspensions are consolidated by a pressure filtration technique. The green density of the compacts consolidated from the well-dispersed suspensions is higher than that from the hetero-coagulated suspensions. The lower density of the latter compacts is improved by a subsequent cold isostatic pressing (CIP) at 400 MPa. The sinterability of the compacts at 1200 °C is greatly affected by the amount of doped silica. The densification and grain growth are hindered by silica doping above 0.3 wt% at 1200 °C. All the compacts are densificated to a relative density of above 99% by sintering at 1300 °C for 2 h.

INTRODUCTION

Yittria-doped tetragonal zirconia polycrystal (TZP) is known as the material that shows a superplasticity of ceramics[1-3]. The characteristic is greatly affected by the microstructure of a sintered body. It comes out under the small grain size of the sintered body less than 1 μm. The extensive tensile elongation up to 800% has been reported for 3Y(3 mol% Y_2O_3)-TZP whose initial grain size is about 0.3 μm[4]. The glassy phase at grain boundaries often promotes the decrease in flow stress and the enhancement of tensile ductility of TZP[5,6]. Kajihara et al. have reported the maximum elongation over 1000% at 1400 °C for 5 wt% SiO_2-doped 2.5Y-TZP[7]. Recently, Hiraga has reported that the increase in fracture strain is observed for >1 wt% SiO_2-doped 3Y-TZP, but the decrease in flow stress is observed for 3Y-TZP with 0.1-0.7 wt% SiO_2[8]. His result shows the importance of silica doping against the mechanical properties of TZP when its doping is between 0.1 and 1.0 wt%.

Pore sizes and their distribution are very important factors that determine the mechanical properties of sintered ceramics. Large pores often become the starting point of fracture. Therefore, uniformity of the initial particle packing of green compacts is demanded. If the TZP is modified with silica uniformly and the defects in the sintered body are decreased, excellent superplasticity of TZP could be expected even though the amount of silica is less than 1 wt%.

Colloidal processing is a powerful method for controlling the density and microstructure of a consolidated compact[9-11]. The particle packing is affected by an interparticle force. The force can be changed from attractive to repulsive by adjusting pH in case of aqueous suspensions. Attractive potential among the particles contributes to the prevention of segregation. On the other hand, repulsive potential among them contributes to the dense particle packing. The aim of the present work is to prepare silica-doped Y-TZP ceramics with (1) uniform modification of zirconia with silica and (2) homogeneous microstructure from nano-sized powders by colloidal processing.

EXPERIMENTAL

The experimental work was performed with 3Y-TZP zirconia powder (Tosoh. Co., TZ3Y; average particle size is 60 nm) and colloidal silica (Nissan Chem. Co., Snowtex O; average particle size is 10-20 nm). The impurities of 3Y-TZP are 0.005 wt% Al_2O_3, 0.004 wt% SiO_2, 0.002 wt% Fe_2O_3 and 0.013 wt% Na_2O. The main impurity of colloidal silica is 0.032 wt% Na_2O. ζ-potential of zirconia and silica powders was measured by a laser electrophoresis ζ-potential analyzer (Otsuka Electronics Co., LEZA-600). 0.01M NaCl was added to the diluted suspensions to control ionic strength. Two types of aqueous suspensions of the zirconia-silica system were prepared according to the following two different conditions; (A) at pH=5.3, solid content is 7 vol% and (B) at pH=8.3, solid content is 28 vol%. For pH adjustment, 1N HNO_3 and 1N NH_4OH were used. Preparation procedure of the suspensions will be described later. The suspensions were ultrasonicated for 10 min and stirred by a magnetic stirrer for 12 h at room temperature. Silica content was changed between 0-1.0 wt% against zirconia. The consolidation of the suspensions was performed by a pressure filtration technique[9-13] at 10 MPa. Before the consolidation, the suspensions were evacuated in a vacuum desiccator to eliminate air bubbles. The schematic diagram of a pressure filtration equipment has been reported elsewhere[12]. A Teflon membrane with pores of 0.1 μm was used as a filter. CIP treatment at 400 MPa was carried out to improve the packing density of the green compacts after the consolidation. The shape of the green compacts was disc and the sizes of diameter and thickness are 50 mmφ and 3-4 mm, respectively. Those compacts were dried overnight at 120 °C, and sintered in air at fixed temperatures for 2 h. The density of the compacts was measured by the Archimedes' method using kerosene. Microstructure of sintered bodies was observed by SEM for the polished and chemically and thermally etched surfaces. Chemical etching was performed by soaking the sintered bodies in a mixed acid of HF: H_2SO_4: H_2O= 1: 4: 20 for 2 min.

RESULTS AND DISCUSSION

Preparation of Suspension

Figure 1 shows the ζ-potential versus pH of 3Y-zirconia and colloidal silica. The isoelectric points of the zirconia and silica are at pH 7.2 and 2.6, respectively. ζ-potential determines the interparticle potentials in a suspension. In case of suspension (A), which was prepared at pH=5.3, the interparticle potentials between ZrO_2-ZrO_2 and SiO_2-SiO_2 are repulsive, but ZrO_2-SiO_2 is attractive. Therefore hetero-coagulated condition, i.e., uniform modification of zirconia particles with silica particles, is expected. In case of

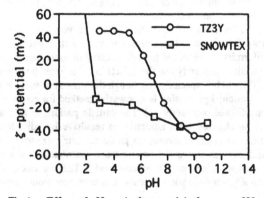

Fig.1. Effect of pH on the ζ-potential of aqueous 3Y-TZP (TZ3Y) and colloidal silica (snowtex) suspensions.

suspension (B), which was prepared at pH=8.3, the interparticle potentials between ZrO_2-ZrO_2, SiO_2-SiO_2 and ZrO_2-SiO_2 are repulsive. Here, an appropriate amount of ammonium polycarboxylate (Toagosei chem. Co., ALON A-6114) was added to improve the negative ζ-potential of zirconia and silica surfaces[14]. Therefore well-dispersed condition, i.e., dense particle packing, is expected.

Figure 2 shows the relative density of the as-pressure filtrated (as-PF) compacts of undoped-TZP

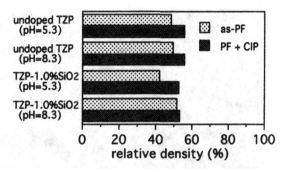

Fig.2. Packing density of undoped TZP and 1.0wt% SiO2-doped TZP consolidated by pressure filtration (PF) at 10 MPa. Their densities were improved by subsequent CIP at 400 MPa.

and 1.0wt%SiO$_2$-doped TZP. The green density of as-PF compacts consolidated from the well-dispersed suspension (pH=8.3) was higher than that from the hetero-coagulated suspension (pH=5.3). The lower packing density of the latter compacts was improved to almost the same density by subsequent CIP treatment at 400 MPa. Therefore we used the compacts prepared from hetero-coagulated suspensions and CIPed at 400 MPa for the following experiments.

Sintering Characteristics

Figure 3 shows a sintering diagram of the PF + CIPed samples. The sinterability of the compacts at 1200 °C was greatly affected by the amount of silica contents.

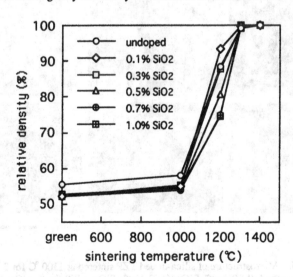

Fig.3. Relative density of silica -doped TZP as a function of sintering temperature.

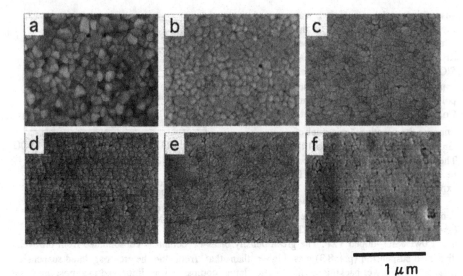

Fig.4. Microstructure of silica-doped TZP sintered at 1200 °C for 2 h;
(a) undoped, (b) 0.1wt% SiO2, (c) 0.3wt% SiO2, (d) 0.5wt% SiO2, (e)
0.7wt% SiO2 and (f) 1.0wt% SiO2.

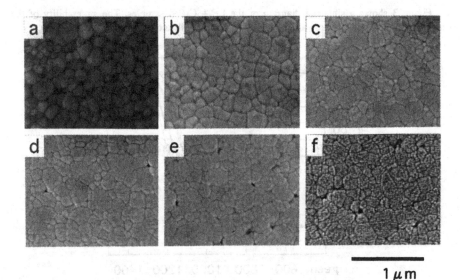

Fig.5. Microstructure of silica-doped TZP sintered at 1300 °C for 2 h;
(a) undoped, (b) 0.1wt% SiO2, (c) 0.3wt% SiO2, (d) 0.5wt% SiO2, (e)
0.7wt% SiO2 and (f) 1.0wt% SiO2.

Density of TZP is increased by the addition of 0.1 wt% silica, but decreased by the addition of >0.3 wt% of silica. The enhancement of densification at 0.1 wt% doped silica is probably related to small amounts of impurities. All the compacts could be densified to a relative density of >99 % by sintering at 1300 °C for 2 h in air.

Figure 4 shows the microstructure of the compacts sintered at 1200 °C for 2 h. The hindrance of grain growth is observed for the large amount of silica-doped samples. Figure 5 shows the microstructure of the compacts sintered at 1300 °C for 2 h. The grain sizes of the sintered compacts are almost the same regardless of the amount of doped silica. For the compacts whose silica contents are ≥0.5 wt%, the excess amount of silica segregates at grain multiple junctions.

Figure 6 shows the log-log plot of grain size measured for TZP- 0.3 wt% SiO_2 against sintering time at 1200 °C. Grain size was determined by the linear intercept method. The slope of the plots is about 1/4. The slope of 1/4 indicates that the mechanism of grain growth is a grain boundary diffusion control[15].

Mechanical properties for the silica-doped TZP prepared from colloidal processing are under investigation. Until now, we investigated the stress-strain relation at 1400 °C for the TZP without silica. The fracture true strain of TZP prepared by colloidal processing increased more than 40 % in comparison with that of TZP prepared by dry processing. A quantitative

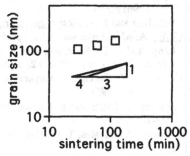

Fig.6   Log-log plot of grain size vs. sintering time of 0.3wt%SiO2-doped TZP at 1200 ℃.

analysis of cavitated volume during the deformation reveals that the damage accumulation is controlled by growth of pre-existent defects and nucleation growth of new cavities. The improvement in the superplasticity of colloidally-processed samples is due to the elimination of pre-existent void.

CONCLUSIONS

We attempt the preparation of 3Y-TZP with (1) uniform modification with silica and (2) homogeneous microstructure from nano-sized powders by colloidal processing. Hetero-coagulated and well-dispersed suspensions are prepared at pH=5.3 and 8.3, respectively, by changing of pH of zirconia-silica aqueous suspension. The green density of as-pressure filtrated compacts consolidated from the well-dispersed suspension is higher than that from the hetero-coagulated suspension. The lower packing density of the latter compacts is improved to almost the same density by subsequent CIP treatment at 400 MPa. The densification and grain growth at 1200 °C are hindered when silica contents are ≥0.3 wt%. The compacts whose silica contents are ≥0.5 wt%, the excess silica segregates at grain multiple junctions. All the compacts are densificated to the relative density of >99% by sintering at 1300 °C for 2 h.

ACKNOWLEDGMENTS

We wish to thank Y. Kaieda and N. Oguro at NRIM for their help with the sample preparation by CIP and C. H. Nelson at Seitoku Univ. for revising the manuscript. This study was performed through Special Coordination Funds (Research on fundamental science of frontier ceramics) of the Science and Technology Agency of the Japanese Government.

REFERENCES

1. F. Wakai, S. Sakaguchi Y. Matsuno, Advanced Ceramic Materials 1, 259 (1986).
2. F. Wakai, S. Sakaguchi and H. Kato, J. Ceram. Soc. Jpn. 94, 721 (1986).
3. F. Wakai, Tetsu-to-Hagané, 75, 389 (1989).
4. T. G. Nieh and J. Wadsworth, Acta metall.mater. 38, 1121 (1990).
5. M. J. Verkerk, A. J. A. Winnubst and A. J. Burggraaf, J.Mat.Sci. 17, 3113 (1982).
6. M. Miyayama, H. Yanagida and A. Asada, Am. Ceram. Soc. Bull. 64, 660 (1985).
7. K. Kajihara, Y. Yoshizawa and T. Sakuma, Acta metall. mater. 43, 1235 (1995).
8. K. Hiraga, H. Yasuda, K. Nakano, E. Takakura and Y. Sakka, Abst. 118th. Meetings Jpn. Inst. Met. (1996) p.250.
9. C. H. Schilling and I. A. Aksay, Engineered Materials Handbook Vol. 4. CERAMICS AND GRASSES, ASM international, (1991) pp.153-160.
10. F. F. Lange and K .T. Miller, Am. Ceram. Soc. Bull. 66, 1498 (1987).
11. F. F. Lange, J. Am. Ceram. Soc. 72, 3 (1989).
12. T. Uchikoshi, Y. Sakka, H. Okuyama and K. Ozawa, J. Jpn. Soc. Powder and Powder Metall. 42, 309 (1995).
13. T. Uchikoshi, Y. Sakka and K. Ozawa, Proc. 5th. World Congr. Chem. Eng. Vol IV (1996) pp. 1007-1012.
14. Y. K. Leong, P. J. Scales, T. W. Healy and D. V. Boger, Colloids and Interfaces A 95, 43 (1995).
15. K. Okada and T. Sakuma, British Ceram. Trans. 93, 71 (1994).

TRANSITION DYNAMICS IN FERROELECTRICS WITH ORDERED NANOREGIONS

I.G. SINY*, R.S. KATIYAR, S.G. LUSHNIKOV*
Department of Physics, University of Puerto Rico, San Juan, PR 00931-3343

ABSTRACT

Raman scattering was used to study two model relaxor ferroelectrics, $PbMg_{1/3}Nb_{2/3}O_3$ (PMN) with the 1:2 stoichiometric composition of Mg^{2+} and Nb^{5+} ions in the oxygen octahedrons and $PbSc_{1/2}Ta_{1/2}O_3$ (PST) with the 1:1 stoichiometric composition of Sc^{3+} and Ta^{5+} ions. In spite of a different stoichiometric ratio the Raman spectra of both materials are consistent with the Fm3m space symmetry which implies the existence of similar 1:1 ordered clusters at least in nanoscale regions. The spectra show some anomalous features in the temperature range preceding a ferroelectric state, namely a broad central peak appears in PMN and a complex structure develops from the initially singlet line in PST. Those phenomena are considered as the dynamic features in course of evolution of the relaxors to a ferroelectric state. The preceding phase is characterized by a breakdown in the selection rules for Raman scattering, so some points in the Brillouin zone can contribute to the light scattering spectra. Comparing all available data, one can assume the determinant role of heterophase fluctuations in that process. The fluctuations in a special preceding phase are caused by a competition between two phases, namely between the ferroelectric phase and an additional nonpolar phase.

INTRODUCTION

Relaxor ferroelectrics with the complex perovskite-type formula $AB'_xB''_{1-x}O_3$ have received considerable attention for many years. Two unlike valance B' and B'' ions in the B sublattice distinguish these mixed materials from the classical ABO_3 perovskites. The arrangement of two different ions in the B sublattice appears to be a determining factor to create a special relaxor behavior. Dynamic features of the evolution to a ferroelectric state in relaxors are far from being clear. At least, relaxors do not exhibit any soft modes which used to be a distinctive feature of the phase transition dynamics in the most "pure" perovskites. Nothing new has appeared in this field since a review book [1] was published in 1977. The recent studies [2] (and Refs. therein) have revealed an important common characteristic feature of relaxors to consist of the nanoscale clusters with the 1:1 B-site order irrelevant to whether a stoichiometric composition for the B ions is 1:1 or 1:2. We assume that the nanoscale arrangement prevents the development of a "normal" ferroelectric transition. One can thus expect a special dynamics of fluctuations with the frustrated transition.

In the present paper, the Raman scattering studies in two relaxors, $PbMg_{1/3}Nb_{2/3}O_3$ (PMN) and $PbSc_{1/2}Ta_{1/2}O_3$ (PST), reveal some unusual features in the spectra which are connected with the evolution of both materials to a ferroelectric state. We believe that a central peak in PMN and a complex structure of the singlet hard mode in PST appear in a preceding phase and have a common nature.

EXPERIMENTAL

The PMN and PST single crystals were grown by spontaneous crystallization from a flux. The samples measured about $5 \times 4 \times 2$ mm^3 and $3 \times 2 \times 1$ mm^3 respectively. The X(ZZ)Y diagonal Raman spectra were measured, with X,Y and Z being along the fourfold cubic axes.

Raman spectra were excited with an argon laser for PMN and with a krypton laser for PST and were analyzed with a Cary-82 triple spectrometer. The instrument was equipped with an Oxford Instruments optical cryostat with a cold stage for low-temperature measurements and with a small furnace for high temperatures. A temperature controller stabilized the temperature in every scanning run to within ±0.5 K.

Both Stokes and anti-Stokes parts of the Raman spectra of PMN were studied. A central part of the spectrum in the limit of ±5 cm^{-1} was supposed to consist of a stray light from the elastic scattering and therefore it was eliminated from our consideration in all of the experimental spectra. In order to reveal a broad central peak, we used the following procedure. First of all, a spectrum at 77 K far below all known anomalies without any visible traces of the central peak was taken as an initial cross-section of light scattering from PMN, $S_0(\nu, 77)$. Then, all necessary spectra at higher temperatures were reconstructed from the initial spectrum at 77 K by using a normal expected temperature dependence for the first order spectra $S(\nu,T) = S_0(\nu,77)[n + 1]$, where the population factor is given by $n = [\exp(h\nu/kT)-1]^{-1}$. The calculated spectra were subtracted from corresponding experimental spectra. Some additional light scattering at the lowest frequencies appeared as wings on the Rayleigh line in a wide temperature interval. Those results are shown in Fig.1. It is clearly seen that the curves obtained by the procedure described above form broad central peaks. The top parts of peaks in the eliminated range ±5 cm^{-1} were obtained by fitting to a Lorentzian line shape.

RESULTS AND DISCUSSION

Local Ordering of PMN

PMN is a well-known model relaxor material with the 1:2 ratio of the two different B ions. High resolution electron micrographs of PMN [2] show the existence of a regular array of ordered clusters about 2 nm in diameter. The composition of these ordered clusters corresponded to the 1:1 ratio of B′ and B″ ions as in various other relaxors like PST with the 1:1 stoichiometric composition of two B ions. The distance between the centers of neighboring clusters is about 2.5 nm. Such nanoscale arrangement of PMN prevents a ferroelectric transition from spreading throughout the crystal. The transition occurs to be frustrated in normal conditions and could be found only in an external electric field above the threshold value of about 1.8 kV.cm^{-1} [3].

A Central Peak in the Range of a Frustrated Ferroelectric Transition

In spite of macroscopically frustrated transition the polarization fluctuations manifest themselves in light scattering from PMN even in the absence of an external field. We found a broad central peak in PMN around 200 K just in the range of a frustrated ferroelectric transition (Fig.1). It is clearly seen in Fig.1 that there is only a weak and extremely broad response in light scattering at lower temperatures. The peaks, at temperatures slightly above 200 K, are more than two times broader in comparison with the exceptional central component in the range of a frustrated transition. This central peak correlates adequately with the sharp anomaly in hypersonic damping in PMN which also is caused by fluctuations [4].

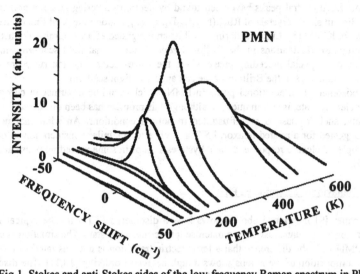

Fig.1. Stokes and anti-Stokes sides of the low-frequency Raman spectrum in PMN on approaching the frustrated ferroelectric transition (T~200 K) [3] and the range of a "diffuse" transition with the main dielectric anomaly (T~270 K) [1]. The hard-mode contribution is eliminated. A narrow central component occurs at T~200K and a broader and more intense component appears in a wide range around T~270 K.

A Central Peak in the Range of a So-called Diffuse Ferroelectric Transition

Besides the "sharp" anomaly at 200 K which is connected in the previous subsection with the polarization fluctuations, there is a main central peak with the maximum intensity and the minimum width around 280 K (Fig.1). This stretched anomaly correlates with the main broad maximum in the dielectric response of PMN [1]. We assume that this main broad central component in light scattering is connected with special heterophase fluctuations. Anomalous light scattering in PMN is very similar to that in a related crystal $Na_{1/2}Bi_{1/2}TiO_3$ (NBT) [5]. A cubic-tetragonal-trigonal sequence of phase transitions in NBT leads to the final ferroelectric state rather like in PMN. A broad central component in light scattering from NBT occurs between two phase transitions in contrast to the ordinary well-known behavior with anomalies in the vicinity of every transition point. This unusual behavior of NBT implies a coupling of two order parameters related to different phase transitions separated by some temperature interval.

Mechanism of Heterophase Fluctuations

One can suppose that the main broad central peak in light scattering from PMN is caused by fluctuations of the coupled order parameters as well. It is important to emphasize that coupled order parameters in a suitable model [6] initiate primary phase transitions in different points of the Brillouin zone. In this case one can expect to find a critical contribution of the heterophase fluctuations from many points on a line in the reciprocal space between the special points of the Brillouin zone. Such central peaks have been found by neutron scattering at some points along the critical Σ-line in single crystals of $Rb_{0.38}(ND_4)_{0.62}D_2PO_4$ [7] or along the R-M line of the cubic Brillouin zone in $KCaF_3$ [8]. Light scattering exhibits an integrated effect summing contributions from all heterophase fluctuations in the Brillouin zone. It seems that light scattering in PMN gives evidence of a special preceding phase where the wave-vector selection rules are broken down and some anomalies in the Brillouin zone can appear in light scattering.

The existence of an additional phase in PMN, which could be a partner in competition with the ferroelectric state, is still in question although this problem has been discussed for a long time. Probably, such a phase is also frustrated in normal conditions. An additional nonpolar phase was suggested for a related relaxor, PST, as well after a similar consideration in order to explain a complex dielectric response, double hysteresis loops in some preceding phase and other anomalies [9].

The Relaxor Behavior of a Disordered PST

The paper [9] mentioned above shows how a disordered PST with the typical relaxor behavior transforms spontaneously into a macroscopic ferroelectric state. The situation is close to the case of PMN with the difference that a ferroelectric transition is not frustrated in PST. PST with the 1:1 composition of the B ions shows a high degree of ordering [10,11]. The disordered sample of PST considered above [9] and studied in the present work implies nanoscale arrangement of the 1:1 ordered clusters in a manner which is close to that in PMN.

Additional Anomalous Structure in the Raman Spectrum of PST

The Raman spectra of PST obtained in our experiments are consistent with the partly ordered complex perovskite belonging to the Fm3m space group. Unlike the "pure" ABO_3 perovskites without any Raman active modes in a cubic phase, the complex $AB'_{1/2}B''_{1/2}O_3$ compounds with the Fm3m space symmetry exhibit a set of Raman active modes: $A_{1g} + E_g + 2F_{2g}$. The A_{1g} mode is a simple motion of the oxygen atoms like the breathing mode of a free oxygen octahedron. However, this mode reflects clearly the effect of subtle changes in the inner structure of PST in course of evolution to a ferroelectric state occurring slightly below room temperature. At high temperatures, far above the transition region, the A_{1g} mode has the shape of a singlet line (Fig.2). An evident structure of the initially singlet line appears when temperature is lowered down to the vicinity of the ferroelectric phase transition (Fig.2). No evidence of a change in the crystal structure in PST above the ferroelectric transition has been published. The structure around the A_{1g} mode is more pronounced in the samples with a higher degree of disorder on the B sites without any macroscopic ferroelectric transition. Thus, the additional structure appears in the PST even if the ferroelectric transition is frustrated as in PMN. Fig.2 shows the behavior of the A_{1g} mode in a sample with the highest degree of order between all studied materials. In this sample, the A_{1g} line takes a singlet shape again in the ferroelectric phase. One can suppose that the complex structure of the A_{1g} mode is connected with a breakdown in the wave-vector selection rules, so some symmetry points along the A_{1g} optical branches in the Brillouin zone contribute to the Raman scattering around the initial singlet line in the zone center.

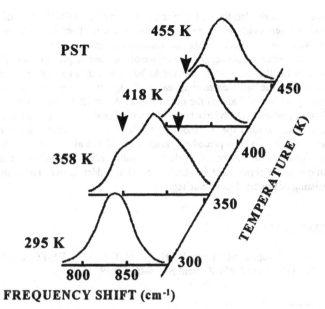

FREQUENCY SHIFT (cm⁻¹)

Fig.2. Appearance of an additional structure around the A_{1g} hard mode in PST on approaching the transition to a ferroelectric state from above (T~300 K) [9]. The arrows show the pronounced structure (T=358 K) and its first emergence (T=418 K) in a wide preceding phase near the low-temperature boundary and near the high-temperature limit respectively.

<u>Heterophase Fluctuations in PST</u>

The light scattering gives evidence of a special state in PST which precedes the transition to a ferroelectric state from above T_c. This result correlates with the existence of some preceding phase in PST with double hysteresis loops and other peculiarities [9]. To explain that unusual behavior of PST, a competition between two phases was suggested, namely between the ferroelectric phase and a postulated nonpolar phase [9]. This suggestion implies intensive heterophase fluctuations between those two phases. The present work gives new experimental evidence in support of such a model. We suppose that the loss of translational symmetry in PST and the breakdown in selection rules occur in a dynamic process initiated by heterophase fluctuations.

CONCLUSION

Two closely related relaxors, PMN and PST, have been studied by Raman scattering. Both materials appear to be constituent of nanoscale ordered clusters with the 1:1 composition of the two different B ions. We assume that such nanoscale arrangement favors the development of fluctuations in course of creation of some new phases, irrespective of whether a transition occurs really or whether it is finally frustrated. Comparing the behavior of PMN, PST and related NBT, we have found enough evidence in support of heterophase fluctuations connected with a competition between the ferroelectric state and additional nonpolar phase.

In any case, Raman scattering gives evidence of a preceding phase in both PMN and PST. The selection rules for Raman scattering occur to be broken down in this preceding phase, so some information from the Brillouin zone appears in the spectra, namely a critical contribution to the broad central peak in PMN and to the initial singlet A_{1g} mode in PST. One should note that our studies showed preliminarily the existence of a central peak in PST as well as some traces of an additional structure around the A_{1g} mode in PMN. Those will be a subject of our further studies. The existence of a special preceding phase is considered as a distinctive characteristic of the transition dynamics in relaxor ferroelectrics with ordered nanoscale clusters. Raman scattering without any electric field is able to reveal a hidden phase transition dynamics in materials consisting of principal nanoscale regions.

ACKNOWLEDGMENTS

This work was supported in part by NASA-NCCW-0088, DE-FG02-94ER75764, and NSF-OSR-9452893 Grants and RFBR Grant No.96-02-17859.

---

*On leave from A.F.Ioffe Physical Technical Institute, Russian Academy of Sciences, St.Petersburg 194021, Russia.

REFERENCES

1. M.E. Lines and A.M. Glass, Principles and Applications of Ferroelectrics and Related Materials, Clarendon, Oxford, 1977

2. C. Boulesteix, F. Varnier, A. Llebaria and E. Husson, J. Sol. St. Chem. **108**, 141 (1994).

3. Z.-G. Ye and H. Schmid, Ferroelectrics **145**, 83 (1993).

4. I.G. Siny, S.G. Lushnikov, C.-S. Tu and V.H. Schmidt, Ferroelectrics **170**, 197 (1995).

5. I.G. Siny, R.S .Katiyar, E. Husson, S.G. Lushnikov and E.A. Rogacheva, Bull. Am. Phys. Soc. **41**, 720 (1996).

6. E.V. Balashova and A.K. Tagantsev, Phys. Rev. B **48**, 9979 (1993).

7. P. Xhonneux, E. Courtens and H. Grimm, Phys. Rev. B **38**, 9331 (1988).

8. C. Ridou, M. Rousseau, P. Daniel, J. Nouet and B. Hennion, Ferroelectrics **124**, 293 (1991).

9. F. Chu, N. Setter and A.K. Tagantsev, J. Appl. Phys. **74**, 5129 (1993).

NANOCRYSTALLINE BaTiO$_3$ FROM THE GAS-CONDENSATION PROCESS

Shaoping Li, J. A. Eastman, L. J. Thompson, Carl. Bjormander, and C. M. Foster

Materials Science division, Argonne National Laboratory, 9700 South Cass Avenue, Argonne IL 60439

ABSTRACT

Nanocrystalline BaTiO$_3$ can be prepared by the gas condensation method at a temperature as low as 700^{0}C, with an average particle size as small as 18nm. The stoichiometry of nanocrystalline BaTiO$_3$ particles can be controlled precisely and reproducibly. Nanocrystalline BaTiO$_3$ powders, fabricated by a novel e-beam evaporation method, show good sintering behavior with a high density at a temperature as low as 1200^{0}C. These samples exhibit a relatively larger dielectric constant than that of coarse-grained BaTiO$_3$. In addition, a thermal analysis has been also carried out to determine the lowest temperature for forming nanostructured BaTiO$_3$ from Ba/Ti oxidized clusters at ambient pressure.

INTRODUCTION

Fine-grained BaTiO$_3$ is an important electronic ceramic widely used in the manufacture of thermistors, multilayer capacitors, and electro-optic devices. Traditional ceramic processing has difficulty in preparing morphologically homogeneous materials with fine grains, resulting in the development of several chemical solution-based methods for preparing well-crystallized submicrometer or nanocrystalline BaTiO$_3$ particles. These processes have the common goal of achieving product formation under mild reaction conditions (low temperatures and short reaction times) in order to limit the extent of grain growth and control particle size. BaTiO$_3$ particles with small and uniform particle size allow for thinner layers of the ceramic to be used in multilayer capacitors without loss of dielectric properties. In addition, small and uniform particle morphology offers the advantage of lower sintering temperature for multilayer devices, which may allow for the use of less expensive electrode materials.

Presently there are several chemical routes for synthesizing nanocrystalline BaTiO$_3$, such as coprecipitation procedures, sol-gel methods, and hydrothermal techniques, in which the coprecipitation procedures and hydrothermal techniques have been used to prepare commercial high purity submicrometer BaTiO$_3$ powders. There are two major shortcomings for coprecipitation procedures. One is the relative difficulty in introducing dopants into BaTiO$_3$. The other is that all the coprecipitated, single phase, complex compounds have been 1:1 for Ba:Ti [1]. As a result, the method involving a unique precursor compound applies to BaTiO$_3$ only and cannot be used to synthesize other compounds that are also of great technical importance in the BaO-TiO$_2$ system, such as BaTi$_4$O$_9$, BaTi$_9$O$_{20}$, and BaTi$_3$O$_{11}$ [2].

On the other hand, hydrothermal techniques also have many disadvantages[3-4] in that they involve several reaction steps and pressures to generate crystalline BaTiO$_3$ particles, and need complicated post-treatment of the powders in order to adjust the stoichiometry.

The purposes of the present work are two fold. One is to identify the feasibility of commercially synthesizing nanocrystalline multicomponent oxides, such as BaTiO$_3$, using the gas condensation method(GC). Unlike chemical synthesis methods, the gas condensation method involves no solution chemistry. The other is to determine the lowest temperature for forming nanostructured BaTiO$_3$ from Ba/Ti oxidized precursors at ambient pressure. This is important because the reaction of Ba/Ti oxidized clusters made by a gas condensation method does not involve a hydrolysis reaction, which creates the possibility of preparing nanocrystalline BaTiO$_3$ at the lowest temperature at ambient pressure condition.

45

SYNTHESIS AND MICROSTRUCTURE

We employed a two-source evaporation process to simultaneously produce a homogenous mixture of partially oxidized Ba/Ti clusters. The processing technique and parameters have been reported somewhere else[5]. One important feature of the process used is that the Ti and $BaTiO_3$ source materials can be evaporated in an oxygen environment rather than the more common inter gas environment typically used with gas condensation.

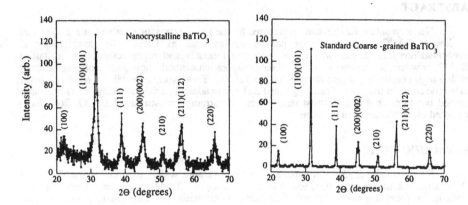

Figure 1 (a)X-ray diffraction scans from nanocrystalline $BaTiO_3$ annealed at 700 ^{0}C in air for 2 hours. (b) X-ray diffraction pattern of standard polycrystalline $BaTiO_3$.

After annealing the mixture of Ba/Ti oxidized powders at 700 ^{0}C in air for 1-2 hours, well crystallized nanocrystalline $BaTiO_3$ was obtained. Figure 1(a) is an x-ray scan for powders after annealing at 700^{0}C for 2 hours, which indicates single phase $BaTiO_3$ is formed. The dominant phase is most likely a pseudo-cubic phase, although peak broadening due to small particle size makes it difficult to distinguish the phase from the tetragonal phase typically found for coarse grained materials. The XRD observations were also reproduced by electron diffraction. Bright-field and dark field TEM images of these nanoparticles are shown in Figure 2. The average particle size is less than 20nm. The nanocrystalline $BaTiO_3$ powders were next pelletized by cold isostatic pressing without use of a binder.

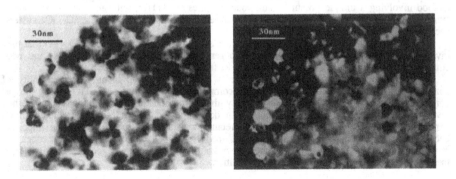

Figure 2 TEM micrographs of nano-$BaTiO_3$: a) Bright-field image; b) Dark-field image.

The pellets were then sintered at temperature ranging from 900 to 1250°C. The densities of the samples sintered at different temperatures are given in Fig.3. Clearly, the well-crystallized nanocrystalline BaTiO₃ particles show good sinterability. Figure 4. is the plot of the dielectric constants and losses as a function temperature of the sample sintered at 1200°C.

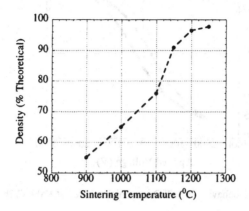

Figure 3 Plot of the density as a function of sintering temperature of nanocrystalline BaTiO₃ pellets.

DIELECTRIC BEHAVIOR

The dielectric constants of samples were determined at 100 kHz during heating. Heating rates were 3°C/min. Quite clearly, the phase transition behavior does not obey the Curie's law, exhibiting a rather diffused phase transition. It can be also found that the orthorhombic-tetragonal phase transformation slightly shifts up to a higher temperature, which is consistent with the recent thermal analyses done by Frey and Payne[6]. The dielectric constants of BaTiO₃ sample made by nanocrystalline powders is larger than polycrystalline BaTiO₃ with coarse grain[7-8] at the room temperature.

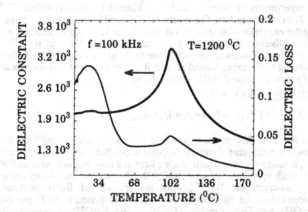

Figure 4 The dependence of dielectric constants and losses on temperature for the sample made by nanocrystalline BaTiO₃ from GC

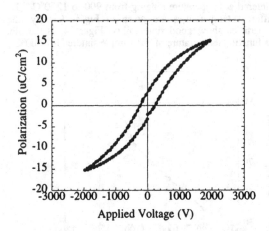

Figure 5 P vs. E hysteresis behavior of BaTiO₃ ceramics sintered at 1200 °C/2h.

However, the dielectric constants reported here are smaller than those of BaTiO₃ ceramics with ultra-fine grains reported in literature[4,8-10]. It is probably due to our inability to fully polarize samples prior to dielectric measurement. Fig.5 shows the polarization vs. applied electric field behavior of the samples sintered at 1200 °C. Apparently, the induced polarization are quite smaller than that of BaTiO₃ with coarse grains, even though the appearance of the classic P vs. E hysteresis is noticeable.

PHASE DEVELOPMENT

Thermogravimetric analysis(TGA) and differential thermal analysis(DTA) were employed to determine the nature of the reactions that led to the formation of nanocrystalline BaTiO₃ particles. The mixture of Ba/Ti precursors was first fully oxidized in air for several months. And then they were analyzed by DTA/TG at different heating rates to disclose the temperature at which exothermic/endothermic reaction took place. Heating rates from 1^0C/min to 20^0C/min were employed in both experiments in order to study the kinetics of nanocrystalline BaTiO₃ particle formation from Ba/Ti oxidized clusters. The results of TGA and DTA experiments are presented in Figure 6, indicating the existence of at least two stages. Stage 1, which extends up to 220°C, was accompanied by an exothermic reaction and a continuous weight loss. In stage 2, at 400°C to 600°C, depending upon heating rates, a sharp decrease in weight with apparent exothermic reaction was observed. According to the experimental results, it is hypothesized that the reactions involved in the production of nanocrystalline BaTiO₃ are as follows:

$$BaO_2 + Ba(OH)_2 \rightarrow BaO_2 + H_2 \uparrow$$

$$BaO_2 + TiO_{2-x}(BaTiO_4) \rightarrow BaTiO_3 + 1/2 O_{2-x} \uparrow$$

It is expected when Ba atom clusters are exposed to air, they could be oxidized as $Ba(OH)_2$ or BaO_2 because of absorbing moisture. It should be pointed out that from thermodynamic and kinetic considerations BaO_2 is relatively more stable oxide phase at a low temperature than BaO, especially for clusters[11], although very little is known about the detailed mechanism of barium oxidation. The formation of nanocrystalline BaTiO₃ from the mixture of Ba/Ti oxidized clusters is hypothesized to proceed along the following path. At low temperature, Ba/Ti precursors consists of a mixture of $Ba(OH)_2$ and TiO_2. Around 100-250 °C, the $Ba(OH)_2$ converts to BaO_2. Such a conversion should result in a 2 % weight loss, which is in close agreement with the observed weight loss.

48

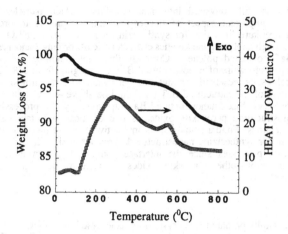

Figure 6 (a) Thermogravimetric analysis and differential thermal analysis curves of the mixture of Ba/Ti oxidized clusters.

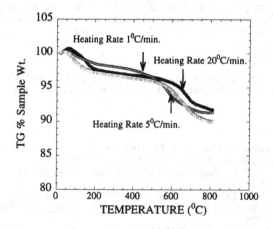

Figure 6 (b) TGA of the mixture of Ba/Ti clusters at different heating rates.

The amorphous mixture of BaO_2 and TiO_2 clusters crystallized to form nanocrystalline $BaTiO_3$ on heating in a temperature range from 400 to 600°C, depending upon the heating rates. Such a reaction is expected to be exothermic with a weight loss of 6-7%, in a good agreement with the experimental observations. Obviously, the faster heat rate leads to the higher temperature for nanocrystalline $BaTiO_3$ formation if the transformation of Ba/Ti precursors to nanocrystalline $BaTiO_3$ in static air is controlled by the oxygen diffusion process. The precise information of kinetics of formation nanocrystalline $BaTiO_3$ from Ba/Ti oxidized clusters can quantitatively obtained from Fig.(6b). From above experimental results, it is quite clearly that under ambient pressure condition well crystallized nanostructured $BaTiO_3$ could not be prepared from Ba/Ti oxidized clusters at a temperature below 400°C within a relatively short period of time. This result is actually consistent with the recent experimental observation by Nourbakhsh et al.[12].

It should be mentioned here that if the mixture of Ba/Ti precursors was not fully oxidized, its DTA/TGA behavior will be different from above presented results, although the mixture of

Ba/Ti precursors can be still converted into nanocrystalline BaTiO$_3$ particles at the similar temperature range. An important aspect of our experimental results is to provide a rough assessment of the temperature limitation for synthesizing nanostructured BaTiO$_3$ under ambient pressure conditions through analyzing the kinetics characteristic of forming nanocrystalline BaTiO$_3$ from the mixture Ba/Ti oxidized powders. Currently there are a number of recent literature reporting synthesizing temperature of nanostructured BaTiO$_3$, ranging from 200^{0}C to 900^{0}C[13-18]. Our experimental results presented here indicate that within a short synthesis time period the synthesizing temperature of nanostructured BaTiO$_3$ should be above 400^{0}C. Otherwise, it is not possible to obtain well crystallized nanostructured BaTiO$_3$. In reality, the processing temperature for synthesizing nanostructured perovskite oxides is one of most important issues for future microelectronics applications. Fundamentally, it is imperative to determine the lowest possible processing temperature for synthesizing nanostructured perovskite oxides in order to use of them with standard Si based processes since the interface compatibility between silicon or other semiconductors and numerous other perovskite oxides is critical for developing new generation microelectronics devices

CONCLUSION

We have successfully prepared well crystallized nanocrystalline BaTiO$_3$ at a temperature as low as 700^{0}C by using a gas condensation method involving evaporation of Ti and BaTiO$_3$ sources in both oxygen and non-oxygen environments. The dielectric properties of sintered BaTiO$_3$ made from nanocrystalline BaTiO$_3$ powders have been reported. The obtained barium titanate powders sinter to high density at a temperature as low as 1200^{0}C, which is favorable for the manufacture of multilayer capacitors. The possible mechanism responsible for forming nanocrystalline BaTiO$_3$ through the mixture of Ba/Ti oxidized clusters has been also discussed.

ACKNOWLEDGMENTS

This work was supported by the U.S. Department of Energy, BES-Materials Science, under Contract W-31-109-Eng-38. We thank Dr Mark Harsh for assisting DTA/TGA measurement.

REFERENCES

1. P.P Phule, and S.H. Risbud, J. Mater. Sci. **25**, 1169 (1990).
2. Zhimin Zhong and K. Gallagher, J. Mater. Res.**10**(4), 945, (1995).
3. J. Menashi, R.C. Reid and L.P. Wagner, Barium Titanate Based Dielectric Compositions, U.S. Patent 4,832,939, May 23, (1989).
4. Y-S Her, E.Matijevic, and M.C. Chou, J. Mater. Res.**10**(12), 3106-14 (1995).
5. Shaoping Li, Jeffy A. Eastman, L.T. Thompson, and P.M. Baldo, Mat. Res. Soc. Symp. Proc. Vol.**400**, 83-88, (1996).
6. M.H. Frey and D.A. Payne, Phy. Rev. B Vol.**54**(5), 3158 (1996).
7. B.W. Lee and K. H. Auh, J. Mater. Res.**10**(6), 1416(1995).
8. K.Kinoshita and A. Yamaji, J.Appl. Phys. **47**, 371, (1976).
9. G. Arlt, D. Hennings, and G.deWith, J.Appl.Phys.**58**, 1619(1985); and J.C. Niepce, Electroceramics **4**, Aachen, 29 (1994).
9. W.R. Buessem, L.E. Cross, and A.K. Goswami, J.Am. Ceram. Soc.**49**, 36 (1966).
10. Technical Report from Cabot Performance Materials (1995).
11. H. J. Schmutzler, M. M. Antony, and K. Sandhage, J. Am.Ceram.Soc.**77**, 721-29 (1994).
12. S. Nourbashsh, I.Vasilyeva, and J.N. Carter, Appl. Phys. Lett. **66**(21), 2804-08 (1995).
13. T. Sonegawa, et al., Appl. Phys. Lett.**69**(15), 2193 (1996).
14. D.L. Kaiser, et al., Appl. Phys. Lett. **66**(21), 2801 (1995); L.A.Wills,et al., Appl.Phys. Lett. 60(1) **41** (1992); and B.S. Kwak, et al., J. Appl. Phys. **69**(2), 767 (1991).
15. G.M. Davis and M.C. Gower, Appl. Phys. Lett. **55**(2), 112 (1989).
16. R.A. Mckee, et al., Appl. Phys. Lett. **59**(7), 789 (1991).
17. K. Iijima, et al., Appl. Phys. Lett. **56**(6), 527 (1990).
18. P.C.V. Buskirk, et al., J. Mater. Res. 7(3), 542 (1992).

THE EFFECT OF SULFATING ON THE CRYSTALLINE STRUCTURE OF SOL-GEL ZIRCONIA NANOPHASES

BOKHIMI*, A. MORALES*, O. NOVARO*, M. PORTILLA**, T. LOPEZ***,
F. TZOMPANTZI***, R. GOMEZ***
*Institute of Physics, UNAM, A. P. 20-364, 01000 México D. F., Mexico,
bokhimi@sysul2.ifisicacu.unam.mx.
**Faculty of Chemistry, UNAM, A. P. 70-197, 01000 México D. F., Mexico.
***Faculty of Chemistry, UAM-I, A. P. 54-534, 09340 Mexico D. F., Mexico.

ABSTRACT

Nanophases of sol-gel zirconia were prepared with HCl, $C_2H_4O_2$ and NH_4OH as hydrolysis catalysts, and sulfated with H_2SO_4. They were analyzed by using X-ray powder diffraction, and their crystalline structure was refined by using the Rietveld method. All samples annealed below 300 °C were amorphous. The non-sulfated samples crystallized around 350 °C, while the sulfated samples crystallized around 600 °C, when they started loosing sulfate ions. In the initial stage of crystallization, both the tetragonal and monoclinic nanophases coexisted, with the tetragonal as the main phase. Annealing the samples at higher temperatures transformed the tetragonal nanophase, stabilized by OH ions, into the monoclinic one.

INTRODUCTION

Zirconia (ZrO_2) has many applications in high technology [1-3]. For example, zirconia-based composites are light, and able to withstand heat, corrosion and wearing. Many of these composites have been used for coating turbine blades [4], and for making engine block, body valves, cylinder liners, pistons, and bearings in internal-combustion engines of automobiles [5]. This is because composites based on stabilized zirconia have a large fracture toughness produced by the martensitic transformation of the zirconia tetragonal phase into the monoclinic one [5]. The mechanical properties of these composites depend on the zirconia crystallite size, in special, for sizes in the range of the nanometers.

Superacidity is another important property of zirconia, which occurs when it is sulfated [6]. In this case, sulfate ions strongly interact with the zirconia matrix. The presence of defects, which are normally found in nanostructured oxides [7, 8], in the zirconia lattice will favor this.

Using the sol-gel technique can make oxide nanophases. Here, hydrolysis catalyst and annealing temperature determine crystallite size, morphology and cation deficiency of the nanophase [7, 8].

In the present paper, we will report the dependence on hydrolysis catalyst and temperature of the crystalline structure of zirconia and sulfated zirconia nanophases obtained by the sol-gel technique and characterized by using X-ray powder diffraction.

EXPERIMENTAL

Sample Preparation

Zirconium n-butoxide in terbutilic alcohol, containing the hydrolysis catalyst HCl, $C_2H_4O_2$, or NH_4OH, was used for preparing the sol-gel zirconia. Gels were dried in air at 100 °C, and annealed at 200, 400, 600, and 800 °C. Sulfated zirconia was prepared by sulfating with H_2SO_4 the dried sol-gel zirconia. Sulfated samples were annealed in air at 200, 400, 600, and 800 °C.

X-ray Diffraction Characterization

The crystalline structure of the nanophases was characterized by using X-ray powder diffraction, and refined by using the Rietveld method. X-ray diffraction patterns were measured with CuK$_\alpha$ radiation. Peak profiles were modeled with a seudo-Voigt function having average crystallite size as one of the profile-breadth fitting parameters [9]. The standard deviations, showing the variation of the last figures of the corresponding number, were given in parenthesis.

TGA Analysis

This analysis was done in a thermoanalyzer Dupont model 950. Both sulfated and non-sulfated samples were analyzed in air from 25 to 1200 °C at the annealing rate of 20 degree/min.

RESULTS AND DISCUSSION

Non-Sulfated Zirconia

Annealing sol-gel zirconia samples from room temperature to 500 °C produced a weight loss (Fig. 1). Evaporation of the residual volatile components used for the sample preparation produced this. Annealing above this temperature did not produce any additional change in the sample weight.

The structure of the samples annealed below 300 °C was amorphous (Fig. 2A) and

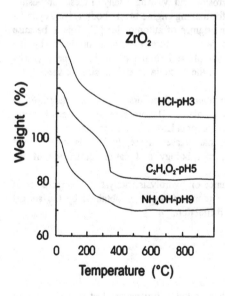

Fig. 1 TGA curve of sol-gel zirconia prepared with different hydrolysis catalysts.

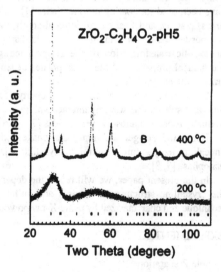

Fig. 2 X-ray diffraction curves of sol-gel zirconia prepared with acetic acid as hydrolysis catalyst. Tick marks correspond to the tetragonal phase of zirconia.

corresponded to $Zr(OH)_4$. The center of the main broad peak of the amorphous phase had the same position as the main peak of tetragonal zirconia (Fig 2B). This suggested that the local order in both phases was the same. Annealing the samples in air between 300 and 350 °C crystallized the amorphous phase.

Crystallized samples had two nanocrystalline phases (Figs. 3 and 4). One phase was tetragonal with space group $P4_2/nmcm$, and the another was monoclinic with space group $P2_1/c$. These nanostructured phases had crystallite sizes that varied between 4 and 34 nm (Table 1).

Table 1. Non-Sulfated Zirconia. Phase Composition and Average Crystallite Size as a Function of Hydrolysis Catalyst and Temperature

hydrolysis catalyst	T (°C)	tetragonal (wt %)	monoclinic (wt %)	tetragonal crystallite size (nm)	monoclinic crystallite size (nm)
HCl	400	78 (5)	22 (4)	17 (1)	4.1 (2)
	600	35 (1)	65 (2)	24 (1)	21 (1)
	800	13 (2)	87 (3)	31 (3)	31.1 (8)
$C_2H_4O_2$	400	100	-------	12.5 (6)	-------
	600	91 (1)	8.6 (2)	22.0 (5)	12.8 (1)
	800	17 (2)	82 (3)	22 (3)	29 (1)
NH_4OH	400	70 (5)	30 (6)	8.4 (3)	5.8 (5)
	600	45 (2)	55 (1)	20.7 (8)	22 (1)
	800	11.7 (4)	88.3 (7)	32 (3)	34 (1)

It is known that doping microcrystalline zirconia with large ions like Y, Ca, or Mg ions stabilizes the tetragonal phase at low temperature [10]. In the sol-gel zirconia reported in the present work, the solutions and the precursors used in the preparation did not include any of these

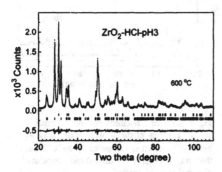

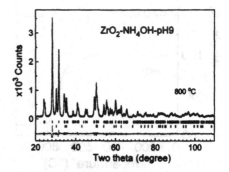

Fig. 3 Rietveld refinement plot for sol-gel zirconia prepared with HCl and annealed at 600 °C. It has the tetragonal (upper tick marks) and monoclinic phases (lower tick marks).

Fig. 4 Rietveld refinement plot of sol-gel zirconia prepared with ammonium hydroxide and annealed at 800 °C. It has the monoclinic (upper tick marks) and tetragonal phases (lower tick marks).

ions, or any similar. OH ions, however, were abundant; therefore, they were the only ions that could stabilize the tetragonal structure. This result agrees with those reported for zirconia obtained by using $Zr(NO_3)_4$ and $Zr(OH)_4$ as precursors [11].

Annealing the samples at even higher temperatures increased the crystallite size, and transformed the tetragonal nanophase into the monoclinic one (Table 1). Leaving of OH ions from the stabilized tetragonal phase caused this transformation.

Sulfated Zirconia

Sol-gel zirconia was prepared with HCl, $C_2H_4O_2$ and NH_4OH hydrolysis catalysts, and sulfated with H_2SO_4. None of the sulfated samples crystallized below 600 °C. This result contrasts with those obtained in the non-sulfated sol-gel zirconia, where crystallization occurred between 300 and 350 °C.

Sulfated samples started loosing part of its weight when they were annealed above 500 °C. They had one transformation around 500 °C and a second one around 600 °C (Fig. 5). After annealing the sample at 800 °C, the total weight loss depended on the hydrolysis catalyst used in the preparation; it was 31.7% for HCl, 29.3% for $C_2H_4O_2$, and 41.5% for NH_4OH. In the transformation at 500 °C, the sulfated samples prepared with HCl and NH_4OH as hydrolysis catalyst only lost 2.4 and 14.7% in weight respectively; the rest of the weight loss occurred above 600 °C. In contrast to this, the samples prepared with acetic acid lost most of their weight (24.1 of 29.3%) in the first transformation. If the weight loss, observed around 500 °C and above 600 °C, was associated to the evaporation of SO_x ions from the sample, then, the above results will suggest that these SO_x ions had a strong interaction with the sol-gel zirconia prepared with hydrochloric

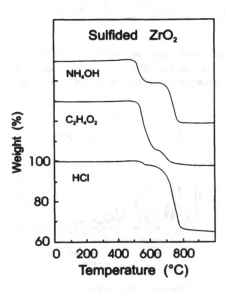

Fig. 5 TGA curve of sulfated sol-gel zirconia prepared with different hydrolysis catalysts.

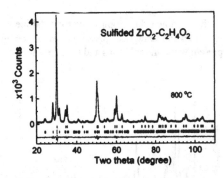

Fig. 6 Rietveld refinement plot of the sulfated sol-gel zirconia prepared with acetic acid as hydrolysis catalyst and annealed at 800 °C. It has the tetragonal (upper tick marks) and monoclinic phases (lower tick marks).

Table 2. Sulfated Zirconia. Phase Composition and Average Crystallite Size
as a Function of Hydrolysis Catalyst and Temperature

hydrolysis catalyst	T (°C)	tetragonal (wt %)	monoclinic (wt %)	tetragonal crystallite size (nm)	monoclinic crystallite size (nm)
HCl	800	3.0 (6)	97 (3)	24 (9)	30.3 (8)
$C_2H_2O_2$	600	91 (3)	8.7 (2)	15.3 (4)	10 (1)
	800	66 (2)	34 (4)	28.4 (1)	32 (2)
NH$_4$OH	800	3.7 (6)	96 (3)	24 (7)	29.5 (7)

acid and ammonium hydroxide, and a weak interaction with the sol-gel zirconia prepared with acetic acid.

In the samples prepared with acetic acid, sulfating stabilized the tetragonal structure (Fig. 6). In the samples annealed at 800 °C, the concentration of the tetragonal phase was 66 (2) wt % (Table 2), while it was only 17 (2) wt % for the respective non-sulfated samples annealed at the same temperature (Table 1).

The sulfated samples annealed at 600 °C and prepared with HCl and NH$_4$OH hydrolysis catalysts had a phase different from those of zirconium oxide. This phase should correspond to a sulfate of zirconium (Zr-O-S). That means that sulfate, zirconium, and oxygen ions reacted between each other to produce a new phase. Annealing these samples at 800 °C transformed this Zr-O-S phase into both the tetragonal and monoclinic zirconia phases (Table 2). This transformation correlated with the observed weight loss between 600 and 800 °C (Fig. 5). The evaporation of the SO$_x$ ions generated the observed weight loss.

CONCLUSIONS

The atomic distribution of sol-gel zirconia annealed below 300 °C was amorphous. Its local order, however, was similar to the local order of the tetragonal crystalline phase. The amorphous phase crystallized into the tetragonal and monoclinic zirconia nanophases, with the monoclinic crystals having a smaller size. Annealing the samples at higher temperatures transformed the tetragonal phase into the monoclinic one.

Sulfating sol-gel zirconia caused a strong interaction between SO$_x$ ions and the zirconia precursor matrix, and stabilized the amorphous phase. When sulfated samples were annealed above 500 °C, SO$_x$ ions left the sample, and produced a large weight loss.

ACKNOWLEDGMENTS

We would like to thank Mr. A. Sánchez for technical support, and the CONACyT (Mexico), the CNRS (France), and the NSF (USA) for financial support.

REFERENCES

1. O. Tatsuya, T. Tomoshiro, S. Masayoshi and N. Hidetoshi, J. of Membrane Science **118**, 151 (1996).
2. C. J. Alexander, Am. Cer. Soc. Bull. **75**, 52 (1996).
3. B. A. Cotton and M. J. Mayo, Scripta Materialia **34**, 809 (1996).
4. B. Nagaraj, G. Katz, A. F. Maricocchi and M. Rosenzweig, Proceedings of the International Gas Turbine and Aeroengine Congress and Exposition (American Society of Mechanical Engineers, ASME, New York, N. Y., USA, 1pp, 1995).
5. J. F. Braza, STLE Tribology Transactions **38**, 146 (1994).
6. B. Li, and R. D. Gonzalez, Industrial & Engineering Chem. Res. **35**, 3141 (1996).
7. Bokhimi, A. Morales, T. López and R. Gómez, J. Sol. State Chem. **115**, 411 (1995).
8. Bokhimi, A. Morales, O. Novaro, T. López, E. Sánchez and R. Gómez, J. Mater. Res. **11**, 2788 (1995).
9. P. Thompson, D. E. Cox, and J. B. Hastings, *J. Appl. Crystallogr.* **20**, 79 (1987).
10. C. J. Norman, P. A. Goulding and Y. McAlpine, Catalysis Today **20**, 313 (1994).
11. R. Srinivasan, and B. H. Davis, *Catal. Lett.* **14**, 165 (1992).

CLUSTER FORMATION BY LASER ABLATION OF ZEOLITES

HIROSHI T. KOMIYAMA, AZUCHI HARANO, TATSUYA OKUBO and MASAYOSHI SADAKATA
Department of Chemical System Engineering, The University of Tokyo
7-3-1 Hongo, Bunkyo-ku, Tokyo 113, Japan. okubo@chemsys.t.u-tokyo.ac.jp

ABSTRACT

Several kinds of zeolites, crystal-SiO_2 (α-quartz), amorphous-SiO_2 (quartz glass and ultrafine particles) and α-Al_2O_3, were ablated by an Nd:YAG laser. Generated positive ions from the targets were measured by TOF-MS (time-of-flight mass spectrometry). In the TOF mass spectra of ablated zeolites, TO_x (x=0-2, T = tetrahedral atom, e.g., Si, Al), T_2O_x (x=1) and T_3O_x (x=4,5) were observed up to m/z=170 (m = mass, z = plus charge of clusters). In the spectra due to the α-quartz, quartz glass and an α-Al_2O_3 plate, smaller species, T^+, TO^+ and TO_2^+ , were mainly detected. These results demonstrate that the clusters from zeolites reflect the characters of the mother structure.

INTRODUCTION

Zeolites are constructed from TO_4 tetrahedra and each apical oxygen atom is shared with an adjacent tetrahedron [1-3]. Silicon atom forms bonds with four neighboring atoms in a

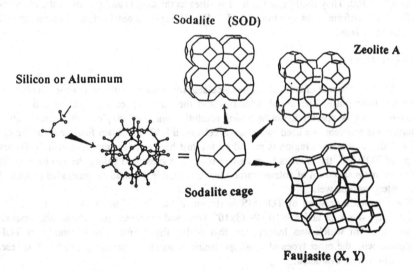

Figure 1   Schematics of zeolite and sodalite cage frameworks.

57

Mat. Res. Soc. Symp. Proc. Vol. 457 °1997 Materials Research Society tetrahedral geometry. Based on this character, zeolites have varieties of structures enclosing micropores and each zeolite has original pore structure and network. For example, LTA(Zeolite A) and FAU (Zeolite X and Y) consist of sodalite cages which are constructed from the 24 T atoms and 48 oxygen atoms. The shape of sodalite cages and the structures of these zeolites are shown in the Figure 1. The sodalite cage is one of the most common cages in zeolites which consist of 4-member rings and 6-member rings. The window size of the 6-member ring is near 3 Å.

The objective of this study is to extract the characteristic unit of zeolite structure into gas phase, and thus to generate new clusters. Structured clusters generated could be precursors for new microporous materials. The information on the cluster is also useful to understand the structure and the properties of the zeolites. Recently, great attention has been paid to the fullerenes [4]. If we can isolate the sodalite cage from the zeolites, it might be another "fullerene". In order to isolate the characteristic structures unique in zeolites, an laser ablation was applied. As the first step, physical laser ablation using an Nd:YAG laser was tested. To measure the mass of the clusters, TOF-MS (time-of-flight mass spectrometer) system was used [5,6]. The main chamber was pumped by a diffusion pump (DPF-6z, DIAVAC LIMITED, 2000L/min).

Recently, some reports on the ablation of zeolites were published [7-10]. For example, Peachey et al.[9] tried to make zeolitic thin film by pulse laser deposition (PLD). Two kinds of zeolites, Mordenite and Ferrierite, were ablated and the mass spectra of the species involved in the PLD process were measured. But the mass spectra reported were restricted to m/z=300 - 400. Jeong et al.[10] ablated and measured the mass number of prepared positive and negative species up to m/z = 360. They used three kinds of zeolites as the targets and measured the mass spectra by Fourier transform mass spectrometry (FTMS). The spectra obtained in this research differed from those of Jeong.

EXPERIMENTAL

In TOF-MS, the m/z of cations can be calculated by measuring the delay time of cation arrival from the laser pulse. The initial velocities and the initial space positions of cations have a distribution, so the spectra do not have high resolution under the single constant acceleration. To conquer this problem, we used two stage acceleration [11]. To obtain finer spectra, the electric field of the acceleration region is pulsed by the fast high voltage transistor switch. To test the prepared TOF-MS, the clusters from carbon rods were measured before the zeolite experiment. Since the mass spectrum of ablated carbon was reasonable, it could be concluded that the TOF-MS system worked well.

The schematic drawing of TOF-MS is shown in Figure 2. The pressure in the flight tube (1.2m long) was kept under 4×10^{-4} Pa (3×10^{-6} Torr) and the mean free path at this pressure was longer than 5m which was longer than that of the flight tube. The advantage of TOF-MS compared with the other types of mass spectrometry its short measuring time. In this research, only cations were evaluated.

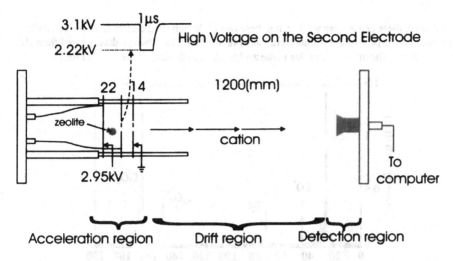

3.1kV ⎯⎯⎯ 1μs

2.22kV ⎯⎯⎯⎯⎯

High Voltage on the Second Electrode

22 14

1200(mm)

zeolite

cation

To computer

2.95kV

Acceleration region   Drift region   Detection region

Figure 2.   Schematic drawing of TOF-MS

Six kinds of zeolites (Zeolite A, X, Y, Mordenite, ZSM-5 and Ferrierite supplied by TOSOH), crystal-SiO_2 (α-quartz), amorphous-SiO_2 (quartz glass and ultrafine particle), and α-Al_2O_3 were ablated as the targets by an Nd:YAG laser (the second harmonic wave 532nm and the third harmonic wave 355nm). The structures of zeolites and the Si/Al ratios are listed in TABLE I. The zeolites and SiO_2 pellet were 20mm in diameter and ~5mm in thickness. After sintered in an electric furnace at 523K for 10 hours, these pellets were set on the electric plate of the acceleration region and ablation was performed. Other materials were ground and the pellets were prepared.

TABLE I.   The Si/Al ratios of zeolites which were used in this experiment

	Si/Al ratio
Zeolite A	1.00-1.03
Zeolite X (Faujasite)	1.27
Zeolite Y (Faujasite)	2.82
Mordenite	4.90
Ferrierite	8.82
ZSM-5	11.9

RESULTS AND DISCUSSION

The TOF mass spectra of ablated zeolites and other materials are shown in Figures 3 - 6. The spectrum of ablated zeolite A is shown in Figure 3. Five peaks are obtained up to m/z = 200. The peaks at m/z = 28, 44, 70, 148 and 163 were assigned to T', TO', T_2O', T_3O_4' and T_3O_5', respectively. The TOF mass spectrum of ablated Ferrierite is shown in Figure 4. The differences between Figures 3 and 4 are the peaks at m/z = 100 and 116. When Zeolite A and other zeolites were ablated, these peaks were weaker, but these two peaks appeared clearly from the Ferrierite. The characteristic peaks from zeolite were m/z=148 and 163. These two peaks appeared from all ablated zeolites. The mass spectra due to the zeolites did not depend on the Si/Al ratios (see

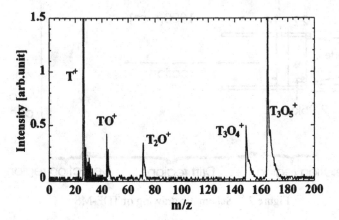

Figure 3.   A TOF mass spectrum of ablated Na-Zeolite A (Si/Al=1.00)

Laser : Nd:YAG 532nm : Laser power : 255mJ pulse^{-1} cm^{-2}

Zeolite A was ablated by Nd:YAG laser and five peaks are obtained up to m/z = 200. The peaks at m/z = 28, 44, 70, 148 and 163 were assigned to T^+, TO^+, T_2O^+, $T_3O_4^+$ and $T_3O_5^+$, respectively.

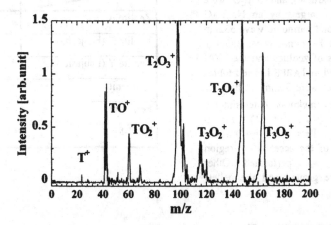

Figure 4.   A TOF mass spectrum of ablated Ferrierite

Laser : Nd:YAG 532nm : Laser power : 255mJ pulse^{-1} cm^{-2}

Ferrierite was ablated and seven peak were observed. Each peak was assigned to T^+, TO^+, T_2O^+, $T_2O_3^+$, $T_3O_2^+$, $T_3O_4^+$ and $T_3O_5^+$.

Zeolite X and Y. They have the same topology.) In this experiment, three kinds of Zeolite A (The cations in the structure were different, K^+, Na^+ and Ca^{2+}) were used, but the cations in the structures have no influence on the spectra.

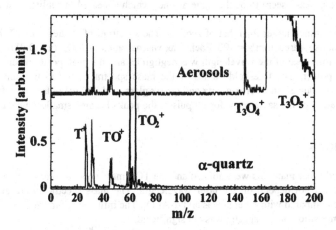

Figure 5.   TOF mass spectra of ablated α-quartz and aerosol
Laser : Nd:YAG 532nm : Laser power : 550mJ pulse^{-1} cm^{-2}

This figure shows the two mass spectra of ablated α-quartz and aerosol. Clear differences between two spectra were observed. From the aerosol, peaks at m/z = 148 and 163 were observed, which seem to be the same ones in the spectra from zeolites.

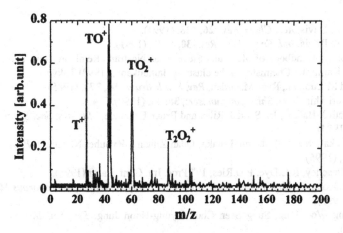

Figure 6.   A TOF mass spectrum of ablated Al_2O_3
Laser : Nd:YAG 532nm : Laser power : 450mJ pulse^{-1} cm^{-2}

Peaks appeared at m/z = 100. Each peak was assigned to AlO_x^+ (x = 0-2) or $Al_2O_x^+$ (x = 2, 3). These peaks were different from that of zeolites and SiO₂.

The spectra of ablated SiO_2 are shown in Figure 5. From the α-quartz and quartz glass, small mass peaks were detected, Si^+, SiO^+ and SiO_2^+. The spectrum of quartz glass was the same as from α-quartz. But from the ultrafine particles, the peaks at m/z = 148 and 163 were mainly observed. These peaks seem to be the same as ones which appeared in zeolites, but the FWMH of the peak
at m/z = 163 is much larger than that of zeolites. The spectrum of ablated α-Al_2O_3 is shown in Figure 6. Peaks appeared at m/z = 100. Each peak was assigned to AlO_x^+ (x = 0-2) or $Al_2O_x^+$ (x = 2, 3). The influence of the wavelength was negligible, and the peak positions did not depend on the laser power. The threshold values of the mass spectra were 145 mJ pulse^{-1} cm^{-2} for zeolites, 210 mJ pulse^{-1} cm^{-2} for α-quartz and quartz glass, and 210 mJ pulse^{-1} cm^{-2} for α-Al_2O_3. After the ablation of the same point for 20 pulses, the peaks became smaller due damage to the pellet surfaces.

CONCLUSIONS

Zeolites and other materials were ablated and the TOF mass spectra were obtained. The TOF mass spectra did not depend on the laser wavelength. The mass spectra due to zeolites were characteristic compared with other silicas and α-Al_2O_3. The influence of the zeolite structure and chemical composition on the spectra was not significant.

ACKNOWLEDGMENTS

We are grateful to TOSOH for the donation of zeolite samples.

REFERENCES

1.   Mark E. Davis, *Acc. Chem. Res.*, **26**, 438, (1993).
2.   Mark E. Davis, *Ind. Eng. Chem. Res.*, **30**, 1675, (1991).
3.   R.Szostak, Handbook of Molecular Sieves, Van Nostrand Reinhold, New York.
4.   Koua Kajimoto, Chemistry of the clusters, Baihuukann, (1992) Tokyo.
5.   David M.Lubman, Russ M.Jordan, *Rev. Sci. Instrum.*, **56**, 373, (1985).
6.   Hisanori Shinohara, *Shituryou bunnseki*, **38**, 43, (1990).
7.   Kenneth J. Balkus, Jr., Scott J. Riley and Bruce E. Gande, *Mat. Res. Soc. Symp. Proc.* **351**, 437, (1994).
8.   Kazutaka Ishigoh, Katsumi Tanaka, Quan Zhuang, Ryouhei Nakata, *J. Phys. Chem.*, **99**, 12231, (1995).
9.   N.M.Peachey, R.C.Dye, P.D.Ries, 17th Proc. Int. Conf. Laser (1995).
10.   Seijin Jeong, Keith J. Fisher, Russell F. Howe, Gray D. Willett, *Microporous Materials*, **4**, 467, (1995).
11.   Kwang Woo Jung, Sung Seen Choi, Kyung-Hoon Jung, *Rev. Sci. Instrum.*, **62**, 2125, (1991).

ELECTRICAL CONDUCTIVITY OF PURE AND DOPED NANOCRYSTALLINE CERIUM OXIDE

E.B. Lavik, and Y.-M. Chiang,
Department of Materials Science and Engineering,
Massachusetts Institute of Technology, Cambridge, MA 02139

ABSTRACT

We have previously shown that dense nanocrystalline CeO_{2-x} of approximately 10 nm grain size exhibits enhanced electrical conductivity and an enthalpy of reduction that is more than 2.4 eV lower than that for conventional ceria [1, 2]. These effects were attributed to preferential interface reduction. In this work, we investigated the relationship between interfacial area, heat treatment conditions, and conductivity by varying the grain size of dense samples through annealing at various temperatures. It is shown that the conductivity does not scale in direct proportion to interfacial area. Moderate temperature (700 °C) anneals which change the grain size by only a few nanometers reduce the conductivity by three orders of magnitude. It is suggested that atomistic relaxation occurs at the interfaces, and eliminates many low energy defect sites.

INTRODUCTION

Cerium oxide is an important catalytic material for oxidation and reduction of gas phase species such as carbon monoxide and sulfur dioxide [3, 4]. Nanocrystalline cerium oxide exhibits significantly improved catalytic properties, including the ability to achieve greater conversion at lower temperatures than its coarse-grained counterpart for the carbon monoxide and sulfur dioxide reactions [5].

It has been proposed that surface oxygen defects participate in the redox process, and that the energies of these reactions vary with the surface orientation. We have sought to understand the defect thermodynamics of nanocrystalline ceria though characterization of the electrical properties. Previously, we have shown that dense nanocrystalline CeO_{2-x} exhibits a lower enthalpy of reduction and higher conductivity than the equivalent coarse-grained counterpart, and we have attributed this behavior to interfacial reduction [1, 2]. In the present work, we have annealed samples to produce a range of grain-sizes to explore the relationship between the excess conductivity and the interfacial area.

EXPERIMENTAL

Freeze-dried acetate precursors were used to prepare homogeneous powders which were subsequently densified via hot-pressing in WC-Co dies at approximately 650 °C and 0.8 GPa to produce pellets of approximately 6.3 mm in diameter and 1 mm in thickness [1, 2]. Sample densities and average grain sizes were determined by the Archimedes method and by x-ray line broadening using Scherrer's equation, respectively. Grain sizes were confirmed in selected samples via high-resolution electron microscopy (HREM). Annealed samples were heated to the desired temperature at a rate of 10°/minute. Treatment temperatures are listed in Table I. The grain size for the coarsest sample was estimated from field emission scanning electron microscopy (FESEM) of a fracture surface, since the grain size is too large to be determined by x-ray line broadening.

For electrical characterization, platinum electrodes of at least 1 μm in thickness were sputtered onto the faces of the pellets, and impedance spectroscopy was performed using a Hewlett Packard 4192-LF impedance analyzer with an oscillating voltage of 50 mV over the frequency range of 5 Hz to 13 MHz. Measurements were conducted in air, oxygen, and oxygen-argon mixtures to obtain oxygen partial pressures between 10^5 Pa and 1 Pa. The temperature of the samples was kept below 550 °C to avoid *in situ* coarsening.

RESULTS

The Archimedes measurements showed that all of the samples have densities greater than 90% of the theoretical value. Table I summarizes the heat treatments and grain-sizes. Annealing at

700 °C for 3 hours (sample a-CeO$_{2-x}$) increases the grain size only slightly from 13 nm to 16 nm. The coarsest sample, c-CeO$_{2-x}$, exhibits a bimodal grain size distribution with fine grains of 100 nm interspersed with larger grains of approximately 1 μm in diameter.

Table I: Samples and Heat Treatments

Sample	Treatment	Grain size
n-CeO$_{2-x}$	as densified	d$_g$~13 nm
n-Ce$_{0.9823}$Gd$_{0.0177}$O$_{2-x}$	as densified	d$_g$~13 nm
a-CeO$_{2-x}$	700 °C for 3 hours	d$_g$~16 nm
c-CeO$_2$	1200 °C for 4 min	d$_g$~ 100 nm-1 μm

Figure 1 shows a representative impedance plot for n-CeO$_{2-x}$. Using the Voigt model as an equivalent electrical circuit, one can deconvolve the plot into two arcs representing the RC components for the bulk and boundary arcs. The "bulk" arc is large in comparison to the "boundary" arc.

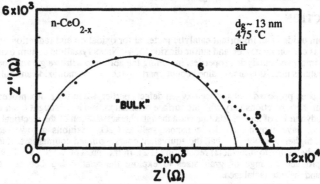

Figure 1: Representative impedance spectrum for n-CeO$_{2-x}$. Numerical labels represent the logarithm of the measurement frequency (Hz).

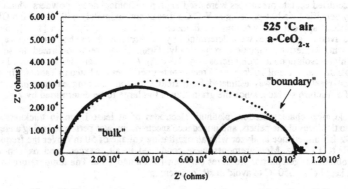

Figure 2: Representative example of an impedance spectrum for a-CeO$_{2-x}$

Annealing causes the relative size of the boundary arc to increase as shown in Figure 2. The coarsest sample, c-CeO$_2$, has a still larger boundary arc (Figure 3) which is typical of polycrystalline ionic conductors in which grain boundary impedance has been attributed to impurity segregation [6, 7, 8].

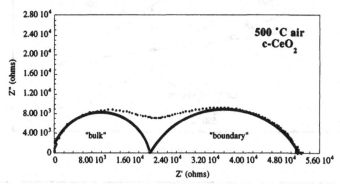

Figure 3: Representative example of an impedance spectrum for c-CeO$_2$

The variation in grain boundary impedance with grain size is most likely due to size-dependent impurity segregation [8, 9, 10]. In order to understand changes in defect thermodynamics with size scale and heat treatment, we focus on the high-frequency arc.

Since CeO$_{2-x}$ is a small-polaron conductor [11], electronic conductivity is given by

$$\sigma_e = ne\mu_e = ne\left(\frac{\mu_o}{T}\right)\exp\left(-\frac{E_h}{kT}\right) \tag{1}$$

where σ_e is the electronic conductivity, n is the carrier concentration, e is the charge on the carrier, and μ_e is the carrier mobility. It exhibits an activated mobility with a hopping energy, E_h, of 0.4 eV [11]. Reduction at high PO$_2$'s and low temperatures occurs by the formation of doubly-ionized oxygen vacancies [12], implying the following defect reaction:

$$O_O^x + 2Ce_{Ce}^x \Leftrightarrow \tfrac{1}{2}O_2(g) + V_O^{\bullet\bullet} + 2Ce_{Ce}' \tag{2}$$

which has an equilibrium constant of the form:

$$K_1(T) = K_1^o \exp\left(-\frac{\Delta H_1}{kT}\right) = [V_O^{\bullet\bullet}]n^2 PO_2^2 \tag{3}$$

where K_1^0 is a constant and ΔH_1 is the enthalpy of reduction per $V_O^{\bullet\bullet}$. When this is the dominant mechanism, electroneutrality is given by $n=2[V_O^{\bullet\bullet}]$ and the log σ- log PO$_2$ plot is expected to exhibit a slope of -1/6. Furthermore, in this regime, the slope of the log (σT)-T^{-1} plot (Figure 4b) gives the activation energy $E_a=(\Delta H_1/3)+E_h$. However, if there are sufficient background impurities to pin the concentration of oxygen vacancies, reduction can still occur, but the electroneutrality condition becomes $[A_{Ce}']=2[V_O^{\bullet\bullet}]$, the PO$_2$ slope is expected to be -1/4, and the activation energy is $E_a=(\Delta H_1/2)+E_h$ [2].

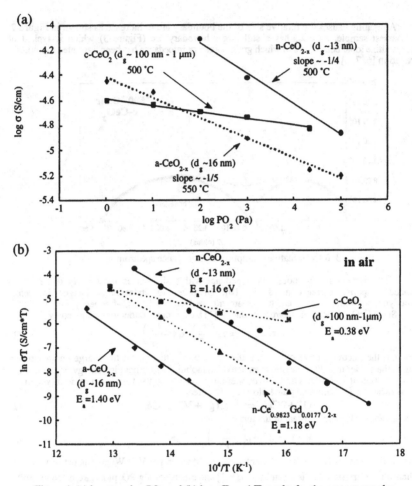

Figure 4: (a) log σ vs. log PO₂ and (b) log σT vs. 1/T results for the present samples.

Figure 4(a) shows the conductivity versus PO$_2$ results. Both the n-CeO$_{2-x}$ and the a-CeO$_{2-x}$ exhibit a PO$_2$ dependence between -1/6 and -1/4. The coarse sample, c-CeO$_2$, however, exhibits a very shallow slope of approximately -1/20. The nearly PO$_2$-independent behavior suggests extrinsic, ionic behavior. The slight PO$_2$ dependence may result from reduction of the smaller grains which are only a small fraction of the whole and therefore have a minor effect on the total conductivity. The c-CeO$_2$ sample behaves nearly extrinsically under the same conditions where the nanocrystalline materials show behavior consistent with reduction.

Figure 4 (b) shows the temperature dependence of the conductivity for these samples along with their activation energies. The coarsened sample, c-CeO$_2$ shows an activation energy of 0.38 eV which is consistent with the hopping energy for ceria [11]. This further supports the interpretation that the coarse sample is not reduced at these temperatures and PO$_2$'s. The relatively high conductivity of this sample is believed to be an artifact of heat treatment which quenched in a population of oxygen vacancies formed at high temperature.

The nanocrystalline, as-densified sample, n-CeO$_{2-x}$, shows an activation energy of 1.16 eV. The annealed sample, a-CeO$_{2-x}$, shows a slightly higher activation energy of 1.40 eV, but more significantly, it has a conductivity that is three orders of magnitude lower than that for n-CeO$_{2-x}$. The respective grain sizes are 13 nm and 16 nm which implies that the surface to volume ratio for the 13 nm sample (n-CeO$_{2-x}$) is 1.23 times that of the 16 nm sample (a-CeO$_{2-x}$). Clearly the conductivity does not scale with grain-boundary area.

As a further comparison, the lightly-doped sample n-Ce$_{0.9823}$Gd$_{0.0177}$O$_{2-x}$ shows an activation energy of 1.18 eV, almost identical to that of n-CeO$_{2-x}$. The sample also exhibits a PO$_2$ dependence of approximately -1/6. The sample appears to be an electronic conductor despite the doping, which we have attributed to size-dependent grain-boundary segregation of gadolinium causing an exhaustion of the solute within the grains [9]. This sample, processed identically to n-CeO$_{2-x}$, also exhibits a conductivity that is approximately two orders of magnitude higher than that for the annealed sample.

Table II shows the enthalpies of reduction which have been calculated from the observed activation energies for both extrinsic and intrinsic reduction models. The as-densified samples, n-CeO$_{2-x}$ and n-Ce$_{0.9823}$Gd$_{0.0177}$O$_{2-x}$ show enthalpies of reduction that are more than 2 eV lower than that for conventional ceria. The annealed sample, a-CeO$_{2-x}$, shows a higher enthalpy of reduction, but the enthalpy is still more than 1.5 eV lower than literature values for the single crystal or conventional polycrystalline sample.

Table II: Volume Heat of Reduction for CeO$_{2-x}$

Sample	E$_a$(eV)	ΔH_R (eV per V$_o^{\cdot\cdot}$)	
		intrinsic, $n = 2[V_o^{\cdot\cdot}]$	extrinsic, $[A_{Ce}'] = 2[V_o^{\cdot\cdot}]$
n-CeO$_{2-x}$	1.16	2.28	1.52
n-Ce$_{0.9823}$Gd$_{0.0177}$O$_{2-x}$	1.18	2.34	1.56
annealed CeO$_{2-x}$	1.40	3.00	2.00
coarsened CeO$_2$	0.38		
Reduced single crystal [12]	1.96	4.67	
Acceptor-doped polycrystals [13]	2.37		3.94

To understand why annealing at 700 °C has such a large impact on the conductivity, we return to the model for interfacial reduction proposed previously [1, 2]. Disorder at the grain boundaries can result in low energy defect sites for which the enthalpy of reduction is lower than within the perfect crystal. For a nanocrystalline material, the high interfacial area leads to domination of the overall enthalpy of reduction by these low energy sites.

The annealed sample's behavior suggests that low energy sites remain, but their density has decreased markedly. Calculations [15] indicate that different crystallographic surfaces exhibit different defect formation energies. Thus, in a polycrystalline material which has a spectrum of grain boundary types, one also expects a spectrum of sites of various defect formation energies. The observed enthalpy of reduction is a function of both the number of sites and their formation energies. Since defect formation is thermally activated, the lowest energy sites are sampled first. In the as-densified sample, n-CeO$_{2-x}$, there are approximately 10^{17}-10^{18} cm^{-3} such sites on a volume averaged basis [1]. Annealing for even short times at 700 °C appears to decreases the density of these low energy sites by a factor of 10^3. The slight increase in the enthalpy of reduction is consistent with reduction now taking place at higher-energy sites. The as-densified samples were annealed for several days at 500 °C during electrical measurements, and showed completely reproducible behavior with no variation in the observed conductivity or activation energy after cycling in both temperature and PO$_2$. Therefore, it appears that the atomistic relaxation to which we attribute the change in defect energies occurs rapidly in the 700 °C temperature range.

CONCLUSIONS

Nanocrystalline cerium oxide annealed for long times at 500 °C shows enhanced conductivity consistent with preferential interfacial reduction. However, annealing for short times at 700 °C results in a three order of magnitude decrease in the conductivity, with only a slight increase in the heat of reduction. This effect is attributed to a decrease in the density of low energy sites for defect formation at the interfaces due to atomistic relaxation upon annealing. In all cases, the heat of reduction of nanocrystalline ceria remains over 1.5 eV lower per oxygen vacancy than in the single crystal or coarse-grained counterparts.

Acknowledgments

Support for this work was provided by the National Science Foundation under Award No. DMR94-00334. E. B. L. gratefully acknowledges the support of an NSF Graduate Fellowship.

References

1. E. B. Lavik, Y. -M. Chiang, I. Kosacki, and H. L. Tuller, *Mat. Res. Symp. Proc.* **400**, Pittsburgh, PA, 1996, pp. 359-364.

2. Y. -M. Chiang, E. B. Lavik, I. Kosacki, H. L. Tuller, and J. Y. Ying, *Appl. Phys. Lett.*, **69** 185 (1996).

3. A. Tschoepe and J. Y. Ying, p. 781 in Nanophase Materials, G. C. Hadjipanayis and R. W. Siegel, eds., Kluwer Academic Publishers, Netherlands, 1994.

4. W. Liu, and M. Flytzani-Stephanopoulos, submitted to J. of Catalysis.

5. A. Tschoepe, W. Liu, M. Flytzani-Stephanopoulos, and J. Y. Ying, *J. of Catalysis*, **157** 42 (1995).

6. R. Gerhardt and A. S. Nowick, *J. Am. Ceram. Soc.*, **69** 641 (1986).

7. R. Gerhardt, A. S. Nowick, M. E. Mochel, and I. Dumler, *J. Am. Ceram. Soc.*, **69** 647 (1986).

8. M. Aoki, Y. -M. Chiang, I. Kosacki, L. J. -R. Lee, H. Tuller, and Y. Liu, *J. Am. Ceram. Soc.*, **79** 1169 (1996).

9. C. D. Terwilliger and Y. -M. Chiang, *Acta Metall. Mater.*, **43** 319 (1995).

10. Y. -M. Chiang, E. B. Lavik, and D. A. Blom, to appear in Nanostructured Materials, Vol. 9, 1997.

11. H. L. Tuller and A. S. Nowick, *J. Phys. Chem. Solids*, **38** 859 (1977).

12. H. L. Tuller and A. S. Nowick, *J. Electrochem. Soc.*, **126**, 209 (1979).

13. H. L. Tuller and A. S. Nowick, *J. Electrochem. Soc.*, **122**, 255 (1975).

14. T. X. T. Sayle et al., *Surf. Sci.*, **316**, 329 (1994).

LOW TEMPERATURE HYDROTHERMAL SYNTHESIS OF NANOPHASE BaTiO$_3$ AND BaFe$_{12}$O$_{19}$ POWDERS

Fatih DOGAN, Shawn O'ROURKE, Mao-Xu QIAN, Mehmet SARIKAYA
University of Washington, Department of Materials Science and Engineering, Seattle, WA 98195,
fdogan@u.washington.edu

ABSTRACT

Nanocrystalline powders with an average particle size of 50 nm has been synthesized in two materials systems under hydrothermal conditions below 100°C. Processing variables, such as temperature, concentration and molar ratio of reactants and reaction time were optimized to obtain particles of reduced size and stoichiometric compositions. Hydrothermal reaction takes place between Ba(OH)$_2$ solution and titanium/iron precursors in sealed polyethylene bottles in the BaTiO$_3$ and BaFe$_{12}$O$_{19}$ systems, respectively. While crystalline BaTiO$_3$ forms relatively fast within a few hours, formation of fully crystalline and stoichiometric BaFe$_{12}$O$_{19}$ require considerably longer reaction times up to several weeks and strongly dependent on the Ba:Fe ratio of the precursors. The structural and compositional evaluation of the nanophase powders were studied by XRD and TEM techniques.

INTRODUCTION

Considerable attention has been paid to materials composed of nanometer-sized phases over the past two decades. The properties of these mesoscopic atomic ensembles in the range 1-100 nm are often different than those of conventional grain sized (>10 μm) polycrystalline materials [1,2]. To improve the performance of the existing materials, several novel processing methods have been developed such as physical vapor deposition [3], chemical vapor deposition [4], plasma processing [5], spray conversion processing [6] etc.. These processes lead to formation of metastable crystalline or quasicrystalline phase of a number of materials, which, in turn, result in considerable improvement in the properties of these materials.

Synthesis of ceramic powders by hydrothermal techniques allow formation of nanometer sized and crystalline particles at relatively low temperatures [7,8]. This study demonstrates the hydrothermal synthesis as a cost effective and high-yield process to prepare nanophase BaTiO$_3$ and BaFe$_{12}$O$_{19}$ powders as multicomponent systems.

Crystalline and stoichiometric nanophase BaTiO$_3$ can be obtained by a reaction between barium hydroxide solution and titanium oxide powders at the temperatures below 100°C. This reaction does not require the use of high temperature autoclaves so that a high yield and continuous production of powders can be achieved without using of high cost processing equipment. Nanophase hydrothermal barium titanate particles form soft agglomerates which can easily be redispersed for further consolidation processes.

In recent years barium ferrite, BaFe$_{12}$O$_{19}$, due its platelet shape, reduced size, high magnetic anisotropy, and reliability, has drawn considerable attention as an advanced magnetic recording material. Current synthesis routes for barium ferrite include glass leach synthesis [9], organometallic precursor method [10] as well as hydrothermal reaction at temperatures exceeding 200°C [11,12]. All of these routes with the exception of hydrothermal synthesis are high temperature methods in which it is difficult to control the size and morphology of particles.

In this study, nanophase BaTiO$_3$ and BaFe$_{12}$O$_{19}$ powders were both synthesized under hydrothermal conditions below 100°C. Processing parameters, such as the temperature, concentration and molar ratio of the reactants and reaction times were controlled in order to reduce the particle size and to achieve stoichiometric compositions of the materials. Structural characterization of the samples were performed by using transmission electron microscopy (TEM) imaging, diffraction, and spectroscopy (electron energy loss spectroscopy, EELS) techniques.

69

The low temperature (95°C) as-prepared samples after a period (8 weeks) of hydrothermal processing, as well as those heat treated at high temperature (e.g., 850°C) were examined for the purposes of (i) evaluating the size and shape of the crystalline particles as well as the amorphous phase(s) that are formed under these conditions; (ii) crystal structure of the particles, and (iii) elemental composition of different phases within the sample.

EXPERIMENTAL PROCEDURE

Barium Titanate, BaTiO3

Nanophase BaTiO$_3$ powders were synthesized using a clear aqueous solution of 2M Ba(OH)$_2$.8H$_2$O (Baker Chemicals) and titanium oxide soot (Typ P-25, DeGussa Co.) with Ti:Ba molar ratio of 1:1.5. The reaction took place at 95°C for 48 hours in sealed polyethylene bottles. After completion of the reaction, the suspension was washed by repeated centrifugation to remove excess Ba(OH)$_2$ solution using boiled CO$_2$-free water. The suspension was redispersed and freeze-dried to obtain soft agglomerates of nanophase BaTiO$_3$

Barium Ferrite, BaFe12O19

Hydrothermal synthesis conditions of nanophase BaFe$_{12}$O$_{19}$ were similar to that of barium titanate whereas longer reaction times and a more precise control of the concentration of starting materials were necessary. A 1.5M solution of barium hydroxide and 1M solution of FeCl$_3$.6H$_2$O were mixed and titrated with NaOH solution until the pH reached 11. The suspension was kept at 95°C for various incubation time from 10 minutes up to 8 weeks. Following centrifuging and washing, the powders were dispersed in methanol and dried.

The samples for TEM analyses were prepared using the suspension technique. The samples in the powder form were suspended in an aqueous solution and a drop of the suspension was placed onto a holey carbon film on a Cu-grid. After drying, the samples were examined in a LaB$_6$-equipped Philips 430 TEM that operated at 200 keV. EELS was done in the diffraction mode (image-coupled) using a GATAN 666 parallel-recording EEL spectrometer with a 3 mm entrance aperture in the range from 450 to 950 eV to incorporate the O K (532 eV), the Fe L$_{2,3}$ (708 eV) and the Ba M$_{4,5}$ (780 eV) edges. During the EELS experiments it was ensured that the sample thickness was < 0.3 mfp (mean free path) of the electrons within the sample at the operating conditions to make quantitative analysis possible.

RESULTS AND DISCUSSION

Low temperature hydrothermal synthesis is a cost effective and relatively simple process to obtain nanophase multicomponent ceramic powders such as barium titanate and barium ferrite. In contrast to other high temperature techniques, hydrothermal powders are crystalline and free of hard agglomerates. Figure 1 shows the TEM bright field image of BaTiO$_3$ particles with an average particle size of 60 nm and cubic crystal structure. Figures 1c and 2 reveal the x-ray diffraction (XRD) pattern of BaTiO$_3$ and BaFe$_{12}$O$_{19}$ obtained by different reaction times. While a fully developed crystalline structure of BaTiO$_3$ was obtained after a reaction time of 48 hours, crystallization of BaFe$_{12}$O$_{19}$ required significantly longer reaction times. Barium ferrite hydrothermally heat treated at 95°C at various periods of times displayed x-ray diffraction pattern from very diffuse to faint peaks that barely rose above the background . The effect of long reaction times was the increase in the height and the narrowing of the x-ray peaks. This may be interpreted as small particles forming from an amorphous matrix with time of hydrothermal treatment. High temperature heat treatment of the powders at 850°C lead to formation of fully crystalline BaFe$_{12}$O$_{19}$ phase. This indicates that the reaction kinetics of hydrothermal barium ferrite is significantly slower than that of barium titanate under similar reaction conditions. Increased reaction rates can be achieved by high temperature hydrothermal synthesis of barium ferrite in autoclaves at temperatures exceeding 250°C [11,12].

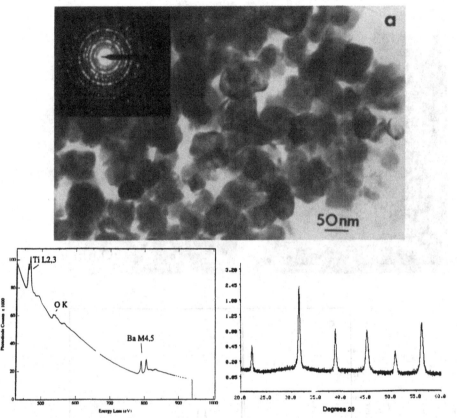

Fig. 1. (a) TEM bright field image and electron diffraction pattern of nanophase BaTiO₃, (b) EEL spectra, (c) XRD pattern

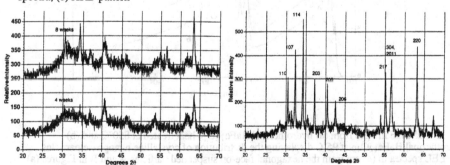

Fig. 2. XRD pattern of (a) hydrothermal BaFe₁₂O₁₉ at 95°C and (b) powders after heat treatment at 850°C

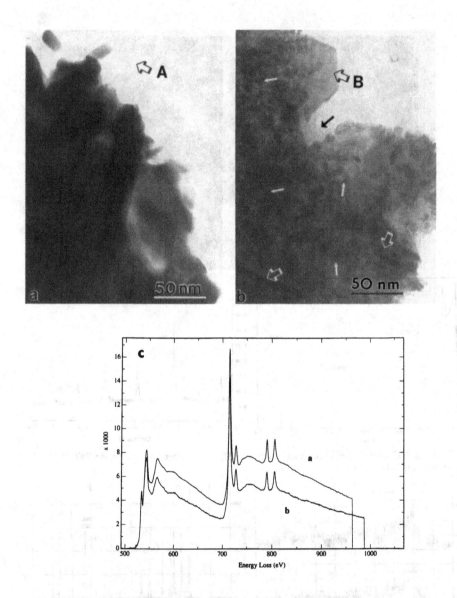

Fig. 3. TEM BF images from (a) hydrothermal BaFe$_{12}$O$_{19}$ heat treated at 850°C and (b) hydrothermal BaFe$_{12}$O$_{19}$ at 95°C displaying the distribution of crystalline nanophase particles. The EEL spectra in (c) are from the hexagonal face-on suspended particles A and B in Figures (a) and (b) respectively.

Barium ferrite samples were studied with TEM after 8 weeks of low-temperature treatment and an example is displayed in Figure 3. The TEM bright field (BF) image in Figures 3(a and b) display characteristic regions of this sample that reveal three types of contrast features. The first one is due to large (about 20-50 nm edge-size) hexagonal shaped crystalline particles (fat arrows), and a high density of small particles (1-5 nm length) that are loosely crystalline (thin arrows) and the matrix. Although it is difficult to obtain isolated diffraction patterns and EELS spectra, we have attempted to focus on the local regions of the sample that would provide such information. In order to prevent background overlap, we have also obtained EEL spectra from particles that were protruding from the edge of the samples into the hole region of the suspending C-film, as shown in Figure 1(c). Such a spectrum displays the edges from the O K, Fe $L_{2,3}$, and Ba $M_{4,5}$. The spectra displayed are given after background subtraction. The quantitative analysis of the spectra revealed (atomic) compositional ratios in terms of Ba/Fe/O of 0.03/0.38/0.59, very close to the stoichiometric composition of $BaFe_{12}O_{19}$.

TEM experiments were also conducted on the high temperature treated samples similar to those for the low temperatures ones. As it was shown in the XRD patterns earlier, by 850°C heat treatment, the samples already displayed peaks that corresponded to a fully crystalline $BaFe_{12}O_{19}$ phase. In Figure 1(a), a TEM BF image of the sample is shown displaying a high number density particles that are overlapping within the matrix as well as those suspending from the edge. Particle dimension in this case was within a narrow range of size distribution, but still relatively small, i.e., 40 - 50 nm edge length. EELS of the high temperature sample was also conducted, and a result is shown in Figure 3(c). Again, the relevant edges corresponding to the O, Fe, and Ba edges are all prominently displayed in the spectrum. In the same figure, EEL spectrum from a hexagonal crystallite in the low temperature is also displayed as a comparison. One can easily assess that there is virtually no difference between the two spectra (except slight differences in the height of the white-lines near the threshold regions which may be due to differences in the sample-thickness).

One can conclude from these analyses that the $BaFe_{12}O_{19}$ particles start forming at low temperature immediately following the first treatment, containing particles with distinctive hexagonal shapes (with $BaFe_{12}O_{19}$ stoichiometry and crystal structure as ensured via micro diffraction) appearing after several weeks. It is expected that these particles would continue to nucleate as smaller particles and grow from the amorphous phase at longer time periods, still at low temperature, to fully convert the amorphous phase into fully crystalline hexagonal barium ferrite particles [13]. As evidenced by the spectroscopy and diffraction analyses in this work, the effect of high temperature treatment appears to accelerate this transformation in the formation of $BaFe_{12}O_{19}$.

CONCLUSIONS

Nanophase $BaTiO_3$ and $BaFe_{12}O_{19}$,respecttively, were synthesized by low temperature hydrothermal techniques through the reaction of titanium or iron containing precursors with barium hydroxide solution. While fully crystalline and stoichiometric $BaTiO_3$ forms within less than two days at 95°C, formation of crystalline $BaFe_{12}O_{19}$ at the same temperature is a sluggish reaction which can continue up to several weeks before its completion. Hydrothermal low temperature synthesis can be considered as a high yield and cost effective processing route to synthesize multicomponent nanophase oxides for functional ceramics applications.

REFERENCES

1.   R. W. Siegel, MRS Bull. **15**, pp. 60-67 (1990).

2.   C. Suryanarayana, Int. Mater. Rev. **40**, pp. 41-64 (1995).

3.   R. L. Bickerdike, D. Clark, J. N. Easterbrook, G. Hughes, W. N. Mair, P. G. Partridge, and H. C. Ranson, Int. J. Rapid Solidif. **1**, pp. 305-325 (1984-85).

4. W, Chang, G. Skandan, S. C. Danforth, and B. H. Kear, Nanostructured Mater. **4**, pp. 507-520 (1994).

5. K. Upadhya (ed.), Plasma Synthesis and Processing of Materials; Warrandale, PA, TMS (1993).

6. B. H. Kear and L. E. McCandlish, Nanostructured Mater. **3**, pp. 19-30 (1993).

7. W. J. Dawson, Am. Ceram. Soc. Bull. **67**, pp. 1673-1678, (1988).

8. S. Somiya in Advanced Ceramics III, edited by S. Somiya, Elsevier Science Publisher LTD, Amsterdam, 1990, pp. 207-243.

9. O. Kubo and E. Ogawa, J. Magnetism and Magn. Mat. **134**, p. 376 (1994)

10. G. Litsardakis, A. C. Stergio, and J. Georgiou, J. Magnetism and Magn. Mat. **120** pp. 58-60 (1993).

11. H. Kumazawa, H. H. Cho and E. Sada, J. Mater. Sci. **28** pp. 5247-5250 (1993).

12. M. L. Wang, Z. W. Shih and C. H. Lin, J. Cryst. Growth **130**, pp. 153-161 (1993).

13. F. Dogan and M. Sarikaya in Polycrystalline Thin Films: Structure, Texture, Properties and Applications II, edited by H. J. Frost, C. A. Ross, M. A. Parker, and E. A. Holm (Mater. Res. Soc. Proc., Boston, MA 1995)

PRODUCTION OF NANOSTRUCTURED IRON OXIDE PARTICLES VIA AEROSOL DECOMPOSITION

J. JOUTSENSAARI, E.I. KAUPPINEN
VTT Chemical Technology, Aerosol Technology Group, P.O.Box 1401, FIN-02044 VTT, Finland, Jorma.Joutsensaari@vtt.fi

ABSTRACT

Nanostructured iron oxide particles with average size below 100 nm were produced by aerosol decomposition method starting from an aqueous iron nitrate solution. Air, nitrogen, or mixture of hydrogen (7 %) and nitrogen were used as the carrier gas. Gas-phase particle number size distributions were determined with a differential mobility analyzer. Particle morphology and crystallinity were studied with scanning (SEM) and transmission (TEM) electron microscopes. Crystalline phase composition of the particles was studied with X-ray diffraction (XRD). The average gas-phase diameter of particles produced in air or N_2 reduced from 80 to 47 nm when temperature was increased from 500 to 1100 °C. In H_2 rich environment, the reduction of average size was larger, from 80 nm at 500 °C to 45 nm already at 900 °C. SEM results showed that very small crystallites (5-10 nm) were formed on the surface of the particles produced in N_2 at 500 °C. When the processing temperature was increased to 700, 900 and 1100 °C, the crystallites on the particle surfaces were grown to 15-30, 30-60 and 60-180 nm, respectively. TEM results show that very small particles (<50 nm) were single crystals and larger particles were polycrystalline with crystallite size of about 50 nm at 700 °C in H_2/N_2. Magnetite particles were produced from aqueous iron nitrate solution at 500 °C in H_2/N_2 and at 900 °C in N_2 according to XRD results.

INTRODUCTION

Iron oxide particles are used in numerous applications, e.g., as magnetic materials (recording media, magnetic fluids), catalysts for industrial syntheses, gas sensors, pigments for paints, and in medical applications.[1,2]

Aerosol decomposition or spray pyrolysis process is commonly used for the production of a wide variety of powders from liquid phase precursors (see e.g., reviews[3-5]). This process involves the atomization of precursor solution into droplets that are then carried through a furnace by a carrier gas. Inside the furnace the solvent evaporates from droplets, and precursor precipitation occurs to form dry precursor particles. This is followed by reactions of the precursors within the dried particles to form the product powder and gaseous reaction products. Nonagglomerated spherical particles with high purity, controlled size and crystallinity can be produced by aerosol decomposition. When processing nanocrystalline powders, the grain-size of product particles can be kept small by operating at temperatures which are sufficient for complete drying and decomposition of precursors, but not high enough to cause excessive grain growth. The spray pyrolysis route has been used for the production of, e.g., nanophase PdO, TiO_2, V_2O_2 and $YBa_2Cu_3O_{7-x}$ particles, and Rh-Fullerene and $YBa_2Cu_3O_{7-x}$/Ag nanocomposites.

Production of iron oxide particles by aerosol decomposition method has been reported in several papers. α-Fe_2O_3 particles have been synthesized starting from an aqueous nitrate solution using air or nitrogen as a carrier gas.[6-10] Average particle sizes were reported to be from 0.3 µm to the supermicron size range and particles were spherical and hollow in shape. Nafis et al.[11] and

Tang et al.[12] produced spherical iron oxide particles with average size of 100 - 200 nm starting from aqueous solution of ferric or ferrous sulfates. The main phases were α-/γ-Fe_2O_3 and Fe_3O_4 when N_2, and mixture of H_2 and N_2 has been used as carrier gas. In addition, Ramamurthi and Leong[13] generated monodisperse particles with diameter about 2 μm using a vibrating orifice aerosol generator from ferrous sulfate. Cabañas et al.[10] synthesized hollow spherical γ-Fe_2O_3 and Fe_3O_4 particles from an aqueous iron citrate solution using air and nitrogen as carrier gas. González-Carreño et al.[14] produced γ-Fe_2O_3 particles with rather different morphologies from a methanol solution of iron acetylacetonate (aggregates of 5 nm primary particles), ammonium citrate (hollow spherical, diameter 170 nm), nitrate (spherical, diameter 180 nm) or chloride (aggregates of 60 nm monocrystallites) using air as a carrier gas. Also, González-Carreño et al.[15] reported of production of hollow spherical β-Fe_2O_3 particles by spray pyrolysis of a $FeCl_3$ alcoholic solution. In industrial or pilot scale, supermicron (20-400 μm) iron oxide particles have been processed with the Ruthner process starting from iron chloride[16] and nitrate solutions.[17]

In the previous studies, there were not reported a production of nonagglomerated spherical iron oxide particles with average size of below 100 nm by aerosol decomposition. In addition, particle size distributions in the gas phase have not been studied in detail at different processing conditions. In this paper we report a production of nanostructured and nanometer-size (<100 nm) iron oxide particles via aerosol decomposition starting from an aqueous iron nitrate solution. Gas-phase particle size distributions, morphology and crystallinity of iron oxide particles produced in various atmospheres and at processing temperatures are discussed. Our goal is to produce black iron oxide (magnetite) particles smaller than 100 nm for medical applications.[18]

EXPERIMENTAL

The precursor solution was prepared by dissolving 2.1 g of high-purity (99.99+%) iron nitrate ($Fe(NO_3)_3 \cdot 9H_2O$, Aldrich Chemical Co.) in 100 ml of ultra-pure water. The solution was atomized by a constant output atomizer (TSI 3076, TSI Inc.) using air, nitrogen, or mixture of hydrogen (7 %) and nitrogen as the carrier gas. A carrier gas flow rate of 3 l/min was used. The aerosol was then passed to the mullite tube (8 cm I.D.) in a three-zone furnace (Lindberg, heated length 91 cm). Furnace temperatures of 500, 700, 900 and 1100 °C were used resulting in residence times of 34, 27, 23 and 19 s, respectively. The aerosol exiting the hot zone of the reactor tube was diluted with clean dry air or nitrogen (27 l/min at 21 °C, dilution ratio 1:10) before sampling.

A differential mobility analyzer (DMA, TSI 3071) with an ultrafine condensation particle counter (UCPC, TSI 3027) and scanning particle mobility sizer[19] software (SMPS, TSI 3934) were used to determine the number size distribution of iron oxide particles in the gas phase. Diffusion losses in the DMA system[20] and tubing[21] were taken into account. Particle morphology and crystallinity were studied by field-emission scanning electron microscopy (SEM; Leo DSM 982 Gemini) and field-emission transmission electron microscopy (TEM; Jeol 2010 F, 200 kV and Philips CM200 FEG, 200 kV). SEM and TEM samples were collected directly onto holey carbon-coated copper TEM grids by an electrostatic precipitator sampler (Combination electrostatic precipitator, InTox Products). SEM analyses were carried out with low acceleration voltage (2 kV) without any sample coating in order to see structures on particle surface. X-ray diffraction (XRD; Philips MPD 1880) was used to identify the crystalline phases in the particles. Particles were collected onto polycarbonate filters (Nuclepore) for XRD analysis. In addition, the color of the powders was observed visually from the polycarbonate filters.

RESULTS AND DISCUSSION

Particle number size distributions (PSDs) in gas phase were determined with the differential mobility analyzer. Total particle concentrations, average sizes (geometric number mean diameters, d_g) and geometric standard deviations (σ_g, $\sigma_g=1$ for monodisperse particles) of size distributions were calculated from the PSDs.[22] Figures 1a-c present PSDs at different processing temperatures when using air, N_2, or mixture of H_2 (7 %) and N_2 as the carrier gas. All PSDs were unimodal, close to lognormal distribution in shape. The total particle number concentrations were rather constant at $1-1.2 \times 10^7$ #/cm³ for all processing conditions, except the concentrations were 8×10^6 #/cm³ at 700 °C in air and N_2. This reduction was probably due to unstability in aerosol generation. Average particle sizes were reduced when temperature was increased at all processing atmospheres. At 500 °C, the average sizes were around 80 nm at all carrier gases. Average size of particles produced in air and N_2 were reduced to 67-69, 52 and 47 nm at temperatures of 700, 900 and 1100 °C, respectively. The reduction of average size was larger in H_2/N_2 than in air or in N_2, i.e. d_g was 62 nm at 700 °C and 45 nm already at 900 °C. In addition, geometric standard deviation (σ_g) of size distribution was decreased when temperature was increased. When the processing temperature was increased from 500 to 1100 °C, σ_g was reduced from 1.95 to 1.88, and from 1.94 to 1.91 in air and N_2 carrier gases, respectively. In H_2/N_2 carrier gas, σ_g was reduced from 1.93 to 1.83 when the temperature was changed from 500 to 900 °C. The reduction of average particle size and deviation is due to particle densification, and also due to formation of different crystalline phase composition of particles when processing conditions were changed.

Particle morphology and surface structure were observed with SEM. SEM results showed that particles produced at different conditions were rather spherical and nonagglomerated (i.e., they were not formed from many small primary particles) in shape. Very small crystallites (5-10 nm) were observed on the surfaces of the particles produced in N_2 at 500 °C. When the processing temperature was increased to 700 (Fig. 2a), 900 (Fig. 2b) and 1100 °C, the crystallites on the particle surfaces were grown

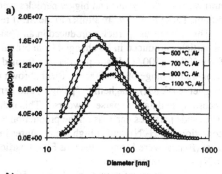

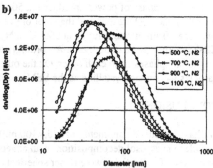

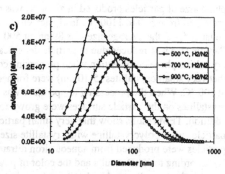

Figure 1. Gas-phase number size distributions of iron oxide particles produced in air (a), N_2 (b) or H_2/N_2 (c) at different temperatures.

to 15-30, 30-60 and 60-180 nm, respectively. This means that the small particles were single crystals and larger ones were polycrystalline. Similar behavior was observed when air was used as the carrier gas. In H_2/N_2 atmosphere, the crystallites on the particle surfaces were larger, 20-40 nm at 500 °C and 40-70 nm at 700°C. Those results indicate that nearly amorphous or nanocrystalline iron oxide particles can be produced at low processing temperatures, and polycrystalline particles with larger crystallites can be formed at higher temperatures.

In addition, the crystal structure of iron oxide particles produced in H_2/N_2 atmosphere at 700 °C was studied further with TEM. The TEM results showed that very small particles (<50 nm) were single crystals and bigger particles were polycrystalline with crystallite sizes of about 50 nm. Those results were similar to the SEM results.

The color of the particles produced in air changed from brown to red-brown, whereas the color of particles produced in N_2 changed from brown to black, when the temperature was increased from 500 to 1100 °C. Black powder was formed when H_2/N_2 was used as the carrier gas at temperature range of 500-900 °C. Black, brown and red powders include probably mainly magnetite (Fe_3O_4), maghemite (γ-Fe_2O_3) and hematite (α-Fe_2O_3) particles, respectively, according to data for coarse powders.[23] This indicates that the crystalline phase composition of iron oxide particles depends on the production temperatures and atmospheres. In reducing conditions, in H_2/N_2, and at high temperatures in N_2, iron oxide particles with a low oxidation state (Fe_3O_4) were formed, whereas Fe_2O_3 particles were formed in oxide rich environment and in N_2 at low temperatures.

The main phase of powder produced at 500 °C in H_2/N_2 and at 900 °C in N_2 was magnetite according the XRD results and the color of powders. This indicates that magnetite particles can be formed from nitrate solution at 500 °C in N_2/H_2 and at 900 °C in pure nitrogen. The peaks in XRD spectrum were quite broad. Accordingly it was difficult to separate magnetite and maghemite phases from each other. On the other hand, the peak broadening indicates that the particles were nanocrystalline as the SEM results showed.

CONCLUSIONS

We have produced nanometer-size (d_g<100 nm) and nanocrystalline (5-200 nm) iron oxide particles via aerosol decomposition of an aqueous iron nitrate solution at different processing conditions. The particles were rather spherical and nonagglomerated in shape. The average gas-phase size of particles produced in air or N_2 was reduced from 80 to 47 nm when temperature increased from 500 to 1100 °C. In H_2 rich environment, the reduction of average size was enhanced, i.e. the average size was 80 nm at 500 °C and it reduced to 45 nm already at 900 °C. Particle size was reduced due to particle densification, and also due to formation of different crystalline phase composition when processing conditions were changed. SEM results showed that very small crystallites (5-10 nm) were formed on the surface of the particles produced in N_2 at 500 °C. When the processing temperature was increased to 700, 900 and 1100 °C, the crystallites on the particle surfaces were grown to 15-30, 30-60 and 60-180 nm, respectively. In addition, TEM results show that very small particles (<50 nm) were single crystals and bigger particles were polycrystalline with crystallite size of about 50 nm at 700 °C in H_2/N_2. Magnetite particles were produced from aqueous iron nitrate solution at 500 °C in H_2/N_2 and at 900 °C in N_2 according to XRD results and the color of powders. More TEM and XRD studies to be carried out in order to understand in detail mechanisms of crystal growth and phase transformations during aerosol decomposition synthesis of iron oxide particles.

a)

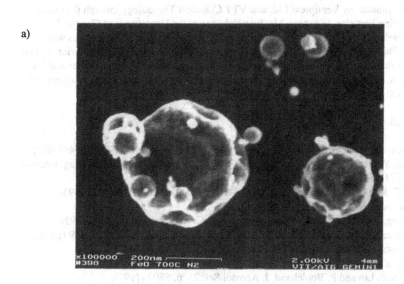

b)

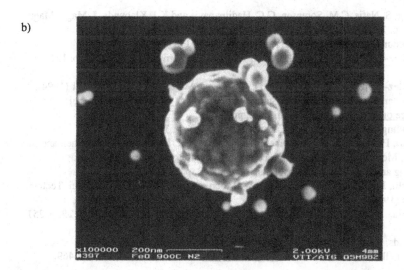

Figure 2. SEM micrographs of iron oxide particles produced in N₂ at 700 °C (a) and 900 °C (b).

ACKNOWLEDGMENT

This work was funded by Ventipress Ltd. and VTT Chemical Technology through the research program Ultrafine Particles. We thank Ms. Erja Nykänen at the Department of Chemical Technology, Helsinki University of Technology for XRD analysis. The TEM analyses were carried out at Philips Electron Optics B.V., Eindhoven, Holland and at Jeol Laboratories, Tokyo, Japan, and we thank staff in those laboratories. In addition, we thank Dr. Unto Tapper and Mr. Petri Ahonen at Aerosol Technology Group for helping with electron microscopy analyses and powder production experiments.

REFERENCES

1. R.M. Cornell and U. Schwertmann, The Iron Oxides, VCH, Weinheim, 1996, pp. 463-482.
2. N. Ichinose, Y. Ozaki and S. Karsu, Superfine Particle Technology, Springer-Verlag, London, 1992, pp. 171-181 and 208-215.
3. A. Gurav, T. Kodas, T. Pluym and Y. Xiong, Aerosol Sci. Technol. **19**, p. 411 (1993).
4. T.T. Kodas, Adv. Mater. **6**, p. 180 (1989).
5. G.L. Messing, S-C. Zhang and G.V. Jayanthi, J. Am. Ceram. Soc. **76**, p. 2707 (1993).
6. T.P. O'Holleren, R.R. Neurgaonkar, D.M. Roy and R. Roy, Ceram. Bull. **57**, p. 459 (1978).
7. D.W. Spronson, G.L. Messing and T.J. Gardner, Ceram. Int. **12**, p. 3 (1986).
8. A.M. Gadalla and H-F. Yu, J. Mater. Res. **5**, p. 1233 (1990).
9. P. Biswas, S.Y. Lin and P. Boolchand, J. Aerosol Sci. **23**, p. S807 (1992).
10. M.V. Cabañas, M. Vallet-Regí, M. Labeau and J.M. González-Calbet, J. Mater. Res. **8**, p. 2694 (1993).
11. S. Nafis, Z.X. Tang, B. Dale, C.M. Sorensen and G.C. Hadjipanayis, J. Appl. Phys **64**, p. 5835 (1988).
12. Z.X. Tang, S. Nafis, C.M. Sorensen, G.C. Hadjipanayis and K.J. Klabunde, J. Magn. Magn. Mater. **80**, p. 285 (1989).
13. M. Ramamurthi and K.H. Leong, J. Aerosol Sci. **18**, p. 175 (1987).
14. T. González-Carreño, M.P. Morales, M. Gracia and C.J. Serna, Mater. Lett. **18**, p. 151 (1993).
15. T. González-Carreño, M.P. Morales and C.J. Serna, J. Mater. Sci. Lett. **13**, p. 381 (1994).
16. M.J. Ruthner, H.G. Richter and I.L. Steiner in Proceedings of the 1st International Conference on Ferrites, Kyoto, Japan, 1970, pp. 75-78.
17. W.F. Kladnig and M.F. Zenger, J. Europ. Ceram. Soc. **9**, p. 341 (1992).
18. P.J. Kolari, P. Maaranen, E. Kauppinen, J. Joutsensaari, K. Jauhiainen, K. Pelkonen and S. Rannikko, Medical & Biological Engineering & Computing **34** S1, p. 153 (1996).
19. S.C. Wang and R.C. Flagan, Aerosol Sci. Technol. **13**, p. 230 (1990).
20. M. Adachi, K. Okuyama, Y. Kousaka, S.W. Moon and J.H. Seinfeld, Aerosol Sci. Technol. **12**, p. 225 (1990)
21. B.Y.H. Liu, D.Y.H. Pui, K.L. Rubow and W.W. Szymanski, Ann. Occup. Hyg. **29**, p. 251 (1985)
22. W.C. Hinds, Aerosol Technology, Wiley, New York, 1982, pp. 69-103.
23. R.M. Cornell and U. Schwertmann, The Iron Oxides, VCH, Weinheim, 1996, p. 469.

NANOPHASE COPPER FERRITE USING AN ORGANIC GELATION TECHNIQUE

D.Sriram*, R.L. Snyder** and V.R.W.Amarakoon*
*NYS College of Ceramics, Alfred University, Alfred, NY 14802
**Dept. of Materials Science and Engineering, The Ohio State University, Columbus, OH 43210

ABSTRACT

Nanocrystalline copper ferrite ($Cu_{0.5}Fe_{2.5}O_4$) was synthesized using a forward strike gelation method with polyacrylic acid (PAA) as a gelating agent. The dried gel was calcined at a low temperature of 400 °C to get the final powder. The effect of pH and the ratio of the cation to the carboxylic group in the initial gel were studied with respect to both the phases and the crystallite size of the final powders synthesized. Phase and crystallite size analysis were done using x-ray diffraction and TEM. Saturation magnetization results were obtained using a SQUID magnetometer. The reactions occurring in the nano-size copper ferrite, in air as a function of temperature, were tracked using a dynamic high temperature x-ray diffraction (HTXRD) system.

INTRODUCTION

In recent years there has been considerable interest in the ceramic community, to synthesize nanoparticles so as to enhance chemical and physical properties of the final product. A chemical route to synthesizing ultra-fine particles of high surface area involves first forming intermediate organometallic materials which are then calcined and sintered to form the high-density ceramic. Organic gelation is one of several methods used to produce nanosized phases. The organic gelation synthesis, first described by Micheli[1] is superior to more common chemical methods like sol-gel synthesis in that 1) the preparation of metal polyacrylate, which is the precursor material, is carried out in an aqueous environment, hence eliminating the need for toxic solvents. This makes the process cost effective and safer. 2) the reaction environment is such that the volatility of the components in the system is minimized and 3) decomposition of the gelatinous precursor is done at low temperatures, leading to uniform, nanophase ceramic oxide powders. The process[2-4] basically involves making two solutions, one containing the metal cations (as nitrates or acetates) and the other containing the polymeric gelating agent. The two solutions are mixed and different precipitates are formed on controlling the pH of the gel. This gel is dried and calcined at low temperatures to obtain the final nanophase powders.

The composition $Cu_{0.5}Fe_{2.5}O_4$ is a soft ferrite that has potential for use as a component in high frequency communication devices. The $Cu_{0.5}Fe_{2.5}O_4$ phase has a spinel structure with copper normally occupying both the tetrahedral (A) site or the octahedral (B) site, depending on its valence.[5] Theoretically the material, with all copper in monovalent state and occupying the A-site of the spinel sublattice, will have a high saturation magnetization of 7.5 μ_B (11,000 Gauss). A high value of saturation magnetization directly translates to its possibility to be used in device applications that require higher operational frequencies.

In a previous publication,[6] we synthesized the ferrite of composition $Cu_{0.5}Fe_{2.5}O_4$ using a standard solid-state route. Soak and quench experiments showed the formation of pure $Cu_{0.5}Fe_{2.5}O_4$ with all copper in the monovalent state, at high temperatures (around 1350 °C). However the copper

81

Mat. Res. Soc. Symp. Proc. Vol. 457 ⁰ 1997 Materials Research Society

was distributed between the A and B-sites of the spinel lattice, giving rise to a saturation magnetization value of 4.9 μ_B. The distribution of the copper between the two sites was understood to be due to the high temperature that creates disorder in the structure. In order to get more copper into the A-site, we proposed that the phase be stabilized at lower temperatures. Also, because of the possibility of small chemical inhomogenity in the powders, inherent to solid-state synthesis technique, we had to look for a synthesis using a more homogeneous chemical route. In line with this, the objective of the present work has been to synthesize nanophase, homogeneous copper ferrite of the composition $Cu_{0.5}Fe_{2.5}O_4$ at low temperature. In this paper we describe the organic gelation method, using polyacrylic acid (PAA), in order to produce phase pure copper ferrite. The effect of pH and the ratio of cation to carboxylic group, in the initial gel, on the final phases and its crystallite size is outlined. Saturation magnetization results are also presented for the as synthesized phase pure powders. Preliminary results of work done on dynamic high temperature x-ray diffraction system is also outlined.

EXPERIMENTAL PROCEDURE

Powder Preparation

The powders were prepared using a forward strike gelation process. In this process, solutions containing the copper and ferric cation were prepared by dissolving their metal nitrates into distilled water. The starting solutions of copper and iron nitrates were 0.0500 M and 0.2517 M respectively. 100ml of each solution was mixed so that the right stoichiometry for copper and iron was achieved in the final solution. In another container the polymer (50% aqueous solution of polyacrylic acid of average molecular weight 1080 g/mole) was dissolved into distilled water. The PAA solution was mixed with the metal ion solution and the pH of the solution was then raised by dropwise addition of concentrated ammonium hydroxide solution. Metal cations complexed with the polyacrylate group and a gelatinous precipitate was formed that changed in color from light brown (at low pH) to dark brown (at higher pH).

In order to study the optimum amount of carboxylic group needed for complete pick-up of the cations in solution, different batches were made with varying amounts of the PAA solution added to the mixed metal ion solution. All these batches were stabilized at a pH of 7.2 and the carboxylic group to cation ratio was controlled between 0.7 to 4.2. So as to study the effect of pH on the various precipitates formed, batches were also made with carboxylic group to cation ratio kept constant at 3.5, but stabilized at different pH values ranging from 1.2 to 8.5.

The as obtained gel from all batches was then vacuum filtered, washed with distilled water and dried at 110 °C for several hours. X-ray diffraction of the dried gel/precipitates revealed them to be essentially x-ray amorphous. These precipitates were then calcined at a temperature of 400°C in a fumehood.

Characterization

X-ray diffraction patterns were obtained using a Siemens θ -2θ configuration diffractometer, with $CuK\alpha$ source and a diffracted beam monochromator, on all of the calcined powders. The XRD patterns of powders stabilized at a pH of 7.2 with varying ratio of carboxylic group to cation in the initial gel is shown in Figure 1. Figure 2 shows the XRD patterns collected for calcined powders whose gels were stabilized with a fixed carboxylic ion to cation ratio of 3.5, but under varying pH values. Phase identification of the powder patterns was performed using program EVA (v.3.3, Diffract-AT software). The average crystallite size of the powders was determined by the x-ray peak broadening method using program SHADOW.[7] The crystallite size of phase pure $Cu_{0.5}Fe_{2.5}O_4$ powders obtained was confirmed using a JEOL-2000FX TEM. The $4\pi M_s$ measurements were made, at

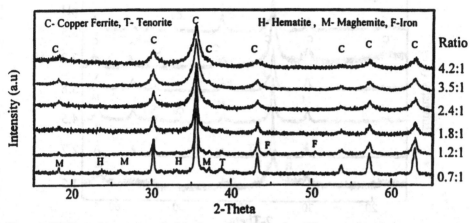

Figure 1: XRD powder patterns of calcined powders obtained at different carboxylic group to cation ratio, for pH =7.2.

the State University of New York at Buffalo, on powders stabilized at different pH, using a SQUID magnetometer.

High temperature x-ray diffraction experiments were done in air on phase pure nano-size powders in order to understand the reactions occurring in the divalent copper ferrite with increasing temperature. The unit consisted of a modified Siemens dynamic hot-stage x-ray diffractometer equipped with chrome radiation and a position sensitive detector (PSD). Detailed information about this set-up can be obtained from reference.[8,9] Continuous scans, lasting a few minutes each, were made between 20 and 80 degrees 2θ, at different temperatures in air.

RESULTS AND DISCUSSION

For the initial gel stabilized at a pH of 7.2, the variation of the average crystallite size of calcined powders with increasing carboxylic group to cation ratio is shown in Figure 3a. It can be seen that for a low ratio, the crystallites are nanosize (about 40nm) and the powders are phase impure, made of hematite (Fe_2O_3), maghemite (Fe_2O_3), tenorite (CuO) and iron. However with increasing carboxylic group in the solution, one obtains not only phase pure $Cu_{0.5}Fe_{2.5}O_4$ powders, but also smaller crystallites in the powder. Finally, at a ratio above 3.5, the nanocrystallites are of sizes less than 10nm. The effect on the crystallite size in the calcined powders, on varying the pH in the initial gel, whose carboxylic group to cation ratio was fixed at 3.5, is shown in Figure 3b. At a low pH of 1.2, the calcined powder is made up of extremely small crystallites of 5nm size but the phases that precipitate are maghemite, tenorite and hematite. As the pH was raised to 2.7, the crystallite size of the powder increases after which there is a decreasing trend in the size of crystallites precipitated with increasing pH. At a pH of 7.2, phase pure copper ferrite with a average crystallite size of 7nm precipitates. Increasing pH toward the more alkaline side leads to precipitation of impurity phases. Figure 4 shows a TEM micrograph of a particle made of a number of nanocrystallites. It can be seen that the phase, precipitated at a pH of 7.2 with the carboxylic

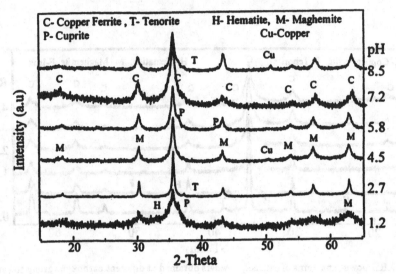

Figure 2: XRD powder patterns of calcined powders obtained at various pH, with carboxylic group to cation ratio kept constant at 3.5.

group to cation ratio held at 3.5:1, shows a average crystallite size of less than 10nm. One has to also note here that for obtaining phase pure powders it was very important that calcining of the dried precipitates not be done at temperatures higher than 400 °C, because of a tendency of the phase to decompose into hematite and another cubic spinel phase ($CuFe_2O_4$).

The results of SQUID saturation magnetization at 10K for three powders stabilized at pH of 7.2, 5.8 and 4.4 are shown in Figure 5. The figure shows the M_s vs. H curve in the first quadrant of the hysterises curve. It can be observed that the saturation magnetization value for the phase stabilized at a pH of 7.2, which is 62 emu/gm, is almost twice that for the precipitates obtained at lower pH . Also, since the powder obtained at pH of 7.2 was single phase copper ferrite of composition $Cu_{0.5}Fe_{2.5}O_4$, conversion of the saturation magnetization value from emu/gm to Bohr magnetron reveals a value of $2.6\mu_B$. This value is comparable to that for pure divalent copper ferrite of composition $Cu_{0.5}Fe_{2.5}O_4$ that has a magnetization of $2.5\mu_B$.[5]

High temperature XRD patterns collected for the nano-sized, $Cu_{0.5}Fe_{2.5}O_4$ phase (with all copper in the divalent state) between room temperature and 1220 °C is shown in Figure 6. It can be noted that, in air, the phase is stable until a temperature of 500 °C . At 500 °C the copper ferrite phase decomposes into hematite and a cubic spinel phase ($CuFe_2O_4$). Increasing temperature leads to further decomposition along with grain growth (which is apparent with decreasing width of the diffraction peaks). At 1200 °C, however the ferrite phase of composition $Cu_{0.5}Fe_{2.5}O_4$ forms again. This phase formed at 1200 °C has copper in both the monovalent and divalent state as our previous studies[6] have indicated. The final temperature of formation of the $Cu_{0.5}Fe_{2.5}O_4$ phase mentioned above was noted to be time dependent. On raising the temperature of the initial nano-phase powders relatively slowly, at the rate of about 10°C /min, the final $Cu_{0.5}Fe_{2.5}O_4$ phase forms at 1200°C. This is the same as the temperature of formation of the phase, for powders, synthesized through solid state route. However on fast ramping of temperature, at a rate of more than 50°C /min, the final $Cu_{0.5}Fe_{2.5}O_4$ phase formed at a much lower temperature of 1050°C. During fast ramping of

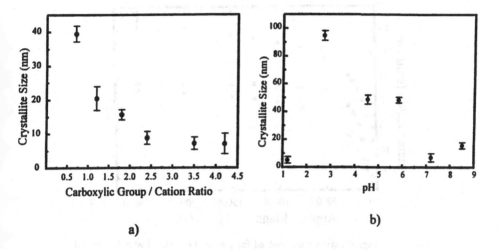

a)

b)

Figure 3: Variation of crystallite size of calcined powder with a) increasing carboxylic group to cation ratio b) increasing pH of the initial gel.

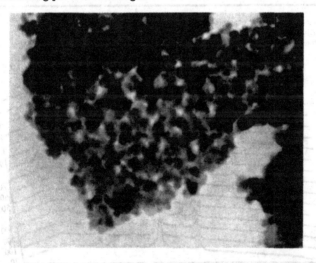

Figure 4: TEM micrograph of calcined powder stabilized at pH of 7.2, with carboxylic group to cation ratio maintained at 3.5, showing nanocrystallites.

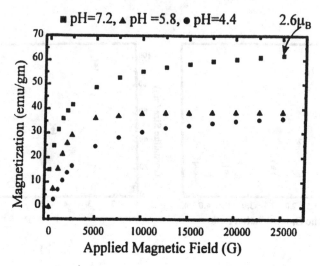

Figure 5: Magnetization data plotted for powders stabilized at different pH.

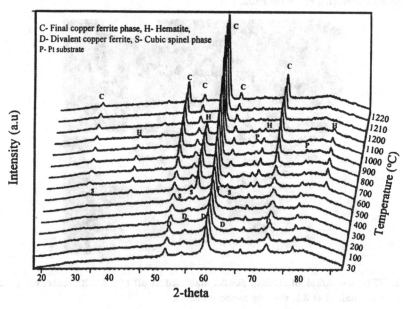

Figure 6: High temperature x-ray plot showing reaction occuring in nanosize divalent copper ferrite with temperature on slow heating.

temperature, the crystallites seem to retain their high surface area because of little time that is allowed for grain growth) and are still very reactive hence forming the final $Cu_{0.5}Fe_{2.5}O_4$ phase at lower temperatures. This is unlike the slow heating of the nano-phase powders, where due to the time one allows for grain growth, reactivity of the powders decrease substantially making the powders almost like powders synthesized by the solid state route, only more homogeneous.

CONCLUSION

X-ray pure nanophase copper ferrite was stabilized at low temperatures using an organic gelation process with PAA as the gelating agent. It was found that changing the carboxylic group to cation ratio and the pH of the the gel leads to different phases of varying crystallite sizes precipitating out. SQUID saturation magnetization of $2.6\,\mu_B$ confirmed that the copper, in the ferrite stabilized at low temperature, was in the divalent state. High temperature x-ray diffraction studies revealed that the divalent copper ferrite decomposes to hematite and a cubic intermediate spinel phase at 500 °C. The two phase composition reacts together to form the $Cu_{0.5}Fe_{2.5}O_4$ phase at elevated temperatures. For the nano-phase powders the final temperature of formation of the $Cu_{0.5}Fe_{2.5}O_4$ phase, where copper is in a mixed valence state, seemed to be dependent on the rate of heat treatment. Future work will be done in the direction of treating the nanosize $Cu_{0.5}Fe_{2.5}O_4$ phase powders at different temperatures under reducing atmosphere, using a controlled atmosphere HTXRD system. The precise temperature and pO_2 parameters will be sought in order to reduce the divalent copper in the nanoferrite to monovalent state and get all monovalent copper ferrite at comparably lower temperatures. Stabilization of monovalent copper in the ferrite at low temperature may encourage copper to occupy the A-site in the spinel lattice leading to a high saturation magnetization ferrite.

ACKNOWLEDGEMENTS

The authors would like to thank Dr. D. Hoelzer, TEM specialist at NYSCC, Alfred and Dr. D. Petrov, Post Doctoral Research Associate at SUNY, Buffalo for their help with TEM and SQUID analysis. The support of NYS Center for Advanced Ceramic Technology (CACT) at Alfred University for funding this research project is also acknowledged.

References

1. A.L. Micheli,U.S. Patent No. 4,627,966 (9 December 1986).
2. P. H. McCluskey, R. L. Snyder and R. A. Condrate, Sr., J. Solid State Chem., **83** (2), 332-339 (1989).
3. P. H. McCluskey, G. S. Fischman and R. L. Snyder, J. Therm. Anal., **34** (5-6), 1441-1448 (1988).
4. J.F. Fagan, PhD thesis, Alfred University, 1995.
5. A. Nagarajan, and A.H. Agajanian, J. Appl. Phys., **41**(4),1642-1647 (1969).
6. D.Sriram and R.L.Snyder, Symp. Proc. Series, Mat. Res. Soc.,**401**, 180-185 (1995)
7. S. A. Howard and R. L. Snyder, J. Appl. Crystallogr., **22**, 238-243 (1989).
8. B. J. Chen, M. A. Rodriguez, S. T. Misture and R. L. Snyder, Physica C, **195**, 118-124 (1992).
9. R. L. Snyder, Advanced Materials & Processes, 8/94, 20-25 (1994).

NANO-SIZED FINE DROPLETS OF LIQUID CRYSTALS FOR OPTICAL APPLICATION

SHIRO MATSUMOTO*, MARTHE HOULBERT*, TAKAYOSHI HAYASHI*, and KEN-ICHI KUBODERA**
*NTT Integrated Information & Energy Systems Laboratories, Musashino, Tokyo 180, Japan, matumoto@ilab.ntt.jp
**NTT Science and Core Technology Laboratory Group, Atsugi, Kanagawa 243-01, Japan

ABSTRACT

Nano-sized fine droplets of liquid crystal (LC) were obtained by phase separation of nematic LC in UV curing polymer. The polymer composite had a high transparency in the infrared region. The fine droplets responded to an electric field causing a change in birefringence. Output power change was brought about by the generated retardation between two polarizations, parallel and perpendicular to the applied electric field. This differs from the composite containing much larger droplets, where output depends on the degree of scattering. The birefringence changed by 0.001 at the applied voltage of 7.5 V/μm.

INTRODUCTION

Nano-sized composition has an important meaning in fiber optics communication system. This is because nano-sized substances are not recognized as scatterers of the light in the infrared wavelength region used in the system and thus the loss of light transmitted through them is kept low. We think that the nano-sized composition of liquid crystal (LC), which is a typical functional material, will be valuable for enlarging LC's territory in optical application. LC essentially has fluctuation in density due to its main characteristics of anisotropy and mobility. The fluctuation causes scattering, which has prevented LC from being applied to waveguides with a long optical path. If the nano-sized composition of LC makes it possible to prevent scattering while maintaining LC's virtuous functions, it will help LC to be used for active optical waveguides.

LC droplets have been energetically investigated in the last ten years [1]. Most of the research has aimed at making use of scattering [2] and diffraction [3], that occurs either at the surface of the droplets or at the surface of a layer consisting of many droplets. The droplets are mostly micro-sized. Our idea, however, is to increase the transmission by using nano-sized droplets whose surfaces are too small to scatter light. The purpose of this work is to make fine LC droplets smaller, to investigate their response to an electric field, and to evaluate the electro-optic effect.

EXPERIMENT

The process we used to prepare the small LC droplets is called photo-polymerization induced phase separation [4]. In this technique, a solution of LC and prepolymer is irradiated with light; the LC droplets are then phase separated according to the polymerization of the prepolymer, because LC solubility in the polymer matrix decreases with increasing molecular weight of the polymer matrix. We investigated several combinations of nematic LC from Merck Industrial Chemicals and UV curable prepolymer from Norland Products.

We fabricated two kinds of samples. A film sample about 400 μm thick was used to measure the transparency and size of the LC droplets. The other sample, used for

89

measuring response to an electric field, is spin coated film about 20 μm thick sandwiched between two electrodes, Si substrate and Au-deposited layer.

We measured the transparency of these film samples with a spectrometer. The size and number density of the LC droplets was evaluated from micrographs taken with a scanning electron microscope (SEM). The film was cracked by bending it in liquid nitrogen, and the liquid crystal was then removed by immersing the film in acetone for several hours. The LC droplets were observed as holes in the polymer.

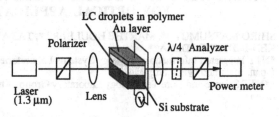

Fig. 1. Experimental setup

The experimental setup we used is shown in Figure 1. A 1.3-μm-wavelength laser beam was polarized at 45 degrees to the plane of the polymer layer. It was collected by an object lens and directly introduced into the polymer layer. The optical path through the polymer was 1 mm. The power passing through the sample was detected by a light power sensor. The analyzer and λ/4 plate were rotated so that the output power was at a minimum. An electric field was applied between the Au layer and the Si substrate and the change in output power was measured.

RESULTS and DISCUSSION

According to general scattering theory for independent scatters, light power T transmitted through a scattering medium of thickness L at input power I_0 is given by

$$T = I_0 \exp. (-NRL) \qquad (1),$$

where N is the density of particles and R is the scattering cross section. In the region of Rayleigh scattering, where the particle size is much smaller than the wavelength, R is expressed as

$$R = 24 \pi^3 ((m^2-1) / (m^2+2))^2 V^2 / \lambda^4 \qquad (2),$$

where m is the relative value of the refractive index of particles to that of the matrix, V is the particle volume, and λ is the wavelength [5]. Figure 2 shows the dependence of scattering loss on particle diameter calculated from this equation, assuming that m is 1.04, volume fraction NV is 10%, and measurement wavelength is 1 μm. It means the particle diameter should be less than 100 nm if the total loss in optical devices with a length of 1 mm to 1 cm is required to be 3 dB or less [6]. Achieving LC droplets of less than 100 nm

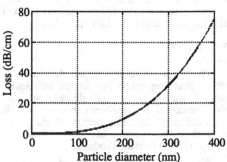

Fig. 2. Calculated scattering loss

has been a target of this work.

Figure 3 shows an example where LC fine droplets smaller than 100 nm were successfully obtained with a combination of prepolymer NOA81 and LC BL24. Transparency was evaluated at the wavelengths of 800 nm and 1300 nm, where absorption peaks do not exist. Below the mixing concentrations of 10 phr, LC droplets were not

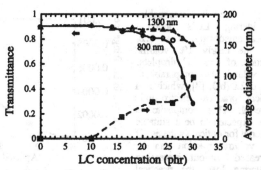

Fig. 3. Transparency and droplet size vs. LC concentration

observed because the LC concentration was not sufficient for the droplets to be phase separated from the polymer matrix. The transmittance decreases slowly as the mixing concentration increases to 26 phr, and then it decreases sharply. Droplet size slowly increases at mixing concentrations below 26 phr. In this region, however, they were smaller than 100 nm.

The polymer sample at the mixing concentration of 20 phr maintained high transparency and was used to study how the fine droplets respond to an electric field and whether they generate birefringence in the polymers. The sample has LC droplets with a diameter of about 50 nm, whose total volume fraction is about 1%. Figure 4 shows the output power change when an electric pulse of 400 ms at ± 5 V/μm was applied. The droplets responded to the pulse. The change in output power increased with applied voltage, reaching a maximum at about 6.3 V/μm, and then decreased at higher voltages. This suggests that a change in output is caused by retardation due to the birefringence generated when the voltage is applied; it is not caused by a change in the extent of scattering, as observed for the much larger LC droplets. The relationship between output power I_o and retardation ϕ is expressed as

$$I_o = I_i \sin^2 (\phi /2) \qquad (3),$$

where I_i is input power. The maximum output at about 6.3 V/μm corresponds to retardation of π. The change in birefringence Δn, is calculated from $\phi = 2\pi \cdot \Delta n \cdot L /\lambda$, where L is optical length and λ is wavelength. Figure 5 shows the dependence of Δn on applied voltage. Δn almost increases in proportion to square of the applied voltage. It is about 0.001 at 7.5 V/μm.

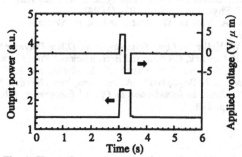

Fig. 4. Change in output power when voltage is applied

This value is reasonable, considering LC's inherent optical anisotropy, about 0.2, and the nearly 1% volume fraction of the LC droplets. The value is also comparable to that in LiNbO$_3$ [7], which is a typical practical electro-optic material. This composite is thus expected to be a suitable material for optical switches. If the volume fraction can be increased without increasing scattering loss, the practical value of this nano-sized LC composite for realizing active LC waveguides will be enhanced.

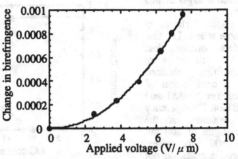

Fig. 5. Change in birefringence vs. applied voltage

CONCLUSION

Liquid crystal fine droplets with a diameter of about 50 nm were prepared in polymer. The composite showed high transparency in the infrared wavelength region. The LC droplets responded to electric fields, causing birefringence. It was about 0.001 at 7.5 V/μm for the composite containing droplets of about 1 vol%. As this composite has the potential to be used for active LC waveguides, further investigation is necessary; for example, increasing the density of droplets, investigating propagation loss, and measuring response time.

ACKNOWLEDGMENT

We wish to express our gratitude to Dr. Tohru Matsuura for his advice in sample preparation. And also to thank Dr. Akira Yamashita and Mr. Seizo Sakata for their help with optical measurements.

REFERENCES

1. L. Bouteiller and P. LE Barny, Liquid Crystals, 21, p. 157(1996).
2. J. L. Fergason, SID Int. Symp. Dig. Tech. 16, p. 68 (1985).
3. R.L. Sutherland, V. P. Tondiglia, L. V. Natarajan, T. J. Bunning, and W. W. Adams, Appl. Phys. Lett., 64, p. 1074 (1994).
4. N. A. Vaz, G. W. Smith, and G. P. Montgomery, Jr., Liquid Crystals, 146, p. 1 (1987).
5. W. Heller, in Light Scattering from Dilute Polymer Solutions, edited by D. Mcintyre and F. Gornick, (Gordon and Breach Science Publishers, Inc., 1964), p. 41.
6. J. E. Watson, C. V. Francis, R. S. Moshrefzadeh, K. M. White, P. Kitipichai, E. M. Cross, G. T. Boyd, and P. A. Pedersen, Proc. SPIE, 2285, p. 328 (1994).
7. A. Yariv, in Introduction to Optical Electronics, (Holt, Rinehart and Winston, Inc., 1971), p. 228.

TRANSMISSION ELECTRON MICROSCOPY AND ELECTRON HOLOGRAPHY OF NANOPHASE TiO₂ GENERATED IN A FLAME BURNER SYSTEM

S. TURNER*, J.E. BONEVICH**, J.E. MASLAR*, M.I. AQUINO*, M.R. ZACHARIAH*
*Chemical Science and Technology Laboratory; **Materials Science and Engineering Laboratory;
National Institute of Standards and Technology, Gaithersburg, MD 20899

ABSTRACT

Nanophase TiO₂ (n-TiO₂) particles were generated in a flame burner system under three experimental conditions. Selected individual nanoparticles were identified and characterized using selected area electron diffraction, bright-field and, in some cases, dark-field imaging to determine morphology and microstructural features. Previously unknown TiO₂ particles with unusual central features were identified as rutile. Electron holography was used to characterize the central features which were found to be consistent with voids. More extensive characterization of individual particles may lead to improved understanding of n-TiO₂ nucleation and growth.

INTRODUCTION

There has been considerable interest in the development and characterization of nanophase TiO₂. It has been found to differ from larger-sized TiO₂ particles in a number of ways including improved catalytic activity [1], sinterability at lower temperatures [2] and increased deformability at lower temperatures [3]. These and other properties have led to a variety of applications for n-TiO₂ ranging from use as an ingredient of sunscreens, to use as a photocatalyst to potential application as a coating on surgical implants [4]. Nanophase TiO₂ has been generated by several methods including inert gas condensation [2], vapor phase deposition [5] and by chemical methods [6]. This work reports on an initial study of n-TiO₂ produced using a flame burner system.

Recent characterization of n-TiO₂ from flame burner systems has shown that at low reaction temperatures, amorphous n-TiO₂ is formed, at intermediate temperatures, mixtures of anatase and rutile are generated, and at high temperatures, spherical particles of anatase are generated [7]. Similar particles were generated in smoke studies in the early 1970's [8-10]. In all of these studies the generated particles were characterized as a whole or in groups of particles by x-ray diffraction, transmission electron microscopy and/or scanning electron microscopy. In the present work, these studies are extended to identification and characterization of individual particles by selected area electron diffraction (SAED), bright-field and dark-field imaging, and, in some cases, by electron holography. Knowledge of the identity and microstructural features of individual particles should aid in understanding nucleation, growth and properties of the particles.

EXPERIMENT

To generate n-TiO₂, titanium isopropoxide was aerosolized from solution and then oxidized in a flame fueled by oxygen and hydrogen. The temperature of the flame was varied by changing the flow rate of the hydrogen gas. Three H₂ flow rates were used giving temperatures ranging from approximately 2000 - 2500 K. Particles were collected for x-ray diffraction (XRD) analysis on a water-cooled cold finger inserted into the flame at 5 and 10 centimeters above the

93

base of the flame. Particles were also collected for characterization by transmission electron microscopy on carbon-coated grids inserted at these heights.

The majority of the transmission electron microscopy was performed on a Philips CM30[1] transmission electron microscope (TEM) equipped with an EDAX energy dispersive analyzer. Bright-field imaging, dark-field imaging and selected area electron diffraction (SAED) were used to identify and characterize individual particles at 300 kV. The twin lens configuration of the CM30 was used to allow tilts of ± 45°. Gatan slow-scan CCD cameras were used to collect images; images were recorded using DigitalMicrograph software. Electron holography was performed on a Philips CM300 TEM equipped with a field emission gun (FEG) and a Mollenstedt biprism. The biprism was maintained at a potential of 65 V giving interference fringes of 0.47 nm. The off-axis holographic images were analyzed using HolograFREE software [11].

RESULTS

Characterization by XRD and low magnification TEM

Initial study of the particles shows that both the ratio of the amount of rutile to anatase and the morphology of the particles generated vary with the temperature of the flame (in agreement with results of previous work [7-10]). At the lowest H_2 flow rate (lowest temperature), both rutile and anatase were generated. Particles occur as isolated grains, as bicrystals and as chains of particles with planar grain boundaries. At the intermediate H_2 flow rate (medium temperature), particles of both rutile and anatase are produced; the particles exhibit spherical or polygonal morphologies and the great majority occur as single grains. At the highest temperature, the majority of particles are anatase which is spherical in morphology. The morphology of the particles generated ranged from polyhedral to rounded to spherical - initial work to characterize some of the particles individually is described below.

Polyhedral particles

The polyhedral anatase particles generated commonly have the morphology shown in Figure 1a. These particles occur so that when oriented on [010], six sides are evident. The platelets can be elongated in one direction or roughly equigranular giving a near hexagonal appearance to the grains.

Faceted rutile particles identified so far occur as twinned grains. Twinning of rutile minerals is common on {011} (more rarely on {092} and {031}) [12]. This {011} twinning occurs in the nanophase particles generated as shown in the image and associated diffraction pattern of Figures 1c,d. The presence of twinning in the n-TiO_2 generated in this study may be diagnostic for identification of rutile. Twinning in these anatase nanoparticles has not yet been established though it does rarely occur in minerals on {112} [12].

[1]Certain commercial equipment, instruments, or materials are identified in this paper to specify adequately the experimental procedure. Such identification does not imply recommendation or endorsement by the National Institute of Standards and Technology, nor does it imply that the materials or equipment are necessarily the best available for the purpose.

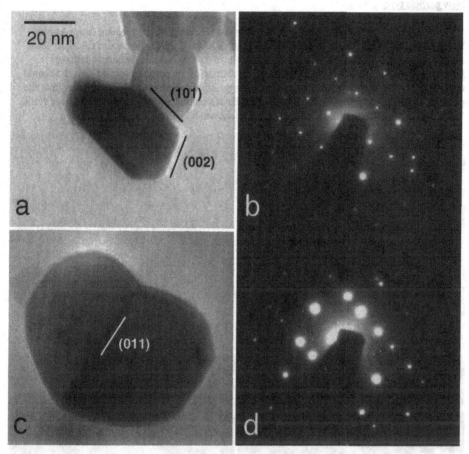

Fig. 1 Images and selected area electron diffraction patterns of polyhedral nanophase TiO$_2$ (scale applies to both images). The anatase particle (a, b) is oriented along [010]. The rutile particle (c, d) is oriented along [111] and contains a twin on (011).

Spherical particles

From XRD and low magnification TEM images, it is deduced that the majority of spherical particles generated in this study are anatase. SAED diffraction patterns obtained from the majority of individual spherical particles examined are consistent with a predominance of anatase. Some spheres do, however, give diffraction patterns consistent with rutile. Although many spherical particles appear to consist of multiple grains, almost all spherical particles examined so far show single crystal diffraction patterns.

<u>New particle type</u>

Unusual particles were noted in two of the sample sets (low and medium temperatures, lower height in flame). These particles are generally 100 nm or larger and are much larger than the majority of particles in the sample sets. They have central features that are rounded or faceted as shown in the examples of Figures 2a,b. Tilting of the particles showed that the central features are in the interior of the particles and energy dispersive x-ray analysis confirmed that the particles are TiO_2. SAED patterns obtained thus far are consistent with an identification of rutile. Rarely, these particles show twinning on {011} as shown in Figures 2b-d.

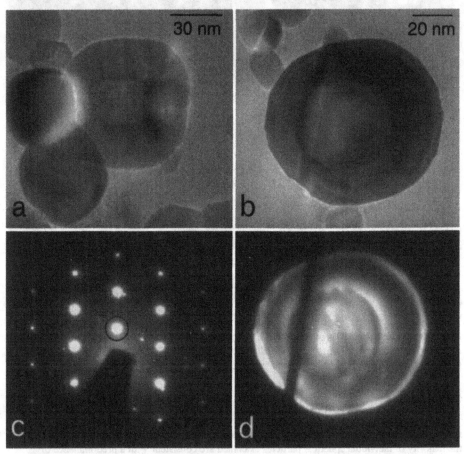

Fig. 2 Images and diffraction patterns obtained from rutile nanoparticles with unusual central features: a) a particle with interior faceting at roughly 90°; b) a particle with a rounded central feature; c) diffraction pattern from particle in b) showing a [111] orientation with weak twinning on (011); d) a dark-field image obtained from the diffracted spot circled in c). The dark band corresponds to the weakly diffracted spots in c).

Electron holography was used to determine the nature of the central feature of the particles. Using electron holography, the phase difference induced in coherent electron wavefronts interacting with matter can be quantitatively measured. In the absence of electromagnetic fields, the phase difference is directly related to the thickness and mean inner potential of the specimen. If the thickness can be deduced, the composition can be modeled to calculate an expected phase change. This procedure has been previously applied to characterize similar features in the interior of palladium nanoparticles [13]. A hologram of a TiO_2 nanoparticle is shown in Figure 3a. A three-dimensional phase image derived from the hologram is shown in Figure 3b. The dip in the top of the phase image corresponds to the central feature.

The phase map was analyzed assuming a roughly spherical TiO_2 particle containing a spherical central feature of unknown composition. Line scan profiles of the experimental phase data were fitted to this model for the following cases of the mean inner potential of the central

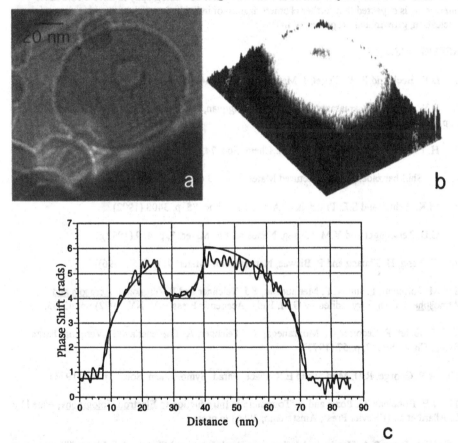

Fig. 3 a) Electron hologram of a TiO_2 particle with an unusual central feature; b) three-dimensional phase image calculated from the hologram. The dip evident in the image corresponds to the central feature; c) trace of the three-dimensional phase image compared with trace calculated for a TiO_2 particle containing a void (smooth line).

feature: greater than that of TiO_2 (mean inner potential of TiO_2 is approximately 12.5 eV), less than TiO_2, and a void (0 eV). In the first case, the central feature caused an increase in the phase difference so as to appear in the data as a phase bump. In the second case, the central feature appeared as a shallow phase dip. The experimental phase data more closely agrees with a central void as shown in Figure 3c.

SUMMARY

This paper has presented initial results of structural characterization of individual n-TiO_2 particles generated in a flame burner system. SAED and bright-field and dark-field imaging have been used to identify and characterize individual polyhedral and spherical particles. A new type of rutile nanoparticle has been identified and shown by electron holography to have a void in its interior. It is expected that further characterization of individual particles will aid in determining nucleation, growth and properties of n-TiO_2.

REFERENCES

1. D.D. Beck and R.W. Siegel, J. Mater. Res. 7, p. 2840 (1992).

2. R.W. Siegel, S. Ramasamy, H. Hahn, L. Zongquan, L. Ting and R. Gronsky, J. Mater. Res. 3, p. 1367 (1988).

3. H. Hahn and R.S. Averback, J. Am. Chem. Soc. 74, p. 2918 (1991).

4. M. Shirkhanzadeh, Nanostructured Mater. 5, p. 33 (1995).

5. M.K. Akhtar and S.E. Pratsinis, J. Am. Ceram. Soc. 75, p. 3408 (1992).

6. C.D. Terwilliger and Y.M. Chiang, Nanostructur. Mater. 1, p. 419 (1992).

7. G. Yang, H. Zhuang and P. Biswas, Nanostructur. Mater. 7, p. 675 (1996).

8. M. Formenti, F. Juillet, P. Meriaudeau, S.J. Teichner and P.Vergnon in Aerosols and Atmospheric Chemistry, edited by G.M. Hidy (Academic Press, NY, NY, 1972) pp. 45-55.

9. F. Juillet, F. Lecomte, H. Mozzanega, S.J. Teichner, A. Thevenet and P. Vergnon, Farad. Symp. Chem. Soc. 7, p. 57 (1973).

10. A.P. George, R.D. Murley and E.R. Place, Farad. Symp. Chem. Soc. 7, p. 63 (1973).

11. J.E. Bonevich, G. Pozzi and A. Tonomura in Introduction to Electron Holography, edited by L. Allard et al. (Elsevier Press, Amsterdam, 1996) in press.

12. W.A. Deer, R.A. Howie, and J. Zussman, Rock-Forming Minerals Vol. 5 Non-Silicates, Longman Group LTD, London, 1962, pp. 34-43.

13. L.F. Allard, E. Volkl, D. Kalakkad and A.K. Datye, J. Mater. Sci. 29, p. 5612 (1994).

COULOMETRIC TITRATION STUDIES OF NONSTOICHIOMETRIC NANOCRYSTALLINE CERIA

O. PORAT, H.L. TULLER, E.B. LAVIK, AND Y.-M. CHIANG
Center for Materials Science & Engineering, Massachusetts Institute of Technology, Cambridge, MA 02139

ABSTRACT

Oxygen nonstoichiometry measurements in nanocrystalline ceria, x in CeO_{2-x} , were performed using coulometric titration. The measurements reveal large apparent deviations from stoichiometry, of the order of 10^{-3} - 10^{-4} at T = 405 - 455 °C and P_{O_2} = 0.21 -10^{-5} atm, as compared to levels of ~ 10^{-9} for coarsened materials under the same conditions. The level of nonstoichiometry is, however, larger then expected from previous electrical conductivity data of nanocrystalline ceria. In addition, $x \propto P_{O_2}^{-1/2}$ while $\sigma \propto P_{O_2}^{-1/6}$. The observed dependence of $x(P_{O_2},T)$ can be explained by either the formation of neutral oxygen vacancies at or near the interface, or by surface adsorption.

INTRODUCTION

Nanostructured materials, polycrystalline solids with crystallites sized in the 1-20 nanometer range, have been receiving increasing attention due to their reported unique structural and physical properties [1,2]. In earlier studies, we and others have found nanocrystalline ceria CeO_2 (N-CeO_2), to exhibit substantially enhanced catalytic activity [3] and electrical properties compared to conventional coarsened CeO_2 (C-CeO_2 , grain size ~ 1-10μm). Specifically, the n-type electronic conduction in N-CeO_2 was found to be ~10^4 times greater than that in C-CeO_2 at low temperatures of 400 - 600 °C [4,5].

This large increase in electronic conductivity was tied to a corresponding large increase in deviations from stoichiometry x in CeO_{2-x} (e.g. x = 10^{-5}-10^{-6} for N-CeO_2 at 500 °C and P_{O_2} = 1 - 10^{-5} atm, vs. x ~ 10^{-9} for C-CeO_2 under the same conditions) by the assumption that each oxygen vacancy formed by reduction ionized to form two electrons or

$$n = 2[V_O^{\bullet\bullet}] = 2[O_O]x \qquad (1)$$

where n, $[V_O^{\bullet\bullet}]$ and $[O_O]$ are electron, doubly ionized oxygen vacancy and oxygen ion densities respectively. Of particular note, was the sharply reduced activation energy for electronic conduction, i.e., 0.99 -1.16 eV for N-CeO_2 versus 2.45 eV for C-CeO_2 [6]. Applying the appropriate defect modeling, the effective enthalpy of reduction ΔH_1, for the reaction

$$O_O \Leftrightarrow V_O^{\bullet\bullet} + 2e' + \frac{1}{2}O_2 \text{ (gas)} \qquad (2)$$

was found to be more than 2.4 eV lower per oxygen vacancy for the nanocrystals compared to the coarse grain counterparts [6].

Given the apparent large changes in thermodynamic properties induced by decreasing grain size down to the nanometer scale and the indirect method used to estimate x in N-CeO_2, we initiated coulometric titration studies on N-CeO_2 capable of measuring deviations from stoichiometry directly in an attempt to confirm the above observations. Our preliminary results are described below.

EXPERIMENTAL

Cerium acetate solutions were atomized by spraying into liquid nitrogen. The product was placed into a freeze dryer under vacuum following which the temperature was raised to sublimate the water, resulting in a fine unagglomerated powder. This was subsequently calcined at 300 ^{0}C for two hours resulting in an oxide powder with ~10nm grain size.

Coulometric titration measurements were carried out using a calcia stabilized zirconia (CSZ) electrochemical cell. Details of the titration cell are given elsewhere [7]. Briefly, the cell consists in part of a CSZ tube. An open circuit EMF is developed on two opposite Pt electrodes painted on both sides of the CSZ tube, given by:

$$EMF = \frac{RT_{CSZ}}{4F} \ln\left(\frac{P_{O_2}(atm)}{0.21}\right) \qquad \text{(Nernst Equation)} \qquad (3)$$

T_{CSZ} is the temperature of the CSZ tube tip. This EMF is a measure of the P_{O_2} over the sample. For coulometric titration, one pumps oxygen into/out of the cell by passing a current, I, through the CSZ tube and the Pt electrodes. The number of O_2 moles added/taken form the sample passing through the CSZ tube in time t_1, is:

$$\Delta n_{O_2}(\text{sample}) = \frac{1}{4F}\int_0^{t_1} I \, dt + \int_0^{t_1+t_2} \dot{n}_{O_2} dt - \frac{1}{R}\left(\frac{V}{T}\right)_{eff}(P_{O_2,f} - P_{O_2,i}) \qquad (4)$$

While the first term on the right hand side of Eq. (4) is the number of moles passing through the CSZ tube during titration, one also needs to take into account the leak rate of oxygen molecules into the cell, $\dot{n}$, and the amount of oxygen taken by/from the dead volume (the right hand term in Eq (4), where t_2 is the time required for equilibration after the titration current is switched off, $P_{O2,i}$ and $P_{O2,f}$ are the oxygen partial pressures over the sample before and after the titration step, respectively. $(V/T)_{eff}$ is the effective dead volume coefficient [8]. Due to the high resolution required for measuring the expected small deviation from stoichiometry in N-CeO$_2$, one needs, prior to the actual measurements, to carefully evaluated the leak rate and the dead volume of the cell [8]. In addition, since the N-CeO$_2$ powder has such a high surface area, one expects to see traces of adsorbed contaminants in the powder, which could interfere with the measurements. Therefore the cell with the sample is mechanically pumped to 10^{-3} mbar at 455 °C for 24 hours prior to the measurements.

RESULTS

Nonstoichiometry x of N-CeO$_2$ is plotted in Fig. 1 as log P_{O_2} vs. x for temperatures of 405, 430, and 455 °C. Measurements were limited to 455 °C for fear of coarsening at higher T and to 405 °C due to difficulties with "dead volume" calibrations at lower temperatures (see experimental section). Note, that while measurable values of Δx initiate only below $P_{O_2}=10^{-3}$ atm, further significant changes are easily observed under more reducing conditions.

The results of Fig. 1 are re-plotted in Fig. 2 as log x vs. log P_{O_2}. The data exhibit a good fit to a linear dependence with a slope of -1/2. A plot of log K_2, vs. 1000/T (not presented), where $x = K_2(T) P_{O_2}^{-1/2}$, results in an Arrhenius dependence characterized by an activation energy of $\Delta H_2=0.36$eV, and a pre-exponential term $K_2^0 = 0.007$ atm$^{1/2}$/molecule fraction.

DISCUSSION

The coulometric titration results do indeed confirm a large apparent deviation from stoichiometry, reaching levels of nearly 10^{-3} at 455 °C and $P_{O_2} \sim 10^{-5}$ atm. This is many orders of magnitude in excess of x values for C-CeO$_2$ extrapolated from high temperature TGA studies [6].

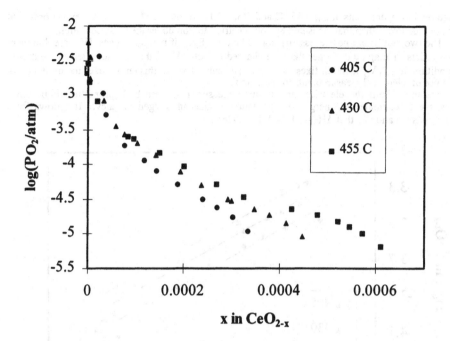

Fig. 1: Nonstoichiometry x, in N-CeO_{2-x} as a function of P_{O_2} at different temperatures

Are these findings necessary in conflict with our earlier electrical measurements? The two types of measurements differ in two important ways. First, while the electronic conductivity follows a P_{O_2} dependence given approximately by $\sigma_e \propto P_{O_2}^{-1/6}$ [4,5], the nonstoichiometry follows a much stronger dependence, $x \propto P_{O_2}^{-1/2}$. While the conductivity results are consistent with the reaction described in Eq. (2), with mass action relation:

$$K_1(T) = [V_O^{\bullet\bullet}]\, n^2\, P_{O_2}^{1/2} \tag{5}$$

and

$$n = 2[V_O^{\bullet\bullet}] = (2K_1(T))^{1/3}\, P_{O_2}^{-1/6} \tag{6}$$

the nonstoichiometry results follow from

$$O_O \Leftrightarrow V_O + \frac{1}{2}O_2(gas)\,, \qquad K_2(T) = [V_O]\, P_{O_2}^{1/2} \tag{7}$$

with

$$[V_O] = x = K_2(T)\, P_{O_2}^{-1/2} \tag{8}$$

The reduction enthalpy, ΔH_1, see Eq. 6 ($K_1(T) = K^0 exp(\Delta H_1/k_B T)$) was reported to be 2.28eV per $V_O^{\bullet\bullet}$, while from this study we find $\Delta H_2 = 0.36eV$. Second, the estimated value of x from the electrical measurements at e.g. 455 °C and $P_{O_2} = 10^{-4}$ atm is ~ 10^{-6}, which is several orders of magnitude lower than that measured by coulometric titration under the same conditions.

The two results are not necessarily contradictory. Even if neutral defects, $[V_O]$ are dominant and orders of magnitude larger than the charged defects $[V_O^{\cdot\cdot}]$, they do not alter the neutrality condition $n = 2[V_O^{\cdot\cdot}]$, which fixes the electron density. Indeed, this emphasizes the limitation in the use of electrical measurements to estimate x.

Assuming both the electrical and titration measurements are bulk properties, it becomes possible to estimate the energy ΔH_3 to doubly ionize an oxygen vacancy. It follows from equations (2) and (7) that $\Delta H_3 = \Delta H_1 - \Delta H_2 = 1.92 eV$.

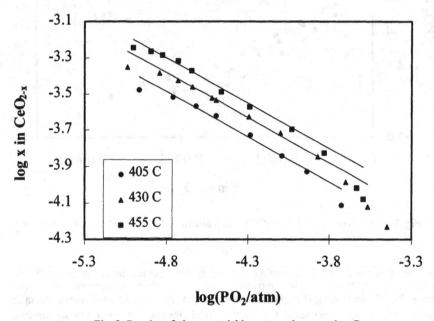

Fig. 2: Re-plot of the nonstoichiometry as log x vs. log P_{O_2}.

An alternative explanation for the titration results is that the oxygen losses detected upon reduction are largely oxygen desorbed from the surface of the particles. In this case one can write

$$O_2 + 2V_{ad} \Leftrightarrow 2O_{ad} \tag{9}$$

where V_{ad} and O_{ad} correspond to empty and occupied surface adsorption sites. Following the formulation of Langmuir adsorption, one can write the mass action relationship as

$$\theta/(1-\theta) = K_{ad}^{1/2} P_{O_2}^{1/2} \tag{10}$$

where θ = fraction of occupied adsorption sites. At high P_{O_2}, $\theta \rightarrow 1$, with

$$(1-\theta) = \text{"x"} = K_{ad}^{-1/2} P_{O_2}^{-1/2} \tag{11}$$

which takes on the same form of Eq. (8)

The obvious way to distinguish between the two possibilities, i.e., bulk vs. surface reduction, is to dramatically reduce the surface area by repeating the titration measurements on dense N-CeO$_2$ as was done in the electrical properties study [5,6]. Such measurements are now under way.

ACKNOWLEDGMENTS

This work was supported by the National Science Foundation under the MRSEC program at MIT, award No. 94000334-DMR.

REFERENCES

[1] H. Gleiter, Progress in Materials Science 33, 1 (1990).
[2] R. Birringer, U. Herr and H. Gleiter, Trans. Jpn. Inst. Met. 27 suppl. , 43 (1986).
[3] A. Tscoepe and J.Y. Ying, in Nanophase Materials: Synthesis-Properties-Applications, edited by G.C. Hadjipanayis and R.W. Siegel (Kluwer Academic Publ., Netherlands, 1994), p. 781.
[4] Y.-M. Chiang, E.B. Lavik, I. Kosacki, H.L. Tuller and J.Y. Ying, Appl. Phys. Lett. 69, 185 (1996).
[5] Y.-M. Chiang, E.B. Lavik, I. Kosacki, H.L. Tuller and J.Y. Ying, J. Electroceramics, in press.
[6] H.L. Tuller and A.S. Nowick, J. Electrochem. Soc. 126, 209 (1979).
[7] O. Porat and H.L. Tuller, J. Am. Ceram. Soc., in press.
[8] O. Porat and I. Riess, J. Electrochem. Soc. 141, 1533 (1994).

SYNTHESIS AND LASER SPECTROSCOPY OF MONOCLINIC Eu³⁺:Y₂O₃ NANOCRYSTALS

BIPIN BIHARI AND BRIAN M. TISSUE*
Department of Chemistry, Virginia Polytechnic Institute and State University
Blacksburg, VA 24061-0212

ABSTRACT

Gas-phase condensation of CO_2 laser-heated Eu^{3+}:Y_2O_3 ceramics produces monoclinic-phase nanocrystalline material. Transmission electron microscopy shows particle diameters in the range 7-30 nm for particles quenched at 60 °C under 400 Torr of nitrogen atmosphere. The optical spectra of nanocrystals produced from 0.1% Eu^{3+}:Y_2O_3 starting material have narrow lines, and the $^5D_0$ lifetimes are 1.8, 1.2 and, 1.3 ms for the three Eu^{3+} cation sites. Nanocrystals obtained from 0.7 -5 % Eu^{3+}:Y_2O_3 starting material show line broadening and the presence of Eu_2O_3 secondary phase.

INTRODUCTION

Nanophase materials can form in new and metastable crystal structures and can exhibit enhanced optical, electronic, and structural properties [1-3]. Nanometer-sized materials have potential as efficient phosphors in display applications, such as in new flat-panel displays with low-energy excitation sources. Decreasing the particle size in conventional phosphors results in decreasing fluorescence quantum efficiency, which is usually attributed to quenching by surface defects [4]. The large surface to volume ratio of constituent atoms can make nanocrystalline materials a suitable model system to study such surface phenomena. The understanding of the surface effects could lead to methods of creating materials with tailored properties for a given application.

The gas-phase condensation of nanocrystalline Y_2O_3 and Eu_2O_3 results in the monoclinic structure being stable under ambient conditions [5,6]. Bulk Y_2O_3, and rare-earth oxides in the middle of the lanthanide series, form in the monoclinic structure only under high pressure and/or high temperatures [7-9]. The stabilization of the high-pressure phase at ambient conditions in nanocrystals has been attributed to an additional hydrostatic pressure component resulting from the Gibbs-Thomson effect [6].

The monoclinic phase of Y_2O_3 is isomorphic to monoclinic Gd_2O_3 and Eu_2O_3 having space group of C2/m. The lattice possesses three crystallographically distinct cation sites each having point group symmetry C_s [10,11]. The coordination of two cation sites (Ln I and Ln II) can be described by six oxygens at the apices of a trigonal prism with a seventh oxygen lying along the normal to a face. The coordination of the third site (Ln III) is described as a distorted octahedron with a seventh oxygen along a three-fold axis at a very long distance [9,11]. Ab initio calculation of crystal-field parameters in monoclinic Eu^{3+}:Y_2O_3 correlate the cation sites Ln I, Ln II, and Ln III to three distinct sets of optical spectra that are labeled sites C, B, and A, respectively [11]. Here, we describe the preparation, site-selective excitation and fluorescence spectroscopy of single-phase monoclinic Eu^{3+}:Y_2O_3 nanocystals.

105

Mat. Res. Soc. Symp. Proc. Vol. 457 © 1997 Materials Research Society

EXPERIMENTAL

The nanocrystalline Eu^{3+}:Y_2O_3 samples were prepared by a gas-phase condensation method using CO_2-laser heating of ceramic pellets [12]. The starting material was prepared by cold pressing mixtures of pre-dried Eu_2O_3 (99.99%, Aldrich) and Y_2O_3 (99.99%, Aldrich) and sintering overnight in platinum crucibles at 1000 °C. The pellets were ground in an agate mortar and pestle, and resintered to improve the homogeneity. The pellets were placed on a slowly rotating platform in a vacuum chamber and the CO_2 laser was focused onto the target pellet to a spot size of approximately 2 mm. The nanocrystals quenched and condensed on a Pyrex cold finger. The distance between target pellet and the end of the cold finger was approximately 3 cm. The CO_2 laser power was 50 ± 2 W, and the chamber atmosphere was 400 Torr of nitrogen. The cold finger was filled with water at 50-60 °C, which kept the quenching temperatures fairly constant. At the end of a synthesis run the nanocrystalline material was scraped from the cold finger.

The phase and particle size of the nanocrystalline powders were characterized by transmission-electron microscopy (TEM) and X-ray diffraction (XRD). Small amounts of nanocrystalline material were dispersed in acetone and dried on thin carbon films on 300-mesh copper grids for TEM or on quartz substrates for powder X-ray diffraction. The TEM instrument was a Philips EM 420 STEM (operated at 100 kV) and the x-ray diffractometer was a Scintag XDS 2000 (using Cu $K\alpha$ radiation). The particle sizes were determined from XRD linewidths using the Scherrer equation [13], and from a survey of TEM micrographs.

For laser spectroscopy, nanocrystalline material was packed in a depression on a copper sample holder, which was mounted on the cold head of a closed-cycle cryogenic refrigerator (Cryomech GB15). The cold head cools to approximately 10 K, although the exact sample temperature was not determined. The fluorescence and excitation spectra were recorded using a Nd^{3+}:YAG-pumped dye laser (using Coumarin 540A dye) as an excitation source, 1-m monochromator (Spex 1000M), GaAs photomultiplier tube (Hammamatsu R-636), gated photon counter (Stanford SR400) or boxcar averager (Stanford SR250), and an in-house written LabView computer data acquisition program. Some of the excitation spectra were recorded using a 0.25-m monochromator with a bandpass of 5.5 nm, a PMT (Hammamatsu P-28) and boxcar averager. Fluorescence transients were recorded with a 350-MHz digital oscilloscope (Tektronix TDS460) with typically averaging 200 laser shots.

RESULTS AND DISCUSSION

The production rate differed slightly from one synthesis run to other, even under similar preparation conditions. We obtained an average production rate of 11 mg/hr under the synthesis parameters described in the experimental section. The sizes of the nanocrystals as determined by the TEM micrograph were in the range 7-30 nm. The average size calculated from XRD linewidths was 23 nm. Figure 1 shows the powder x-ray diffraction of 0.1% Eu^{3+}:Y_2O_3 nanocrystals. The XRD pattern is the same as one reported by Skandan et al. for monoclinic Y_2O_3 nanocrystals [6].

Samples of nanocrystals were prepared from 0.1, 0.7, 2, and 5 % Eu^{3+}:Y_2O_3 starting material. We did not determine the exact concentration of Eu^{3+} in the final nanocrystalline material, and the spectra are labeled according to the percentage Eu^{3+} in the starting material. Figure 2 shows the $^7F_0 \rightarrow {}^5D_0$ excitation spectra for Eu_2O_3 and 5 and 0.1% Eu^{3+}:Y_2O_3

nanocrystals. The excitation spectra in Fig. 2 were obtained by monitoring $^5D_0 \rightarrow {}^7F_2$ fluorescence at 624 nm with the wide-band 0.25 m monochromator. We recorded the excitation spectra with short and long boxcar gate widths and delays to uncover any discrimination among the various sites and phases that might be present. At a boxcar delay of 20 μs the excitation spectrum for 5% Eu^{3+}: Y_2O_3 shows broadened and asymmetric lines with higher intensity towards the Eu_2O_3 transitions. At a boxcar delay of 200 ms, the lines appear symmetric and the line positions shifted

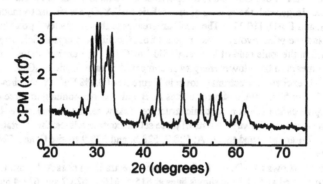

Fig. 1   X-ray diffraction of monoclinic 0.1% Eu^{3+}:Y_2O_3 nanocrystals (average size of 23 nm).

to longer wavelength. The above-mentioned change in the behavior of line profiles for 0.1% Eu^{3+}: Y_2O_3 is negligible, and the excitation lines remain sharp and symmetric. The excitation peaks corresponding to sites B and C are resolved in this spectrum, unlike the spectrum of the

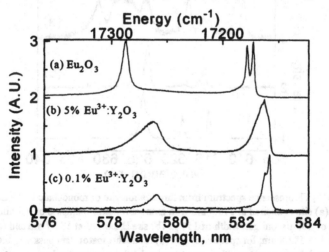

Fig.2. $^7F_0 \rightarrow {}^5D_0$ excitation spectra of nanocrystalline (a) Eu_2O_3 (b) 5.0% Eu^{3+}: Y_2O_3 and (c) 0.1% Eu^{3+}:Y_2O_3 monitoring $^5D_0 \rightarrow {}^7F_2$ fluorescence at 624.0 nm using wide-band 0.25 m

monochromator with a bandpass of 5.5 nm. Boxcar delay and gatewidths were 20 and 150 μs respectively for all three spectra.

5% Eu^{3+}:Y_2O_3. Considering these observations we conclude that nanocrystalline materials prepared by gas-phase condensation from starting materials containing 0.7% or more of Eu^{3+} have mixed phases. We could not detect any secondary Eu_2O_3 phase in the nanocrystals prepared from 0.1% Eu^{3+}:Y_2O_3 starting material, and the resulting nanocrystals appear to be single-phase Eu^{3+}: Y_2O_3.

To the best of our knowledge no spectroscopic data is available for Eu^{3+} in monoclinic Y_2O_3 matrix. In general, the spectra of monoclinic Eu^{3+}:Y_2O_3 are similar to monoclinic Eu^{3+}:Gd_2O_3, and Eu_2O_3 [10,11]. The small differences in the energy level positions between Eu^{3+}:Y_2O_3 and the other two systems originate from the different crystal-field strengths due to the difference in the ionic radii of Y^{3+} versus Gd^{3+} or Gd^{3+}. The excited states in the 0.1% Eu^{3+}:Y_2O_3 appears to have lower energies as compared to monoclinic Eu_2O_3 and Eu^{3+}:Gd_2O_3. The $^7F_0 \rightarrow {}^5D_0$ excitation spectrum (shown in Figure 2) of 0.1% Eu^{3+}: Y_2O_3 show three lines at 579.4, 582.7, and 582.8 nm, assigned to sites A, B and C corresponding to three distinct crystallographic cation site in monoclinic structure. Likewise, the site-selective $^7F_0 \rightarrow {}^5D_1$ excitation spectra (not shown) consists of three set of excitation lines; the excitation lines at 526.2 and 527.4 are assigned to site A; 528.7, 528.8, and 528.9 to site C; and 528.2, 528.6, and 529.3 to site B.

Figure 3 shows the $^5D_0 \rightarrow {}^7F_2$ fluorescence spectra from sites A, B, and C. The fluorescence excited at 582.7 nm shows lines at 615.5, 616.2, 623.7 and 631.4 and a weak peak at 618.3 nm which could be assigned to site B. Excitation at 582.8 nm, corresponding to site C

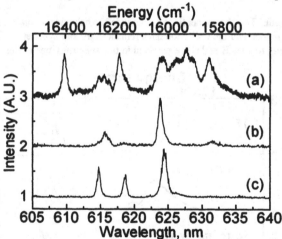

Fig.3. $^5D_0 \rightarrow {}^7F_2$ Fluorescence spectrum from three cation site of monoclinic 0.1% Eu^{3+}: Y_2O_3 from (a) site A excited at 579.4 nm with monochromator bandpass of 0.4 nm, boxcar delay of 100 μs and gatewidth of 150 μs (b) site B excited at 582.7 nm and (c) site C excited at 582.8 nm; for spectra (b) and (c) monochromator bandpass was 0.6 nm, photon counter delays and gatewidths were 10 μs and 1.0 ms, respectively.

gives three strong fluorescence peaks at 614.8, 618.6, 624.4 nm, and a weak shoulder at 526.0 nm. Exciting site A at 579.4 nm results in a rather complex spectrum consisting of more than 10 weak lines. However, monitoring the fluorescence at relatively shorter time (see Fig. 3) the lines at 609.8, 617.7, 624.0, 626.0, 627.7, 628.4, and 630.7 nm appear prominent The transitions at 609.8, 617.7, 627.7, 628.4 and 630.7 nm could be assigned to site A conclusively. The 624.0 line is somewhat broad and overlaps with the strongest $^5D_0 \rightarrow ^7F_2$ transitions at 623.8 and 624.4 nm corresponding to sites B and C. In view of observed energy transfer from A to B and C, there is an uncertainty in assigning 624.0 nm line to site A. Similarly the observed intensity of the line at 626.0 nm makes its assignment ambiguous. In monoclinic Eu_2O_3 and $Eu^{3+}:Gd_2O_3$ some of the longer wavelength lines have been assigned to a vibronic origin [10,11]. However, we do not comment on the origin of these lines in monoclinic $Eu^{3+}:Y_2O_3$ in absence of any theoretical calculations for this system.

Fluorescence decay measurements of 0.1% Eu^{3+}: Y_2O_3 under direct excitation of site A at 579.4 nm showed a decay time of approximately 1.8 ms. The decay time for sites B and C were found to be about 1.2 and 1.3 ms, respectively. These decay times are longer than that of radiative decay time of 1.1 ms for C_2 $(^5D_0)$ site in cubic Eu^{3+}: Y_2O_3. Observations of transitions from $^5D_0$ levels after excitation of $^5D_1$ levels show an initial rise-time followed by exponential decay. $^5D_0$ decay times measured after $^5D_1$ excitation are in agreement with the one obtained after direct excitation of $^5D_0$ and rise times are consistent with the $^5D_1$ decay times. The decay times for $^5D_1$ levels were measured by monitoring $^5D_1 \rightarrow ^7F_3$ transitions in the 585-602 nm spectral range. The fluorescence decay times of $^5D_1$ levels in 0.1% $Eu^{3+}:Y_2O_3$ were measured as approximately 46, 117, and 156 μs for site A, B, and C respectively. The results are summarized in Table 1. The qualitative behavior of the $^5D_0$ lifetimes and energy transfer is similar in $Eu^{3+}:Y_2O_3$ and $Eu^{3+}:Gd_2O_3$ [11]. This indicates the similarity of crystal-field perturbation in Y_2O_3 and Gd_2O_3 host materials.

Table 1 Fluorescence decay times in monoclinic 0.1% Eu^{3+}: Y_2O_3 at 10 K.

	Site A	Site B	Site C
$^5D_1$	46 μs	117 μs	156 μs
$^5D_0$	1.8 ms	1.2 ms	1.3 ms

Contrary to pure Eu systems, the probability of depopulation through ion-ion interaction decreases as the concentration of europium ions decreases in doped systems. At sufficiently low concentration the decay mechanism is mainly radiative. This results in longer and almost exponential decay curves. In the 0.1% Eu^{3+}: Y_2O_3 nanocrystals, although the energy difference between $^5D_0$ levels for site B and C is less than the thermal energy at 10 K interaction between sites B and C appears very weak. In the cubic system also no observable interactions between Eu^{3+} ions have been reported for concentrations below 0.1% of Eu^{3+} [14]. However, we notice significant energy transfer from site A to B and C in nanocrystalline monoclinic Eu^{3+}: Y_2O_3. Phonon-assisted energy transfer could be a possible explanation of this observed behavior [15]. Because of small energy mismatch between the $^5D_0$ levels of Eu^{3+} at sites B and C, the one-phonon assisted energy transfer between these two sites will be inhibited as compared to transfer from site A to B and C. The fluorescence from site A appears to be inefficient as compared to site B and C. The decay characteristic of site A is rather complex and fluorescence from this site appears inefficient as compared to sites B and C.

CONCLUSIONS

Gas-phase condensation of 0.1% Eu^{3+}:Y_2O_3 nanocrystals results in a metastable monoclinic structure. Nanocrystals show three sets of spectral lines corresponding to the three cation sites in the monoclinic lattice. Eu^{3+}:Y_2O_3 nanocrystals obtained from starting materials with 0.7% or more Eu^{3+} in Y_2O_3 show mixed phases. The 0.1% Eu^{3+}: Y_2O_3 shows single-phase spectrum having sharp and symmetric lines at longer wavelength than the corresponding transitions in monoclinic Eu_2O_3. Energy transfer from the higher energy site A to lower energy sites B and C was found significant, while no interaction has been observed between sites B and C. The three cation site have different fluorescence lifetimes, which could be attributed to the difference in the local environment of each site. Experiments are now underway to elucidate the size-dependent spectral behavior in Eu^{3+}:Y_2O_3 and Eu_2O_3.

ACKNOWLEDGMENTS

This work was supported by a National Science Foundation Career award (CHE-9502460).

REFERENCES

1. R. P. Andres, R. S. Averback, W. L. Brown, L. E. Brus, W. A. Goddard, III, A. Kaldor, S. G. Louie, M. Moscovits, P. S. Peercy, S. J. Riley, R. W. Siegel, F. Spaepen and Y. Wang, J. Mater. Res. 4, 704-736 (1989).
2. G. C. Hadjipanayis and R. W. Siegel, eds., Nanophase Materials: Synthesis - Properties - Applications, NATO ASI Series E Vol. 260 (Kluwer, Dordrecht, 1993).
3. H. Gleiter, Prog. Mater. Sci. 33, 223 (1989).
4. T. Hase, T. Kano, E. Nakazawa and H. Yamamoto, Adv. Electronics and Electron Phys. 1990 271.
5. H. Eilers and B. M. Tissue, Chem. Phys. Lett. 251, 74-78 (1996).
6. G. Skandan, C. M. Foster, H. Frase, M. N. Ali, J. C. Parker and H. Hahn, Nanostruct. Mater. 1, 313 (1992).
7. G. Shen, N. A. Stump. R. G. Haire and J. R. Peterson, J. Alloys and Comp. 181, 503 (1992).
8. H. R. Hoekstra, Inorg. Chem. 5, 754 (1966).
9. H. T. Hintzen and H. M. van Noort, J. Phys. Chem. Solids 49, 873 (1988).
10. K. C. Sheng and G. M. Korenowski, J. Phys. Chem. 92, 50 (1988).
11. J. Dexpert-Ghys, M. Faucher and P. Caro, Phys. Rev. B23, 607 (1981).
12. H. Eilers and B. M. Tissue, Mater. Lett. 24, 261-265 (1995).
13. J. Doss and R. Zallen, Phys. Rev. B48, 15626 (1993).
14. Blasse, in Energy transfer processes in condensed matter edited by B. Di-Bartolo (Plenum, New York, 1984), p. 251.
15. T. Holstein, S. K. Lyo and R. Orbach, in Laser spectroscopy of solids edited by W. M. Yen and P. M. Selzer (Springer-Verlag 1986) pp. 39-82.

Part II

Nanophase Metals, Alloys, and Non–Oxides

Part II

Nanostructured Metals, Alloys
and Non-Oxides

PROPERTIES OF NANOPHASE MATERIALS SYNTHESIZED BY MECHANICAL ATTRITION

H.-J. FECHT and C. MOELLE
Technical University Berlin, Institute of Metals Research, Hardenbergstr. 36,
PN 2-3, D-10623 Berlin, Germany

ABSTRACT

Mechanical attrition and mechanical alloying has been developed as a versatile alternative to other processing routes in preparing nanophase materials with a broad range of chemical composition and atomic structure. In this process, lattice defects are produced by "pumping" energy into initially single-crystalline powder particles of typically 50 μm particle diameter. This internal refining process with a reduction of the average grain size by a factor of 10^3 - 10^4 results from the creation and self-organization of small-angle and high-angle grain boundaries within the powder particles during the milling process. This microstructural evolution has been characterized by X-ray, neutron and electron scattering methods revealing the grain refinement and increase in internal stress. As a consequence, a change of the thermodynamic, mechanical and chemical properties of these materials has been observed with the properties of nanophase materials becoming controlled by the grain size distribution and the specific atomic structure and cohesive energy of the grain or interphase boundaries. An analysis of the thermal stability of attrited powder specimen gives the grain boundary energy of non-equilibrium and fully relaxed grain boundaries as well as their mobility. In summary, it is expected that the study of mechanical attrition processes in the future not only opens new processing routes for a variety of advanced nanophase materials but also improves the understanding of technologically relevant deformation processes, e.g. surface wear, on a nanoscopic level.

INTRODUCTION

Nanocrystalline materials have attracted considerable scientific interest in the last decade because of their unusual physical properties (for a review see reference 1). Such materials are characterized by their small crystallite-size which is in the range of a few nanometers. The grains are separated by high-angle grain or interphase boundaries. As such, they are inherently different from glasses (ordering on a scale of < 2 nm) and conventional polycrystals (grain size of > 1 μm).

These materials can be synthesized by a range of different physical, chemical and mechanical methods [1]. Among these, mechanical attrition offers several advantages in comparison with other methods. Here, cyclic mechanical deformation at high strain rates leads to large quantities of nanostructured powder particles which can be compacted to bulk samples. As a result, a wide range of metals, alloys, intermetallics, ceramics and composites can be prepared in an amorphous, nanocrystalline or quasicrystalline state [2]. Due to the broad range of possible atomic structures very different properties in comparison with conventional materials are obtained. For example, nanostructured particles prepared by mechanical attrition can exhibit unusually high values in hardness [3,4], enhanced hydrogen solubility [5], magnetic spin-glass behavior [6] etc.

Whereas these deformation processes within the powder samples are important for fundamental studies of extreme mechanical deformation and the development of nanostructured states of matter with particular physical and chemical properties, similar processes control the deformation of technologically relevant surfaces. For example, the effects of work hardening, material transfer and erosion during wear situations result in microstructures of wear surfaces comparable to those observed during mechanical attrition [7]. In particular, during sliding wear, large plastic strains and strain-gradients are created near the surface [8]. Similar to mechanical attrition of powder particles this is the consequence of the formation of dislocation cell networks, subgrains and grain boundaries with the subgrains becoming smaller near the surface.

Although several problems still have to be solved in order to use these materials for technological applications, mechanical attrition offers interesting perspectives in preparing nanostructured powders with a number of different interface types in terms of structure (crystalline / crystalline, crystalline / amorphous) as well as atomic bonding (metal / metal, metal / semiconductor, metal / ceramic etc.). This opens exciting possibilities for the preparation of advanced materials with particular grain- or interphase-boundary design.

EXPERIMENT

In the metallurgical processes of mechanical attrition or mechanical alloying, powder particles are subjected to severe mechanical deformation from collisions with steel or tungsten carbide balls and are repeatedly deformed, cold welded and fractured. Shaker mills (e.g. SPEX model 8000) which are preferable for small batches of powder sufficient for research purposes, are highly energetic and reactions can take place by one order of magnitude faster compared with other types of mills. A variety of milling devices has been developed for different purposes including tumbler mills, attrition mills, shaker mills, vibratory mills, planetary mills etc. [9].

The powder samples are placed together with a number of hardened steel or WC coated balls in a sealed container which is shaken or violently agitated. Since the kinetic energy of the balls is a function of their mass and velocity, dense materials (steel or tungsten carbide) are preferable to ceramic balls. During the continuous severe plastic deformation, a continuous refinement of the internal structure of the powder particles to nanometer scales has been observed frequently.

Powder samples with high purity (typically 99.95%) and particle size of 320 mesh are generally used. The results presented in the following are based on powder samples sealed in a stainless steel vial together with steel balls under high purity Argon and milled in a Spex 8000 mill. The ball to powder ratio is typically 5:1. Since generally contamination from the milling devices can occur, the experiments discussed in the following are mainly concentrated on iron powder as a model system for mechanical attrition.

RESULTS

Structural Properties

During mechanical attrition the metal powder particles are subjected to severe plastic deformation from collisions with the milling tools. Consequently, plastic deformation at high strain rates ($\sim 10^3 - 10^4$ s^{-1}) occurs within the particles.

The microstructural changes as a result of mechanical attrition can be followed by X-ray diffraction methods averaged over the sample volume. The X-ray diffraction patterns exhibit an increasing broadening of the crystalline peaks as a function of milling time.

The peak broadening is caused by size as well as internal strain effects [10]. The average coherently diffracting domain size (grain or crystal size) and the microstrain as function of milling time are obtained from the integral peak widths assuming Gaussian peak shapes (Fig. 1) [11]. In the very beginning mechanical attrition leads to a fast decrease of the average grain size of 40 - 50 nm. Further refinement occurs slowly to less than 20 nm after extended milling. In addition, the average atomic level strain as calculated from the X-ray broadening exhibits an increase to about 0.7 % for the iron particles. Direct observations of the individual grains within the deformed powder particles by transmission electron microscopy agree with the grain size determination by X-ray diffraction.

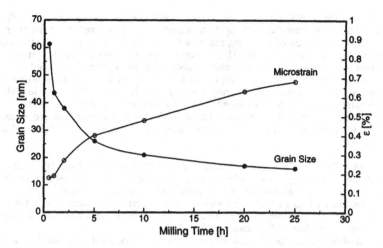

Figure 1: The average grain size and microstrains as determined from X-ray
line broadening as function of milling time for iron powder

From wide angle X-ray spectra, the information about lattice defects (grain boundaries, dislocations etc.) is obtained via their disturbing influence on the coherent superposition of radiation diffracted at the atomic lattice sites which causes the broadening of Bragg peaks. In small angle scattering experiments the lattice defects themselves give rise to a scattering contrast because of the (scattering length) density fluctuations associated with them.

Small angle neutron scattering (SANS) measurements were performed at the Hahn-Meitner-Insitut Berlin. Ballmilled Fe powders were measured in quartz cuvettes at a neutron wavelength λ of 0.60 nm. The cuvettes were filled up with D_2O resulting in a 79% reduction of scattering contrast at the wetted surfaces of the powder particles.

Different positions between the sample and the area-sensitive detector were chosen covering a range of momentum transfer $q=(4\pi/\lambda)\sin\theta$ from 0.045 nm^{-1} to 0.85 nm^{-1} (2θ: scattering angle). During the measurements a homogeneous magnetic field of 0.7 T was applied to the sample in horizontal direction perpendicular to the incoming neutron beam.

For a magnetically saturated sample the scattered intensity as function of the vector of momentum transfer **q** can be written as

$$I(q) = I_N(q) + I_M(q) \sin^2\alpha \qquad (1)$$

where $I_N(q)$ ($I_M(q)$) represents the structure function of nuclear (magnetic) scattering [12] and α is the azimuthal angle between **q** and the magnetic field, projected on the area perpendicular to the incoming beam. The anisotropic intensity distributions were analyzed by radial averaging over angular sectors parallel and perpendicular to the direction of the applied magnetic field. While in the first case ($\alpha=0°$) the spectra thereby obtained represent the nuclear scattering contribution, the perpendicular averaging ($\alpha=90°$) yields a linear combination of nuclear and magnetic scattering according to Eq. 1. For the correction of detector efficiency and the conversion of sample scattering intensities to absolute scattering cross sections additional measurements of the uniform scattering of a water sample were performed. Radial distribution functions (RDF) were calculated from the SANS spectra by the indirect Fourier-transformation method [13].

Fig. 2 shows the SANS spectra of Fe powder samples before milling and after milling for 0.5 hs and 30 hs averaged parallel to the applied magnetic field. Compared to the spectrum of the unmilled sample, an increase of scattering intensity occurs after 0.5 hs milling over the whole q-range covered by the measurements. This increase may be explained by the refinement of microstructure in the early stages of the milling process (i.e. the scattering contribution of grain boundaries, dislocations and triple junctions) which is proved by the x-ray peak broadening. It should be noted that the increase of scattering intensity extends to the high q range implying that structural inhomogenities on a small length scale of a few nm are present after short milling times. This observation underlines that the structural refinement is strongly inhomogeneous and might occurs in shear bands of high dislocation density surrounded by less deformed sample regions.

After 30 hs milling time, a drop of the scattering intensity is observed over the whole q-range compared to the 0.5 hs sample. Especially the drop at high q values is surprising since from X-ray measurements it is known that the average grain size is further reduced to a volume average of about 16 nm which is good agreement with the average grain size derived from the RDF of about 15 nm after 30 hs milling. Furthermore, the RDF also shows a decrease in magnitude without any significant shift of their maxima (Fig. 3).
Obviously the observed change of the SANS intensity cannot be solely explained by a shift of crystallite size distribution to smaller distances in real space during the milling process. Instead it is believed that the scattering contrast due to dislocations plays an important role for the interpretation of the measured data. Since the volume dilatation in the vicinity of dislocations is small their nuclear scattering contrast is small and disappears for pure screw dislocations [14]. Their magnetic scattering contrast due to orientation fluctuations of the magnetic moments of Fe atoms may exceed by factors up to 10-100 their contribution to nuclear scattering [15]. These fluctuations are caused by magnetoelastic coupling between the magnetic moments and the dislocation strain field.

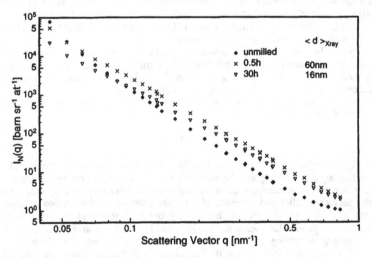

Figure 2:   SANS spectra of ballmilled Fe powder after different milling
times (0 hs, 0.5 hs, 30 hs)

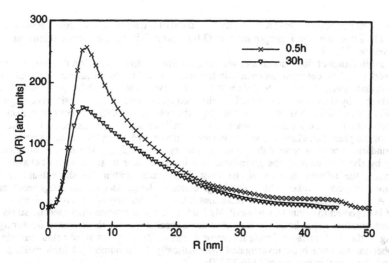

Figure 3: Radial distribution functions (volume weighted) calculated from spectra in Fig. 2 (milling times 0.5 hs, 30 hs)

The SANS data may be interpreted with respect to the magnetic scattering contribution of dislocations as a result of a changing distribution of dislocations in the deformed material. According to this model, the increase of scattering intensity at high q after 0.5 hs milling is caused by an increasing dislocation density with an average dislocation distance of a few nm which is in the range of the final grain size after long milling times. The subsequent decrease of scattering intensity after 30 hs milling time may be caused by the rearrangement of dislocations to form grain boundaries and the absorption of dislocations as secondary grain boundary dislocations. Further experiments on annealed nanocrystalline samples which are planned for the future are expected to give more information on the structural defects and their interactions in heavily deformed metals.

The elemental processes leading to the grain size refinement include generally three stages [16]:

(i) Initially, the deformation is localized in shear bands consisting of an array of dislocations with high density.

(ii) At a certain strain level, these dislocations annihilate and recombine to small angle grain boundaries separating the individual grains. The subgrains formed via this route are already in the nanometer size range with diameters often between 20 and 30 nm. During further attrition, the sample volume exhibiting small grains extends throughout the entire specimen.

(iii) The orientations of the single-crystalline grains with respect to their neighboring grains become completely random. Probably superplastic deformation processes together with grain boundary sliding cause this self-organization into a random nanocrystalline state.

This behavior is typical for deformation processes of bcc metals and intermetallic compounds at high strain rates. However, it is surprising that nominally brittle materials, such as intermetallics, develop considerable ductility under shear conditions.

Similar observations regarding the deformation mechanism have been reported in chips removed during machining [17] and simple metal filings [18,19]. In analogy to the mechanically attrited powder at the early stage, large inhomogeneities have been observed in the filings with the deformation process leading to the formation of small angle grain boundaries. Here, the dislocation cell size dimensions are basically a function of the acting shear stress τ resulting in an average cell size dimension L proportional to G b / τ with G being the shear modulus and b the Burgers vector [20,21].

The annihilation of dislocations can set a natural limit to the dislocation densities which can be achieved by plastic deformation (typically less than 10^{13} m^{-2} for screw dislocations and 10^{16} m^{-2} for edge dislocations). Steady state deformation is observed when the dislocation multiplication rate is balanced by the annihilation rate. This situation corresponds to the transition of stage (i) to stages (ii) / (iii) as described above. In this stage the role of dislocations becomes reduced and further deformation occurs probably via slip of grain boundaries. It is expected that the shear modulus of the grain boundary regions is lowered by about 40% when the volume-fraction of the grain boundaries becomes comparable to that of the crystals [22,23]. Localized deformation then proceeds by the dilatation of the grain-boundary layers similar to superplastic behavior [24]. Furthermore, the relative motion of the crystalline grains within the shear band leads to impingement on one another which should give rise to large, locally inhomogeneous elastic stresses. As a consequence, in order to relax these strains, formation of nanovoids about 1 nm in diameter is expected to occur which inevitably leads to crack formation under tensile stress [25]. Such a deformation mode basically also provides a mechanism for the repeated fracturing and rewelding of the fresh surfaces during mechanical attrition leading to a steady state particle size. These phenomena have been investigated systematically for a number of high melting point metals and intermetallic compounds [26,27,28].

Thermal Properties

Decreasing the grain size of a material to the nanometer range leads to a drastic increase of the number of grain boundaries reaching typical densities of 10^{19} interfaces per cm^3. The large concentration of atoms located in the grain boundaries in comparison with the crystalline part scales roughly with the reciprocal grain size 1 / d.

Consequently, due to their excess free volume the grain boundaries in nanocrystalline cause large differences in the physical properties of nanocrystalline materials if compared with conventional polycrystals. In all cases discussed here, the short range order typical for an amorphous material is not observed as the characteristic structure of grain boundaries. As such, the grain boundary structure in these materials must be different from the structure of the single crystal as well as the amorphous structure of a glassy material. It turns out that the thermodynamic properties of nanostructured materials produced by mechanical attrition can be realistically described on the basis of a free volume model for grain boundaries [29].

Thus, as a result of the cold work, energy has been stored in the powder particles. During heating in the DSC, a broad exothermic reaction is observed for all of the samples starting at about 370 K and being typically completed at 870 K (Fig. 5). Integrating the exothermic signals gives the energy release ΔH during heating of the sample. For comparison, the ΔH values are included in Table I in addition to other characteristic values, such as the average grain size and excess specific heat after 24 hs of mechanical attrition.

The stored enthalpy reaches values up to 7.4 kJ/mole (after 24 hs) and 10 kJ/mole (after 32 hs) for Ru, which corresponds to 30-40% of the heat of fusion ΔH_f. One would expect that the recovery rates during the milling process correlate with the melting point of the specific metal. With the exceptions of Co (due to a large number of stacking faults) and Hf, Nb and W (possibly due to an increased level of Fe-impurities from the milling tools stabilizing the nanostructure) such a relationship is indeed observed. Similar results have been obtained for metals with fcc structure as given in Table 1 [28]. Consequently, most effective energy storage occurs for metals with melting points above 1500 K resulting in average grain sizes between 6 (Ir) and 13 nm (Zr). For the compound phases similar high values for the stored energies are found ranging from 5 to 10 kJ/mole and corresponding to values between 18 and 39% of the heat of fusion for grain sizes between 5 and 12 nm.

Table 1

Structural and thermodynamic properties of metal and intermetallic powder particles after 24 hours ball milling, including the melting temperature T_m, the average grain size d, the stored enthalpy ΔH, and the excess heat capacity Δc_p.

material	structure	T_m (K)	d(nm)	ΔH (% of ΔH_f)	$\Delta c_p(\%)$
Fe	bcc	1809	8	20	5
Cr	bcc	2148	9	25	10
Nb	bcc	2741	9	8	5
W	bcc	3683	9	13	6
Co	hcp	1768	(14)	6	3
Zr	hcp	2125	13	20	6
Hf	hcp	2495	13	9	3
Ru	hcp	2773	13	30	15
Al	fcc	933	22	43	-
Cu	fcc	1356	20	39	-
Ni	fcc	1726	12	25	-
Pd	fcc	1825	7	26	-
Rh	fcc	2239	7	18	-
Ir	fcc	2727	6	11	-
NiTi	CsCl	1583	5	25	2
CuEr	CsCl	1753	12	31	2
SiRu	CsCl	2073	7	39	10
AlRu	CsCl	2300	8	18	13

The final energies stored during mechanical attrition largely exceed those resulting from conventional cold working of metals and alloys (cold rolling, extrusion etc.). During conventional deformation, the excess energy is rarely found to exceed 1-2 kJ/mole and, therefore, is never more than a small fraction of the heat of fusion [30,31]. In the case of mechanical attrition, however, the energy determined can reach values typical for crystallization enthalpies of metallic glasses corresponding to about 40% ΔH_f.

A simple estimate demonstrates that these energy levels can not be achieved by the incorporation of defects which are found during conventional processing. In the case of pure metals, the contribution of point defects (vacancies, interstitials) can be safely neglected because of the high recovery rate at the actual processing temperature [31]. Even taking non-equilibrium vacancies into account, which can form as a consequence of dislocation annihilation up to concentrations of 10^{-3} [32], such contributions are energetically negligible in comparison. On the other hand, for intermetallics point defects are relevant in order to describe the stability of the material [33].

The maximum dislocation densities which can be reached in heavily deformed metals are less than 10^{16} m^{-2} which would correspond to an energy of less than 1 kJ/mole. Therefore, it is assumed that the major energy contribution is stored in the form of grain boundaries, and related strains within the nanocrystalline grains which are induced through grain boundary stresses. Recent estimates suggest that the grain boundary energies in nanocrystalline metals are about twice as high in comparison with high energy grain boundaries in conventional polycrystals which are approximately equal to 1 J/m^2.

Large differences generally also arise in the specific heat c_p at constant pressure. The specific heat of the heavily deformed powder particles was measured in the range from 130 K to 300 K, i.e. at low enough temperatures to prevent the recovery processes from taking place. For all samples, a considerable increase in c_p has been found experimentally after 24 hours milling, reaching values up to 15% for Ru. These data are also included in Table I given as a percentage of heat capacity increase in comparison to the unmilled state at 300 K. For pure metals, a linear correlation between the heat capacity change Δc_p and the stored enthalpy ΔH given as a percentage of the heat of fusion ($\Delta H/\Delta H_f$) after extended mechanical attrition is observed (Fig. 4). Such a relationship is also predicted by the free volume model for grain boundaries.

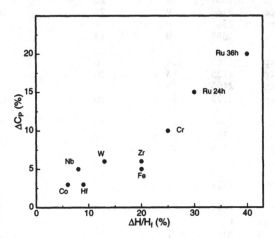

Figure 4:   Specific heat increase at room temperature in comparison to the unmilled state as function of the stored enthalpy ΔH (given as percentage of ΔH_f) for bcc and hcp nanocrystalline metals

Phase Stability

As a result of the cold work considerable energy has been stored in the powder particles (see Table 1). Therefore, thermodynamically these materials are far removed from their equilibrium configuration and a large driving force towards equilibrium exists. The stored energy is released during heating to elevated temperatures due to recovery, relaxation processes within the boundaries and grain growth. As a consequence, during annealing at elevated temperatures, relaxation and grain growth processes will occur leading to a concomitant increase of the grain size.

This behavior has been investigated for iron in detail [34]. For extended periods of milling time a decrease of the average grain size to nanometer dimensions is observed with a stationary average grain size d=16 nm and 0.7% microstrain (see Fig. 1). The enthalpy release during a DSC heating experiment spreads over the entire temperature range of the scan as shown in Figure 5. The very broad signal does not exhibit any distinct peaks but a further increase of the exothermic signal for T > 250 - 300 °C.

X-ray diffraction of powder samples annealed for 80 min at each temperature revealed the evolution of grain size and strain as function of annealing temperature as shown in Figure 6. The microstrain is decreasing rapidly below 200 °C while the grain size remains nearly constant. As such, the enthalpy release during the first exotherm in Figure 5 is only related to relaxation and not to grain growth. Grain growth starts to become significant above about 300 °C. Furthermore, it has been found that after a fast increase at early times the average grain size d changes from 16 nm to about 30 - 40 nm. The average grain size remains constant for t ≥ 2400 sec and reaches values of 100 - 200 nm at temperatures about 600 °C.

As such, two regimes with and without grain growth can be distinguished. However, since the influence of lattice point defects and lattice dislocations is negligible, the enthalpy release can be clearly assigned to the existence of grain boundaries. The reduction of the microstrains are probably caused by grain boundary relaxation and annihilation of secondary grain boundary dislocations. Based on elastic theory it is estimated that this contribution to the overall energy is less than about 5%. On this basis, the grain boundary energy can be estimated.

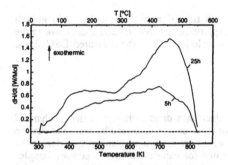

Figure 5: DSC heating scan at 10 K/min of iron powder after mechanical attrition for 5 and 25 hs

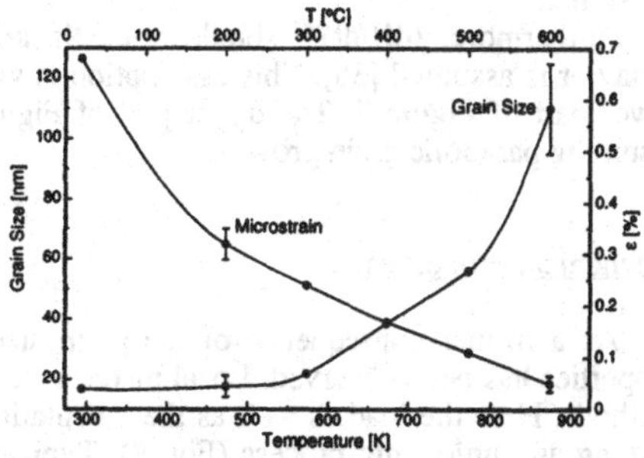

Figure 6: Dependence of stationary grain size and microstrain on annealing temperature

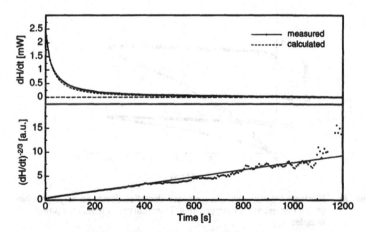

Figure 7: Isothermal exothermic DSC curve at 500 °C of nanocrystalline iron (upper part) and plot of $(dH/dt)^{-2/3}$ versus time (lower part)

By simple geometric considerations [35,36] the specific grain boundary excess enthalpy is estimated to be about 2.1 J/m^2. This would correspond to a value for non-equilibrium unrelaxed grain boundaries, whereas after relaxation, the grain boundary energy is reduced to 1.5 J/m^2. Values resulting from computer simulations suggest excess enthalpies between 1.2 and 1.8 J/m^2

[37]. Therefore we conclude that grain boundaries in the as prepared state are characterized by increased values of about 25% due to their unrelaxed atomic structure.

Further isothermal DSC measurements allow to analyze the grain growth processes in nanocrystalline Fe. For example, the isothermal DSC curve shown in the upper part of Figure 7 was measured at 500 °C after annealing the sample at 400 °C and heating to 500 °C at a rate of 50 °C/min. A monotonically decreasing signal typical for grain growth is observed. Similar signals are observed at 200 °C, 300 °C and 400 °C and clearly differ from those measured in isothermal recrystallization processes controlled by nucleation and growth in conventional polycrystalline metals which are described by Johnson Mehl Avrami type models [36]. Figure 7 does not exhibit the expected maximum related to an incubation time for nucleation, but shows only a decrease in the signal.

Furthermore, $(dH/dt)^{-2/3}$ should scale linearly with time if normal parabolic grain growth behavior is assumed [35]. This assumption is well approximated for $t < 1200$ s as shown in the lower part of Figure 7. The upper part of Figure 7 includes a fit to the measured DSC signal assuming parabolic grain growth.

<u>Mechanical Properties</u>

As a further consequence of the grain size reduction a drastic change in the mechanical properties has been observed. Local mechanical properties can be measured by nano-indentation methods. Here the load as well as the indentation depths are monitored continuously during the loading and unloading process (Fig. 8). Typical results for nanocrystalline Fe powder samples exhibit an increase in hardness by a factor 7 (9.3 GPa for nx-Fe with d about 16 nm versus 1.3 GPa for annealed px-Fe). In general, the hardness follows a trend similar to the Hall-Petch relationship though the deformation mechanism in the nanocrystalline regime remains unclear.

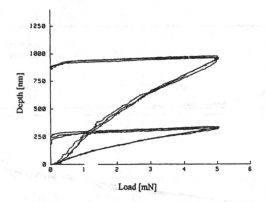

Figure 8: Hardness measurement using a nanoindentation device on polycrystalline (upper curves) and nanocrystalline (lower curves) attrited iron powder samples

Furthermore, the Youngs modulus can be measured by this method as well and shows a decrease by 10 - 20 % in comparison with the polycrystal. Therefore, it is suggested that the mechanical properties of nanophase materials prepared by mechanical attrition after extended periods of milling are not being controlled by the plasticity of the crystal due to dislocation movement anymore but rather by the cohesion of the nanocrystalline material across its grain boundaries. From the considerable increase of hardness and the principal changes of the deformation mechanisms improved mechanical properties can be expected as attractive features for the design of advanced materials.

CONCLUSIONS

By mechanical attrition in ball mills a refinement of microstructure to a nanometer scale can be achieved. In iron a volume-averaged stationary grain size of 16 nm has been determined. The correlation of structural and thermal analysis reveals that the defect structure after long milling times is comprised of a network of unrelaxed large angle grain boundaries. Acccording to TEM observations and SANS data the grain boundaries are formed by reorganisation of dislocations produced in the initial step of the deformation process. Due to the high density of grain boundaries and the large amounts of stored enthalpy, relaxation processes of the heavily deformed structure and grain growth occur at rather low temperatures. Therefore the control of the thermal stability of the nanostructure is important for the compaction of nanocrystalline powders and future applications as bulk materials. For example the enormous increase of microhardness by factors in the range of 5-8 with respect to conventional polycrystals is one of the promising properties of nanocrystalline materials which have to be retained in further processing of milled powder samples.

ACKNOWLEDGEMENTS

The financial support by the Deutsche Forschungsgemeinschaft (grant Fe 313/1-2) is gratefully acknowledged. We would like to thank U. Keiderling for his support in the SANS measurements.

REFERENCES

1. H. Gleiter, Prog. Mat. Sci. **33**, p. 223 (1989).
2. Mechanical Alloying, edited by P.H. Shingu, Mat. Sci. Forum **88-90** (1992).
3. C.C. Koch, Nanostructured Materials **2**, p. 109 (1993).
4. A. Kehrel, C. Moelle and H.J. Fecht in Nanophase Materials, edited by G.C. Hadjipanayis and R.W. Siegel, Kluwer Acad. Publ., 1994, p. 287.
5. C. Moelle and H.J. Fecht, Nanostructured Materials **3**, p. 93 (1993).
6. G.F. Zhou and H. Bakker, Phys. Rev. Lett. **72**, 2290 (1994).
7. D.A. Rigney, L.H. Chen, M.G.S. Naylor and A.R. Rosenfield, Wear **100**, p. 195 (1984).
8. D.A. Rigney, Ann. Rev. Mater. Sci. **18**, p. 141 (1988).
9. W.E. Kuhn, I.L. Friedman, W. Summers and A. Szegvari, ASM Metals Handbook, Vol. 7, Powder Metallurgy, Metals Park (OH) (1985), p.56.
10. C.N.J. Wagner and M.S. Boldrick, J. Mat. Sci. Engg. **A133**, p. 26 (1991).
11. C. Moelle, Ph.D. Thesis, Technical University Berlin (1995).
12. J.B. Hayter, J. Appl. Phys. **21**, p. 737 (1988).
13. O. Glatter, J. Appl. Cryst. **10**, p. 415 (1977).
14. H.H. Atkinson and P.B. Hirsch, Phil. Mag. **3**, p. 213 (1958).
15. H. Kronmüller, A. Seeger and M. Wilkens, Zeitschrift für Physik **171**, p. 291 (1963).
16. H.J. Fecht in Nanophase Materials, edited by G.C. Hadjipanayis and R.W. Siegel, Kluwer Acad. Publ., 1994, p.125.
17. D. Turley, J. Inst. Metals **99**, p. 271 (1971).
18. C.N.J. Wagner, Acta Met. **5**, p. 477 (1957).
19. B.E. Warren, Prog. Metals Phys. **8**, p. 147 (1956).

20. D. Kuhlmann-Wilsdorf and J.H. Van der Merwe, J. Mat. Sci. Engg. **55**, p. 79 (1982).
21. U. Essmann and H. Mughrabi, Phil. Mag. A **40**, p. 40 (1979).
22. J.J. Gilman, J. Appl. Phys. **46**, p. 1625 (1975).
23. P.E. Donovan and W.M. Stobbs, Acta Metall. **31**, p. 1 (1983).
24. M. Hatherly and A.S. Malin, Scripta Metall. **18**, p. 449 (1984).
25. S.M. Goods and L.M. Brown, Acta Metall. **27**, p. 1 (1979).
26. H.J. Fecht, E. Hellstern, Z. Fu and W.L. Johnson, Metall. Trans. A **21**, p. 2333 (1990).
27. H.J. Fecht, E. Hellstern, Z. Fu and W.L. Johnson, Adv. Powder Metallurgy **1**, p. 111 (1989).
28. J. Eckert, J.C. Holzer, C.E. Krill III, and W.L. Johnson, J. Mat. Res. **7**, p. 1751 (1992).
29. H.J. Fecht, Phys. Rev. Lett. **65**, p. 610 (1990).
30. W.L. Johnson, Prog. Mater. Sci. **30**, p. 81 (1986).
31. M.B. Bever, D.L. Holt and A.L. Titchener, Prog. Mater. Sci. **15**, p. 5 (1973).
32. U. Essmann and H. Mughrabi, Phil. Mag. A **40**, p. 40 (1979).
33. H.J. Fecht, Nature **356**, p. 133 (1992).
34. C. Moelle and H.J. Fecht, Nanostructured Materials **6**, p. 421 (1995).
35. A. Tschöpe, R. Birringer, H. Gleiter, J. Appl. Phys. **71**, p. 5391 (1992).
36. L.C. Chen and F. Spaepen, J. Appl. Phys. **69**, p. 679 (1991).
37. D. Wolf, Phil. Mag. A**62**, p. 447 (1990).

RAPID SYNTHESIS OF NANOSTRUCTURAL INTERMETALLICS AND THEIR BULK PROPERTIES

S.M. PICKARD AND A. K. GHOSH
Materials Science and Engineering Department, The University of Michigan, Ann Arbor MI 48109

ABSTRACT

A rapid physical vapor deposition process (PVD) utilizing a high speed rotating substrate and small substrate-to-source spacing has been used to produce bulk sheet of Ti-Al alloys in the compositional range Ti-12% Al to Ti-75% Al[1] at a rate of 1-3 μm/minute. Microstructural architectures produced by the method comprise of either fully homogenous phase mixtures of nano-grains, or nanolaminated material, depending on the substrate rotational rate, with lower rotational rate producing a layered microstructure. Defect populations within the as-deposited material are characterized by TEM and SEM, and hot pressing consolidation of the as-deposited material, which retains a grain size < 1000 nm, has been investigated. While indentation hardness of $\alpha_2 + \gamma$ (2 phase) alloys exceeded 7 GPa, brittle failure occurred in the elastic regime at nominally lower tensile stress than that for conventionally produced alloys containing Nb and Cr as solute elements. $\alpha_2 + \gamma$ alloys can exhibit tensile elongations of more than 100% at 850°C with retention of fine grain size. Elevated temperature failure occurs by the formation of voids in regions of compositional variability in the composite where single phase α_2-Ti_3Al structure was present.

INTRODUCTION

Nanoscale processing of metals produces ultrafine grained homogenous microstructures with compositional variations confined to the nanometer size of the microstructure. Property combinations possible through the nanophase processing route are higher strength and superplastic forming capability at a much lower temperature than for conventional grain size material [1]. Our aim has been to explore a rapid physical vapor deposition (PVD) method for nanoscale processing of materials which otherwise show compositional inhomogeneities in the cast microstructure and low room temperature ductility. Interest exists in using this approach on Ni and Ti based aluminides which offer a combination of good oxidation resistance, high temperature strength and lower density than nickel base superalloys [2]. Optimum nanophase processing has to address the opposing considerations of strength and ductility needs for practicality, especially for those materials which are prone to grain boundary rupture such as Ni_3Al [3].

In this study, TiAl has been the alloy selected. The potential of nanophase processing of strategic intermetallics of Ti-40 to 50% Al is critically assessed. These $\alpha_2 + \gamma$ alloys can have excellent potential to be economically shaped and retain their high temperature properties, when processed by such methods

EXPERIMENTAL

A modified TEMESCAL 2550 Evaporator unit which contained 2 independent evaporant sources heated by two 15 kW electron guns, using a 10 kV high voltage supply was utilized for the PVD work (Fig. 1). Evaporant was deposited onto a rapidly rotating substrate plate which revolved at a rate of 500-2000 r.p.m.. To increase the rate of deposition, the substrate-to-source spacing was reduced to 5 cm. No active heating of the substrate was employed, the temperature rise being solely from the radiant heating due to E-beam sources being in close proximity to the substrate. The evaporants comprised of a pure titanium sponge (98.8%) and commercial purity Al which were placed in the 156 c.c. Cu hearth (water cooled) of each of the two sources without the use of ceramic liners. A Ti-6Al-4V (wt%) alloy source was also utilized instead of Ti sponge for some of the deposition runs.

The chamber temperature was monitored by a thermocouple placed mid-way up the chamber wall. The temperature during a given deposition run reached a plateau value of 270-300°C after 45-50 mins., although the substrate temperature is expected to be much higher.

---

[1] All compositions given in at.% unless otherwise stated

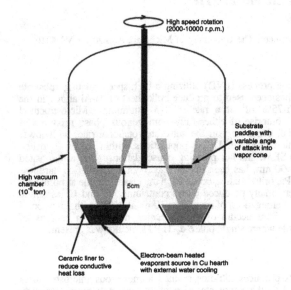

High speed rotation
(2000-10000 r.p.m.)

Substrate
paddles with
variable angle
of attack into
vapor cone

High vacuum
chamber
(10⁻⁶ torr)

5cm

Ceramic liner to
reduce conductive
heat loss

Electron-beam heated
evaporant source in Cu hearth
with external water cooling

Fig. 1. Modified coevaporation unit operated with two 15 kW electron-beam guns. Note the reduced substrate/source spacing and the high rotational speed of the substrate which spins above the evaporant sources.

The Ti-Al alloy deposits produced by the evaporator were between 50 μm and 400 μm in thickness after deposition runs of between 45 mins and 5 hrs, and were deposited at a rate of 1-3 μm per minute.

Consolidation of the as-deposited material was achieved by vacuum hot pressing (10^{-6} torr) in the temperature range 700-950°C and under pressure of 40 MPa. Tensile testing of the deposits were conducted at both room temperature and elevated temperature of 850°C after consolidation of the material. Ribbons of the thin sheet deposit, approximately 1 cm long and 4-5 mm wide and 230-400 μm thick were cut using high speed precision diamond blade sectioning. Ti-6Al-4V (wt%) mill annealed sheet was spot welded to both grip ends of the specimen ribbons, to secure a reasonable bond which would be effective at both room temperature and elevated temperature. Testing was performed on an Instron screw driven test machine. Rupture strength values for thick consolidated materials (0.3-1.5 mm thick) were also obtained using 3 -point bending at room temperature.

RESULTS

Microstructure of as-deposited alloys

As-deposited Ti-Al alloys of composition range between Ti-8% Al and Ti-75% Al were examined in the SEM. The PVD deposits showed a variety of void-like defects, the nature of which were found dependent on the speed of substrate rotation, substrate temperature, and the phase region of the Ti-Al phase diagram. A common source of defects in all the deposits resulted from either a cold or insufficiently heated substrate at the start of a run, before the equilibrium chamber temperature is reached, or on interruption of a deposition run to replenish the sources, which allowed the chamber to cool down. In either case, the material deposited at the start of the run often constituted a porous seam in the deposit as shown in Fig. 2.

The fine layered structure at slow rotational rates, appeared to result from the sequential accumulation of alternate thin Ti and Al evaporant layers, less than 0.1 μm thick on each rotation of the substrate. The coarser banded regions shown in the deposit for the Ti-40% Al alloy indicates the variability of the source output during the run, which is detected by compositional BSE imaging, and is directly related to compositional fluctuations. TEM observations on a Ti-40% Al alloy, showed a grain size of 20-50 nm in the as-deposited condition.

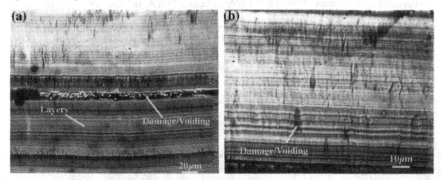

Fig. 2. As-deposited Ti-40 to 50% Al alloy. (a) and (b) show the deposit produced at low rotation rate (500 r.p.m). Note the defect content within the material and layered nature of its appearance in the SEM (BSE image).

Selected area diffraction patterns from the α_2 grains from planar sections parallel to the growth direction showed a nearly (0001) basal orientation diffraction pattern indicative of the c-axis orientation of the hexagonal α_2 phase parallel to the growth axis (Fig. 3). Voids, 5-10 nm size, were also seen in the TEM. The grain boundary 'pinning' from these and the 2-phase contributes to retaining fine grain size in these alloys.

Fig. 3. In-plane TEM views of the Ti-40 to 50% Al deposit. The section of the foil is within a region of reduced Al content (Ti-30% Al) and shows a single phase α_2 microstructure. The diffraction pattern from the nanograins, shows [0001] is aligned parallel to the growth axis.

Effect of Heat-treatment and Consolidation

Ordered α_2 and γ phase mixtures, showed fine retained microstructure after 2 hrs vacuum (10^{-6} torr) heat treatment at 700°C, with fine second phase dispersions (Fig. 4), which varied due to compositional variation during deposition. Both the phases appeared either as a continuous matrix phase or as well-dispersed eqiuaxed grains.

Hot pressing of the deposit for 4 hrs at 850°C under 40 MPa pressure was used to eliminate the porous interlamellar defects, resulting in a pore-free material, which retained fine grain microstructure of the deposit. Heat treatment of as- consolidated deposit for 10 hrs at 800°C resulted in an increase in the average eqiuaxed grain size to 1.5 μm. Porous defects, found in the as-deposited 2-phase $\alpha_2+ \gamma$ alloys, now appeared to be healed after heat treatment, with affected regions apparent by disruption of the ordered microstructure with Al rich deposits at the sites of the

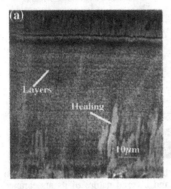 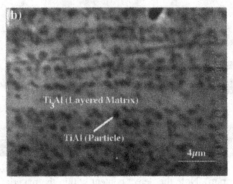

Fig. 4. Ti 40 to 50% Al (500 r.p.m), heat treated for 2 hrs at 700°C. Note that phase separation of fine eqiuaxed gamma particles has occurred in (a) and (b). Note void healing and disruption of the normal microstructure in the regions which contained defective material.

starting voids, with surrounding Ti rich zones.

Texture measurement of the 2-phase $\alpha_2+ \gamma$ deposits conducted using electron backscattering patterns showed a strong basal (0001) orientation of the α_2-phase parallel to the growth direction, as also seen in the as-deposited alloy. However, the γ (face-centered tetragonal) phase showed only a weak texture for [$\bar{1}$11] growth direction and poor correlation with the α_2 orientation. Best matching of the pole figures for the two phases is for (111)//(0001) hexagonal. This finding indicates a less strong crystallographic relationship between the phases in the PVD material than found for twin lamella $\alpha_2+ \gamma$ cast microstructures, where the usual crystallographic relationship between $\alpha_2+ \gamma$ is (0001)//(111)/[2110]//[110] [2].

Hardness

The Ti-Al deposits were evaluated for Vickers microhardness (100 gram load). As-deposited alpha Ti-Al solid solution has the lowest hardness value of 2.2 GPa. $\alpha_2+ \gamma$ phase nanograined regions (20-50 nm size) show much higher hardness values of up to 7 GPa in the as-deposited condition, which compares with literature values of up to 12 GPa for nanocrystaline near-gamma alloys of similar grain size produced by mechanical alloying [2]. The reason for the lower hardness of deposits produced by PVD could be due to lower intrinsic impurities and second phase particle content arising from the cleaner high vacuum process of PVD compared to the mechanically milled material which may be particle strengthened.

For the heat treated two-phase $\alpha_2+ \gamma$ deposits, elastic modulus was measured using nanoindentation which showed modulus values ranging from 190-212 GPa, with the highest values resulting from indentations in the Ti$_3$Al phase. The range of modulus values recorded are 10-20% higher than those reported in the literature for near alpha and near gamma alloys, the reason for this discrepancy is unclear at present. Variability of hardness values between deposits and within a single deposit probably reflects the range of defect densities and composition ranges sampled by the indentor tip.

Flow Strength and Ductility

In the as-consolidated form, the thin sheets of $\alpha_2+ \gamma$ alloy were extremely brittle and required delicate handling. Tensile testing was conducted on the consolidated thin sheet $\alpha_2+ \gamma$ alloys of 250 μm thickness, which showed strength values in the range from 200-350 MPa, with linear-elastic loading to catastrophic failure and no indication of plasticity. Examination of the fracture faces showed a rough surface appearance with intergranular facets observed on the fracture surface at high magnification (Fig. 5). A fast fracture region on crack initiation was seen close to the surface of the specimen and was characterized by a relatively featureless appearance, indicative

 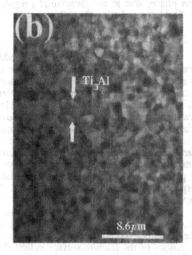

Fig. 5. Fracture surface observations of consolidated Ti-40 to 50%Al. (a) shows a general view of the fracture surface (b) shows the predominantly intergranular nature of the failure, with cleavage of occasional thin layers of brittle α_2- Ti$_3$Al.

that a surface flaw had initiated failure. Isolated thin bands of material which had failed by formation of cleavage steps were occasionally observed, due to compositional variations within the material which resulted in formation of thin layers of brittle Ti$_3$Al. Compression testing of stack consolidated deposits revealed that initial plastic yielding of the material occurred at approximately 900 MPa, which is higher than for conventionally processed near gamma material [2].

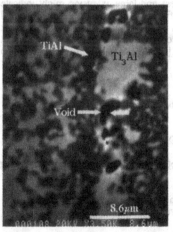

Fig. 6. Polished section (SEM) close to the fracture plane of the high temperature failure of Ti-40 to 50% Al at 850°C. Note the voided regions within the α_2 rich layers of the alloy.

High temperature tensile tests were conducted on the two phase alloy at a temperature of 850°C and stain rate 5×10^{-5} s^{-1}. The material showed a flow strength of about 20 MPa, and failure elongation of about 90% engineering strain. This was a lower limit failure strain since rupture invariably occurred at the grips. Microstructural examination of cross-sections of the tested material sectioned normal to the loading direction revealed that voiding had occurred within the regions devoid of gamma phase, due to microstructural irregularities in the deposit, which consisted of the more brittle Ti_3Al ordered alloy (Fig. 6).

CONCLUSIONS

$\alpha_2 + \gamma$ two-phase TiAl alloys produced by rapid PVD, were microstructurally and mechanically evaluated in this study. These materials provide a highly stable microstructure at elevated temperature, due to the mutual pinning of the fine eqiuaxed grain structure in both phases. This material showed a high resistance to coarsening with $\alpha_2 + \gamma$ grain size of 1.5 μm after 12 hrs at 850°C. The microstructure contains less Al than is optimum for ductility in a cast twin lamellar microstructure exhibiting maximum tensile ductility at Ti-48% Al [1]. It is interesting to note that a twin lamellae as-cast microstructure is not produced in the PVD deposit because the processing temperature is below the α_2 to $\alpha_2 + \gamma$ lamellae transformation temperature [1]. Interestingly, the low temperature processing route by PVD also eliminates the strong parent-matrix orientation relationship seen in the as-cast material, however the PVD deposit shows a strong texture in the α_2-Ti_3Al phase, with the growth direction of the deposit parallel to the basal (0001) plane, with a corresponding weak texture in the coexistent γ-TiAl. Such strong texture component is not retained in conventionally cast or powder metallurgy derived deposits produced by mechanical milling.

Somewhat disappointedly, the room temperature tensile strength of the nanophase processed binary Ti-40% Al of 300-500 MPa was less than for an optimally processed cast alloy, inspite of higher room temperature hardness of more than 7 GPa exhibited by the PVD material. Observations of the fracture surface indicated that flaw dominated surface crack nucleation might occur in the material. Propagation of the crack and failure of the fine grained deposits occurred by intergranular separation, suggesting weak grain boundaries. Hence, methods of grain boundary engineering aimed at increasing the cohesive strength should be investigated [5]. Additions of Nb, Cr, V can be successfully incorporated into these PVD deposits. Grain boundary weakening in our PVD deposit may be attributed to impurity segregation or Al enrichment. Elevated temperature ductility in the 2-phase alloy showed approximately 100% extension exhibited at 850°C in the consolidated Ti-40% Al at a strain rate of 10^{-4} s^{-1}. The high temperature strength of the fine grained material of about 20 MPa was much less than is reported for conventional alloys which indicates that processing of these alloys at moderately elevated temperature is extremely attractive. Elongations of more that 100% are seen which is greater than those expected for conventional processed near γ material at 850°C [1,3]. The low flow stress and large extension might indicate the grain boundary sliding mechanism at low stresses in the fine grained material and the potential for superplastic extension. Ball milled nanophase material exhibits similar low flow stress at elevated temperature, yet the stress exponent measured on compression testing, n=2, implies that the extent of superplasticity may be limited in that material [4]. Additional testing is required to determine elevated temperature rate sensitivity.

REFERENCES

1.   F. H. Froes, C. Suryanarayana, G.-H. Chen, Abdulbaset Frefer, and G. R. Hyde, JOM, **44** (May 92), p. 26.

2.   Y.-W. Kim and D. M. Dimiduk, JOM, **43** (August 91), p. 40.

3.   D. A. Kaibyshev in Superplasticity of alloys, Intermetallics and Ceramics, Springer-Verlag New York 1992, Chapter 9.

4.   R. S. Mishra and A. K. Mukerjee, to be published in Advances in the Science and Technology of Titanium alloys, edited by I. Weiss, R. Srinivassan, P. Bania and D. Eylon, TMS, Warrendale, 1996.

5.   C. T. Liu, Scr. Metall. Mater., **25**, p. 1231 (1991)

PREPARATION OF NANOMETER SIZED ALUMINUM POWDERS

CURTIS E. JOHNSON AND KELVIN T. HIGA
Chemistry and Materials Branch, Research and Technology Group,
Naval Air Warfare Center, Weapons Division, China Lake, CA 93555

ABSTRACT

Nanometer aluminum powders have been prepared from the catalytic thermal decomposition of aluminum hydride adducts in organic solvents. This process provides excellent control of particle size and uniformity with batch particle sizes ranging from 50 to 600 nm. Variables affecting particle size were explored, including reaction temperature, catalyst concentration, and adducting amine concentration. Passivation techniques of the reactive aluminum powders were established for safe handling in air. Samples were mainly characterized by scanning electron microscopy and thermogravimetric analysis.

INTRODUCTION

Aluminum powders are used in a broad range of applications including rocket propellants, paints, and powder metallurgy parts for aircraft and automobiles.[1] Since the reactivity of aluminum increases as the particle size decreases, small particles are desirable for aluminum used in propellants, explosives, and powder metallurgy processes. Commercially available aluminum powders are generally several microns or larger in size. We present here the synthesis of aluminum powders in controlled sizes ranging from 50 to 600 nm. The process is a modification of the previously known decomposition of the trimethylamine adduct of alane ($AlH_3 \cdot N(CH_3)_3$) to elemental aluminum.[2]

EXPERIMENT

The compound $AlH_3 \cdot (N(CH_3)_3)_2$ was prepared by a literature procedure.[3] The ALEX aluminum powder was obtained from The Argonide Corporation. The H-3 and H-5 aluminum powders were obtained from Valimet. Toluene, xylene, and tetramethylethylenediamine (TMEDA) were distilled from sodium under argon. Reactions were conducted under argon atmosphere using standard Schlenk techniques. In a typical aluminum powder preparation (B in Table I), 0.2 g of $AlH_3 \cdot (N(CH_3)_3)_2$ was dissolved in 10 mL of toluene, and 0.5 g of TMEDA was added. The solution was heated to 110°C, and a solution of 1 µL of titanium isopropoxide in 2 mL of toluene was rapidly added via syringe. A gray precipitate formed in five seconds and decomposition was complete within a few minutes (evolution of H_2 ceased). After cooling to room temperature, the precipitate settled out and the solution was removed via cannula. The solid was then washed twice with 5 mL of toluene, each time removing the liquid via cannula after the solid settled out. The solid was then dried under vacuum. After refilling the flask with argon the solid was slowly exposed to air by opening a stopcock and allowing air to diffuse into the flask over 30 min. Scanning electron micrographs were recorded on an Amray Model 1400 instrument. Thermogravimetric analysis was conducted on a TA Instruments 2950 Thermogravimetric Analyzer with a DuPont 2100 Thermal Analyst controller.

RESULTS AND DISCUSSION

Aluminum powders were prepared by decomposing tertiary amine adducts of alane (AlH_3) in organic solvents. The reaction was catalyzed by addition of tetravalent titanium compounds.[4] The reaction conditions and characterization results are collected in Table I. Several variables were examined for their influence on particle size, including temperature, solvent, adducting species, catalyst, and catalyst concentration. For comparison, we obtained two fine powders prepared by alternative methods, "LANL" (prepared by aluminum vapor condensation at Los Alamos National Laboratory[5]) and "ALEX" (prepared by an exploded aluminum wire technique).

131

Table I. Reaction conditions and characterization results for aluminum powders prepared from $AlH_3 \cdot (N(CH_3)_3)_2$.

Powder[a]	Solvent	Temp.,°C	%Ti Catalyst vs. Al	TGA Calc'd. Size (nm)	SEM Size Est. Avg.	Wt % Active Al
Ti-1[b]	Toluene	110	0.25	135	200	78.8
A[c]	Toluene	110	0.25	230	300	86.9
B	Toluene	110	0.25	127	200	77.6
B2	Toluene	110	0.25	264		88.5
B3	Toluene	110	0.25	228	200	86.8
E	Toluene	110	0.023	409	500	92.4
F	Toluene	110	2.2	103	100	73.1
G	Toluene	82	0.21	64	250	60.5
H	Xylene	141	0.22	100	200	72.5
I	Xylene	82-138	0.25	149	150	80.5
J[d]	Xylene	82-130	0.24	520	350	94.0
K	TMEDA	110	0.22	66	<100	61.5
L[e]	Toluene	110	0.25	225		86.6
M[f]	Toluene	110	0.40	320	350	90.4
N[g]	Toluene	110	0.25	195	300	84.8
A-50	Xylene	50	0.23	111	150	74.9
B-65	Xylene	65	0.23	96	125	71.6
C-80	Xylene	80	0.25	120	125	76.4
D-95	Xylene	95	0.25	213	200	86.0
E-110	Xylene	110	0.25	167	150	82.5
F-125	Xylene	125	0.25	165	150	82.2
G-140	Xylene	140	0.25	175	150	83.1
LANL				61	TEM 48[h]	59.3
ALEX				137	TEM 142[h]	79.1

[a]Each preparation was conducted by adding titanium isopropoxide catalyst to a heated solution containing $AlH_3 \cdot (N(CH_3)_3)_2$ plus three equivalents of tetramethylethylenediamine (TMEDA), except where otherwise indicated.
[b]Reverse of normal addition procedure: alane adduct added to a heated catalyst solution.
[c]No tetramethylethylenediamine was used.
[d]Reaction scale and concentration increased ten-fold compared to other runs.
[e]Alane adduct was $H_3Al \cdot N(CH_3)_2(CH_2CH_3)$.
[f]Catalyst was $TiCl_4$.
[g]Catalyst was $Ti(N(CH_3)_2)_4$.
[h]Data from C. Aumann and J. Martin, Los Alamos National Laboratory.

The powders were mainly characterized by scanning electron microscopy (SEM) and thermogravimetric analysis (TGA). In addition, x-ray powder diffraction was conducted on powder A which showed it to be crystalline aluminum. Figure 1 shows SEM micrographs of three of the powders along with a micrograph of a commercial powder. The solution-prepared powders have a much smaller particle size and a much narrower particle size distribution compared to the commercial powder which contains particles ranging from 0.2 to 8 μm in diameter. Increasing the amount of catalyst led to a reduction in particle size (compare samples B and F in Figure 1), while the use of $TiCl_4$ catalyst in place of $Ti(OCH(CH_3)_2)_4$ led to larger particles and a wider distribution of sizes (sample M in Figure 1). Figure 2 shows TGA data for some of the powders. By completely oxidizing the aluminum powder to Al_2O_3 the weight increase on oxidation was used to calculate a particle size by assuming an initial oxide thickness of 3.68 nm on uniform spherical aluminum particles. Each TGA curve is normalized to the minimum weight to account for the weight loss due to oxidation and volatilization of organic impurities below 400°C. The particle sizes from TGA data generally agreed well with SEM particle sizes as shown in Table I. The TGA data was also used to calculate the weight percent of active aluminum for each powder where the balance of the weight is due to the aluminum oxide coating. Figure 3 shows the weight percent active aluminum for the various powders and how it depends on particle size (a 3.68 nm oxide thickness was assumed for all powders). The figure shows that the active aluminum content drops rapidly as the particle size drops below 100 nm, indicating a diminishing return at some point on the increasing reactivity of aluminum powders as particle size decreases.

The particle size is most sensitive to catalyst concentration among the variables examined (see entries E and F in Table I). The particle size roughly doubles for an order of magnitude decrease in catalyst concentration. Thus, the number of particles, which varies as the inverse third power of particle size, appears to be roughly proportional to catalyst concentration over the range 0.023-2.2% Ti(O-i-Pr)$_4$. This indicates that the catalyst affects the nucleation of aluminum particles.

Fig. 1. Scanning electron micrographs of aluminum powders. H-3 is a commercial aluminum powder, while the other three powders were prepared from $AlH_3 \cdot (NMe_3)_2$ at 110°C using 0.25% Ti(OCH(CH$_3$)$_2$)$_4$ (B), 2.2% Ti(OCH(CH$_3$)$_2$)$_4$ (F), and 0.40% TiCl$_4$ (M). All pictures originally at 20,000X magnification.

When no catalyst is used, very slow decomposition of the alane occurs, even after some aluminum is precipitated. Thus, the catalyst affects both the nucleation and the growth of aluminum particles.

Addition of TMEDA to the reaction solution helped prevent agglomeration of the aluminum particles. TMEDA binds strongly to alane and replaces the coordinated trimethylamine. When the reaction was conducted in neat TMEDA (run K in Table I), the rate was much slower with aluminum precipitate appearing 150 s after catalyst addition compared to about 5 s when a 1:20 ratio of TMEDA to toluene is used. This result supports reversible amine dissociation as the first step in alane decomposition. Using TMEDA as the solvent is also effective at reducing particle size.

The as-formed aluminum powders are very reactive to air, and can ignite if not passivated. Techniques we have used for passivation are (1) slow exposure of dry powder to air, and (2) addition of air while the powder is still suspended in the reaction solvent.

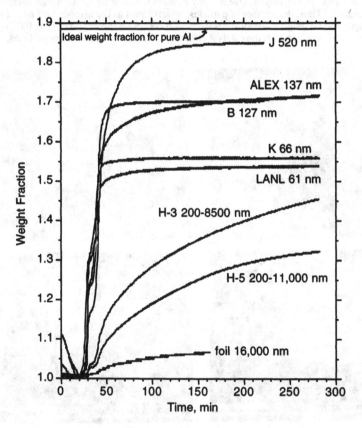

Fig. 2. Thermogravimetric analysis of aluminum powders. Samples were heated in air at 20°C/min to 850°C, then held at 850°C. The sample weight was normalized to the minimum weight.

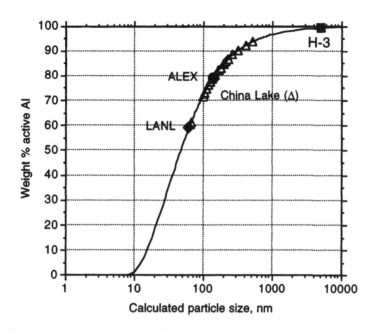

Fig. 3. Dependence of active aluminum content on particle size. The curve was calculated based on uniform spherical particles with a 3.68 nm oxide layer. TGA data was used to determine the weight percent of active aluminum for each powder.

CONCLUSIONS

The solution process for preparing fine and ultra-fine aluminum powders is a convenient and versatile method for producing well controlled particle sizes in the 50-600 nm range. Parameters that favor small particles are high ratio of catalyst to alane, high TMEDA concentration, dilute alane solution, and low temperature. While the mechanism is not yet well understood, it involves dissociation of the amine adduct followed by particle nucleation and growth that are promoted by a titanium compound.

ACKNOWLEDGMENTS

We thank Richard Scheri and Michael Dowd for SEM characterization. We also thank Chris Aumann and Joe Martin for providing the LANL aluminum powder and communicating unpublished results. This work was supported by the Naval Air Warfare Center Core Science and Technology program.

REFERENCES

1. T. B. Gurganus, Adv. Mater. Process. **148**, 57-59 (1995).
2. O. Stecher and E. Wiberg, Chem. Ber. **75**, 2003-2012 (1942).
3. J. K. Ruff, Inorg. Synth. **9**, 30 (1967).
4. D. L. Schmidt and R. Hellmann, U. S. Patent No. 3 462 288 (19 August 1969).
5. C. E. Aumann, G. L. Skofronick, and J. A. Martin, J. Vac. Sci. Technol. B **13**, 1178-1183 (1995).

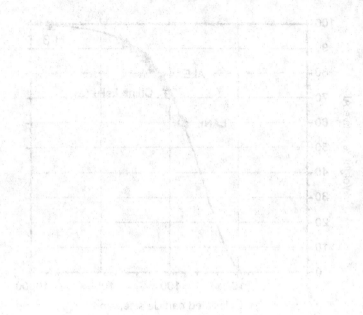

THREE-DIMENSIONAL SUPERLATTICE PACKING OF FACETED SILVER NANOCRYSTALS

S.A. HARFENIST*, Z. L. WANG**, M.M. ALVAREZ***, I. VEZMAR* and R.L. WHETTEN***

* School of Physics, Georgia Institute of Technology, Atlanta, GA. 30332-0430
** School s of Physics, Chemistry, and Microelectronics Research Center, Georgia Institute of Technology, Atlanta, GA. 30332-0430
*** Schools of Materials Science and Engineering, Georgia Institute of Technology, Atlanta, GA. 30332-0245

ABSTRACT

Orientational ordering of faceted nanocrystals in nanocrystal arrays has been directly observed for the first time, by use of transmission electron microscopy imaging and diffraction to resolve the structure of thin molecular-crystalline films of silver nanocrystals passivated by alkylthiolate self-assembled monolayers. The type of ordering found is determined by the nanocrystal's faceted morphology, as mediated by the interactions of surfactant groups tethered to the facets on neighboring nanocrystals. Orientational ordering is crucial for the understanding of the fundamental properties of quantum-dot arrays, as well as for their optimal utilization in optical and electronic applications.

INTRODUCTION

Materials composed of metal nanocrystals are of great current interest because of the enhancement in optical and electrical-conductance properties that arises from confinement and quantization of conduction electrons within a small volume [1]. Recent investigations have revealed interesting and potentially useful properties, including extremely large optical polarizabilities [2], nonlinear electrical conductance of nanocrystal arrays that also have small thermal activation energies (~ 0.1 eV) [3], and Coulomb-blockade and -staircase phenomena in the conductance of single nanocrystals persisting to unusually high temperature [4,5]. The optimal manifestation of these properties depends on the nanocrystal material's quality, encompassing uniformity in size, shape (or crystallite morphology), internal structure, surface passivation, and spatial orientation of the nanocrystals. The significance of the last of these is apparent from the fact that nanometer-scale crystallites preferentially have well-faceted polyhedral shapes, in which the vertices and the edge- and face-centers point along crystallographic axes of the nanocrystal's internal lattice [6,7]. Consequently, all transport and optical properties are orientation-dependent for a single nanocrystal, and, in the case of a condensed nanocrystal array or superlattice, are sensitive to orientational order.

High quality nanocrystal solids have become possible because of a remarkable set of facts [8], namely that nanocrystals passivated by suitably matched surfactants can retain their ideal core morphology, and yet behave as stable macromolecules, i.e. can reversibly form solutions in solvents appropriate to the surfactant tail-groups and be evaporated rapidly to form a molecular vapor [9], that they can be separated from mixtures into highly purified substances composed of a single size and shape, and that they can condense to form crystalline *molecular* solids with high translational (center-to-center) order [10,11]. It is in such molecular crystals that the question of orientational order becomes crucial. Generally, crystals formed from high-symmetry molecules are orientationally ordered, i.e. the molecular axes are co-oriented with respect to each other and with respect to the crystallographic axes of the lattice, only below an orientational ordering temperature [12]. Despite their distinctly non-spherical, polyhedral shape, it is not clear whether passivated nanocrystals interacting mainly through their surfactant groups will form highly oriented molecular crystals.

Recently, limited evidence for nanocrystal *alignment* has been provided by x-ray diffraction patterns obtained on films of diverse nanocrystal materials [11], and by direct observation of co-aligned nanocrystal lattice planes in a very thin film [12]. These results are insufficient to reveal the packing of faceted particles and their surfactant groups, and even in the most favorable interpretation would only establish alignment, rather than orientation. Here we report clear evidence of high molecular orientational order in nanocrystal lattices, in the form of direct observations of co-oriented shapes, lattice fringes, and surfactant density profiles, in thin crystalline films of nanocrystal silver (*Ag*) molecules. Our results also suggest a directionality in the distribution of surfactant chains, which serve as the *inter*molecular bonds connecting nanocrystals.

EXPERIMENT

The molecular silver nanocrystals under investigation here are from the same class of materials that has enjoyed rapidly growing interest in the past several years. They are charge-neutral entities comprised entirely of an elemental metal (*Au, Ag, Pt*) core and a dense mantle of alkylthiol(ate) surfactant groups. Elemental *Ag* is evaporated at temperatures ranging from 1200 to 1500 K into a flowing, preheated atmosphere of ultrahigh purity helium. The flowstream is cooled over a short distance (and flow-time) to ca. 400 *K*, stimulating growth of nanocrystals in the desired size-range, here ca. 4 - 6 *nm*. Growth is abruptly terminated by expansion through a conical funnel accompanied by dilution in a great excess of cool helium. Depending on conditions, most of the material produced is soluble in non-polar solvents, from which excess surfactant can be removed. The soluble material is found to consist entirely of nanocrystal *Ag* molecules (NCAM s). A Hitachi HF-2000 cold field emission gun TEM (200 kV) has been used to perform imaging and diffraction experiments. Each TEM specimen was prepared by depositing a highly concentrated drop of toluene solution of *NCAMs* on an ultrathin amorphous carbon film substrate, and allowing it to dry slowly in air prior to TEM observations. Nanocrystal superlattices, such as those shown in figure 1, appear on the carbon film, in the form of highly-oriented thin crystalline films, whose lateral dimensions can be as large as 1 micrometer.

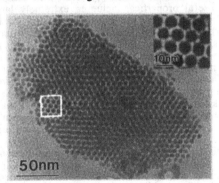

Figure 1. Transmission Electron Micrograph of Ag Nanocrystal Superlattice. Inset is a magnified image of the boxed region.

RESULTS

Our first objective has been to establish the packing structure of the nanocrystal superlattice. Figure 2a shows a TEM image of an ultrathin (monolayer) film of 5.0 *nm* diameter *NCAMs* (6.4±0.2 *nm* center-to-center spacing), in which 2-D close-packing is deduced from the clear six-fold symmetry. The electron diffraction pattern obtained from this region reveals the periodic structure of the nanocrystal lattices, and also provide an absolute calibration of the superlattice constant, because the *fcc*-Ag {111} ring is recorded simultaneously in the same photograph (figure 2b). Figures 2c and 2d show a TEM image and the corresponding diffraction pattern corresponding to the [110] orientation, thus confirming that the superlattice has the *fcc* structure. Such [110]$_S$-oriented films are the most frequently observed structure in our specimens, and are very advantageous, as described previously [11], in that the projection from individual columns of nanocrystals can be viewed without obstruction by neighboring columns, allowing one to clearly view a projected shape of the nanocrystals, as we now describe. [The subscript *s* denotes the nanocrystal superlattice, to distinguish it from the reflections of atomic lattice of elemental *Ag*.]

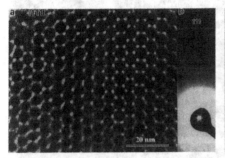

Figure 2 (a) 2-D close-packing and (c) $[110]_s$ TEM images and (b) and (d) corresponding electron diffraction patterns, respectively, obtained on thin films of nanocrystal Ag molecules, establishing the fcc packing structure of the multilayer superlattices. The Ag {111} reflection ring is seen in (b), which serves as an absolute calibration for determining the lattice constant of the superlattice, $a_s = 10.4 \pm 0.3$ *nm*. The solid line in (c) indicates the $[110]_s$ projected unit cell of the superlattice

Our second objective has been to establish a relationship between this 3-D superlattice structure and the orientation of the NCAMs within it. *Figures 3a* and *c* are two TEM images recorded from the same specimen region under differing defocus conditions. Three striking features are observed. First, the superlattice contains "coherent" $\{200\}_s$ twin (T) and $\{111\}_s$ stacking fault (SF) structures. Close examination shows that the (projected) nanocrystal shape is modified across the twin boundary, accompanying the transformation in superlattice orientation. Second, profiles of the faceted structure of the *NCAM* cores can be seen in the image recorded at the condition close to in-focus (*Figures 3a* and *b*), and the co-aligned arrangement of these projected shapes is apparent. The majority of the nanocrystal profiles observed are consistent with the truncated octahedral (*TO*) morphology, which exposes {100} and {111} facets, as described previously [13]. Viewed along [110], two {100} facets and four {111} facets are projected *edge-on* (*Figure 3b*). This establishes the orientational relationship between the *NCAM* core's lattice and the superlattice as [110] ∥ $[110]_s$ and [002] ∥ $[1\bar{1}0]_s$. Third, the image recorded at the out-focus condition exposes a web-like pattern of directional "contacts" among neighboring *NCAM*s, as indicated by dotted lines in *Figure 3d*, where the *Ag* cores are in dark contrast and the bright spots are channels enclosed by regions of enhanced density of surfactant chains. By comparison with *Figure 3b*, one can see that these dense regions interconnect between *facets* of *Ag* cores, rather than edges or vertices. It is therefore natural to suppose that the overlap of surfactant groups emanating from facets on neighboring nanocrystals serves as the prime contact which holds, or "bonds", and orients the particles in their 3-D superlattice. We further propose that materials of special type be called "highly oriented molecular (*Ag*) nanocrystal arrays", *HOM(A)NA* s.

A precise orientational relationship between nanocrystal core lattices and the superlattice has been determined from a high-resolution lattice image, *Figure 4*, recorded on a very thin film (two layers) for which the electron beam is parallel to $[110]_s$, and the {111} lattice fringes of the *Ag* cores are clearly resolved. By tracing the $\{111\}_s$ planes of the superlattice, one finds a striking feature: the {111} lattice fringes of the *Ag* cores (as indicated by arrowheads) consistently show a small deflection (in the same sense) from the $\{111\}_s$ planes of the superlattice, thus defining a preferred orientational angle, α. A histogram of the experimentally measured angles is given in the inset, from which a value of $\alpha = 16 \pm 5^o$ is obtained.

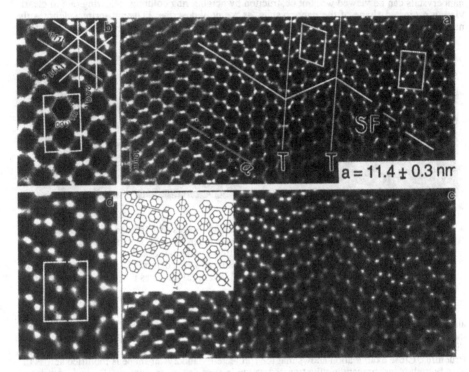

Figure 3 [110]$_s$ TEM images recorded from the a single region of a film under (a) in-focus and (c) out-focus conditions, showing the faceted shape and the directional intermolecular bonds of the particles. The white lines indicate twin (T) planes, stacking faults (SF) and the rotation of the {111}$_s$ plane across the twin plane. An angle, a, is indicated which measures the projected orientation between the {111} facet of a nanocrystal (dotted line) and the {111}$_s$ plane of the superlattice. A structural model for the formation of {100}$_s$ twins in the fcc superlattice is inset in (c) according to the structure model to be given in Figure 5b. (b) and (d) are enlarged TEM images selected from (a) and (c), respectively, showing the relationship of the {111} and {100} facets of the truncated octahedron particles to the projected unit cell (white lines) of the superlattice. The dotted lines in (d) indicate the directional intermolecular contacts formed by the groups of surfactant chains from facets on neighboring nanocrystals.

Working from all these experimental observations, we have constructed a 3-D structural model (*Figure 5b*) of the packing and orientational relationships in the *HOMANAs* observed, where each *NCAM* core is represented by a truncated octahedral (*TO*) shape, consistent with the information provided by *Figure 3(a,b)*. Any variation in particle size and shape may introduce significant strain to the superlattice, and so this model is only an idealized representation of the practical situation. The model is generated by orienting the *TO* polyhedra according to the relationships established by the TEM images in *Figures 3-4*. Clearly, this orientation of objects of cubic symmetry within a cubic lattice breaks the lattice symmetry; hence, the faceted structure of the *Ag* cores could result in different lengths of the *a, b* and *c* axes of the superlattice. Because of this loss of symmetry, the faceted structure of the nanocrystals produces inequivalent projected images when viewed along

$[110]_s$ and $[1\bar{1}0]_s$. As viewed along $[110]_s$, the [110] orientation of the Ag nanocrystal is parallel to the $[110]_s$ of the superlattice (*Figure 5ca*, where four {111} and two {100} facets of the particle are parallel to the electron beam, in correspondence to the experimental image shown in Figure 3b. Considering also the strong Fresnel fringes near the edge of the particles, these profiles will have almost circular shapes in the TEM images, such as those shown in *Figure 2b*.

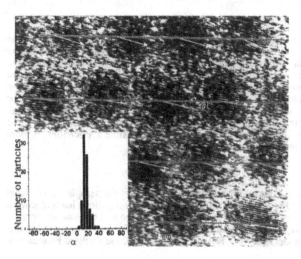

Figure 4   A high-resolution TEM image recorded from a thin *NCAM* film viewed in the $[110]_s$ direction. The markings indicate the angular relationship α between the {111} lattice fringes of the Ag particles with the $\{111\}_s$ planes of the superlattice (indicated by white-dark solid lines). The inset is a histogram of α values measured from the experimental images.

(a) $[110]_s$ projection   (b) NCAM fcc superlattice unit cell

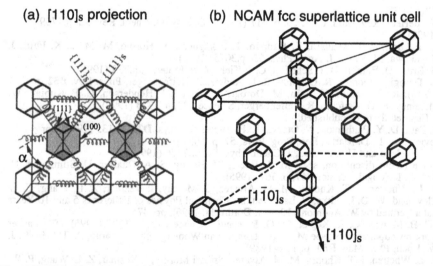

Figure 5. (b) A schematic model of the *fcc* packing of the TO Ag cores and following the orientational relations of [110] ‖ $[110]_s$ and [002] ‖ $[1\bar{1}0]_s$.   (a) The $[110]_s$ projection of the model shown in (b). The shaded polyhedra are out-of-plane by a distance $(\sqrt{2}/4)a$ and the angle a is indicated corresponding to that shown in Figure 3a. The intermolecular interactions are represented by coils that interconnect the {111}-facets with {111}-facets on neighboring nanocrysals and {100} with {100} facets.

From the 2-D projected model in *Figure 5b* an angle $\alpha = 19.4°$ is calculated between the {111} facet (as indicated) of the nanocrystal with the {111}$_s$ plane of the superlattice. This is the angle implicated from core orientations in the images shown in *Figure 3b* and d established, by the high-resolution TEM image shown in *Figure 4*, to be $16 \pm 5°$. By contrast, the $[hkl]$ ||$[hkl]_s$ orientation of nanocrystal to superlattice axes would have necessitated $\alpha = 0$.

CONCLUSIONS

Orientationally ordered molecular Ag- nanocrystal lattices have been observed, their structures have been determined, and a model consistent with the nanocrystal's geometry and bonding (polyhedral shapes and the grouping and interdigitation of surfactant chains) has been proposed to account for the principal structure and its details. These observations have important consequences: orientational ordering, which is desirable for so many purposes, occurs readily in such materials, and its microscopic origins can be comprehended. Electronic transport may be described as taking place within an ordered crystalline structure, rather than in a granular material. Furthermore, regular channels interconnecting a lattice of uniform voids are intrinsic to these materials; these have a lengthscale that is interesting with regard to possible applications. A more detailed discussion of this topic is reported elsewhere [14].

ACKNOWLEDGMENTS

The authors thank Dr. Joseph T. Khoury for assistance in preparing the samples, Prof. Janet Hampikian and Dr. Srihari Murthy for assistance with preliminary investigations leading toward this research, and Prof. Uzi Landman and Dr. W. David Luedtke for frequent discussions on their unpublished work on surfactant layers on passivated nanocrystals. Financial support for this research has been provided by the Georgia Tech Research Foundation, the Packard Foundation, and the Office of Naval Research.

REFERENCES
1. W. deHeer, Rev. Mod. Phys. **65**, p. 611 (1993); G. Schmid, Chem. Rev. **92**, p.1709 (1992).
2. K. Fukumi, A. Chayahara, K. Kadono, T. Sakaguchi, Y. Horino, M. Miya, K. Fujii, J. Hayakawa, M. Satou, J. Appl. Phys. **75**, p.3075 (1994).
3. M. Brust, D. Bethell, D. J. Schriffin, C. J. Kiely, Adv. Mater. **7**, p.795 (1995).
4. M. Dorogi, J. Gomez, R. Osifchin, R. P. Andres, R. Reifenberger, Phys. Rev. **B52**, p.9071 (1995); R. P. Andres, T. Bein, M. Dorogi, S. Feng, J. I. Henderson, C. P. Kubiak, W. Mahoney, R. G. Osifchin, R. Reifenberger, Science **272**, p. 1323 (1996).
5. P. First, et al., to be published.
6. A. Patil, D. Y. Paithankar, N. Otsuka, R. P. Andres, Z. Phys. **D26**, p. 135 (1993); for review, see L. D. Marks, Rep. Prog. Phys. **57**, p. 603 (1994).
7. C. Cleveland and U. Landman, J. Chem. Phys. **94**, p. 7376 (1991).
8. For a recent discussion, see: M. T. Reetz, W. Helbig, S. A. Quaiser, U. Stimmung, N. Breuer, R. Vogel, Science **267**, p. 367 (1995).
9. R. L. Whetten, J. T. Khoury, M. M. Alvarez, S. Murthy, I. Vezmar, Z. L. Wang, C. L. Cleveland, W. D. Luedtke, U. Landman, in "Chemical Physics of Fullerenes 5 and 10 Years Later", edited by W. Andreoni (Kluwer, Dordrecht, 1996), pp. 475.
10. C. B. Murray, C. R. Kagan, M. G. Bawendi, Science **270**, p. 1335 (1995). For earlier work on 'supercrystals', see M. D. Bentzon, J. van Wonterghem, S. Mørup, A. Thölén, C. J. W. Koch, Phil. Mag. **B60**, p. 169 (1989).
11. R. L. Whetten, J. T. Khoury, M. M. Alvarez, Srihari Murthy, I. Vezmar, Z. L. Wang, P. W. Stephens, C. L. Cleveland, W. D. Luedtke, U. Landman, Adv. Mater. **5**, p. 428-433 (1996).
12. For example, see: P. A. Heiney, J. E. Fischer, A. R. McGhie, W. J. Womanow, A. M. Denenstein, J. P. McCauley, A. B. Smith, Phys. Rev. Lett. **66**, p. 2911 (1991).
13. P.-A. Buffat, M. Flüeli, R. Spycher, P. Stadelmann, J.-P. Borel, Faraday Discuss. **92**, p. 173- (1991).
14. S. A. Harfenist, Z. L. Wang, M. M. Alvarez, I. Vezmar, R. L. Whetten, J. Phys. Chem. **100** p. 13904 (1996).

APPLICATION OF MeV ION IMPLANTATION IN THE FORMATION OF NANO-METALLIC CLUSTERS IN SILICA

D. ILA*, Z. WU*, R. L. ZIMMERMAN*, S. SARKISOV*, C.C. SMITH[b], D. B. POKER[c] and D. K. HENSLEY[c]
a Center for Irradiation of Materials, Alabama A&M University, Normal AL 35762
b NASA-MSFC, Marshall Space Flight Center AL 35898
c Solid State Division, Oak Ridge National Laboratory, Oak Ridge TN

ABSTRACT

The implantation of metal ions into photorefractive materials followed by thermal annealing leads to an increase in resonance optical absorption as well as an enhancement of the nonlinear optical properties. We have implanted ions of Au (3.6 MeV), Ag (1.5 MeV) and Cu (2.0 MeV) into pure silica followed by careful heat treatment. Using optical absorption spectrophotometry and rutherford backscattering spectrometry we have measured the cluster size for each heat treatment temperature and determined the activation energies for their formation. The third order electric susceptibility for silica with 2 nm gold clusters has been determined by Z-scan to be 65×10^{-8} esu.

INTRODUCTION

Introducing metal colloids into a glass matrix has long been used to change the color of glasses for decoration and recently for fabricating optical filters. The production of these devices was motivated mostly by desired changes in the linear properties of glasses. In recent years, attention has been directed to nonlinear properties of the material caused by the surface plasmon resonance optical absorption which depends on the index of refraction of the host substrate and the electronic properties of the metal colloids. These glasses were prepared classically by mixing selected metals powders with molten glass then cooling to form a homogeneous glass. In recent years, ion implantation has been used to introduce nonlinear optical properties [1-6] in layers near the surface of optical materials. To form nanoclusters after ion implantation the material must be heat treated either by thermal annealing, by laser annealing or by inert ion bombardment. An attractive feature of ion implantation is that the linear and nonlinear properties occur in a well defined space in an optical device, and by using focused ion beams, point quantum confinement may be accomplished.

EXPERIMENTAL PROCEDURES

The silica glass used in this work is Suprasil-1 provided by Heraeus Amersil, Inc. It contains 150 ppm OH, 0.05 ppm of Ti, Na, Ca and Al and less than 0.01 ppm other metals. Samples 10 x 10 x 0.5 mm were implanted with 3.6 MeV gold, 1.5 MeV silver or 2.0 MeV copper ions at a current density less than 2 $\mu A/cm^2$ to avoid the premature formation of metal clusters due to the ion beam heating. Ion energies were chosen to implant at similar depths in the silica samples. Ion fluences were chosen to give a guest atom to silicon ratio near 1:10. Table 1 summarizes the characteristics of the implanted layers.

The heat treatment for the samples was done in air at temperatures between 500°C and

Table 1 Specie, energy and fluence of the ions implanted in Silica. The dimensions of the implanted layers were calculated with SRIM computer code [7] and verified by RBS.

Ion Specie	Ion Energy (MeV)	Ion fluence (cm^{-2})	Depth / Width (nm / nm)	Volume Fraction
gold	3.6	1.2×10^{17}	860 / 280	0.023
silver	1.5	1.38×10^{17}	700 / 370	0.034
copper	2.0	5.0×10^{17}	1440 / 770	0.020

1250°C for 1 hour at each heat treatment temperature. After each heat treatment, an optical absorption spectrum was measured. From these spectra we calculate the size of the metallic clusters which are responsible for the optical absorption band.

The average radius of metal spheres small compared with the wavelength of light is determined [10] from the resonance optical absorption spectrum according to the equation

$$r = \frac{v_f}{\Delta\omega} ,$$

where v_f is the Fermi velocity of metal and $\Delta\omega$ is the full width at half maximum of the absorption band due to the plasmon resonance in small metal particles.

A frequency doubled mode locked Nd:YAG laser (532 nm) with 70 ps width at a 76 MHz repetition rate with a peak power density of 2.57 GW/cm^2 was employed for Z-scan measurements of the third order susceptibility [11]. Prism coupling of light into the implanted surface was used to investigate the waveguide that is formed as a consequence of the layered structure.

RESULTS and DISCUSSIONS

The higher the atomic number and fluence of the bombarding ion, the more the damage and modification of the optical properties [8,9] of the silica glass. This was observed by absorption spectrometry during all of our bombardments. Except for the effects attributable to the metal clusters, these other effects were reduced with heat treatments above 700° C.

The fluences used in these implantations were chosen to reach a concentration of guest atoms about 10 times less than that of the host atoms; i. e., a ratio of Au to Si of 0.1 for example. For such ratio the mean random separation of the guest atoms is barely three times that of the host atoms. Large persistent mechanical strains and electric charge separation occur in the implantation layer, beyond it and from it to the surface. Not surprisingly, the resulting energies are partially reduced by clustering of the implanted ions at ambient temperature during and after bombardment. We used low current densities such that no significant temperature increase occurred during implantation. When clusters of high density metal form, the volume in the silica occupied by metal atoms decreases.

Heat treating the sample further reduces these strains and charge imbalances. Moreover, heating increases the diffusion coefficient and the implanted atoms move to the lower energy metallic state of the clusters whose volume fraction approaches those shown in Table 1. For the ion energies used in these experiments, the implanted layers were about 1000 nm from the surface with full thickness at half maximum concentration half that value. With an initial mean separation of implanted atoms only a few nanometers virtually all will diffuse to a cluster and not reach the surface of the layer. Moreover other experiments have shown that the damaged region

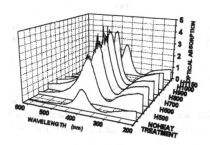

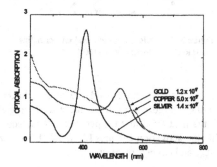

Figure 1   Resonance optical absorption in silica implanted with 1.38x10¹⁷ cm⁻² silver and heat treated for 1 hour to the temperatures shown.

Figure 2   Comparison of the resonance optical absorption in silica implanted with the metal ions shown and heat treated to 1100°C

in the host may itself offer lower energy states to individual guest atoms and inhibit diffusion loss from, and enhance diffusion into, that region. With heat treatment the near neighbor clusters coalesce and the host accommodates to the volume reduction. RBS measurements confirm that the depth profile of the metal clusters formed after heat treatment is almost identical to, or slightly narrower than, that of the atoms initially implanted by ion bombardment.

Figure 1 shows the optical absorption spectra for 1.5 MeV silver implanted in silica at 1.38x10¹⁷ Au/cm² at various heat treatment temperatures. The prominent resonance optical absorption at 410 nm increases until it disappears for heat treatment temperatures above 1150°C. The copper and gold implanted samples behave similarly. Figure 2 compares the optical absorption spectra for three samples, implanted either with gold, silver or copper and heat treated at a temperature near that of the respective bulk material melting point. For a heat treatment temperature beyond a critical temperature the resonance optical absorption disappears as metal atoms evaporate from the clusters as shown in Fig. 1 for silver implanted silica. The critical temperature correlates only roughly with the bulk metal melting temperature.

Using the Doyle theory [10] and the resonance optical absorption bandwidths, average cluster radii in the three implanted samples are calculated for each heat treatment temperature.

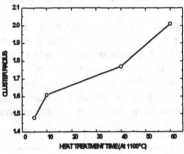

Figure 3 Dependence of the metal cluster radius on time of heat treatment at 1100°C

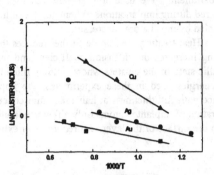

Figure 4 The natural logarithm of metal cluster radii versus reciprocal heat treatment temperature. The lines determine the activation energies given in the text.

Figure 3 shows the dependence of the cluster radius of gold on the time of heat treatment at 1000°C. Figure 4 presents the natural logarithm of all radii as a function of the reciprocal of the heat treatment temperature, all for equal heat treatment times. From the slopes, we find that the activation energies for metal cluster formation in silica are 80, 83 and 290 meV for gold, silver and copper, respectively.

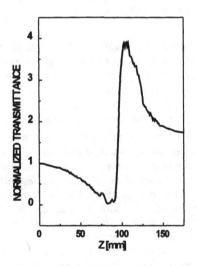

Figure 5 The Z-scan data for silica implanted with 1.2×10^{17} cm^{-2} 3.6 MeV gold heat treated for one hour at 1200°C. They determine the third order susceptibility to be 65 x 10^{-8} esu.

CONCLUSION

Samples of silica implanted with MeV ions of Au, Ag or Cu and subjected to careful heat treatment contain metal nanoclusters whose radii may be obtained from the resulting characteristic resonance optical absorption bands. By measuring the dependence of the metal cluster radii on the heat treatment temperature, the activation energies for the formation of gold, silver and copper clusters in silica are determined to be 80, 83 and 290 meV, respectively. The third order susceptibility for 532 nm light passing perpendicularly through a layer of gold clusters with 2 nm average radius distributed in a gaussian layer 280 nm full width half maximum at a distance of 860 nm below the surface of silica is observed to be 65×10^{-8} esu (Fig. 5). A waveguide for 632 nm light between the gold implanted layer and the surface has been observed and indicates that the refractive index in the guiding region is more than that in or near the implanted layer.

ACKNOWLEDGMENTS

This project was supported by the Center for Irradiation of Materials at Alabama A&M University and Alabama EPSCoR-NSF/ALCOT Grant No. OSR-9553348. The work at ORNL was sponsored by the Division of Materials Science, U.S. Department of Energy, under Contract DE-AC05-96OR22464 with Lockheed Martin Energy Research Corp.

*Corresponding author: Tel +1 205 851 5866, FAX +1 205 851 5868, e-mail ila@cim.aamu.edu

REFERENCES

1. G. W. Arnold, J. Appl. Phys. **46**, 4466 (1975).
2. G. W. Arnold and J. A. Bordes, J. Appl. Phy. **48**, 1488 (1977).
3. R. H. Magruder III, R. A. Zuhr, D. H. Osborne, Jr., Nucl. Inst. & Meth. in Phys. Res. **B99**, 590 (1995).
4. Y. Takeda, T. Hioki, T. Motohiro, S. Noda and T. Kurauchi, Nucl. Instr. Meth. **B91**, 515-519 (1994).
5. C. W. White, D. S. Zhou, J. D. Budai, R. A. Zuhr, R. H. Magruder and D. H. Osborne, Mat. Res. Soc. Symp. Proc. **Vol. 316**, 499 (1994).
6. K. Fukumi, A. Chayahara, M. Adachi, K. Kadono, T. Sakaguchi, M. Miya, Y. Horino, N. Kitamura, J. Hayakawa, H. Yamashita, K. Fujii and M. Satou, Mat. Res. Soc. Symp. Proc. **Vol. 235**, 389-399 (1992).
7. F. Ziegler, J. P. Biersack and U. Littmark, The Stopping and Range of Ions in Solids (Pergamon Press Inc., New York, 1985).
8. E. R. Schineller, R. P. Flam and D. W. Wilmot, J. Opt. Soc. Am. **58**, 1171 (1968).
9. P. D. Townsend, Nucl. Instr. Meth. in Phys. Res. **B46**, 18 (1990).
10. W. T. Doyle, Phys. Rev. **111**, 1067 (1958).
11. D. Ricard, Ph. Roussignol and Chr. Flytzanis, Optics Letters **10**, 511-513 (1985).

PREPARATION OF METAL NANOSUSPENSIONS BY
HIGH-PRESSURE DC-SPUTTERING ON RUNNING LIQUIDS

M.WAGENER*, B.S. MURTY**, B. GÜNTHER*
* Fraunhofer-Institut für angewandte Materialforschung, Lesumer Heerstr.36,
  D-28717 Bremen, Germany
** on leave from Indian Institute of Technology, Kharagpur -721 302, India

ABSTRACT

A modified VERL-process (vacuum evaporation on running liquids) employing high pressure magnetron sputtering has been used for the preparation of suspensions with metal nanoparticles. The method has been tested for Ag- and Fe-suspensions by varying the pressure of the Argon sputtering atmosphere in the range of 1 to 30 Pa. A narrow particle size distribution with a mean particle size ranging from 5-18 nm has been found. The mean particle size increases with increasing Argon pressure in the pressure range under investigation. A descriptive model for the process of particle formation as a function of sputtering gas pressure is given.

INTRODUCTION

Vacuum evaporation on running liquids (VERL) is an established method for the preparation of metal nanosuspensions suspended in a non-aqueous, low vapour pressure liquid, like various oils and resins [1]. In this method a metal is evaporated onto a liquid surface that is continuously renewed by the action of a rotating drum or spinning disk. If the background pressure is below about 10^{-2}Pa, then a PVD-like process is realized onto the liquid surface. In this case particle formation takes place at the surface or in a subsurface layer of the liquid substrate. In this method evaporation of the metal melt is usually performed by Joule heating.

Recently, the formation of particulate nanomaterials on solid substrates by using the sputtering method has also been reported [2-4, 7-9]. In this case elevated gas pressures have been used in order to induce nucleation and grain growth within the sputtering gas atmosphere. According to [2,3] particle growth occurs at gas pressures exceeding about 20 Pa.

Sputtering combined with the usual VERL-technique has the following specific advantages in comparison to evaporation by Joule heating:

- High melting point materials can be evaporated;
- Evaporation conditions can be held more stable;
- Heat load of the liquid substrate is reduced.

In this report we describe a new VERL-geometry including high pressure magnetron sputtering enabling the production of Ag- and Fe-nanosupensions in various organic liquids. The possibilities and limits of controlling the particle size is elucidated by varying the sputtering conditions such as sputtering gas pressure and input power.

Mat. Res. Soc. Symp. Proc. Vol. 457 © 1997 Materials Research Society

EXPERIMENTAL

In the experimental setup (figure 1) a planar 400 x 130mm magnetron sputtering cathode is used that has been especially designed for high pressure sputtering conditions (up to 50 Pa). The maximum input power is 10kW. Experiments have been performed at a fixed input power of 1000W. The substrate consists of a rotating steel drum dipping into a reservoir of the

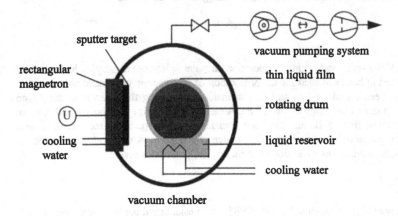

Fig. 1   Outline of the modified VERL process used for the preparation of metal nano-suspensions. The vacuum chamber has a diameter of 700mm.The size of the rotating stainless steel drum is ⌀200mm x 400mm.

respective liquid. The speed of rotation is varied between 0 and 10 rpm. The distance between the target and the drum was held fixed at 8cm.

Before starting the experiments the vacuum chamber was pumped down to $2*10^{-4}$ Pa. After flooding the chamber with Argon to the desired gas pressure (1 to 30 Pa) a constant gas flow ranging from 5 to 150 sccm was adjusted. Both the liquid reservoir and the sputtering target have been water cooled.

The sputter parameters have been varied in the following range:
- gas pressure (Ar) 1 to 30 Pa,
- sputtering time 0 to 4 hours.

Vacuum pumping oils, resins, and polymeric precursors have been used as liquid matrices exhibiting vapour pressures at the respective processing temperature below 10^{-2} Pa.

The metal filling factor of the resulting suspensions have been determined from thermogravimetric analysis (TGA). Particle size distributions have been derived from bright-field images obtained in a transmission electron microscope (Philips CM30).

For these TEM examinations a droplet of the suspension was diluted with acetone using an ultrasonic bath. A TEM Cu-grid was dipped into the suspension and - after drying - inspected in the transmission mode.

RESULTS and DISCUSSION

Ag-suspensions

Sputtering of Ag has been performed at a constant input power of 1kW and at Ar gas pressures ranging from 1 to 30 Pa. The mean particle sizes have been determined from fits of the respective size histograms to log-normal size distributions. Fig.2 gives an idea of the quality of these fits. The results are plotted as a function of Ar gas pressure in fig. 3.

$$f_{LN}(d) = \frac{1}{\sqrt{2\pi}} \exp(-\frac{(\ln d - \ln \overline{d})^2}{2 \ln^2 \sigma}) \tag{1}$$

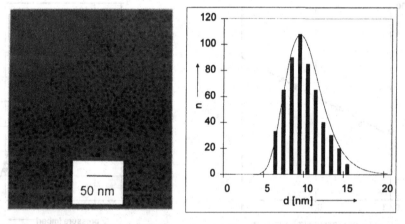

Fig. 2 TEM bright field image (left) and respective particle size histogram (right) of Ag-nanoparticles in mineral oil prepared at a sputtering pressure of 5 Pa and a sputtering input power of 1 kW. The size histogram was fitted to a log-normal distribution function (equ.(1)) using a mean particle size of <d> =9.5 nm and a standard deviation of σ =1.28.

Obviously in the pressure range under investigation the mean particle size increases steadily with increasing gas pressure. This behaviour is similar to the results obtained in the inert gas condensation process employing thermal evaporation [6]. There the statistical coalescence mechanism of particle growth applies: with increasing background gas pressure the mean-free-path for the metal atoms decreases and consequently both the cooling rate and the statistical coalescence probability increase leading to larger particles that are log-normally distributed. It is interesting to note that the dependence on background pressure of the mean particle size given in the paper of Granqvist [6] - d/[nm] ∝ (p/[Pa])$^{0.35}$ - is very close to that observed here: d/[nm] ∝ (p/[Pa])$^{0.40}$.

Furthermore Granqvist and Buhrmann have shown [6] that in the steady state of a coalescence type particle growth mechanism the width σ of the resulting particle size distribution is more or less independent of the mean particle size and ranges at $1.3 < \sigma < 1.6$. The fact that from our sputtering experiments the particle sizes are also log-normally distributed with similar pressure dependences and with widths in the same range as given above suggest that here also a coalescence type mechanism for the growth of particles within the gas phase dominates.

In contrast to thermal evaporation in the case of sputtering a second concurrent process may affect the resulting particle sizes: Keeping both sputtering current and voltage constant the sputtering rate strongly decreases with increasing gas pressure due to a reduction of the average kinetic energy of the impinging Ar atoms [3]. Here this effect shows up in a reduction of metal content of the liquid suspension with increasing gas pressure (fig. 4).

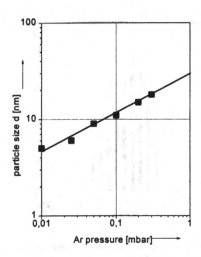

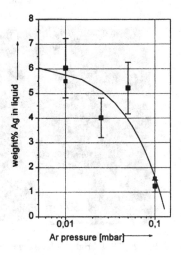

Fig. 3: Mean size of Ag particles in the prepared suspensions as a function of sputtering gas pressure. The data have been derived from fits of the respective size histograms (TEM) to log-normal distribution functions. The solid line fits the data with a slope of 0.40.

Fig. 4: Ag concentration in the suspensions as a function of sputtering pressure. The sputtering input power and the sputtering time were held constant at 1000W and 1 hour, respectively.

The weight-% of silver in the suspensions was determined by thermogravimetric analysis.

Further, it is to be expected that this decrease in sputtering rate would lead to a reduced density of sputtered metal atoms in the gas phase and thus to a decrease of the coalescence probability. However, under the present sputtering conditions employing fairly low gas pressures (< 30 Pa) this effect does not seem to dominate the growth mechanism (fig. 3).

It should be mentioned that for technical reasons the maximum content of metal in suspension is limited to about 3% by volume (which in the case of silver equals about 25% by weight) due to increasing viscosity of the suspensions.

Fe - Nanosuspensions

Some VERL experiments have also been performed using a 4.5mm thick iron sputter target (purity 99.99%). The TEM images of the resulting suspensions show that in contrast to silver the sputtered iron forms facetted particles with cubic shape. However, the particles are nearly completely oxidized. At this point we cannot discriminate wether oxidation of Fe particles has taken place within the liquid matrix already or during the preparation procedure of the TEM specimens.

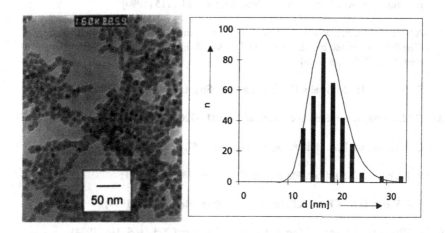

Fig. 5 TEM bright field image (left) and respective particle size histogram (right) of Fe nanoparticles in silicone oil prepared at an Ar sputtering pressure of 10 Pa and a sputtering input power of 1 kW. The particle size histogram was fitted to equ.(1) with a mean particle size 18 nm and a standard deviation of $\sigma=1.4$.

SUMMARY

It has been shown that it is possible to produce suspensions of metal nanoparticles in organic liquids via the sputtering technique. By changing the sputtering pressure it was possible to vary the mean particle size in the range of about 5 to 20nm. The variation of particle size with sputtering gas pressure is consistent with particle formation within the gas phase. Such metal nanosuspensions may be useful, e.g. as additives for functional metal/polymer composites or for metal based magnetic liquids.

ACKNOWLEDGMENTS

Financial support of the german federal ministry of education and research (BMBF) under contract no 03N 2004 D is gratefully acknowledged. One of us (BSM) would like to thank the german DFG for funding his research fellowship in Germany.

REFERENCES

1. S. Yatsua, Y. Tsukasaki, K. Mihama, R. Uyeda, J. Cryst. Growth, **45**, pp 490-494 (1978)

2. H. Hahn, R.S. Averback, J. Appl. Phys., **67**, pp 1113-1115 (1990)

3. A.S. Edelstein, F. Kaatz, G.M. Chow, J.M. Peritt, in Studies of Magnetic Properties of Fine Particles edited by J.L. Dormann, D. Fiorani (Elsevier Science Publishers, New York, 1992, pp 47-54

4. S. Terauchi, N. Koshizaki, H. Umehara, NanoStruct. Mater., **5**, pp 71-78 (1995)

5. G.M. Chow, Mater. Res. Soc. Proc., **206**, pp 315-320 (1991)

6. C.G. Granquist, R.A. Buhrmann, J. Appl. Phys., **47**, pp 2200- (1976)

7. S. J. Choi, M. J. Kushner, Mater. Res. Soc. Proc., **206**, 283 (1991)

8. F.H. Kaatz, G.M. Chow, A.S. Edelstein, J. Mater. Res., **8**, 995 (1993)

9. D. Zhaojing, W. Qiao, L. Jian, S. Qiang, J. Coll. Interf. Sci., **165**, 346 (1994)

SELF ORGANIZED GROWTH AND ULTRAFAST ELECTRON DYNAMICS IN METALLIC NANOPARTICLES

A. STELLA*, P. CHEYSSAC**, S. DE SILVESTRI***, R. KOFMAN**, G. LANZANI***, M. NISOLI***, P. TOGNINI*
*Istituto Nazionale per la Fisica della Materia, Dipartimento di Fisica "A. Volta", Università di Pavia, Via Bassi 6, 27100 Pavia, Italy
**Laboratoire de Physique de la Matière Condensée, URA 190, Université de Nice Sophia Antipolis, Nice Cedex, France
***Centro di Elettronica Quantistica e Strumentazione Elettronica-CNR, Dipartimento di Fisica, Politecnico, Milano, Italy.

ABSTRACT

We present ultrafast transient reflectivity measurements performed on metallic tin nanoparticles with an average radius between 20 and 60 Å in amorphous matrix. The samples, grown using an evaporation-condensation technique, were characterized by a relatively low nanocrystals size dispersion and a negligible clusters-matrix interaction. The excitation decays exhibited a size-dependent behaviour, which is interpreted in terms of the important role played by the electron-surface interactions.

INTRODUCTION

The growing interest for the nanoscale systems as it has been developing in the last few years is strongly motivated not only by the peculiarity of their basic properties, which show remarkable deviations from bulk behaviour, but also by their potential applications in the field of the electronic and electro-optic devices [1]. Among these systems, the nanoparticles, which are confined in all of the three space dimensions, are going to play a major role.

The main problems related to their investigation can be attributed to the difficulty in obtaining structures with a low size dispersion, without being limited by the characteristics of the substrate or of the matrix. A large size dispersion (more than 20-30 %) can lead to a blurring of the effects due to the small size (e.g. blueshift of the spectral structures or reduction of the melting temperature [2]), and, in the worse cases, to their disappearance.

Different growth techniques for nanoclusters have been developed. The main ones are here summarized [3]:
- patterning and etching, using a lithographic process, that allows to control the form and the distribution, but with a considerable perturbation of the physical properties;
- heat controlled precipitation of semiconductor solution dispersed in a glass matrix;
- chemical synthesis of semiconductor nanocrystals leading to colloidal solutions, with a low size dispersion and a careful size distribution control; the main limitation is the need for an organic matrix;
- stress driven, self-organized growth based on a lattice mismatch, that allows a low size dispersion and a good control of the shapes, but a limited possibility of varying the size in a large range.

Mat. Res. Soc. Symp. Proc. Vol. 457 © 1997 Materials Research Society

TIN NANOPARTICLES

The samples described were grown using an evaporation-condensation self-organization technique in ultra-high vacuum, on substrates of sapphire [3]. After a previous evaporation of amorphous Al_2O_3, metallic tin vapor was condensed. Since tin only partially wets the amorphous matrix, and since the temperature of the substrate is kept high enough, there was nucleation of tin nanoparticles at the liquid state, with a truncated spherical shape determined only by the balance of the surface tension. Being the Sn-Al_2O_3 contact angle about 114°, the volume of the truncated spheres was 80% of that of ideal spheres with the same radius [4]. Due to the amorphous character of the matrix and to the peculiar conditions of growth, the interaction between matrix and nanoparticles was minimized, so that any strain or stress effect was negligible.

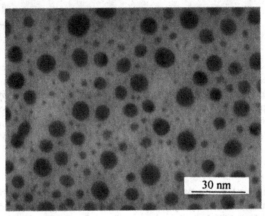

Fig. 1 TEM picture of a sample containing metallic tin nanoparticles in amorphous dielectric matrix (Al_2O_3).

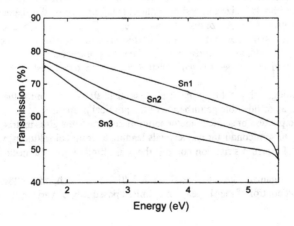

Fig. 2 Transmission spectra of the three samples investigated.

The presence of the nanoparticles, their crystalline character and their size dispersion was verified by Transmission Electron Microscopy. The size distribution of the particles was bimodal, 95% peaking around the average radius, and could be varied over a large range (between about 10 Å and more than 1000 Å) with a size dispersion ≤20%. This is one of the main advantages of this technique, together with the resulting symmetrical shape of the nanocrystals.

The micrographs were obtained with three tin samples with average radii 20 (Sn1), 40 (Sn2) and 60 (Sn3) Å. The corresponding transmission spectra are reported in fig. 2; the high energy absorption of the sapphire substrate prevents from observing the plasmon peak, which is expected to fall around 6 eV [4].

TRANSIENT REFLECTIVITY

The sample characteristics (i.e. relatively low size dispersion, large size range, weak interaction between nanoparticles and matrix, near-spherical shape) allow a careful investigation of the size dependence of the ultrafast transient reflectivity [4].

These measurements were performed in a pump-probe configuration, using a Ti:sapphire laser with chirped pulse amplification, which provides pulses of 150-fs duration at 780 nm with energy up to 750 μJ at 1-kHz repetition rate. In the experiment, the pump and probe wavelengths were 390 and 780 nm, respectively. The pump beam was obtained by frequency doubling a fraction of the laser beam in a LiB_3O_5 crystal. The pump pulse duration was 180 fs and the energy used in the measurements was in the range 6-60 nJ. Degenerate transient reflectivity measurements were performed on a subpicosecond time scale by using ultrashort pulses (10 fs) at 780 nm, with energy up to 240 μJ. Pump and probe beam polarizations were made orthogonal to avoid thermal grating buildup.

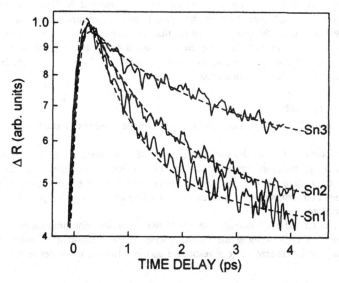

Fig. 3 Transient reflectivity changes in the three samples as a function of the probe time delay. Dashed lines: fitting curves.

The transient reflectivity changes, ΔR, in the samples are shown in fig. 3 as a function of the probe delay. The time decay can be separated in three distinct time scales, each one characterized by a time constant τ resulting from a fit of the curves using three exponential functions. The values of these time constants τ_1, τ_2, τ_3, given in Table 1, are largely different, being related to three different relaxation phenomena.

Sample	Radius (Å)	τ_1 (ps)	τ_2 (ps)	τ_3 (ps)
Sn1	20	0.62	12	580
Sn2	40	1.05	16.7	300
Sn3	60	1.42	22.7	460

Table 1 Time constants obtained from the fitting of the transient reflectivity measurements

These time constants can be associated to the processes of hot-electron thermalization by electron-electron (e-e), electron-phonon (e-p) and electron-surface (e-s) collisions [4],[5]. The electron thermalization usually occurs on a subpicosecond timescale, with the formation of a Fermi distribution with $T_e > T_0$ (where T_0 is the initial sample temperature). The hot electron system thermalizes with the lattice by e-e, but mainly by e-p and e-s collisions on a time scale of a few picoseconds [6]. The excess energy is then dispersed toward the matrix, in tens of picoseconds [7],[8], and the whole system decays to the original temperature in a time scale of hundreds of ps.

τ_1 can be associated to the e-p and e-s collisions, but its size-dependence demonstrates that it cannot be explained in bulk-like terms. From the collision time and the Fermi velocity of the electrons it is possible to estimate the mean free path l_∞ of the electrons in bulk tin [9]. In the nanoparticles, when R is of the order or smaller than the mean free path, collisions with the surface become important. In such case, the electrons oscillate inside the potential well with a frequency $\omega = v_F/R$ [10], and the time constant τ_1 can be written, taking into account the surface contribution to the electronic population relaxation, as:

$$\frac{1}{\tau_1} = \frac{1}{\tau_0} + \frac{v_F}{\alpha R} \qquad (1)$$

where τ_0 is the bulk time constant and α is the average number of inelastic collisions with the surface.

The dependence of τ_2 on the particle size can be understood if one considers that the energy absorbed by the nanoparticles is proportional to its volume, while the energy dissipated is proportional to its surface, so that one obtains longer τ_2's for the bigger particles. τ_3 is not expected to depend on size, since it is related to the cooling of the ensemble particles+matrix to the initial temperature.

The subpicosecond e-e thermalization [4] shows a transient rise time characterized by a time constant of 100 fs, describing the thermalization of the electron distribution, which is substantially constant in the size range investigated and in a fluence range between 30 and 300 $\mu J/cm^2$.

CONCLUSIONS

The quality and the characteristics of the nanoparticles allow a careful investigation of the size dependence of the ultrafast electron dynamics. The analysis here carried out accounts for the transient behavior of the systems under the effects of ultrafast pulses. The size range here explored is suitable for detecting and analysing important modifications of the nanoparticles properties with respect to the bulk, without reaching however the strong confinement regime that would take place when the number of costituent atoms of each nanoparticle is about 100. In this case an appreciable discretization of the electronic levels is expected to show up, with a strong blueshift [5]. We plan to study this kind of systems in the next future, with particular care to their transport and nonlinear optical properties.

REFERENCES

1. C.B. Murray, C.R. Kagar and M.G. Bawendi, Science **270**, 1335 (1995); A.P.Alivisatos, Science **271**, 933 (1996).

2. A. Stella, P. Cheyssac, R. Kofman, P. G. Merli, A. Migliori, Mat. Res. Soc. Symp. Proc. **400**, 161 (1996); P. Tognini, L.C. Andreani, M. Geddo, A. Stella, P. Cheyssac, R. Kofman, Il Nuovo Cimento D **18**, 865 (1996).

3. E. Sondergard, R. Kofman, P. Cheyssac and A. Stella, Surf. Sci. **364**, 467 (1996).

4. A. Stella, M. Nisoli, S. De Silvestri, O. Svelto, G. Lanzani, P. Cheyssac and R. Kofman, Phys. Rev. B **53**, 15497 (1996).

5. U. Kreibig and M. Vollmer, Optical Properties of Metal Clusters, Springer-Verlag, Berlin, 1995.

6. J. Shah, Ultrafast Spectroscopy of Semiconductors and Semiconductor Nanostructures, Springer-Verlag, Berlin, 1996.

7. R.W. Schoenlein, W.Z. Lin, J.G. Fujimoto and G.S. Eesley, Phys. Rev. Lett. **58**, 1680 (1987).

8. C.-K. Sun, F. Vallée, L. Acioli, E.P. Ippen and J.G. Fujimoto, Phys. Rev. B **48**, 12365 (1993).

9. N.W. Ashcroft and N.D. Mermin, Solid State Physics, Holt-Saunders, Orlando, 1976.

10. U. Kreibig and C.V. Fragstein, Z. Phys. **224**, 307 (1969).

Synthesis of Nanophase Noble Metal Systems
Utilizing Porous Silicon

I. Coulthard *, T.K. Sham *
* Chem. Department, University of Western Ontario, London, Canada, N6A 5B7,
icoul@julian.uwo.ca.

ABSTRACT

Apart from its well known ability to luminesce very intensely at room temperature in the visible range, porous silicon is also an effective reducing agent. We report the formation of several noble metal (Pd, Ag, Au, Pt) nanostructures by reductive dispersion of metal ions from aqueous solutions onto the surface of porous silicon. The nanophase systems produced by reductive deposition vary with the element deposited and the metallic salt utilized in the process. The resulting nanophase systems were studied using a variety of techniques including: scanning electron microscopy (SEM), X-ray photoelectron spectroscopy (XPS), and spectroscopic methods using synchrotron radiation.

INTRODUCTION

The ability of porous silicon to produce room temperature visible luminescence [1] has largely overshadowed some of its other interesting properties. One of these properties in particular is the ability of porous silicon to act as a modest reducing agent [2]. Any aqueous system with a standard reduction potential of greater than zero can be reduced by porous silicon which results in the liberation of hydrogen gas from the surface of the porous silicon. If the end product of this reduction is a solid, it is subsequently deposited onto the surface of the porous silicon.

In this study, several metal ions in aqueous solution were reduced using porous silicon resulting in the deposition of the metal onto the surface of the porous silicon. It was found that the deposited metal formed nanometre scaled clusters on the surface of the porous silicon.

Metal clusters of this size are very interesting in that their properties are in between those of condensed matter and atoms [3]. These clusters also provide excellent models for the study and understanding of surfaces and catalysis. The size, morphology, and electronic properties of the clusters deposited onto porous silicon were studied. The effect of choice of metal and preparation conditions is discussed.

EXPERIMENT

The porous silicon samples were prepared with a method described in detail previously [2]. Samples were typically prepared utilizing a p$^+$ (Boron) type Si (100) wafer with a resistivity of 1- 10 Ohm.cm. A current density of 20 mA/cm^2 was utilized for a period of 20 minutes. The electrolyte was composed of 1:1 48% by weight HF: absolute ethanol. The end

161

product was typically a reddish-brown film many microns in thickness which produces an orange-red luminescence under UV irradiation.

The reductive deposition of the metals was carried out immediately after preparation of the porous silicon layer. This allowed for minimum time for the porous silicon layer to oxidize in the ambient atmospheric conditions. The oxidation of the porous silicon surface has been shown to hinder and/or block the reductive deposition [4]. In cases where the deposition could not be performed immediately after preparation of the porous silicon layer, the layer was "refreshed" with a drop of HF which effectively removed all the oxide from the surface while leaving the hydrogen passivated layer intact. In each case, the porous silicon layer was treated with 1mL of a metal ion/complex solution with a concentration of 0.001M. The salts utilized to prepare the solutions were: $AgNO_3$, $Pd(NO_3)_2$, $K_2[PtCl_4]$, and $Na[AuCl_4]$.

Scanning electron microscope (SEM) images were recorded utilizing a Hitachi S-4500 Field Emission Scanning Electron Microscope (FESEM). X-ray photoelectron spectra (XPS) were recorded using a Surface Science Laboratories SSX-100 ESCA Spectrometer. X-ray absorption spectroscopy (XAS) measurements were carried out at the Double Crystal Monochromator soft X-ray beamline of the Canadian Synchrotron Radiation Facility at the University of Wisconsin-Madison 800 MeV Synchrotron Radiation Centre (SRC), and at beamline X11A of the National Synchrotron Light Source (NSLS) 2.5 GeV x-ray ring at Brookhaven National Laboratory. Measurements at both beamlines were carried out in total electron yield (TEY) mode.

RESULTS AND DISCUSSION

Figure 1. shows the Pd L_3 XANES (X-ray absorption near edge structure) for Pd reductively deposited onto porous silicon as compared to bulk Pd. An increase in the intensity of the white line (intense spike arising from p–d dipole transition whose intensity is a measure of the population of the unoccupied d states) is evident in the Pd/PS sample. This is indicative of small cluster and/or a chemical change such as a change in oxidation state. However, evidence shows that this effect is in fact due to small Pd clusters on the surface of the porous silicon. Previous results [2] using Pd K-edge EXAFS (extended X-ray absorption fine structure) showed that locally, the Pd on the surface of the porous silicon was metallic Pd. The only differences were a reduction in the average coordination number and a reduction of long range order. Both these results are consistent with the presence of small Pd clusters on the surface of the porous silicon. This was confirmed using scanning electron microscopy (SEM) as seen in Figure 2. Apart from the porous areas on the surface of the porous silicon, the sample is covered with Pd clusters of the order of 20-150 nm in diameter.

The change in the intensities of X-ray absorption white lines is most evident when samples of Pt on porous silicon are examined. The L_2 and M_2 edges of Pt are unusual in that no white lines are evident in the bulk metal. The lack of white line in the Pt L_2/M_2 edges has been attributed to the uneven distribution of $d_{5/2}$ and $d_{3/2}$ holes [5] in Pt. An increase in white line intensity indicates a redistribution of d charge. It is expected as the number of like nearest neighbours decrease as in the case of a small cluster, the d-d interaction is reduced leading to an increase of the d hole count at the Pt site [5].

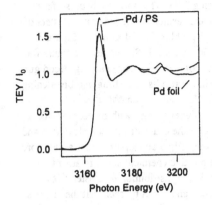

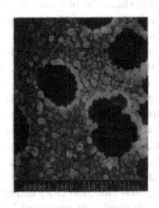

Figure 1. Pd L$_3$ XANES of Pd on porous silicon

Figure 2. SEM of Pd clusters on porous silicon

Figure 3. shows the L$_2$ edge for Pt on porous silicon compared with that of the bulk metal. An intense white line is clearly evident where it was absent in the bulk metal. This spectroscopic change is present in all the metal on porous silicon samples examined. Figure 4. shows the SEM of a gold cluster on porous silicon. The cluster is an aggregate of yet smaller clusters. At several hundred nanometres in diameter, Au clusters are much larger (approx. one order of magnitude), than the clusters of other metals. This is consistent with the X-ray absorption spectra where the difference in intensity between bulk Au and Au on porous silicon is very small.

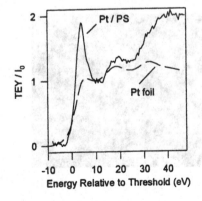

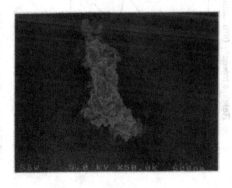

Figure 3. Pt L$_2$ XANES of Pt on porous silicon

Figure 4. SEM of aggregate of Au clusters on porous silicon

The effect of cluster size can also be seen utilizing XPS. The porous silicon surface oxidizes readily under ambient atmospheric conditions and results in isolated small clusters of metal on an insulating substrate. This results in charging problems in all samples. Figure 5. shows the XPS results for Ag on porous silicon at the 3d core level. Shown are the signal for bulk silver, and Ag on porous silicon, with and without compensation for charging (flood gun vs. no flood gun). The charging of the Ag on porous silicon results in shifting and broadening of the peaks which makes them unsuitable for analysis. However, when charging is compensated for, the resulting peak is shifted 0.4 eV to higher energy with respect to bulk silver and is considerably broader. Silver oxides cannot be the cause of this chemical shift and broadening since there is almost no chemical shift between silver and silver oxides at this core level [6]. It has been shown that small cluster produce positive chemical shifts relative to the bulk material [7]. We attribute the broadness of the peak to a distribution of cluster sizes. Figure 6. shows the SEM of Ag clusters on porous silicon and it can be seen that the clusters have a wide range of diameters from 20-120 nm consistent with the interpretation of the XPS results. Consistent behaviour for XPS have also been observed for Au and Pt on porous silicon. XPS experiments of Pd on porous silicon will be carried out in the near future.

It is also interesting to note that the choice of salt used in the preparation of the aqueous solution had an effect upon the reductive deposition. For silver and palladium nitrates, the deposition occurred very rapidly and a metallic sheen was readily observable with the naked eye. In these cases the clusters were deposited uniformly on the surface. However, in the case of the square planar, tetrachloro gold and platinum complexes, the deposition occurred more slowly. This is likely due to the reduction of the complexed metals being a multi step process with a rate limiting step. In these cases the clusters tended to be less uniformly distributed on the surface of the porous silicon but tended to be found with greater frequency near pores where there is a higher density of high energy sites (dangling bonds). In these two cases, the

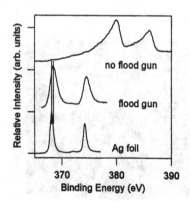

Figure 5.   Ag 3d XPS of Ag on
            porous silicon

Figure 6.   SEM of Ag clusters on
            porous silicon.

smaller clusters tended to aggregate into larger clusters. Other factors including pH of the solutions and the concentrations of spectator ions can also have an effect on metal deposition [6] making this system potentially very complex.

CONCLUSIONS

We have reported the preparation of metal clusters on the nanometre scale utilizing porous silicon as a reducing agent/substrate. The size, morphology, and electronic properties of the clusters were studied with a variety of methods. The metal / porous silicon system was found to be very complex and dependent upon a great number of different conditions. Further study is required to determine the effect of these conditions upon the deposition and to what degree of precision the deposition can be controlled.

ACKNOWLEDGMENTS

We would like to acknowledge J.W. Lorimer for discussion and the use of electrochemical equipment. Also to be thanked are R. Davidson and M. Biesinger at Surface Science Western, and K. Tan, and B. Yates at CSRF for technical assistance. This work received financial support from NSERC of Canada and the University of Western Ontario ADF Fund.

REFERENCES

[1]   L.T. Canham, *Appl. Phys. Lett.* **57**, p. 1046 (1990).

[2]   I. Coulthard, D.T. Jiang, J.W. Lorimer, and T.K. Sham, *Langmuir* **9**, p. 3441 (1993).

[3]   H. Gleiter, *Prog. Mater. Sci.* **33(4)**, p. 223 (1989).

[4]   I. Coulthard, and T.K. Sham, *Appl. Phys. Lett.* (submitted).

[5]   A.N. Mansour, J.W. Cook Jr., and D.E. Sayers, *J. Phys. Chem.* **88**, p. 2330 (1984).

[6]   M.J. Scaini, G.M. Bancroft, J.W. Lorimer, and L.M. Maddox, *Geochim. Cosmochim. Acta* **59(13)**, p. 2733 (1995).

[7]   S.B. DiCenzo, S.D. Berry, and E.H. Hartford Jr., *Phys. Rev. B.* **38(12)**, p. 8465 (1988).

SURFACE MELTING OF PARTICLES: PREDICTING SPHERULE SIZE IN VAPOR-PHASE NANOMETER PARTICLE FORMATION[1]

Y. XING[2] and D.E. ROSNER[3]

Department of Chemical Engineering, High Temperature Chemical Reaction Engineering Lab., Yale University, New Haven, CT 06520-8286, USA

ABSTRACT

There is still no reliable method to predict the all-important size of the 'primary' spherules found in combustion-synthesized particulate products. Toward this end, we introduce surface melting concepts in developing a coagulation-coalescence model for nanoparticles in the low temperature regions of diffusion flames. The associated surface self-diffusivity, which controls the rate of spherule sintering at temperatures well below the equilibrium melting point, is therefore modified to include an important size effect. This formulation is used to calculate the sintering rate of two adjacent particles in a flame coagulation environment corresponding to a condensed phase volume fraction of $ca.$ 10^{-1} ppm. Predicted spherule sizes show encouraging agreement with our experimental data for $ca.$ 10 nm diameter Al_2O_3 particles synthesized on the fuel side of a laminar CH_4/O_2 diffusion flame seeded with $Al(CH_3)_3$ (TMA)[1].

1. INTRODUCTION

Nanometer particles synthesized from the vapor phase are now attracting considerable attention in the research community because of their growing technological importance. Synthesis techniques include using liquid (or gaseous) chemical precursors in a combustor (*e.g.* pigment industry and our experiments [1]). Generally speaking, the particle formation processes involve chemical reaction (*e.g.* hydrolysis and oxidation), nucleation and growth stages; and fast chemical reaction and nucleation occur for materials such as refractory oxides or low vapor pressure metals. Thus, the growth process offers the greatest opportunity to control the final specific surface area and particle morphology.

In our system particle growth occurs primarily by two simultaneous processes: coagulation and coalescence. Coagulation kinetics are well-described by the Smoluchowski Brownian motion formulation (see below), while coalescence after binary contact is often described using simple macroscopic sintering models. However, for nanoparticles, sintering evidently occurs much more rapidly since it has been found that the spherule sizes predicted using such macroscopic sintering models are orders of magnitude smaller than those found experimentally [1, 2]. Therefore, for nanoparticles their size- (curvature-) dependent properties must be taken into account. However, until now no rational yet practical treatment has been proposed that can be used to predict the spherule size in flame-generated materials. In this work we initiate development of such a theory.

Our formulation is based on the observation that for a particle in an environment with increasing temperature the surface will tend to become disordered well before the bulk phase [3]. Thus, an enhanced surface self-diffusion coefficient is to be expected. This would be especially important for nanoparticles because a significant fraction of their atoms reside at/near the surface. The surface melting temperature, dependent on particle size and material properties, is

---

1. Paper #V5.36, MRS Fall Meeting , 1996 Boston, MA USA.
2. Graduate Research Assistant, PhD program.
3. Prof. ChE; Director, Yale HTCRE Lab.; Email: rosner@htcre.eng.yale.edu

167

Mat. Res. Soc. Symp. Proc. Vol. 457 ⁰1997 Materials Research Society

incorporated in the self-surface diffusivity which we use to calculate the coalescence rate of two touching particles. As expected, metal oxide particle sizes formed from vapor phase are generally controlled by Brownian coagulation when the particle coalescence rate is sufficiently rapid; however, when the particles grow larger, spherule growth effectively stops because of the slower spherule coalescence rate at larger sizes. Thus, as emphasized in recent studies [4], by comparing the characteristic times [1,5] of these rate processes we are able to predict terminal spherule sizes and compare them to those measured from TEM images of our thermophoretically-sampled particles [1]. Our rational procedure for modifying the surface properties of nanoparticles (see below) produces reasonable agreement for Al_2O_3 synthesized from our TMA-seeded counterflow diffusion flame environment [1]. The comparison discussed here was made for Al_2O_3 because its many physical properties are well known. In our future work our methods will be tested refined as necessary using data for other materials (e.g. TiO_2 ...) of industrial interest.

2. THEORY

For simplicity, we illustrate our methods here for the simplest case of the Brownian coagulation of a population of small spheres (Knudsen regime) in an isothermal gaseous environment, using elapsed time, t, as the relevant independent variable. We assume there are enough monomers to represent the distribution function by a continuous function. We ask when the mean time between such coagulation events will become comparable to the time to have such spherules coalesce by the mechanism of surface diffusion, here allowing for the phenomenon of *surface melting*.

Coagulation Rate

From the Smoluchowski's population balance equation, we can obtain [6]

$$\frac{dN}{dt} = -\frac{\alpha}{2}\left(\frac{3}{4\pi}\right)^{1/6}\left(\frac{6kT}{\rho_p}\right)^{1/2} \phi^{1/6} N^{11/6},$$ (1)

for long enough elapsed time to achieve the self-preserving PSD shape, where N is the total number density of particles, α a dimensionless collision integral ($= 6.67$ in free molecule regime) ρ_p the intrinsic particle density, T the temperature, k the Boltzman constant and ϕ the solid particle volume fraction. Therefore, the following relaxation time for coagulation, t_C,

$$t_C \equiv 2/(\beta N),$$ (2)

can be defined [1] to characterize the coagulation rate process (in comparison to the sintering rate process described below), where $N(t)$ is easily found from Eq. (1) and the collision frequency function, β, is evaluated using the mean particle size.

Surface Melting

Consider a particle undergoing surface melting process, as illustrated in Fig. 1. As the temperature is increased, a liquid layer is formed on the particle. The corresponding free energy change can be shown [7] to have the form:

$$\Delta f = L\tau(T_{mp} / 3T_{mb})(\delta^3 - 3\delta^2 + 3\delta) - (1 - \Theta)[(\gamma_{sg} - \gamma_{lg}) - \gamma_{sl}(1 - \delta)^2], \qquad (3)$$

where Δf is the specific surface free energy difference between the surface melted particle and the solid particle, L the latent heat of melting (J/m^3), T_{mb} the bulk melting temperature, $\delta = \xi / r$ dimensionless length with ξ the surface liquid layer thickness and r particle radius, T_{mp} the particle melting temperature given by [8] $T_{mp} / T_{mb} = 1 - 3[\gamma_{sg} - \gamma_{sl}(\rho_s / \rho_l)^{2/3}] / Lr$,

$\tau = (T_{mp} - T) / T_{mp}$ the dimensionless temperature, γ_{sg}, γ_{sl}, γ_{lg} the surface energies of the solid-gas, solid-liquid and liquid-gas interfaces, respectively, and $\Theta = \exp[-\delta r / (1 - \delta)\zeta]$ the partic

surface order parameter [9], with ζ the liquid correlation length, which here takes account of the size effect. The (finite) equilibrium liquid layer thickness at any given temperature T can be obtained by minimizing the free energy difference, i.e. by setting $\partial \Delta f / \partial \xi = 0$, satisfying the conditions $\partial^2 \Delta f / \partial \xi^2 > 0$ and $\Delta f \leq 0$. The surface melting temperature, T_{ms} (of the topmost surface molecules) can be determined by setting ξ to the surface layer thickness ξ_0.

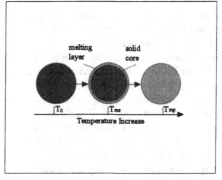

Fig. 1. Schematic diagram of the surface melting process.

Coalescence Rate

Many experiments and simulations have shown that surface diffusion is the dominant mechanism for nanometer particles [10,11,12], especially at temperatures well below the equilibrium melting point. Accordingly, for flame-generated nanoparticles of interest to us [1] we assume that surface diffusion is the dominant coalescence mechanism. Then, the characteristic time for complete coalescence of two touching spherules can be obtained by formally extrapolating the short time behavior of surface diffusion sintering model [13], giving the size-sensitive sintering time [1]:

$$t_S = \frac{kTr^4}{225wD_s\Omega\gamma_{sg}}, \qquad (4)$$

where w is the surface layer width, Ω is the molecular volume and D_s is the relevant surface self-diffusion coefficient.

Now the surface self-diffusion coefficient can be written as [14] $D_s = D_{s0} \exp(-Q_s T_{mb} / RT)$, where the ratio $Q_s = E_s / T_{mb}$ has been found to have an approximately constant value for similar structure materials [14] (E_s being the familiar activation energy for *surface diffusion*). However, for nanoparticles considered in this work the particle surface will melt earlier so that T_{mb} is now replaced by the surface melting temperature T_{ms}. If we further postulate that Q_s is explicitly size-independent, then we have the surface self-diffusion coefficient for small particles:

$$D_s = D_{s0} \exp(-Q_s T_{ms} / RT). \qquad (5)$$

This equation is used below to predict Al_2O_3 spherule sizes in our flame environment.

3. RESULTS AND DISCUSSION

Surface Melting and Associated Diffusivity

In Fig. 2 we present the surface melting temperature calculated using the formulation stated above. The model material is alumina, whose familiar physical properties are listed in Table I. For comparison in Fig. 2 we also display the small particle melting temperature and corresponding bulk melting temperature contours. For small particles, as anticipated, the surface melting temperature deviates from the particle melting temperature as well as the bulk melting point. The surface melting temperature (for the topmost layer here) is lower as the particle sizes become smaller.

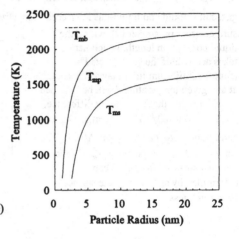

Fig. 2. Predicted surface melting and melting temperatures of alumina nanoparticles. Also shown (dashed) is the bulk melting temperature.

Table I. Property Values of Alumina Used in Calculations

Property	Value	Reference
ρ_s	3970 kg/m^3	[15]
T_{mb}	2327 K	[15]
L	4.32 GJ/m^3	[15]
γ_{sg}	2.3 J/m^2	[16]
γ_{lg}	0.7 J/m^2	[17]
γ_{sl}	0.15 J/m^2	[18][a]
ζ	0.7 nm	[19][a]
ξ_0	0.51 nm	[20][b]
E_s	0.518 MJ/mol	[21]
D_{s0}	1.0 x 10^5 m^2/s	[21]
w	1.0 nm	[21]
Ω	0.022 (nm)^3	[13]

a. estimated from the method proposed in the references.
b. taken as the lattice spacing of α-alumina.

The corresponding surface self-diffusivity is plotted in Fig. 3, from which we see that the enhancement of this property is especially impressive for small particles. Of course, for a sintering process *via* surface diffusion we are only concerned with the surface structure, not the state inside each nanoparticle.

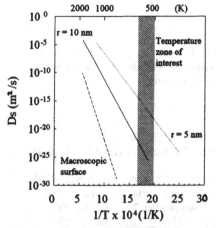 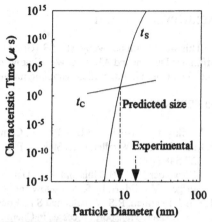

Fig. 3. Predicted size-dependent surface self-diffusivity for Al_2O_3 particles, showing the enhancement for small particles. Also shown is the macroscopic diffusivity.

Fig. 4. Coagulation and sintering characteristic times used to determine Al_2O_3 spherule sizes. Beyond the crossover particle size, t_S becomes much longer than t_C and particle growth is expected to stop at /near the indicated crossover size.

Comparison with Experiments [1]

Here we consider two conditions, *viz.* precursor TMA seeding levels 0.4 mL/hr and 0.8 mL/hr, corresponding to initial particle loadings (ϕ in ppm) 0.14 and 0.28. The calculated characteristic times for coagulation and sintering are plotted in Fig. 4 for the lower loading condition. We see that the predicted crossover occurs at a particle diameter of 8.2 nm. This is taken to be the 'arrest' size for particle growth since, beyond this crossover, coalescence by sintering becomes *much* slower (note the steep increase of the time with particle size). The experimental spherule diameter, 13 nm, is also shown in the plot for comparison. For the higher loading the predicted spherule sizeis 9.5 nm, while our experiments give 16 nm. Although there are still deviations, they are rather small when compared to the results of using literature-values from macroscopic properties [2]. Moreover, preliminary application of our method to the larger spherule Al_2O_3 data of Ref. [2] also reveals promising agreement [7].

4. CONCLUSIONS

We have proposed a rational theory to predict spherule size (and hence, specific surface area) in vapor phase nanoparticle synthesis. While based on the competition between the rate processes of Brownian coagulation and particle sintering, we propose an enhanced surface diffusivity for spherules in this important size range ($O(10$ nm) diameter). These results, when compared with our diffusion flame experiments on Al_2O_3 nanoparticles, show encouraging agreement. We are now in a position to test and refine this formulation using TiO_2 spherule size data from aggregates thermophoretically-sampled from our flames [1]. We expect that this approach could also be used to anticipate sintering time requirements for other types of nanophase materials, such as nanocomposite powders.

ACKNOWLEDGEMENT

This work is supported by AFOSR (Grant No. 94-1-0143) and Yale HTCRE lab industrial affiliates: DuPont and ALCOA. We acknowledge the valuable contributions of Drs. D Albagli and U.O. Koylu in our particle sampling/image analysis experiments.

REFERENCES

1. Y. Xing, U.O. Koylu and D.E. Rosner, Combust. & Flame **107**, 85-102 (1996).
2. M.K. Wu, R.S. Windler, C.K. Steiner, T. Boris and S.K. Friedlander, Aerosol Sci. Tech. **19**, 527-548 (1993).
3. J.F. van der Veen, B. Pluis and A.W. Denier van der Gon, in Chemistry and Physics of Solid Surfaces VII, Edited by R. Vanselow and R.F. Howe, Springer-Verlag, New York ,1988.
4. K.E.J. Lehtinen, R.S. Windler and S.K. Friedlander, J. Aerosol Sci. **27**, 883-896 (1996).
5. D.E. Rosner, Transport Processes in Chemically Reacting Flow Systems, Butterworth-Heinemann, Stoneham, MA, 1986 (Third printing ,1990).
6. S.K. Friedlander, Smoke, Dust and Haze, Wiley, New York, 1977.
7. Y. Xing and D.E. Rosner, to be submitted to J. Colloid Interface Sci. (1997).
8. P.R. Couchman and W.A. Jesser, Nature **269**, 481-483 (1977).
9. H. Lowen, Phys. Rep. **237**, 249-324 (1994).
10. J.E. Bonevich and L.D. Marks, J. Mater. Res. **7**, 1489-1500 (1992).
11. M. Kusunoki, K. Yonimitsu, Y. Sasaki and Y. Kubo, J. Am. Ceram. Soc. **76**, 763-765 (1993).
12. M.R. Zachariah and M.J. Carrier, Mater. Res. Soc. Symp. Proc. **351**, 343-348 (1994).
13. W.S. Coblenz, J.M. Dynys, R.M. Cannon and R.L. Coble, Mater. Sci. Res. **13**, 141-157 (1980).
14. G.A. Somorjai, Introduction to Surface Chemistry and Catalysis, Wiley, New York, 1994.
15. JANAF Thermochemical Tables, 3rd ed., ACS and AIP, 1986.
16. W.C. Mackrodt, Phil. Trans. Roy. Soc. Lond. **A341**, 301-312 (1992).
17. E.T. Turkdogan, Physicochemical Properties of Molten Slags and Glasses, The Metal Society, London, 1983.
18. E.S. Machlin, An Introduction to Thermodynamics and Kinetics Relevent to Materials Science, Giro Press, Croton-On-Hudson, 1991.
19. B. Pluis, D. Frenkel and J.F. van der Veen, Surf. Sci. **239**, 282-300 (1990).
20. Handbook of Chemistry and Physics, Editor-in-Chief, D.R. Lide, CRC Press, 1994.
21. E.G. Seebauer and C.E. Allen, Prog. Surf. Sci. **49**, 265-330 (1995).

SYNTHESIS AND CHARACTERIZATION OF SMALL GRAIN SIZED MOLYBDENUM NITRIDE FILMS

S.L. Roberson*, D. Finello**, R.F. Davis
Department of Materials Science and Engineering, North Carolina State University, Box 7907
Raleigh, NC 27695-7907

* U.S. Air Force Palace Knight student attending North Carolina State University
** U.S. Air Force, WL/MNMF, Eglin AFB, FL 32542

ABSTRACT

Polycrystalline, small grain sized, 15 μm thick Mo_xN (x=1 or 2) films, void of detectable concentrations of molybdenum oxides, have been prepared on 50 μm thick nitrided Ti substrates via conversion of precursor MoO_3 films in a programmed reaction with NH_3. The latter films were produced via liquid spray pyrolysis of an $MoCl_5$/methanol mixture in air at 500° C. The reaction of MoO_3 films with NH_3 resulted in a two-phase Mo_xN mixture consisting of γ-Mo_2N and δ-MoN. The change in density of MoO_3(ρ=4.69 g/cm^3) to γ-Mo_2N(ρ=9.50 g/cm^3) and δ-MoN (ρ=9.05 g/cm^3) produced grains with a calculated average size of 10 nm without losing adherence to the substrate. The composition of the Mo_xN films was determined by X-ray diffraction (XRD) and Auger electron spectroscopy (AES) to be $\approx$ 60% γ-Mo_2N and 40% δ-MoN. The results of scanning electron microscopy (SEM) showed the surface morphology of the Mo_xN films to be highly porous.

INTRODUCTION

The numerous potential applications of high surface area (HSA) nanoparticle materials have prompted significant research into their production and development [1-3]. Recently, HSA Mo_xN powders have been investigated as hydrodenitrogenation, dehydrogenation, and desulfurization catalysts [4-6]. Previous studies regarding the preparation of HSA Mo_xN powders have focused on the reduction of MoO_3 in NH_3 using temperature programmed process routes [1, 2, 7, 8]. The conversion of MoO_3 to Mo_xN is a topotactic transformation [7] in which HSA particle formation occurs accompanied by the retention of the crystallographic relationships between MoO_3 and Mo_xN. Thus a substantial change in density results without the collapse of individual particles [7].

Because of their high electrical conductivity (1/ρ $\approx$ 10 ^{4} ohm^{-1}cm^{-1}), Mo_xN films have been prepared and investigated in the present research for use as electrodes in high energy density storage devices. The preparation of small grain sized Mo_xN films with low oxygen concentrations on Ti substrates via conversion reactions between granular MoO_3 films and NH_3 has invariably resulted in the formation of MoO_2. The formation of TiO_2 and TiO on Ti substrates during the deposition of MoO_3 inhibits the complete conversion of MoO_2 to Mo_2N at the interface. In addition, during the conversion of MoO_3 to Mo_xN the adhesion of the film to the substrate oxides reduces the amount of densification that can occur. Another potential problem in the conversion process is the occurrence of unstable Mo_xNCl compounds as a result of the incomplete pyrolysis of the $MoCl_5$ precursor.

In the current research, we have observed that the preparation of small grain sized Mo_xN films can be accomplished by using nitrided Ti substrates and lengthy conversion reaction processes. The use of nitrided Ti substrates both substantially reduced the amount of TiO_2 and TiO formed and, as a result, reduced the presence of MoO_3 (which converts to MoO_2 at temperatures above 400° C in NH_3) in the converted Mo_xN films. Lengthy conversion reaction processes involving the diffusion of NH_3 into the MoO_3 films were also developed to eliminate MoO_2 near the interface region.

173

EXPERIMENTAL

As-received, 50 μm thick polycrystalline Ti substrates, were cut into 1" squares, cleaned sequentially in trichloroethylene, acetone, and methanol, etched in 1.0 M HCl at 90° C for ten minutes to remove the surface oxide, rinsed in methanol, and dried. Surface nitridation was accomplished by heating the prepared Ti substrates to 700° C for 1 hour in NH_3 flowing at 4 l/min in a quartz lined stainless steel tube furnace, which had been previously evacuated to 10^{-3} torr.

The production of MoO_3 films was accomplished in a resistively heated, pancake-style, atmospheric spray pyrolysis deposition system. This system incorporated a thin layer chromatography sprayer (TLC) to deliver the liquid molybdenum source (5g of $MoCl_5$/100 ml DI H_2O +50 ml methanol) to the substrate. The substrates were heated to the MoO_3 deposition temperature of 500° C, and the process sequence initiated by flowing N_2 at 14 l/min over the liquid inlet of the TLC sprayer. At this flow rate, the liquid source was siphoned up the inlet tube, dispersed into fine droplets, and deposited on the substrate. The pyrolysis of the liquid source in air produced MoO_3. The process route for the growth of MoO_3 films employed an on/off cycle involving the deposition of MoO_3 for thirty seconds and the cessation of the flow of N_2 for thirty seconds. This process was repeated 15 times to prepare 15 μm thick MoO_3 films.

To achieve the conversion of the MoO_3 films to Mo_xN, the former were placed in a quartz lined stainless steel tube furnace and evacuated to 10^{-3} torr. The system was seven times sequentially back-filled with UHP (99.999%) N_2 to atmospheric pressure and evacuated to 10^{-3} torr to reduce the oxygen level in the system to a minimum. To eliminate any detectable MoO_2 in the Mo_xN films, the following temperature programmed reaction sequence was developed by the authors and conducted at 760 torr and an NH_3 flow rate of 7 l/min : 325° C/hour from room temperature to 325° C, 20° C/hour from 300 - 580° C, 160° C/hour from 580 - 740° C, and a final soak at 740° C for 2 hours. The resulting Mo_xN films were passivated for 24 hours in N_2 flowing at 2 l/min to prevent rapid oxidation due to their high surface areas.

The structural, microstructural, and chemical properties of the Mo_xN films were determined using several techniques. Scanning electron microscopy (SEM) and chemical analysis were performed using a JEOL 6400FE at 5 kV and equipped with an energy dispersive X-ray microanalyzer (EDX) capable of light element detection. X-ray diffraction patterns were obtained using a Rigaku Model A operating at 27.5 kV and 15 A. The average particle size of the Mo_xN films was determined using the Debye-Scherrer equation

$$d= K\lambda/B_{hkl}\cos\theta \tag{1}$$

where θ is the Bragg angle; K is a correction factor, taken as unity; λ is the X-ray wavelength in Å; and B_{hkl} is the full width at half the maximum peak height in radians corrected for instrument broadening. Auger electron spectroscopy depth profiling was conducted using a JEMP-30 at an accelerating voltage of 5 kV.

RESULTS AND DISCUSSION

Prior research in the authors' laboratories showed that polycrystalline Mo_xN films converted from MoO_3 previously deposited on Ti metal substrates contained $\leq$ 20% MoO_2. In contrast, Mo_xN films, void of detectable amorphous or crystalline MoO_2, have been prepared in the current research on nitrided Ti substrates using the temperature programmed conversion reaction described above. The slow heating rates employed reduced hydrothermal sintering of the individual grains, prevented rapid water evolution and the associated loss of adherence to the substrate, and resulted in nitrogen-rich and nitrogen-deficient planar regions in the upper and lower sections, respectively, of the films. It was discovered that MoO_2 converted to Mo_2N in NH_3 at temperatures above 730° C. Thus, to eliminate MoO_2, samples were held at a soaking temperature of 740° C for 2 hours. This procedure also reduced the surface area and increased the grain size of the films.

The XRD pattern presented in Figure 1a of a representative low oxygen concentration Mo_xN film shows no peaks associated with MoO_2. The two phase Mo_xN structure consisted of $\approx$ 60% γ-Mo_2N and $\approx$ 40% δ-MoN. An XRD pattern of an MoO_2 contaminated Mo_xN film is shown in Figure 1b for comparison. The peak broadening in both films is due to the small size of the grains in the films as discussed below.

A representative SEM micrograph of a porous Mo_xN film having an average grain size of $\approx$ 10 nm and with no detectable MoO_2 is shown in Figure 2a. An SEM micrograph of a typical MoO_2- contaminated Mo_xN film is shown in Figure 2b. The films contaminated with the oxide possessed a large variation in grain size which is believed to be due to the incomplete conversion of MoO_3 to Mo_xN. The Mo_xN films with no detectable oxide had a more uniform grain size.

As shown in Table I, the calculated grain size of low oxygen concentration Mo_xN films below 19° was $\approx$ 10 nm. Good agreement with this value was observed in all of the peaks below 19°. Although not shown, the average grain size of oxide contaminated Mo_xN films was $\approx$ 15 nm. The calculated particle size decreased with increasing θ values due to the 0.001 Å line width of $K\alpha$ radiation. This line width resulted in an increase in the peaks of 0.08 ° over the width of a true monochromatic beam at 22.5 °. Peak broadening due to this phenomena increases with increasing theta. Therefore, the average grain sizes within the films are best represented in the theta values that are less than this angle.

The nitridation of the Ti substrates reduced the MoO_2 present in the converted Mo_xN films. The nitrided titanium layer made up only 5-10% of the bulk, but it was concentrated at the surface of the substrate. Due to their excellent oxidation resistance, the Ti nitrided substrates did not oxidize to the extent of the Ti substrates during the preparation of MoO_3 films. It was discovered that once titanium oxides (TiO_2 and TiO) were formed on the substrate, it was thermodynamically impossible at the reaction temperatures and pressures employed to convert these oxides to Ti_2N or TiN. As the Gibbs free energy of formation of MoO_2 possesses a more negative value than TiO at temperatures below 1020 K, the presence of TiO was found to inhibit the conversion of MoO_3 to

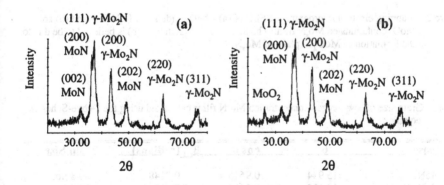

Figure 1. X-ray diffraction patterns of (a) a low oxygen concentration Mo_xN film and (b) an MoO_2 contaminated Mo_xN film. The peak broadening is due to small grain size.

Mo_xN. Above this temperature, TiO is thermodynamically more stable than MoO_2. However, to reduce the presence of MoO_2 in the Mo_xN films, the conversion process had to be conducted at 1040 K. The effects of using nitrided Ti substrates are shown in Figure 3. In Figure 3, an XRD pattern of the same temperature programmed reaction [1] conducted on substrates of Ti and nitrided Ti indicate a significant reduction in oxide contamination by using nitrided Ti substrates. The Ti, TiO_2, and TiO peaks have been removed for clarity.

To monitor contamination in the Mo_xN films, AES and EDX was conducted to determine the presence of oxygen and chlorine, respectively. The AES depth profile shown in Figure 4a of an Mo_xN film on a nitrided Ti substrate indicates that oxygen levels were near background concentrations adjacent to film/substrate interface. Conversely, in Figure 4b, oxygen is indicated near the interface region in an Mo_xN film on Ti substrates prepared by using the same temperature programmed reaction. The EDX data (not shown) of these two films did not indicate the presence of chlorine. Therefore at the MoO_3 deposition temperatures employed, complete pyrolysis of $MoCl_5$ was achieved.

(a)   **(b)**

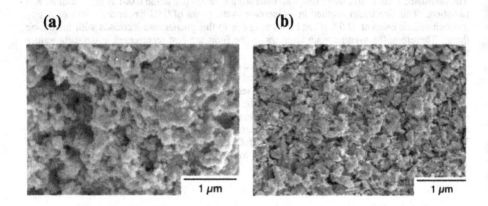

Figure 2. Scanning electron micrographs (SEM) of (a) a bulk oxide free Mo_xN film and (b) an MoO_2 contaminated Mo_xN film. The particle size variation in (b) is believed to be due to the formation of MoO_2 in place of Mo_xN.

Table I. Grain size of low oxygen concentration Mo_xN films calculated using the Debye-Scherrer equation.

Phase	θ	$\cos\theta$	B_{hkl} (radians)	Grain Size
MoN	15.944	0.9615	0.0148	10.8 nm
MoN	18.103	0.9504	0.0153	10.6 nm
γ–Mo_2N	18.688	0.9472	0.0153	10.6 nm
γ–Mo_2N	21.725	0.9289	0.0206	8.1 nm
MoN	24.506	0.9099	0.0239	7.1 nm
γ–Mo_2N	31.555	0.8521	0.0257	7.0 nm

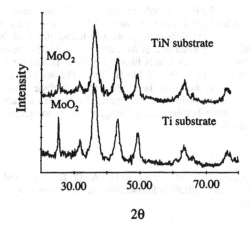

Figure 3. X-ray diffraction patterns of Mo$_x$N converted from MoO$_3$ using the conversion process developed by Choi et. al [1] on different substrates. Molybdenum nitride (Mo$_x$N) films on nitrided Ti substrates had less oxide contamination than on Ti substrates. The TiO, TiO$_2$, TiN, and Ti$_2$N peaks have been removed for clarity.

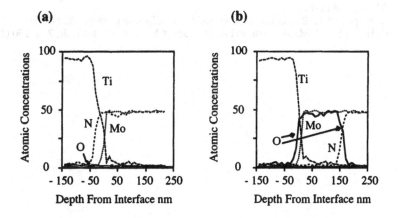

Figure 4. Auger electron spectroscopy depth profiles of the interface region of (a) an Mo$_x$N film on a nitrided Ti substrate and (b) an MoO$_2$ contaminated Mo$_x$N film on a Ti substrate. Both (a) and (b) were converted using the process described in the text. Oxygen was indicated in the Ti substrate in (b).

CONCLUSIONS

The preparation of 15 μm thick polycrystalline Mo_xN films having a low oxygen content and very small grain size has been achieved on nitrided Ti substrates via conversion of previously deposited MoO_3 in NH_3. X-ray diffraction patterns indicated that peak broadening was due to small grain sizes. The average grain size of the films determined by Debye-Scherrer calculations was ≈ 10 nm. The grain size of the Mo_xN films on nitrided Ti substrates varied only slightly in size when compared to oxide contaminated Mo_xN films. The use of nitrided titanium substrates greatly reduced oxide contamination near the film/substrate interface. The AES depth profile data indicated that the oxide contamination was concentrated near the interface in Mo_xN films on Ti substrates. The use of nitrided Ti substrates and lengthy conversion procedures reduced the oxide contamination in the Mo_xN films.

ACKNOWLEDGMENTS

The authors express their appreciation to Dr. A. D. Batchelor and David Ricks for their assistance. One of the authors (SLR) acknowledges the continuing support via U.S. Air Force Palace Knight Program.

REFERENCES

1. J. G. Choi, R. L. Curl and L. T. Thompson, Journal of Catalysis, **146**, p. 218 (1994).
2. R. S. Wise and E. J. Markel, Journal of Catalysis, **145**, p. 344 (1994).
3. J. D. Wu, C. Z. Wu, Z. M. Song, L. H. Wu and F. M. Li, Applied Surface Science, **90**, p. 81 (1995).
4. H. Abe and A. T. Bell, Catal. Lett, **18**, p. 1 (1993).
5. K. S. Lee, H. Abe, J. A. Reimer and A. T. Bell, Journal of Catalysis , **139**, p. 34 (1993).
6. J. C. Schlatter, S. T. Oyama, J. E. Metcalf and J. M. Lambert, Ind. Eng. Chem. Res., **27**, p. 1648 (1988).
7. L. Volpe and M. Boudart, Journal of Solid State Chemistry, **59**, p. 332 (1985).
8. C. H. Jaggers, J. M. Michaels and A. M. Stacy, Chemistry of Materials, **2**, p. 150 (1990).

SINTERING OF SPUTTERED COPPER NANOPARTICLES ON (001) COPPER SUBSTRATES

M. YEADON, J.C. YANG, M. GHALY, D.L. OLYNICK, R.S. AVERBACK, J.M. GIBSON
Materials Research Laboratory, University of Illinois at Urbana-Champaign, Urbana IL 61801
myeadon@uiuc.edu

ABSTRACT

The sintering of sputtered copper nanoparticles with a thin film copper substrate has been studied in real time using a novel *in-situ* UHV TEM. Particles were generated by dc sputtering in argon and transferred directly into the microscope along a connecting tube. The particles were deposited onto a clean (001) copper substrate of thickness 40nm. As-deposited particles assume a random orientation on the copper substrate as evidenced by electron diffraction patterns. Upon annealing, however, the particles were observed to reorient and assume the orientation of the substrate. Reorientation of individual, isolated particles occurred at ~200°C, primarily by grain-boundary motion and not by surface diffusion.

INTRODUCTION

Nanophase materials can exhibit a number of novel properties and are thus of great interest for a number of engineering applications. A nanophase material is typically defined as one having grain sizes of less than ~50nm and may be produced by Gleiter's method [1] of inert gas condensation followed by in-situ compaction of the particles to form a nanocrystalline pellet. Sintering of this pellet leads to the formation of a compact whose density depends on the degree of agglomeration of condensed particles in the gas phase prior to collection and compaction. Highly agglomerated particles typically produce compacts of lower density by decreasing particle coordinations and introducing pores [2]. There is therefore considerable interest in understanding the processes involved in the growth, collection and sintering of nanoparticles. Frequently, collected particles are of necessity in contact with a bulk surface, and consequently it is important to understand the sintering processes occurring both between particles in contact with other particles, and particles in contact with a substrate. Moreover, nanoparticle/film composites will be useful in the development of magnetic storage media where magnetic nanoparticulates are required to remain insoluble with a thin film substrate. The stability and sintering properties of the nanoparticles as a function of composition, processing and temperature will be key issues in the development of such applications. Deposition of clean, small particles on a clean single crystal substrate is also an ideal experiment to test theoretical models of the sintering processes occurring in small particles [3].

In the present work we have addressed the issue of the sintering of nanoparticles in contact with a thin film substrate. Using copper as a model system, we have investigated the sintering of sputtered clean copper nanoparticles in contact with a clean single-crystal copper substrate using a novel in-situ UHV TEM. The experiments were performed in parallel with molecular dynamics simulations of the interactions between particles and a substrate upon initial contact.

Mat. Res. Soc. Symp. Proc. Vol. 457 ©1997 Materials Research Society

EXPERIMENT

Copper nanoparticles were generated by dc magnetron sputtering of a copper target (99.999% purity) in 1.3 Torr of ultraclean argon (<1ppb impurities) and a baked UHV chamber as described elsewhere [4]. Individual particles and agglomerates of these particles were then transferred through a pipe directly into the microscope in the gas phase. The instrument, a JEOL 200CX TEM modified for UHV capability [5] and nanoparticle introduction [4], is equipped for resistive heating of substrates and selective introduction of gases. The base pressures of the chamber and column are 10^{-10} and 10^{-9} Torr respectively.

Copper films 40nm in thickness and of predominantly (001) orientation were produced by e-beam evaporation of copper(99.999% purity) onto (001) rocksalt substrates held at 250°C. Following flotation of the copper films in deionized water the films were mounted on specially prepared Si supports and transferred into the microscope. In order to produce a clean copper surface the native oxide (Cu$_2$O) on the surface of the film was removed by heating to between 300 and 350°C in flowing methanol at a partial pressure of ~5.10^{-5} Torr for 20 minutes. Methanol decomposes on the film surface to yield water, formaldehyde (CH$_2$O), carbon monoxide and carbon dioxide. The reaction products desorb from the surface leaving an oxide-free film as confirmed by electron diffraction. Prior annealing of the copper film also led to a reduced step density on the surface.

RESULTS

The morphology and orientation of as-deposited particles was examined using bright field and dark field imaging in combination with electron diffraction. A typical area of a film following particle deposition is shown in figure 1 (a). The presence of both agglomerates of particles and individual particles can be seen in the figure. Electron diffraction patterns indicate the particles to be of random orientation at this stage.

Upon heating, in ~10°C increments, at a rate of ~40°min^{-1}, the particles exhibited a stable orientation until, above ~200°C, a proportion of the smaller particles (5-10nm diameter) began to reorient. At this stage the particles became invisible in bright field images (figure 1(b)), however examination of DF images using a Cu(220) reflection revealed local thickness changes of similar dimensions to the particle diameters prior to annealing. Depending upon the angle of the incident illumination, and hence the positions of thickness fringes in the image, the particles either appeared bright on a dark background, or dark on a bright background. Examples of this contrast reversal are shown in figure 2. In this region of the film the particles at 'A' and 'B' can be seen to be of opposite contrast in figures 2(a) and (b). Further tilting of the beam led to contrast reversal of other particles. This contrast change implies that the particles have 'joined' the crystalline material of the substrate, and so contribute to the Cu(220) diffraction contrast of the substrate. At this point they become almost invisible in bright field. From these contrast changes the height of these epitaxial particles is estimated to be ~5nm. Dynamic studies of the realignment show that it typically occurs by grain boundary motion through the particle.

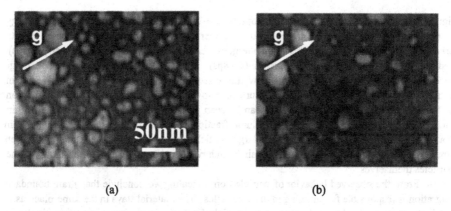

(a)   (b)

Figure 1   Two bright field images (g=220) of the same region of a sample (a) immediately after particle deposition and (b) after heating to 200°C.

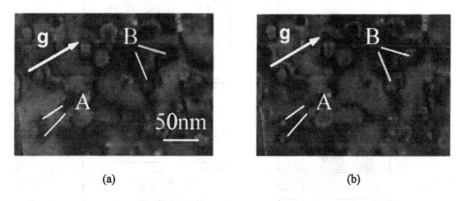

(a)   (b)

Figure 2 (a) Weak-beam dark-field image (g=220) of the film; the particles at A and B show dark contrast on a bright background. (b) Weak-beam dark-field image of the same region of the film with illumination tilted relative to (a). The particles at A and B now show a bright contrast on a dark background, indicating a local thickness increase and confirming their epitaxial orientation. At this point the particles are almost invisible in bright-field.

Room temperature sintering of copper nanoparticles with one another has been observed previously in our system on amorphous SiN substrates [6] and it is expected that a grain boundary will form between the particle and copper substrate at the point of contact. In order to minimize its free energy a 'notched' boundary between the particle and substrate is the most probable form of this interface. Indeed, molecular dynamics simulations of particles ~7nm in diameter confirm this and indicate that the first several monolayers of the particle in contact with the substrate are of epitaxial orientation. Very small copper particles (≤2nm diameter) are predicted to sinter in less than a nanosecond and an example of such a simulation is presented in figure 3(a). The figure shows a copper nanoparticle, ~1nm in diameter and with a kinetic energy of 0.2eV, experiencing a 'soft' impact with a copper surface. From the sequence of events predicted by the simulation the particle undergoes recrystallisation after initial impact (t=15ps) and becomes fully epitaxial with the surface (t=18ps). Simulations of particles in the size range accessible to our experiments do not show this effect, and in figure 3(b) a particle ~3nm in diameter is shown incident upon the same surface, also with a kinetic energy of 0.2eV. A fraction of the particle, closest to the substrate, can be seen to be epitaxial with the substrate within picoseconds of impact. The majority volume fraction does not become epitaxial and a grain boundary is formed within the particle. In larger particles the epitaxial volume fractions are even smaller, with the projected areas of the grain boundaries also becoming lower than those of the particles themselves.

From the observed behavior of particles on annealing we conclude that grain boundary migration is responsible for sintering in these particles. The material stays in the same place as it reorients, as evidenced by the disappearance in bright field and simultaneous appearance in the

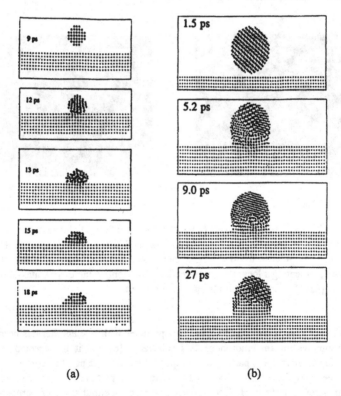

(a)   (b)

Figure 3 Molecular dynamics simulations of the atomic positions at various time instants for soft landings of copper particles on a copper substrate. The particles are (a) 1nm and (b) 3nm in diameter and possess 0.2eV kinetic energy before impact.

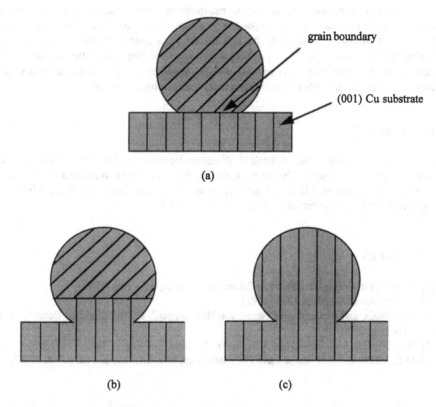

Figure 4 Cross-sectional illustration of the reorientation process occurring upon annealing an as-deposited particle. (a) Particle prior to annealing, showing the presence of a grain boundary close to the particle: bulk interface, (b) fluctuation of the position of the grain boundary upon heating and (c) epitaxial reorientation of the particle. Shaded lines represent copper (100) planes.

Cu(220) dark field image. This process is illustrated in figure 4. Although surface diffusion is occurring, as evidenced by necking in agglomerates, it is not fast enough to allow particle dissolution by a ripening process involving the substrate as the largest particle. This does occur for dirtier depositions on unclean substrates, in which grain boundary motion is presumably impeded. In this case particle dissolution at high temperatures occurs by surface diffusion. Even for 'clean' depositions, some particles do not reorient at high temperatures, which is possibly a consequence of special orientations (e.g. twins) or impurities.

CONCLUSIONS

We have demonstrated that copper nanoparticles with a size distribution of 4-20nm assume a random orientation when deposited on a clean (001) copper thin film. Upon heating, the particles become epitaxial with the film mainly by a grain boundary migration process. Reorientation occurs at ~200°C for particles of between 5 and 10nm, whilst the majority of larger particles and agglomerates become epitaxial at higher temperatures. Our studies provide valuable controlled experimental data for modeling of this important sintering process.

ACKNOWLEDGEMENTS

The authors would like to thank H. Birnbaum for access to the UHV e-beam evaporator used during the experiments. The use of facilities in the Center for Microanalysis of Materials at the University of Illinois at Urbana-Champaign is acknowledged. This work was funded by the Department of Energy under contract DEFG02-96ER45439.

REFERENCES

[1] H. Gleiter in *Progress in Materials Science* (Pergamon, New York, 1990)
[2] E. Arzt, Acta Met **30**, p. 1883 (1982)
[3] 'Synthesis and Processing of Nanocrystalline Copper,' Ed. by David L. Bourell, (TMS, Pennsylvania, 1996)
[4] D.L. Olynick, J.M. Gibson and R.S. Averback, Mater. Sci. Eng. A **204**, p. 54 (1995)
[5] M.L. McDonald, J.M. Gibson and F.C. Unterwald, Rev. Sci. Instrum. **60**, p. 700 (1989)

Part III

Nanophases: Simulation Studies

FRACTURE OF NANOPHASE CERAMICS: A MOLECULAR-DYNAMICS STUDY

AIICHIRO NAKANO, RAJIV K. KALIA, ANDREY OMELTCHENKO, KENJI TSURUTA, PRIYA VASHISHTA
Concurrent Computing Laboratory for Materials Simulations
Department of Computer Science and Department of Physics & Astronomy
Louisiana State University, Baton Rouge, LA 70803
Email: nakano@bit.csc.lsu.edu, kalia@bit.csc.lsu.edu, omeltch@rouge.phys.lsu.edu,
           kenji@rouge.phys.lsu.edu, priyav@bit.csc.lsu.edu
URL: http://www.cclms.lsu.edu

ABSTRACT

New multiscale algorithms and a load-balancing scheme are combined for molecular-dynamics simulations of nanocluster-assembled ceramics on parallel computers. Million-atom simulations of the dynamic fracture in nanophase silicon nitride reveal anisotropic self-affine structures and crossover phenomena associated with fracture surfaces.

INTRODUCTION

Light-weight materials with high-temperature strength are urgently needed in aeronautical and automotive industries for future-generation, energy-efficient engine components [1]. Silicon nitride has been at the forefront of research, since the combination of low thermal expansion and high strength makes it one of the most thermal-shock-resistant materials currently available [2].

However, the brittleness of silicon nitride hinders its practical applications. In recent years a great deal of progress has been made in synthesizing nanophase ceramics containing ultrafine microstructures in the range of a few nanometers [3]. The synthesis of these materials involves the generation and sintering of nanometer-size clusters. Generally speaking, the physical properties of nanophase solids are much superior to those of ordinary coarse-grained materials. For example, it has been found that nanophase ceramics are more ductile than conventional ceramics with the same constituents.

Understanding the fracture mechanisms of nanophase ceramics is essential for structural applications. While it has been speculated that the improved toughness is due to the deflection and multiple branching of cracks at interfaces, there is little understanding about the influence of interfacial structures on fracture dynamics.

New computational approaches are augmenting the time-consuming search-and-test methods traditionally used for the development of new materials. Molecular dynamics (MD) approach provides the phase-space trajectories of particles through the solution of Newton's equations, thereby shedding light on how atomistic level processes lead to macroscopic phenomena such as fracture [4-10]. However, simulations of nanophase ceramics involve at least multimillion atoms [7]. In addition, characteristic time for many technologically important processes such as sintering is many orders-of-magnitude longer than the atomic time scale [8,9]. The required large-scale, long-time simulations are beyond the scope of current simulation technologies. To extend the scope of MD simulations, we have developed new simulation methods including i) space-time multiresolution schemes [11], ii) hierarchical dynamics via a rigid-body/implicit-integration/normal-mode approach [12], and iii) adaptive curvilinear-coordinate load balancing for parallel computing [13].

Using large-scale molecular-dynamics (MD) simulations, we have investigated the influence of ultrafine microstructures on dynamic fracture in nanophase silicon nitride [7]. The simulations reveal significant crack branching and pinning of the crack front by microstructures. As a result, the system is able to sustain an order-of-magnitude larger strain than crystalline Si_3N_4. The morphology of the crack surfaces exhibits significant anisotropy and crossover phenomena, in agreement with recent experiments on intergranular brittle fracture [14-16].

In this paper we report i) the development of multiscale MD techniques, ii) million-atom MD simulations of dynamic fracture in nanophase Si_3N_4 on parallel computers, and iii) the roughness analysis of MD fracture surfaces.

MULTISCALE MOLECULAR DYNAMICS

The compute-intensive part of MD simulations is the calculation of interparticle interactions. Highly efficient algorithms have been designed to compute these interactions on parallel machines [11]. The long-range Coulomb interaction is calculated with a divide-and-conquer scheme, called the fast multipole method (FMM), which reduces the computational complexity from $O(N^2)$ to $O(N)$. For short-ranged interactions, we have employed a multiple time-step (MTS) approach in which a significant reduction in computation is achieved by exploiting different time scales for different force components.

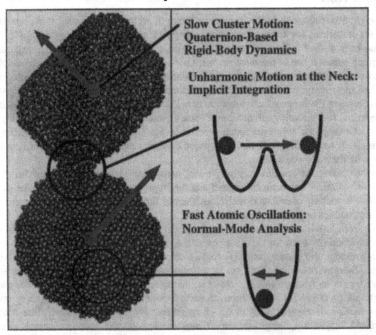

FIG. 1: Various physical processes involved in the sintering of nanoclusters: i) relative rotation of clusters is included through rigid-body dynamics; ii) unharmonic atomic motions responsible for surface diffusion are included by implicit integration of Newton's equations; iii) thermal atomic motions assist the diffusion processes, and these high-frequency modes are dealt with normal-mode analysis.

We have also developed a new algorithm for long-time MD simulations of nanocluster-assembled materials [12]. The rigid-body/implicit-integration/normal-mode (RIN) method combines: i) quaternion-based rigid-body dynamics for global conformational changes; ii) symplectic implicit integration of Newton's equations for the coalescence of clusters; and iii) normal-mode analysis of fast atomic oscillations (Fig. 1). The RIN scheme using a time step Δt of 10^{-12} sec speeds up a conventional explicit integration scheme with $\Delta t = 2 \times 10^{-15}$ sec by a factor of 28 without loss of accuracy. A parallel implementation of the scheme achieves an efficiency of 0.94 for a 1.28-million-atom nanocrystalline silicon nitride solid on 64 nodes of the IBM SP computer at Argonne National Laboratory.

Simulation of nanophase materials is characterized by irregular atomic distribution. Uniform spatial decomposition on parallel computers would result in unequal partition of workloads among processors and low efficiency. Recently, we have added a dynamic-load-balancing capability to an existing parallel MD code which was based on uniform mesh decomposition [13]. The new load-balancing scheme uses adaptive curvilinear coordinates to represent partition boundaries. Workloads are partitioned with a uniform 3-dimensional mesh in the curvilinear coordinate system (Fig. 2). Simulated annealing is used to determine the optimal coordinate system which minimizes load-imbalance and communication costs. Periodic boundary conditions are naturally incorporated, and the new scheme requires no change in the data structures of the MD program. The computational overhead to perform load balancing is only 3.7% of the total MD execution time for a 633,696-atom SiO_2 system on 32 nodes of the SP machine. Inclusion of load balancing reduces the execution time by a factor of 4.2.

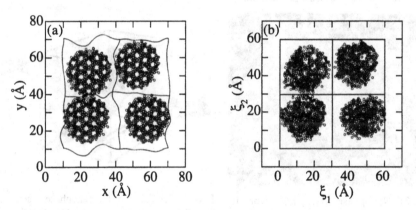

FIG. 2: Partition of 2,964 atoms into $2 \times 2 \times 1$ processors: (left) curved partition boundaries in the Euclidean space; (right) uniform mesh partition in the curvilinear space.

PARALLEL MOLECULAR DYNAMICS SIMULATIONS OF NANOPHASE Si_3N_4

We have performed MD simulations of nanophase Si_3N_4 on a parallel computer [7]. The interatomic interactions are characterized by steric repulsion, Coulomb and charge-dipole interactions, and three-body covalent interactions [17]. First 108 spherical α-Si_3N_4 clusters each containing 10,052 atoms were randomly packed in a box with periodic boundary conditions. An external pressure of 15 GPa was applied to consolidate them at 2,000K. The consolidated system was cooled to 300K, and the external pressure was reduced to zero.

To study fracture in the consolidated system, periodic boundary conditions were removed and the system was relaxed with the conjugate-gradient approach. Using the MD method the system was thermalized at room temperature and then subjected to an external strain by displacing atoms in the uppermost and lowermost layers (5.5Å thick) normal to the x direction. Initially a tensile strain of 5% (at the rate of 0.25% per picosecond) was applied. After relaxing the system for several thousand time steps with the MD approach, a notch of length 50Å was inserted in the y direction.

At this strain we observed the growth of a few crack branches in addition to slight progression of the notch in the y direction. These local branches and nanoclusters decelerated the crack front. When the motion of the crack front subsided, we increased the strain along x by displacing the same boundary-layer atoms by 1% and then we relaxed the system for 10 ps. The crack front then advanced significantly in the y direction. We also observed that interfacial regions allowed the crack front to meander and cause local crack branches to sprout in the system. At a strain of 14%, a secondary crack with several local branches merged with the primary crack. With further increase in the strain the secondary front advanced toward the initial notch and finally when the strain reaches 30% the two fronts joined and the system was completely fractured. Figure 3 shows a snapshot of the nanophase system just before it fractures. It should be noted that the critical strain (30%) at which the nanophase system fractures is enormous compared to what the crystalline Si_3N_4 system can sustain (at an applied strain of only 3% the crystal undergoes cleavage fracture).

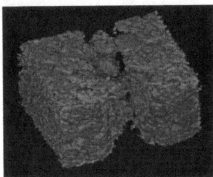

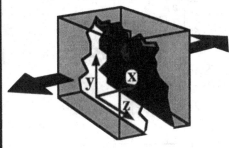

FIG. 3: Nanophase Si_3N_4 just before it fractures under an applied strain of 30%. Evidently the crack-front advances along disordered interfacial regions in the system.

FIG. 4: Schematic of the nanophase system with the applied strain in the x (the direction of the arrows) and the notch in the y direction (the crack propagates along y).

ANISOTROPIC ROUGHNESS OF FRACTURE SURFACES

There are considerable current interests in the correlation between the roughness of fracture surfaces and materials properties such as toughness [18]. It is now well established that fracture surfaces are self-affine, i.e., they remain invariant under the transformation $(x, y, z) \rightarrow (a^\zeta x, ay, az)$ where ζ is known as the roughness exponent [19]. Recent experimental findings include:

- **Universality**: Various experiments on brittle as well as ductile materials have found that for the fracture profile, $x(z)$, perpendicular to the direction of crack propagation, y, the roughness exponent has a universal value independent of material and mode of fracture, $\zeta_\perp \approx 0.8$, above a certain domain of length scales [20,21].
- **Crossover**: At smaller length scales the roughness exponent crosses over to a value close to 0.5 which corresponds to quasi-static crack propagation [14].
- **Anisotropy**: Recently Schmittbuhl *et al.* pointed out that the self-affine correlation length scales with the distance to the notch as $\xi \sim y^{1/1.2}$ [13]. As a result, the roughness exponent, $\zeta_\parallel$, for the out-of-plane fracture profile, $x(y)$, should be $\zeta_\perp/1.2$.
- **Crack-front morphology**: Measurements of in-plane fracture profile, $y(z)$, have revealed the existence of a new exponent, ζ. In Al-Li and Superα_2 Ti$_3$Al-based alloys the observed values are 0.60 ± 0.04 and 0.54 ± 0.03, respectively [12].

To investigate the nature of self-affine fracture surfaces in the nanophase Si$_3$N$_4$, we calculate the height-height correlation functions both in and out of the fracture plane y-z (see Fig. 4). Figure 5a shows the best fit to the out-of-plane height-height correlation function $g_{xx}(z)$ ($= <[x(z+z_0) - x(z_0)]^2>^{1/2}$) for the fracture profile $x(z)$ requires two roughness exponents: $\zeta_\perp = 0.84\pm0.12$ above a certain length scale (64Å) and $\zeta_\perp = 0.58\pm0.14$ otherwise. The inset shows the MD results for the other out-of-plane height-height correlation function $g_{xx}(y)$. In this case the best fit to the results gives a roughness exponent $\zeta_\parallel = 0.75\pm0.08$. The MD results for $\zeta_\perp$ and $\zeta_\parallel$ are very close to experimental values [14]. We have also determined the in-plane roughness exponent, ζ. Figure 5b shows that the best fit to the corresponding height-height correlation function gives $\zeta = 0.57\pm0.08$, which is in good agreement with experiments [14].

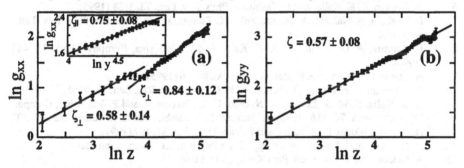

FIG. 5: (a) Log-log plot of the height-height correlation function, $g_{xx}(z)$, versus z for the out-of-plane fracture profile $x(z)$ (the inset contains the corresponding results for the out-of-plane profile $x(y)$); (b) the variation of $g_{yy}(z)$ for the in plane fracture profile $y(z)$.

SUMMARY

We have simulated dynamic fracture in nanophase Si$_3$N$_4$ ceramic using a multiscale molecular dynamics approach on parallel computers. Deflection and multiple branching of cracks at interfaces make critical strain an order-of-magnitude larger than crystalline systems. Fracture surfaces exhibit anisotropic self-affine structures and crossover phenomena, which is in good agreement with recent experiments.

ACKNOWLEDGMENTS

This work was supported by the U.S. Department of Energy, Grant No. DE-FG05-92ER45477, National Science Foundation, Grant No. DMR-9412965, Air Force Office of Scientific Research, Grant No. F 49620-94-1-0444, Army Research Office, Grant No. 36347-EL-DPS, Louisiana Education Quality Support Fund (LEQSF), Grant No. LEQSF96-99-RD-A-10, and USC-LSU Multidisciplinary University Research Initiative, Grant No. F 49620-95-1-0452. A part of these simulations were performed on the 128-node IBM SP computer at Argonne National Laboratory. The computations were also performed on parallel machines in the Concurrent Computing Laboratory for Materials Simulations (CCLMS) at Louisiana State University. The facilities in the CCLMS were acquired with the Equipment Enhancement Grants awarded by the Louisiana Board of Regents through LEQSF.

REFERENCES

1. The Federal Research and Development Program in Materials Science and Technology (National Science and Technology Council, Washington, D. C., 1995).
2. Silicon Nitride Ceramics. Scientific and Technological Advances, edited by I.-W. Chen, P. F. Becher, M. Mitomo, G. Petzow, and T.-S. Yen (Mat. Res. Soc. Symp. Proc. **287**, Pittsburgh, PA, 1993).
3. J. Karch, R. Birringer, and H. Gleiter, Nature **330**, 556 (1987); R. W. Siegel, in Physics of New Materials, edited by F. E. Fujita (Springer-Verlag, Heidelberg, 1994) p. 65.
4. P. Vashishta, R. K. Kalia, and I. Ebbsjö, Phys. Rev. Lett. **75**, 858 (1995); C.-K. Loong, P. Vashishta, R. K. Kalia, and I. Ebbsjö, Europhys. Lett. **31**, 201 (1995).
5. A. Nakano, R. K. Kalia, and P. Vashishta, Phys. Rev. Lett. **73**, 2336 (1994).
6. A. Nakano, R. K. Kalia, and P. Vashishta, Phys. Rev. Lett. **75**, 3138 (1995).
7. R. K. Kalia, A. Nakano, A. Omeltchenko, K. Tsuruta, and P. Vashishta, Phys. Rev. Lett., submitted.
8. K. Tsuruta, A. Omeltchenko, R. K. Kalia, and P. Vashishta, Europhys. Lett. **33**, 441 (1996).
9. H. Zhu and R. S. Averback, Mat. Sci. Eng. A **204**, 96 (1995).
10. P. Keblinski, S. R. Philpot, D. Wolf, and H. Gleiter, Phys. Rev. Lett. **77**, 2965.
11. R. K. Kalia, S. W. de Leeuw, A. Nakano, D. L. Greenwell, and P. Vashishta, Comput. Phys. Commun. **74**, 316 (1993); A. Nakano, P. Vashishta, and R. K. Kalia, *ibid.* **77**, 302 (1993); A. Nakano, R. K. Kalia, and P. Vashishta, *ibid.* **83**, 197 (1994).
12. A. Nakano, Int. J. Supercompter Appl. High Performance Comp., submitted.
13. A. Nakano and T. Campbell, Para. Comp., submitted.
14. P. Daguier, E. Bouchaud, and G. Lapasset, Europhys. Lett. **31**, 367 (1995)
15. J. Schmittbuhl, S. Roux, and Y. Berthaud, Europhys. Lett. **28**, 585 (1994).
16. E. Bouchaud and S. Navéos, J. Phys. I (France) **5**, 547 (1995).
17. P. Vashishta, R. K. Kalia, A. Nakano, W. Li, and I. Ebbsjö, in Amorphous Insulators and Semiconductors, edited by M. F. Thorpe and M. I. Mitkova (Kluwer, Dordrooht, 1996).
18. V. Y. Milman, N. A. Stelmashenko, and R. Blumenfeld, Prog. Mat. Sci. **38**, 425 (1994).
19. B. B. Mandelbrot, D. E. Passoja, and A. J. Palma, Nature **308**, 721 (1984).
20. E. Bouchaud, G. Lapasset, and J. Planes, Europhys. Lett. **13**, 73 (1990).
21. K. J. Måløy, A. Hansen, E. L. Hinrichsen, and S. Roux, Phys. Rev. Lett. **68**, 213 (1992).

MOLECULAR DYNAMIC COMPUTER SIMULATION OF ELASTIC AND PLASTIC BEHAVIOR OF NANOPHASE Ni

H. Van Swygenhoven, A. Caro
* Paul Scherrer Institute, Villigen, CH-5236 Switzerland, helena.vs@psi.ch
** Centro Atomico Bariloche, 8400 Bariloche, Argentina, caro@cab.cnea.edu.ar

ABSTRACT

Molecular dynamics computer simulations of low temperature elastic and plastic deformation of Ni nanophase samples with several mean grain size in the range 3-5 nm are reported. The samples are polycrystals nucleated from different seeds, with random locations and orientations. Bulk and Young modulus are calculated from stress-strain curves and the onsett of plastic deformation is discussed. At higher loads substantial difference in the plastic behaviour with respect to the coarse grain counterpart is observed: among the mechanism responsible for the deformation, grain boundary sliding and motion, as well as grain rotation are identified. An interpretation in terms of grain boundary viscosity is proposed and a linear dependence of strain rate with the inverse of the grain size is obtained.

INTRODUCTION

Many of the distinctive properties of nanophase materials come from the large number of grain boundaries compared to the coarse grained polycrystals. In particular these materials exhibit mechanical properties which are considerably different from their coarser-grained counterparts. Several mechanisms have been proposed to explain the deformation behavior based on the extensive number of triple junctions, the lack of dislocation activity, grain boundary sliding, grain rotation, the presence of a softer phase in the boundaries, etc. [1-5]. Conventional physical models for crystal plasticity have been revised to include size effects due to the presence of a large density of grain boundaries. The influence of these interfaces is crucial; while in conventional models of plasticity, interfaces represent obstacles for the deformation processes, contributing to the strengthening, in nanophase materials they are probably responsible for all of the observed plasticity.

Despite of the continuously increasing body of experimental evidences, a definite picture of elasticity and plasticity in this new class of materials is still missing, in part due to difficulties in reproducing the experimental conditions and controlling crucial parameters such as density and grain boundary structure, which translates into a significant dispersion of data [3,6]. Reported experiments on elastic and plastic behaviour concentrate mainly on determination of Young modulus, hardness measurements and creep tests in the temperature regime corresponding to Coble creep and Nabarro - Herring creep.

Large scale molecular dynamics computer simulations can help understanding the relationship between grain boundary structure and overall properties. Computer simulations of nanocrystalline materials consist of two parts. The first part is the modeling of the material itself; the second part is the simulation of the physical property of interest and the analysis of the representative character of the results in relation to experimentally observed behavior.

In the case of nanocrystalline materials, where the stochastic nature of the interface structure is relevant, this is not only a matter of the correct choice of the potential, but also of the number

193

of grains simulated and their mean grain size, i.e. of the total number of atoms included in the sample. Approaches from different perspectives have lead to different atomic modeling ranging from the aggregation and consolidation of different randomly oriented clusters [7] to a deterministic aggregation of grains where position, shape, and orientation of each grain of the system is specified [8]. Between them, samples have also been constructed by filling up an assigned volume with a polycrystal nucleating from different seeds [9] or by cooling down a liquid containing pre-oriented crystalline seeds [10]. Careful analysis of grain boundaries in computer generated nanosamples has been reported [8], interpreting their structure in terms of amorphous regions by comparison of the similar energetics involved.

This paper reports on the elastic and plastic behaviour of nanophase Ni with mean grain sizes between 3.2 and 5.4 nm. Plastic deformation is modeled in the framework of viscous plasticity of non-crystalline materials, combining the concept that nearly all plastic activity is concentrated at the grain boundaries and that these grain boundaries behave as amorphous regions.

COMPUTATIONAL

Results on computer simulations of plastic deformation of nanoscale metallic samples are reported: five samples containing approximately 100.000 atoms each, with 15, 20, 25 and 50 grains (Ni_15, ...Ni_50), which represent average grain sizes ranging from 3.2 to 5.4 nm, and a test sample containing two grains and consequently two glide systems consisting of parallel interfaces at 45 degrees of the applied stress. They are constructed by filling the simulation cell volume with nanograins nucleated from stochastically chosen seeds with random crystallographic orientations. Subsequently, those atoms in the grain boundaries closer than 2 A to each other are removed. The samples are relaxed to a minimum energy, using a parallel molecular dynamics code with a Finnis Sinclair potential in the Parinello-Raman approach, and periodic boundary conditions, as described in [11]. The final density of the samples at 0 K is in all cases above 97% of the perfect crystal value. Samples are then loaded and subsequently unloaded with an uniaxial stress between 10 MPa and 3 Gpa, the latter being ~ 2 % of the bulk modulus, below the range assumed for the theoretical strength of perfect crystals. In the lower load region the elastic properties are calculated from stress-strain curves. The highest load is applied in order to study the mechanism responsible for plastic deformation. The need of large strain rates imposed by computer running-time limitations, determine these large values of applied stresses, clearly above the experimental range. However, this stress level gives strain rates in the range 10^9 sec^{-1} which, for this sample size, is still considered as stationary creep (in the sense that no transient stress gradients develop). Simulations were done in the microcanonical ensemble, in such a way that the mechanical work was converted into both heat and defect energy, which were then easily monitored.

Despite this procedure, sample temperature in all runs was below 70 K after 10% deformation, for a starting temperature of 0 K. These stresses and temperature ranges gave us information on essentially a-thermal plastic deformation mechanisms. Some runs were continued until deformations of 50% or more, but these runs show artificial departures of the steady state induced by the periodic boundary conditions (these samples never break; deformation stops when the sample is a long rod with all interfaces perpendicular to the applied stress, i. e. a bamboo-like structure Deformation has been followed using strain-time (creep) curves, visual inspection of slabs and energy contour plots. Grain rotation has quantitatively been determined by Fourier transforming the atomic positions, and measuring the displacement of the diffraction spots on the Ewald sphere.

RESULTS

Figure 1a shows histograms of potential energy for the two extreme sizes, Ni_15 and Ni_50, in the undeformed state at 0 K. Bin width is 40 meV, which is approximately the value corresponding to potential thermal energy at room temperature. In this scale, it is clearly seen the relevance of the interfaces in terms of the number of atoms that have potential energy above the perfect crystal value, within a range of thermal energies: these numbers are 50% for Ni_15, and 70% for Ni_50, with 5.4 and 3.2 nm respectively; if we count the atoms with energy larger than the latent heat of melting, these numbers reduce to 12% and 18% respectively. This figure not only tells us about the significant perturbation to the crystalline structure that these grain boundaries introduce, but also shows that the average potential energy those defected atoms store is comparable to the latent heat of melting of this Ni model potential, 180 meV/atom. This gives strong support to the interpretation of boundaries in terms of amorphous (or liquid-like) structures. Also shown in this figure is the grain size distributions for Ni_50 and Ni_15.

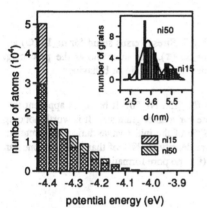

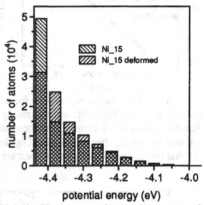

Fig. 1.a: Histograms of potential energy of each atom for Ni_15 and Ni_50 Inset: Grain size distribution of Ni_50 and Ni_15, together with a log-normal fit.

Fig. 1b: Histograms of potential energy of each atom of another sample Ni_15 before and after 10% deformation and subsequent relaxation.

Figure 2 shows the stress-strain curve for Ni_15 and Ni_50. At stresses lower then 80 MPa the deformation can be said to be elastic and the curve fits to a straight line within a standard deviation of 1%. On the plot only the linear fit for ni_15 is shown, because of the very little difference. Unloading the sample leads to the original sample size. From linear fitting of the strain versus stress in longitudinal and transverse direction, the elastic constants C_{11} and C_{12} can be calculated. For Ni_15 the values are 170 and 90 Gpa and for Ni_50, 150 and 60 GPa respectively. This means a reduction of about 30 % in Young modulus for a sample with mean grain size of 5.4 nm and a reduction of 40 % for a sample with a mean grain size of 3.2 nm relative to the value of a single crystal calculated with this potential (C_{11} :250 Gpa, C_{12} : 155 Gpa). The reduction in E-modulus is less then the measured values in consolidated samples and is in the range of the value (50%) reported in other computer simulations on Fe with mean grain size of 1nm [12]. The reduction is however larger then the predicted values from the three-compound model described in reference[6]. At higher loads the curve start to deviate from linear behaviour. The inset in fig.2 shows the deviation from linear behaviour expressed in %. The exact onset of plastic deformation is difficult to determine because of the very slowly or even maybe never reached maximum deformation. In order to reduce error, values given are all strain values loading during 30 psec, when strain rate was very slow. As can be observed, the strains reached in Ni_50 are higher then those reached in Ni_15, which means that there is more plastic deformation at the lower grain size. This indicates a lowering in yield stress or a reduced hardness when grain size is reduced from 5.4 to 3.2 nm. At 5.4 nm the yield stress is between 0.1 and 0.2 Gpa, at 3.2 nm the yield stress is below 0.1 Gpa.

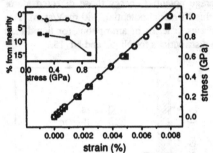

Fig.2 Stress versus strain for ni_15 (O) and ni_50 (■). The inset shows the percentage deviation from linear behaviour.

Figure 3 shows strain versus time for all samples (Ni_15.. Ni_50). It becomes apparent that they show a linear behavior, with increasing strain rate for smaller grain size. It is worth noticing that the density of all samples is reduced to about 98% of the initial value during deformation, and that relaxation after ultimate loading does recover density to 99% of that value, so one can say that deformation takes place at constant volume (i. e. no pore formation).

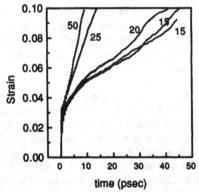

Fig. 3 strain versus time for two Ni_15 samples and Ni_20, Ni_25 and Ni_50.

An evaluation of the energy distribution after 10% deformation (fig.2b) followed by a short unloading, reveals that deformation takes place increasing slightly the total number of defected atoms; 13% more atoms join the grain boundary region, but less then 1% of them end in an amorphous-like position. The strain field induced by the deformation extends from the boundary region into the crystallites. Relaxation during longer times would certainly reduce the number of atoms joining the grain boundary.

All samples continue to be nanocrystalline after deformation. Some grains have been followed individually during deformation revealing the presence of several accommodation mechanisms such as grain boundary sliding, grain rotation and grain boundary motion, as is demonstrated in (13,14). As far as we have been able to detect by visual inspection of slabs and grain boundaries, and by using energy contour plots, we have not seen any evidence of dislocation activity.

Creep experiments are usually performed at high temperatures, where thermally activated migration of point defects at the interfaces (Coble creep) or at the grains (Nabarro - Herring creep) are the main source of plasticity. In our simulations at low temperature diffusion is excluded; however we still observe that the major contribution comes from grain boundary a-thermal sliding, a viscous-like behavior, with additional contributions from rotation and grain boundary motion.

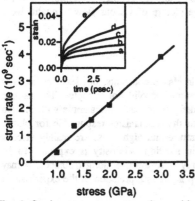

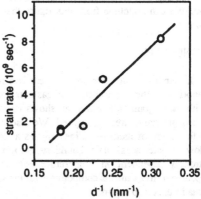

Fig. 4: Strain rate versus stress for an ideal planar interface with random missorientation, showing linear viscosity. Data is obtained from linear fits to strain-time curves shown in inset; labels a-e indicte increasing stress from 1.0 to 3.0 GPa.

Fig. 5: Strain rate versus average grain diameter, together with a linear fit according to equation (1).

For a better understanding of this viscous behavior, we deformed a sample with a particularly simple geometry, namely, two grains. Because of periodic boundary conditions these grains are infinite rods along the z axis, with square cross section in the x-y plane, which produce a regular tiling with grain boundaries at 45 degrees of the tensile x-axis. This design presents two possible glide systems, at +45 degrees and -45 degrees with respect to the direction of the applied stress. However, because of the random nature of the misfit orientation, and the constrains imposed by the geometry, these two systems are not equivalent and therefore only one of them will be activated. For the very early stages of deformation, this configuration allows us to directly measure the relation between stress and strain, keeping constant the interface geometry. The results are presented in Figure 4; in the inset we report ε vs. time at five different values of applied stress, between 1 and 3 GPa; while in the main section of the figure a linear fit to strain rates indicates that the salient feature of this selected planar interface is linear viscosity with a coefficient b = 1.6 Pa sec. Having characterized one interface, we can simply estimate the behavior of samples with different amount of grain boundaries by making the two following assumptions: i- a 3-dimensional array of interfaces is self similar as a function of the scale d and therefore the fraction of interfaces that are activated to contribute to plastic deformation is the same at all scales; and ii- increasing the interface surface increases the slip channels that contribute to total strain in the same way as parallel dashpots would do in a mechanical model of visco-elasto-plastic solid. By further noting that in a sample with grain size d, the total surface per unit volume is proportional to d-1, we can write the following relation for the strain rate as a function of grain size in the stationary state:

$$de / dt = (a / d \, b) \, s \qquad (1)$$

where a is a constant such that a/d characterize the sample in terms of the ideal planar interface used to determine b. Figure 5 shows the results for strain rate vs. d-1 obtained from Fig. 3, together with a linear fit whose slope, which accounts for geometric factors, is a = 1.1 nm. We can conclude from this figure that the simple picture of viscous flow is qualitatively correct.

CONCLUSIONS

In summary, we have shown there is a reduction in elastic constants, but less drastic as is reported in the experiments on gas-condensation/vacuum consolidation samples. The sample with mean grain size of 5.4 nm shows larger plastic deformation then the sample with 3.2 nm when the same stress is applied. We also showed that the mechanisms responsible for plastic deformation of nanocrystalline samples at low temperature and high stress are mainly viscous flow of interfaces, grain boundary motion, and grain rotation. Viscosity is expected to be strongly dependent on temperature and does not imply diffusion over long distances; it may become a mechanisms competing with the Coble creep in the interpretation of the experimental data in the creep regime.

REFERENCES

1. J.R. Weertman, Materials Science and Engineering, A166, p 161, (1993).
2. V.Y. Gertsman, M. Hoffman, H. Gleiter, R. Birringer, Acta Metall. Mater. 42, p 3539, (1994).
3. R.W.Siegel, G.E. Fougere, Nanostruct. Mater., 6,p 205, (1995).
4. Papers in "Superplasticiy in advanced Materials", ICSAM-94, Materials Science Forum,170-172 (1994).
5. M. Ke, S.A. Hackney, W.W. Milligan, E.C. Aifantis, Nanostruct. Mater. 5,p 689, (1995).
6. T.D.Shen, C.C. Koch, T.Y. Tsui, G.M. Pharr, J. Mater. Res., 10(11), p 2892, (1995)
7. M. Celino, G. D´Agostino, V. Rosato, Nanostuct. Mater., 6,p 751, (1995).
8. D. Wolf, J. Wang, S.R. Phillpot, H. Gleiter, Phys. Rev. Letters, ,p 4686, (1995).
9. Da Chen, Computat. Mater. Science, 3, p 327, (1995).
10. S.R. Phillpot, D. Wolf, H. Gleiter, J. Appl. Phys. 78, p 847, (1995).
11. D'Agostino, H. Van Swygenhoven, to appear in Proc. MRS Fall Meeting, 1995
12. Da Chen, Mater,. Sci. and Eng., A190, p 193, (1995)
13. H. Van Swygenhoven. A. Caro, proceedings of NANO'96, Kona, Hawai, July 2, 1996.
14. H. Van Swygenhoven, A. Caro, submitted .

THREE DIMENSIONAL SIMULATION OF THE MICROSTRUCTURE DEVELOPMENT IN Ni-20%Fe NANOCRYSTALLINE DEPOSITS

HUALONG LI, F. CZERWINSKI and J.A. SZPUNAR
Department of Metallurgical Engineering, McGill University, Montreal, PQ, Canada, H3A 2A7

ABSTRACT

A Monte-Carlo computer model was applied to simulate a development of the three dimensional microstructure during electrodeposition of nanocrystalline alloys. The driving force for this process was the minimization of free energy of the system. For a particular deposit of Ni-20%Fe, the influence of the overpotential and current density on the grain size was tested. A strong decrease in grain size with increasing overpotential and current density obtained from the simulation is in qualitative agreement with the experimental data.

INTRODUCTION

Nanocrystalline solids are a new class of materials which have recently attracted an increasing engineering interest. A large number of technologies are involved in their manufacturing. A relatively effective way of obtaining nanocrystalline solids is the electrodeposition from aqueous solutions [1-3]. It is obvious that the properties of nanocrystalline materials depend on their microstructure and texture. Hence, the computer simulation of the influence of deposition parameters on the formation of nanocrystalline deposits may contribute to developing the new materials with the microstructure precisely designed for some particular applications. For example, designing the specific character of grain boundaries between nanocrystalline grains is known to be critical for controlling the diffusion processes inside the material [1].

The previous simulations, based on the energy minimization assumption, [4,5] were effective in qualitatively predicting the microstructure and texture of pure iron deposits. In this work, that model is improved and applied for predicting the microstructure development in nanocrystalline nickel-iron alloys.

DESCRIPTION OF THE SIMULATION

During electrodeposition, metal atoms are ionized at the anode; then the ions are hydrated and move to the cathode through the electrolyte. When reaching the cathode, the ions are dehydrated, discharged and then adhered to the metal substrate. Some of the deposited atoms can join each other to form small clusters, and some of the clusters will, after reaching a stable size, become nuclei with a new orientation. Other deposited atoms can diffuse on the surface in order to find a minimum energy position. Once reaching that position, they have a tendency to stay there. The driving force for the electrodeposition is the overpotential, which is defined as a potential difference between the cathode potential and the equilibrium value.

The flow chart of the computer program used to simulate the electrodeposition process is shown in Fig. 1. The essential parts of this program are described below.

199

Mat. Res. Soc. Symp. Proc. Vol. 457 ⁰1997 Materials Research Society

Initialization

An 80×80×80 array was defined in the program which represents a three dimensional structure with a 80 sites length, a 80 site width and 80 layers height. At the beginning, all the sites were empty.

Deposition stage

In this subroutine, n blocks were deposited one by one into the defined structure and n sites in the structure were occupied from the bottom to the top, as shown in the Fig. 2. The number n was proportional to the current density I. The block was defined as a group of metal atoms whose position and orientation are described by three coordinates (x,y,z) and three Euler angles (φ_1, ϕ, φ_2). When a block was deposited into the structure, its position parameters x, y were randomly assigned between 1 and 80 and z was increased layer by layer. The orientation of this block was assigned according to a random number p' and

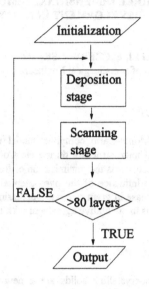

Fig. 1 Flow chart of computer programm simulating microstructure development

the nucleation probability (p_{nuc}). If $p' < p_{nuc}$, the block was labeled as a nucleus and the orientation (φ_1, ϕ, φ_2) was determined randomly. Otherwise, it was labeled as an ad-block and assigned with the same orientation as one of its neighbours.

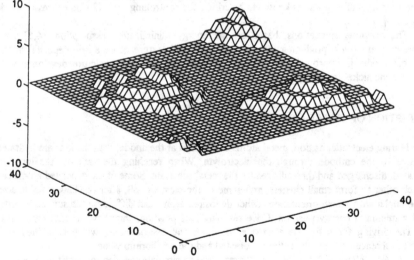

Fig. 2 Simulated initial stage in the electrodeposition

<u>Scanning stage</u>

At this stage, the energy minimizing process was performed for 80x80 times by the surface diffusion and rotation of the block. The block was located at the top layer with a random x and y from the following ranges: $1 \leq x \leq 80$ and $1 \leq y \leq 80$. Then, the selected block was permitted to change either its orientation or its position with equal possibility. This change was determined by the following probability equation:

$$p = \exp(-\Delta E_i / kT) \tag{1}$$

where k is Bolzman's constant, T is a temperature in Kelvin, and ΔE_i is the energy difference before and after the rotation or surface diffusion of the block. The ΔE_i was calculated according to the equation:

$$\Delta E_i = (\sum_{j=1}^{6} E_{ij})_{after} - (\sum_{j=1}^{6} E_{ij})_{before} \tag{2}$$

where E_{ij} is the energy of the grain boundary between block i and its neighbour j. If the neighbour position was empty, the E_{ij} was replaced with the surface energy, $E_{surface}$, of block i. The deposition- and scanning stage formed a main loop in the program, which was repeated until the desired layer was reached.

<u>Output</u>

The microstructure was derived from the orientation and position of each block. The blocks with the same orientations were labeled as one grain. The grain boundary was obtained by drawing a line between two blocks with different orientations. Moreover, through statistical analysis of the orientation of each block, the orientation distribution function for polycrystalline material can be derived. However, this is not analyzed in this paper.

CALCULATION OF THE INPUT PARAMETERS (n, p, E_{gb}, E_{sur})

The number of blocks n deposited on the cathode during certain period of time is proportional to the current density. However, the exact form of this dependence is unknown. Therefore, this simulation offers, so far, qualitative results only. During simulation, the nucleation probability (p_{nuc}) was calculated using the following equation:

$$p_{nuc} = \exp(-\Delta G / kT) \tag{3}$$

where ΔG is the activation energy of nucleation. Assuming a semi-sphere nucleus shape, the ΔG is expressed as [6]

$$\Delta G = \frac{4H^3 \sigma^3}{27 z^2 e^2 \eta^2} \tag{4}$$

$$H = (4 \pi r^2) * 2^{-1/3} \tag{5}$$

Table I. Anisotropic Surface Energies E_{sur} (eV/atom)

	(111)	(100)	(110)	(311)	(210)	(211)	(221)	Ran.
Ni-20%Fe	1.34	1.52	1.82	1.90	2.09	1.95	1.95	1.97

where z is the number of electrons transferred per atom or molecule, e is the electron charge, η is the overpotential, σ is the surface free energy and r is the radius of the atom or molecule.

The value for σ as 1000 mJ/m² was estimated for Ni-20%Fe from the anisotropic surface energies listed in Table I. The activation energy of nucleation ΔG as a function of overpotential η was calculated using Eq. 4 and is shown in Fig.3.

The anisotropic surface energies and grain boundary energies were calculated with a L-J potential at the atomistic level [7] and are shown in Table I and Fig. 4. In order to simplify the simulation procedure, the average data from <111>, <110> and <100> tilt grain boundary energy were used and those are represented by the solid line in Fig. 4.

RESULTS AND DISCUSSION

The input parameters used in the simulation are specified in Table II. The microstructure obtained from simulation for five values of overpotential, 400 500, 600, 700 and 800 mV, are shown in Fig. 5a-e. The grain size distribution as a function of grain radius, which was derived from Fig.5e, is shown in Fig. 5f. It is seen, that the grain size decreases rapidly as the overpotential increases. During electrodeposition at lower overpotentials, columnar grains are formed (Fig. 5a). Conversely, at higher overpotentials, smaller and equi-axis grains are formed. Both the overpotential and current density were found to contribute to the microstructure development.

An analysis of Equation 4 indicates that increase in overpotential results in a decrease in activation energy. Hence, the nucleation probability increases. During simulation, nucleation takes place at the deposition stage. A block at this stage has the possibility of becoming a nucleus or ad-block. Increasing the nucleation probability results in a higher chance for a block to become a nucleus with a new orientation rather than remain an ad-block. Therefore, the resultant grain growth process is depressed. Thus, for relatively high values of the overpotential, the grain growth along the deposition direction can be effectively restricted and smaller equi-axis grains in the microstructure should be expected.

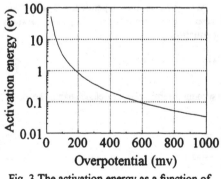

Fig. 3 The activation energy as a function of overpotential for Ni-20%Fe

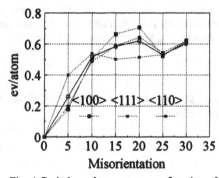

Fig. 4 Grain boundary energy as a function of misorientation

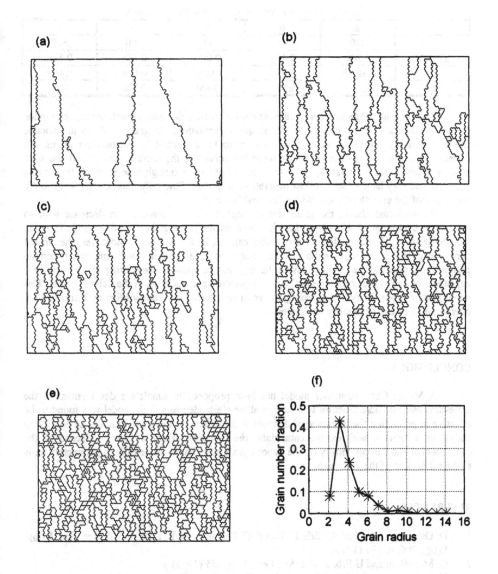

Fig. 5 Simulated evolution of microstructure for different overpotentials. (a) η=400mv;
(b) η=500mv; (c) η=600mv; (d) η=700mv; (e) η=800mv; and (f) the grain size
distribution for microstructure shown in Fig. 5 (e)

Table II. The input parameters used in the simulation

	a	b	c	d	e
η (mv)	400	500	600	700	800
ΔG (eV)	0.208	0.133	0.092	0.068	0.520
p	0.0008	0.011	0.04	0.098	0.169
n	2000	2500	3000	3500	4000

The scanning stage for energy minimization is, in fact, the grain growth process controlled by surface diffusion and rotation. At this stage of simulation, the grains with an unfavorable energy status are unstable and have a tendency to be consumed by surrounding grains. An increase in overpotential shortened this stage by increasing the current density and the total number of blocks n. As a result, some deposits do not have enough time to reach their stable positions and they have to stay in their unstable energy state. Physically, this process leads to the restriction of the growth of grains with a favourable energy.

As discussed above, the grain size of deposit shows a tendency to decrease with an increase in overpotential. However, there is a limit value for the grain size of a specific material. According to Eq. 3, the nucleation probability can not be larger than 1. Moreover, the current density cannot be increased without limits. For specific deposition conditions, the maximum current density, called the limiting current density, may be applied. The limiting current density is dependent on ion's diffusion properties, bath temperature and thickness of the diffusion layer. This implies that the possibility of decreasing the grain size during electrodeposition by increasing the overpotential or current density are limited.

CONCLUSIONS

A Monte Carlo computer model has been proposed to simulate a development of the three-dimensional microstructure in nanocrystalline electrodeposits. This model was found to be effective in predicting the influence of deposition conditions on the microstructure of Ni-20%Fe alloys. In agreement with experimental data, the modeling shows a strong influence of the overpotential and current density on the deposit grain size. The increase of both parameters leads to the refinement of the grain size.

REFERENCES

1. D. Osmola, E. Remaud. U. Erb, L. Wong, G. Palumbo and K.T. Aust, Mat. Res. Soc. Symp. Proc., 286, p. 191 (1993)
2. G. McMahon and U.Erb, J. Mat. Sci. Lett., 8, p. 865 (1993)
3. G. McMahon and U. Erb, Microstr. Sci., 17, p. 447 (1989)
4. D.Y. Li and J.A. Szpunar, Electroch. Acta, 42, 37, 47 (1989).
5. D. Y. Li and J. A. Szpunar, Mat. Sci. Forum, 157-162, p. 1827-1838 (1994)
6. Fleischmann and H.R.Thirsk, Advances in Electrochemistry and Electrochemical Engineering, 3, p. 123 (1963)
7. Hualong Li, F. Czerwinski and J.A. Szpunar, Nanostructured Materials, (in press)

STRUCTURE, MECHANICAL PROPERTIES, AND DYNAMIC FRACTURE IN NANOPHASE SILICON NITRIDE VIA PARALLEL MOLECULAR DYNAMICS

Kenji Tsuruta, Andrey Omeltchenko, Aiichiro Nakano, Rajiv K. Kalia, and Priya Vashishta

Concurrent Computing Laboratory for Materials Simulations
Department of Physics & Astronomy
Department of Computer Science
Louisiana State University, Baton Rouge, LA 70803-4001

E-mail: kenji@rouge.phys.lsu.edu

http://www.cclms.lsu.edu

ABSTRACT

Million-atom molecular-dynamics (MD) simulations are performed to study the structure, mechanical properties, and dynamic fracture in nanophase Si_3N_4. We find that intercluster regions are highly disordered: 50% of Si atoms in intercluster regions are three-fold coordinated. Elastic moduli of nanophase Si_3N_4 as a function of grain size and porosity are well described by a multiphase model for heterogeneous materials. The study of fracture in the nanophase Si_3N_4 reveals that the system can sustain an order-of-magnitude larger external load than crystalline Si_3N_4. This is due to branching and pinning of the crack front by nanoscale microstructures.

INTRODUCTION

Nanophase ceramics are rapidly gaining an edge on conventional ceramics for high-temperature applications. The former have larger fracture toughness, higher strength and sinterability than conventional ceramics [1, 2]. Various experimental observations indicate that these properties of nanophase ceramics are due to unusual microstructures in intergranular (intercluster) regions. A quantitative understanding of the intercluster structure and its influence on mechanical properties are not yet understood. Computer-simulation approach is crucial to obtaining quantitative information on the intercluster structure [3, 4]. Until recently, an atomistic simulation of nanophase materials was a difficult task because such a simulation requires a large number of atoms ($\sim 10^6$). Recent advances in parallel computers and algorithms have made it possible to carry out multimillion atom MD simulations for real materials [5].

In this paper we present the results of large scale (up to 1,085,616 atoms) molecular-dynamics (MD) simulations of nanophase silicon nitride (Si_3N_4). We have investigated the structure of intercluster regions, the effect of consolidation on mechanical properties, and fracture in the consolidated nanophase Si_3N_4.

The MD calculations for Si_3N_4 are based on an effective interatomic potential consisting of two-body and three-body terms [6]. The two-body terms account for: charge-transfer effects through screened Coulomb potentials; charge-dipole interaction due to the large electronic polarizability of nitrogen; and steric repulsion between atoms. Three-body bond-bending and bond-stretching terms take into account covalent effects. This interatomic potential for Si_3N_4 has been validated by comparing MD results with various experimental measurements: The MD results for bond-length and bond-angle distributions for both α-crystal and amorphous Si_3N_4, and the static structure factor of amorphous Si_3N_4 are in excellent agreement with x-ray and neutron scattering measurements [6, 7]. The MD calculations of elastic constants, phonon density-of-states, and specific heat of crystalline α-Si_3N_4 also agree well with experiments [6-8].

Parallel MD simulations reported in this paper were performed with the domain decomposition scheme [9]. The physical system was divided into subsystems of equal volume, and each subsystem was assigned to a processor. The atomic trajectories were followed by integrating the equation of motion for each atoms with the velocity-Verlet algorithm [10] using a time step of 2 femto seconds. The reversible multiple time-scale algorithm [11] was used in the MD calculations. The simulations were carried out on the IBM SP computer at Argonne National Laboratory and on Digital's 40-node Alpha system with two Gigaswitches in our Concurrent Computing Laboratory for Materials Simulations.

CONSOLIDATION OF NANOPHASE Si3N4

Nanophase Si_3N_4 system with 1,085,616 atoms was consolidated with the variable-shape MD approach [10]. Initially, 108 clusters of diameter 60Å were positioned randomly in a cubic box of length 288.49 Å. (Each cluster contained 10,052 atoms.) The mass density of the starting system was 1.0 g/cc. We first thermalized the system at zero pressure and 2,000K. Subsequently an external pressure 1 GPa was applied and the system was thermalized at 2,000K. We then increased the pressure to 5 GPa and the system was again well thermalized at 2,000K. Repeating this procedure, we obtained well-thermalized systems at 1, 5 and 15 GPa at 2,000K. All of these systems were subsequently cooled to 1500, 1000, 700 and 300K. At each temperature and pressure, we thermalized the system for several thousand time steps. After cooling the system to room temperature, we gradually reduced the external pressure. The consolidated nanophase systems with mass densities 2.24, 2.67, and 2.94 g/cc (corresponding to applied pressures of 1, 5, and 15 GPa) were thereby obtained.

Figure 1 depicts snapshots of the nanophase Si3N4 before consolidation (mass density 1.0 g/cc) and after consolidation under an external pressure of 15 GPa (the corresponding mass density is 2.94 g/cc). The density of the well-consolidated system is 92% of the density of the crystalline Si3N4 system. The consolidated system contains a few small pores in intercluster regions.

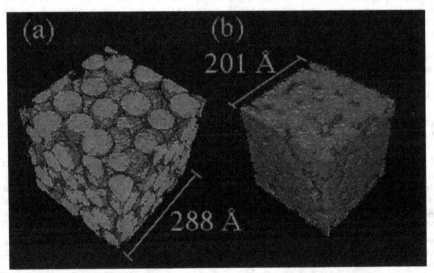

Fig. 1 Nanophase Si3N4 with 1.08 million atoms (a) before consolidation (1.0 g/cc) and (b) after consolidation (2.94g/cc) .

STRUCTURAL ANALYSIS

We have calculated various correlation functions to determine the structure of the consolidated nanophase Si_3N_4 at mass density 2.94 g/cc. In Fig. 2 (a), we depict partial (Si-N) pair-distribution functions (PDF) for atoms inside the nanoclusters (dashed curve) and in intercluster regions (solid curve) at 5 K. The sharp peaks in the PDF for atoms inside the clusters indicate that the interior regions of these clusters remain crystalline. On the other hand, the PDF for intercluster regions has much broader peaks (except the first one) than those inside the nanoclusters. The inset in Fig. 2(a) shows that the height of the first peak in the PDF for intercluster regions is only one fourth of the height of the first peak for interior regions. It should also be noted that the position of the first peak for intercluser regions is shifted to a lower value relative to that for the interior region of nanoclusters. This shift implies a decrease of the nearest-neighbor coordination for Si atoms in intercluster regions. Lower Si coordination in intercluster regions is also evident in Fig. 3. The figure shows that the average Si coordination in intercluster regions is approximately 3.5 which means that there are 50% three-fold coordinated Si atoms while the remaining Si atoms are four-fold coordinated. (In crystalline Si3N4 the coordination of Si is four [12].) The figure also shows that there are a few pores (black regions) inside the system. We have also examined N-Si-N and Si-N-Si bond-angle distributions (BAD) inside the nanoclusters and in intercluster regions. We find that the BADs for intercluster regions are much broader than those corresponding to the interior region of nanoclusters. This also indicates that the local structure in the intercluster regions is highly distorted [13].

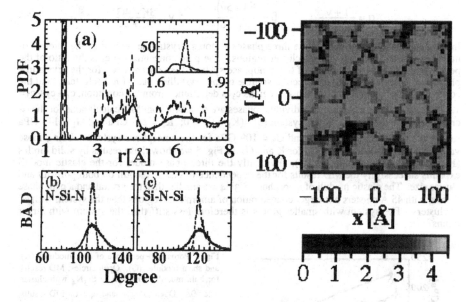

Fig. 2 (a) Si-N pair-distribution functions (PDF); (b) N-Si-N and (c) Si-N-Si bond-angle distributions (BAD). (Dashed and solid curves correspond to the interior regions of nanoclusters and the intercluster regions of the nanophase system, respectively.) The inset in (a) shows the first peaks in the PDF.

Fig. 3 Spatial distribution of the average Si coordination, projected onto the x-y plane, in the nanophase system at 2.94 g/cc.

MECHANICAL PROPERTIES

We have also investigated mechanical properties of nanophase Si3N4 using the MD method. In addition to the systems with cluster size 60Å we have simulated nanophase systems with smaller cluster size (diameter=45Å). Those systems were prepared following the procedure described in the previous section. The mass density of those systems are 2.28, 2.64, 2.84, and 2.95 g/cc (corresponding to external pressures of 1, 5, 10, and 15 GPa).

Figure 4 shows the porosity dependence of the bulk modulus, K, and the shear modulus, G. The open circles are the MD results for the bulk modulus of the three nanophase systems with 60Å clusters. The open triangles and squares are the bulk and shear moduli, respectively, of systems with 45Å clusters. The figure shows the strong dependence of elastic moduli on both porosity and cluster size. These results of elastic moduli can be understood in terms of a multicomponent model for heterogeneous materials [14]. In this model, the elastic moduli of a heterogeneous material are given by,

$$\sum_{i=1}^{n} \frac{c_i}{1 - \alpha\left(1 - K_i/K\right)} = \sum_{i=1}^{n} \frac{c_i}{1 - \beta\left(1 - G_i/G\right)} = 1 \; , \tag{1}$$

where n is the number of phases. c_i, K_i, and G_i denote the concentration, bulk modulus, and shear modulus for the ith phase, respectively. The quantities, α and β, and the Poisson's ratio v are calculated from the relations:

$$\alpha = \frac{1 + v}{3(1 - v)} \; ; \qquad \beta = \frac{2(4 - 5v)}{15(1 - v)} \; ; \qquad v = \frac{3K - 4G}{6K + 2G} \; . \tag{2}$$

In the nanophase systems, there are three phases -- pores, crystalline regions in the interior of nanoclusters, and amorphous intercluster regions. The pore concentration c_1 is the ratio of the pore volume to the total volume of the nanophase system; the concentration for the crystalline phase c_2 is calculated from the effective volume of the crystalline part of nanoclusters; and the concentration of amorphous intercluster regions c_3 is determined from the condition, $c_1 + c_2 + c_3 = 1$. The bulk and shear moduli of individual phases are obtained from MD calculations for the α-crystal and the amorphous Si3N4 system [8]. ($K_1 = G_1 = 0$; $K_2 = 289$ GPa and $G_2 = 145$ GPa for the α-crystal; $K_3 = 181$ GPa and $G_3 = 109$ GPa for the amorphous system.) Using those values we solved Eqs. (1) and (2) for K and G. In Fig. 5 we show these results by solid circles (K), triangles (K) and squares (G). Evidently the three-phase model for the elastic moduli explains successfully the MD results for the dependence of elastic moduli on both porosity and cluster size. The elastic moduli of amorphous Si3N4 are smaller than the crystal and a nanophase system with 45 Å clusters has larger concentration of amorphous regions than the system with 60 Å clusters. The system with smaller grains is therefore less stiff than the system with larger grains.

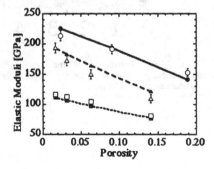

Fig. 4 Porosity dependence of bulk modulus (K) and shear modulus (G). Open circles: MD results for bulk moduli of nanophase Si3N4 with cluster size 60Å. Open triangles and squares: MD results for bulk and shear moduli of nanophase systems with 45Å clusters. Solid circles and triangles denote bulk moduli and solid squares represent shear moduli calculated from the three-phase model. The solid, dashed, and dotted lines are drawn to guide the eye.

DYNAMIC FRACTURE

We have also investigated fracture in the consolidated nanophase system. We first removed periodic boundary conditions, and then relaxed the system with the conjugate-gradient approach. The system was thermalized at room temperature using the MD method. We then applied an external strain to the system by displacing the position of atoms in the top and bottom layers (5.5Å thick) normal to the x direction. First, a tensile strain of 5% (at the rate of 0.25% per picosecond) was applied and the system was relaxed for several thousand time steps with the MD approach. We then inserted a notch of length 25Å in the y direction. The external strain was increased gradually by 1%. Subsequently the system was relaxed for 10 ps. Following this procedure, we stretched the system until it fractured.

Figures 5(a) and 5(b) show snapshots of the crack front at 5% and 14%, respectively. The crack front shown in the figure was identified by finding empty voxels of size 4Å connected to the initial notch. As the notch progresses in the y direction, we observe several microbranches sprouting off the crack front. Figure 5(b) shows a snapshot of the crack front in the nanophase system under 14% strain. Evidently the crack front has advanced significantly in the y direction. We also observe that the crack front meanders along the intercluster regions and a secondary crack (top right hand corner of the figure) with several local branches merges with the primary crack. The secondary front advances toward the initial notch as the strain is increased. When the strain reaches 30%, the system fractures completely. To show contrast, we display in Fig. 5(c) the fracture surface in crystalline Si_3N_4 at 3% strain. (This system also contains 1.08 million atoms.) The figure shows atomically sharp cleavage-like crack surface in the crystalline system. From Fig. 5(b) with Fig. 5(c), it is evident that multiple branching and crack-front pinning by nanoscale microstructures allow the nanophase system to sustain much larger external strain then the crystalline Si_3N_4 solid [15].

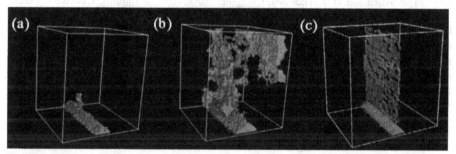

Fig. 5 Snapshots of fractured Si_3N_4: (a) The consolidated nanophase Si_3N_4 under an applied strain of 5%; (b) the nanophase system under an applied strain of 14% strain; and (c) the crystalline Si_3N_4 under an applied strain of 4%.

CONCLUSION

In conclusion, large-scale MD simulations of nanophase Si_3N_4 reveal: i) intercluster regions are amorphous and have 50% undercoordinated Si atoms; ii) the dependence of elastic moduli on porosity and grain size can be understood in terms of a three-phase model for heterogeneous materials; and iii) the critical strain at which the nanophase system fractures is much larger than that for the crystalline Si_3N_4 system.

ACKNOWLEDGMENT

This work was supported by the U.S. Department of Energy, Grant No. DE-FG05-92ER45477, National Science Foundation, Grant No. DMR-9412965, Air Force Office of Scientific Research, Grant No. F 49620-94-1-0444, Army Research Office, Grant No. 36347-

EL-DPS, Louisiana Education Quality Support Fund (LEQSF), Grant No. LEQSF96-99-RD-A-10, and USC-LSU Multidisciplinary University Research Initiative, Grant No. F 49620-95-1-0452. A part of these simulations were performed on the 128-node IBM SP computer at Argonne National Laboratory. The computations were also performed on parallel machines in the Concurrent Computing Laboratory for Materials Simulations (CCLMS) at Louisiana State University. The facilities in the CCLMS were acquired with the Equipment Enhancement Grants awarded by the Louisiana Board of Regents through LEQSF. Discussions with Dr. R. W. Siegel are gratefully acknowledged.

REFERENCES

1. S. Komarneni, J. C. Parker, and G. J. Thomas, Nanophase and Nanocomposite Materials (Mater. Res. Soc. Symp. Proc. 286, Pittsburgh, PA, 1993).

2. R. W. Siegel in Materials Interfaces: Atomic-Level Structure and Properties, edited by D. Wolf and S. Yip, Chapman and Hall, London, 1992, p. 431.

3. H. Zhu and R. S. Averback, Mater. Sci. Eng. **A204**, p. 96 (1995).

4. J. Wang, D. Wolf, S. R. Philpot, and H. Gleiter, Phil. Mag. **73**, p. 517 (1996).

5. P. Vashishta et al., Comput. Mater. Sci. **2**, p. 180 (1994).

6. P. Vashishta, R. K. Kalia, and I. Ebbsjö, Phys. Rev. Lett. **75**, p. 858 (1995).

7. C.-K. Loong, P. Vashishta, R. K. Kalia, and I. Ebbsjö, Europhys. Lett. **31**, p. 201 (1995).

8. A. Nakano, R. K. Kalia, and P. Vashishta, Phys. Rev. Lett. **75**, p. 3138 (1995).

9. A. Nakano, R. K. Kalia, and P. Vashishta, Comput. Phys. Commun. **83**, p. 197 (1994).

10. M. P. Allen and D. J. Tildesley, Computer Simulation of Liquids, Clarendon Press, Oxford, 1987.

11. M. Tuckerman and B. J. Berne, J. Chem. Phys. **97**, p. 1990 (1992).

12. L. Cartz and J. D. Jorgensen, J. App. Phys. **52**, p. 236 (1981).

13. P. Keblinski, S. R. Philpot, D. Wolf, and H. Gleiter, Phys. Rrev. Lett. **77**, p. 2965 (1996).

14. B. Budiansky, J. Mech. Phys. Solids **13**, p. 223 (1965).

15. M. Marder and J. Fineberg, Physics Today **49**, p. 24 (1996).

Part IV

Magnetic and Metal Nanocomposites

FERROMAGNETIC NANOCOMPOSITE FILMS FROM THERMALLY LABILE NITRIDE PRECURSORS

L. MAYA*,M. PARANTHAMAN*, J.R. THOMPSON*, T. THUNDAT*, and R.J. STEVENSON**
* Oak Ridge National Laboratory, P.O. B. 2008, Oak Ridge TN 37831.
** K-25 Plant, P.O.B. 2008, Oak Ridge TN 37831.

ABSTRACT

A series of nanocomposite films containing nickel or cobalt nitride dispersed in a ceramic matrix of aluminum nitride, boron nitride or silicon nitride, were prepared by reactive sputtering of selected alloys or compounds such as nickel aluminide or cobalt silicide. Thermal treatment of the nitride composites in vacuum at $\leq 500\ ^0C$ leads to selective loss of nitrogen from CoN or Ni_3N to generate dispersions of the metal in the ceramic matrix. This treatment may be performed in a localized manner by means of a focused laser beam to generate microscopic features that are imaged by magnetic force microscopy. The films are potentially useful for data storage with superior chemical and mechanical stability provided by the ceramic matrix and high encoding density made possible because of the size of the magnetic particles of less than 10 nm generated in the thermal treatment. The films were characterized by chemical and physical means including FTIR, TEM, MFM and magnetic measurements. Preliminary results on similar iron composites are also described.

INTRODUCTION

Nanocomposite films, described initially as granular solids, have been the subject of many studies since the pioneering work of Abeles et al[1] who studied transport properties of very small gold particles in silica as a function of metal loading. Interest in this area of research is driven by theoretical and practical considerations since these materials may show optical, electrical, magnetic or mechanical properties that depart from those shown by conventional solids containing particles with dimensions ≥ 50 nm. A promising area for development is the use of ferromagnetic nanocomposites as data recording media. The properties of such media may be manipulated by controlling the size and volume fraction of the ferromagnetic particles in the matrix. An additional advantage inherent in the configuration of nanocomposite films is that the matrix provides, if properly chosen, chemical and mechanical protection for the metal particles. A recent review by Chien[2] gives a description of the principles controlling the complex relationship of particle size and temperature on the magnetic characteristics of ferromagnetic nanocomposites as well as a phenomenological account of a variety of systems. Nanocomposites, in general, are conveniently prepared by sputtering of multiple homogeneous targets as done by Abeles et al[1] or through reactive sputtering of alloys and compounds[3]. Alternate approaches are sol-gel techniques[4] and ion implantation[5].

EXPERIMENT

Film deposition was conducted in a parallel plate glow discharge apparatus previously described[6]. In a typical experiment, a target such as nickel aluminide was subjected to reactive sputtering for a period of about 20 hours in a nitrogen plasma generated by a potential drop of

213

500 V and a current density of 1.8 mA/cm^2. Films, a few micron thick, were deposited on a variety of substrates by passing 10 ml/min of nitrogen while maintaining a pressure of 2 torr in the system. Characterization was performed with a variety of physical and chemical techniques including temperature programmed thermal decomposition (TPTD), conducted under vacuum using a 10 ^{0}C/min ramp with a mass analyzer on line to derive a profile of volatile evolution as a function of temperature. Also performed were direct chemical analyses, XRD, TEM, and FTIR. Magnetic characterization was conducted with a superconducting quantum interference device (SQUID) magnetometer in the temperature range of 5-300 K using a field of up to 65 kOe; in addition to this, microscopic imaging was also obtained by magnetic force microscopy (MFM).

RESULTS AND DISCUSSION

Thermal properties

Central to the development of the nanocomposite systems examined in this study are the thermal properties of the ferromagnetic metal nitrides utilized as precursors. Their relatively limited stability allows selective decomposition at moderate temperatures, ~ 600 ^{0}C, while the ceramic matrix is unaffected. The thermal decomposition of pure CoN and FeN was examined by Suzuki et al [7,8] who conducted annealing treatments in a stepwise manner; on the other hand, the thermal decomposition of pure Ni$_3$N was examined[6] using a continuous incremental ramp. Pure Co and Fe foils were sputtered in the present study to confirm the stoichiometry of the products formed under the experimental conditions described above. The nitrides produced in both cases were the ones with a 1:1 stoichiometry. Both elements form nitrides with lower nitrogen content, Co$_2$N and Co$_3$N in the case of cobalt, and in the case of iron, Fe$_4$N as a additional phases. The thermal decomposition profile of CoN shows two nitrogen evolution events at 340 and 400 ^{0}C. The integrated areas for the two events are approximately equal, suggesting a stepwise decomposition to Co$_2$N in the first step and formation of metallic cobalt in the second step. The profile is given in Fig. 1 along with the result of a run that was quenched halfway to establish weight loss and to examine the crystalline phase present at that stage.

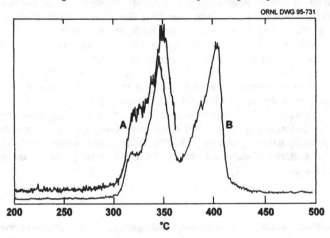

ORNL DWG 95-731

Fig. 1. Thermal decomposition profile, showing mass analyzer response to ion m/e = 28. Trace A decomposition of pure CoN quenched. Trace B pure CoN.

Weight loss in the quenched experiment corresponded to 9.7 wt% of the initial which is approximately half of the theoretical nitrogen content of CoN of 19.2%. XRD however showed both Co_2N and Co_3N. The decomposition profile of Ni_3N, (no higher nitrides are formed) also exhibits two events at 300 and 380 °C. The intermediate step corresponds to the formation of Ni_4N. Finally, FeN also decomposes in two steps to yield initially Fe_2N with a maximum volatile evolution at 480 °C and a second step which reaches a maximum at 565 °C but in this case the evolution tails off to temperatures higher than 650 °C reflecting the higher thermal stability of Fe_4N.

The decomposition profiles of all the nanocomposite films studied, Ni_3N/AlN, CoN/Si_3N_4, CoN/BN, and FeN/Si_3N_4 exhibit similar patterns to the pure metal nitride compounds. The x-ray diffraction of the nanocomposite films identified the metal nitride as the only crystalline phase. There was significant line broadening which was used to calculate an average particle size that turned out to be < 10 nm in every case. A very significant characteristic of these systems is that the particle size of metal produced by the thermal treatment does not show any crystal grain growth, probably due to the ceramic matrix which prevents aggregation. This is illustrated in Fig. 2 for the cobalt nitride system. The presence of the ceramic matrix in each composite was revealed by their corresponding typical infrared spectra and also confirmed by direct chemical analysis.

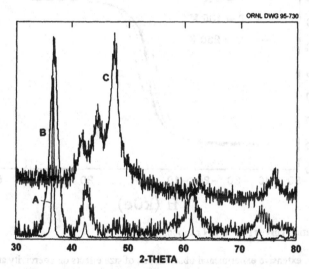

ORNL DWG 95-730

Fig.2 Diffraction patterns of (A) pure crystalline CoN, (B) CoN/BN nanocomposite, and (C) pyrolysis product, pattern corresponds to hcp cobalt.

Magnetic properties

The magnetic properties of the nitride nanocomposites were examined. Nickel nitride, Ni_3N, dispersed in aluminum nitride, is paramagnetic, the slope dM/dH increased from 2×10^{-5} cm^3/g of Ni at 200 K to 6×10^{-5} cm^3/g of Ni at 5 K. Cobalt nitride, CoN, dispersed in boron nitride, is also paramagnetic. The corresponding values for 200 K and 30 K were 2.1×10^{-5} and 1.1×10^{-4} cm^3/g of Co, respectively. Suzuki et al[7] found paramagnetic susceptibilities for pure

CoN that are significantly smaller (~1/10) than the values given above; this difference may stem from the nanostructure of the composite. Iron nitride, is non-ferromagnetic at room temperature and antiferromagnetic at temperatures below 100 K[8]. Our study of the iron nitride composite is still in progress.

Saturation magnetization of the ferromagnetic nanocomposites derived from the thermal treatment were to about 74% of that for pure metal for Ni/AlN. The corresponding figures are 94 % for Co/BN and 50 % for of Fe/Si₃N₄. The lower saturation magnetization is common for nanostructured metals and has been explained as arising from the oxidation of the clusters and disorder at interfaces. The iron case is somewhat more complicated since the pyrolysis has to be conducted at temperatures above 700 °C to insure that all of the nitride is converted to the metal. Under those conditions, however reaction between the metal and the ceramic matrix is possible and silicides were formed which accounts in part for the lower saturation magnetization for iron.

In common with nanostructured ferromagnetic metals, the nanocomposites examined in this study showed hysteresis upon reversal of the field in a given temperature range. Typical results are illustrated in Fig 3 for the Co/BN system.

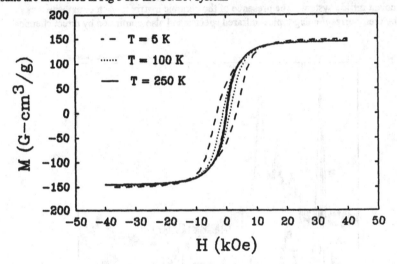

Fig. 3 Magnetization curves for Co/BN nanocomposite

There are extensive experimental observations of size effects on coercivity and its temperature dependence that have been explained[9] in relation to the size of single-domain particles. Coercivity reaches a maximum for single domain particles but decreases both for larger particles that are multidomain and for smaller particles which fall within the realm of superparamagnetic behavior. Superparamagnetism is dependent on temperature and is evident above a critical temperature, the blocking temperature, T_B, above which thermal motions control the alignment of the assembly with the applied field and the system is magnetically reversible(coercive field, $H_C=0$). The assembly is stable at temperatures lower than T_B and the system shows hysteresis. Experimentally we derive T_B from plots of H_C vs $T^{1/2}$. T_B values are 392 K for Ni/AlN, 206 K for Co/BN, 50 K for Co/Si₃N₄ and 100 K for Fe/Si₃N₄. The coercive field at room temperature for nickel is 35 Oe; on the other hand, the other systems should not show any coercivity at room temperature under ideal conditions since the T_B values are much lower than ambient temperature, and indeed this is the case for the iron nanocomposite but the cobalt showed residual coercivity of about 30 Oe in the silicon nitride matrix and 88 Oe in the boron nitride matrix. Similar departures from ideal behavior have been observed, among others, for nanocrystalline nickel[10] and are interpreted as being due to a transition from a regimen where crystal anisotropy energy dominates at low temperature to a regimen where shape anisotropy dominates at higher temperatures.

Microscopy

Examination of our films by Atomic Force Microscopy, AFM, and Magnetic Force Microscopy, MFM, reveals a botryoidal morphology, consisting of aggregates of smaller clusters, about 15 nm in diameter. The AFM and MFM images are similar, other than subtle differences in contrast, since the MFM responds to a field gradient generated by the presence of the ferromagnetic particles. This is illustrated in Fig. 4. The AFM and MFM produce an image of the outer ceramic matrix and thus complement the XRD and TEM observations which are sensitive to the denser crystalline metal component in the nanocomposite.

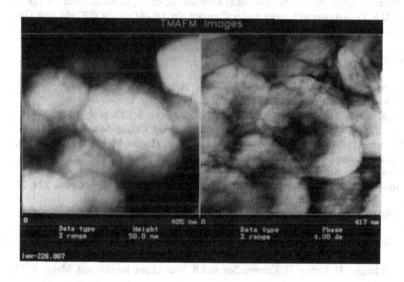

Fig. 4 Atomic force (left side) and magnetic force (right side) micrographs of Ni/AlN composite. Images are about 400 nm on the side.

Data storage

Localized heating, produced with a focused argon ion laser beam, was utilized to generate features a few microns in width which then could be imaged by MFM. This takes advantage of the fact that the areas not affected by the laser, containing the nitride which is paramagnetic, do not interact in the same manner with the magnetic tip of the microscope and produce a marked contrast. This constitutes a proof of principle that these systems may be utilized for ultrahigh density data storage once a method to generate smaller features is developed. One possible approach is to use the tip of the microscope as the heating source.

The long term stability is of concern for a successful data storage medium. Exposure to ambient air for a few months have shown that the aluminum nitride and silicon nitride matrixes remain unchanged and in the case of nickel nanocomposite no nickel oxide was detected by XRD. On the other hand, boron nitride was found to undergo partial hydrolytic decomposition.

CONCLUSIONS

The ferromagnetic nitrides in the form of nanocomposites in a ceramic matrix provide a media that through the use of localized heating may provide a useful data storage. Extended chemical stability for some of these systems seems to be adequate for practical use as recording media.

ACKNOWLEDGMENT

Research sponsored by the Division of Materials Sciences, Office of Basic Energy Sciences, U.S. Department of Energy, under Contract No. DE-AC05-96OR22464 with Oak Ridge National Laboratory managed by Lockheed Martin Energy Research Corp.

REFERENCES

1. B. Abeles, P. Shoeing, M.D. Coutts, and Y. Arie, Adv. Phys. **24**, 407 (1975).
2. C.L. Chien, J. Appl. Phys. **69**, 5267 (1991).
3. L. Maya, W.R. Allen, A.L. Glover, and J.C. Mabon, J. Vac. Sci Technol. B **13**, 361 (1995).
4. D. Kundu, I. Honma, T. Osawa, and H. Komiyama, J. Am. Ceram. Soc. **77**, 1110 (1994).
5. K. Fukumi, A. Chayahara, K. Kadono, T. Sakaguchi, Y. Horino, M. Miya, J. Hayakawa, and M. Satou, Jpn. J. Appl. Phys. B **30**, L742 (1991).
6. L. Maya, J. Vac. Sci. Technol. A **11**, 604 (1993).
7. K. Suzuki, T. Kaneko, H. Yoshida, H. Morita, and H. Fujimori, J. Alloys Comp. **224**, 232 (1995).
8. K. Suzuki, H. Morita, T. Kaneko, H. Yoshida, and H. Fujimori, J. Alloys Comp. **201**. 11 (1993).
9. D. Cullity, Introduction to Magnetic Materials, Addison-Wesley, Reading, MA, 1972, pp. 386-417.
10. H. E. Schaefer, H. Kisker, H. Krommuller, and R. Wurschum, NanoStruct. Mater. **1**, 523 (1992).

CARBON COATED NANOPARTICLE COMPOSITES SYNTHESIZED IN AN RF PLASMA TORCH

JOHN HENRY J. SCOTT[1], SARA A. MAJETICH[1*], ZAFER TURGUT[2], MICHAEL E. MCHENRY[2], and MAHER BOULOS[3]
[1]Department of Physics, [2]Department of Materials Science and Engineering, Carnegie Mellon University, Pittsburgh, PA 15213; [3]Plasma Technology Research Center (CRTP), University of Sherbrooke, Sherbrooke, Quebec, CANADA
* Author to whom correspondence should be addressed. email: sm70@andrew.cmu.edu.

ABSTRACT

FeCo alloy nanoparticles are synthesized in an RF plasma torch reactor and characterized using X-ray powder diffraction (XRD) and transmission electron microscopy (TEM). Bare, uncoated particles exhibit a chain-like agglomeration morphology marked by large ring- and bridge-like structures surrounding open voids. Acetylene was used to generate large numbers of carbon-coated nanoparticles similar to those produced in carbon arc reactors. Conventional TEM of this powder revealed numerous particles below 50 nm in diameter embedded in a carbonaceous matrix. These results establish RF plasma torch processing as a well-characterized, scalable alternative to carbon arc synthesis of encapsulated nanoparticles.

INTRODUCTION

Carbon-coated nanoparticles [1-4] represent one of the most promising classes of magnetic nanocomposites. Composed of nanophase ferromagnetic material encapsulated by graphitic or amorphous carbon, these systems often display interesting physical or magnetic properties because of their small size and reduced dimensionality [5]. In addition to providing an effective barrier to oxidation and corrosion [6], the encapsulating carbon overcoat can act as a form of "nanolamination". This versatile nanostructure can perform several important functions simultaneously: preventing coarsening and particle coalescence, attenuating interparticle magnetic interactions, and reducing eddy current losses in high frequency environments. Numerous technological applications have been suggested for magnetic nanoparticles, including use as data storage media [7], magnetic inks and ferrofluids, xerographic toners [2], and biomedical imaging contrast agents. Here we discuss FeCo nanoparticles, whose very low magnetocrystalline anisotropy and large saturation magnetization make them valuable for applications requiring magnetically soft materials.

If commercial applications for nanoparticles are to be realized, synthesis routes capable of producing large quantities of product (kilograms/day) must be developed. Early research relied on modified Huffman-Kratschmer carbon arc reactors to produce gram quantities of soot, but scaling this technology to kilogram yields has proven difficult. Recently, second-generation synthesis routes have begun to appear, including tungsten-arc and blown-arc techniques [8]. These new reactors permit large amounts of soot to be produced quickly, but more importantly they afford a greater degree of control over the processing parameters. In this work, radio frequency (RF) plasma torch synthesis is presented as an efficient, highly scalable, electrode-less synthesis route using pure metal starting materials and a gaseous carbon source. The flow conditions in the reactor are uniform and well-characterized [9], a key ingredient in understanding the resulting particle morphology and coating mechanisms.

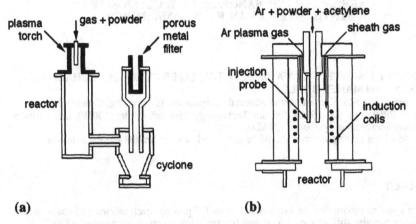

(a)   **(b)**

Fig. 1a. Overview of RF plasma torch reactor showing plasma torch head, expansion and reaction vessel, cyclone separator, and porous metal filters. **1b.** Detailed schematic of plasma torch head used to produce nanoparticles.

Here we report the successful synthesis of FeCo nanoparticles in an RF plasma torch using metal powder starting materials and acetylene as the carbon source. After a brief description of the experimental apparatus and production parameters, X-ray powder diffraction and transmission electron micrographs are presented for coated and uncoated nanoparticles. The performance of the RF plasma is compared to existing nanoparticle generation methods and future extensions of this fruitful technique are discussed.

EXPERIMENTAL PROCEDURE

Nanoparticles were synthesized in a radio frequency (RF) plasma reactor (Fig. 1a) consisting of a plasma torch head (Tekna Model PL-50), a gas expansion and reaction vessel, a cyclone separator, and two 5 μm porous metal filters. A 60 kW RF power supply (Lepel) was used to energize the torch head with 6.6 A delivered at 8.8 kV and 3 MHz. The torch head (Fig. 1b) consists of a inner quartz tube surrounded coaxially by a ceramic heat shield. Water-cooled copper coils outside the larger ceramic tube inductively couple the RF energy to the plasma gas in the inner quartz tube. A high flow-rate gas sheath between the inner and outer tube minimizes heat transfer to the torch body. Argon flowing at 40 standard liters per minute (slpm) was used as the plasma gas while the sheath gas consisted of 80 slpm of Ar mixed with 9 slpm of hydrogen. Metal feedstock powders were entrained in a 3 slpm Ar flow and injected axially into the gas stream just above the plasma via a tubular injection probe. A screw-driven vibratory powder feeder dispatched a mixture of 6-10 micron Fe powder and 1.6 micron Co powder to the carrier gas at a rate of 2 g/min. This ability to continuously feed starting material to the plasma has clear advantages over carbon arc reactors where synthesis must be stopped periodically to install a new consumable electrode. Acetylene could also be mixed with the powder feed gas between the vibratory feeder and the injection probe. The torch was operated nominally for 30 minutes before powder collection, producing approximately 50 g of product in this time. This compares favorably with the yields achieved using carbon arc reactors, typically only a few grams per hour in optimized systems. During each 30 minute session the pressure in the reactor vessel climbed steadily from 300 torr to 600 torr as the powder collected on the filters and the effective speed of the vacuum system decreased.

Two synthesis protocols were used to produce the nanoparticles reported here. The first used a mixture of 50 wt. % Fe and 50 wt. % Co. The goal of this run was to produce uncoated FeCo alloy in the absence of carbon. The second protocol used a mixture of 70 wt. % Fe and 30 wt. % Co with a 1.5 slpm acetylene flow as the carbon source. In both cases a black, powdery reactor product was separately collected from the reactor vessel walls and the porous metal filter traps by gently cleaning the surfaces with a brush.

X-ray powder diffraction was performed on the nanoparticle-containing soots using a Rigaku diffractometer fitted with a fixed tube Cu target and a bent graphite monochromator. TEM was used to characterize the soot structurally and examine the coating morphology. Micrographs were recorded on a Philips EM-420T operated at 120 kV using Cu grids with amorphous carbon substrates.

RESULTS AND DISCUSSION

Fig.2 is a conventional TEM micrograph of bare FeCo nanoparticles produced in the plasma torch reactor using the first processing protocol. With only argon and hydrogen gases present in the plasma during particle production (no carbon source), the particles nucleate and agglomerate without carbon coatings. Surface tensions acting during the liquid phase of particle growth have spheroidized many of the particles in this micrograph. Examples of chain-of-spheres agglomerates are also seen in this field of view (upper left), possibly caused by magnetic interactions between the particles during flocculation. Chain-like morphology has also been seen in FeCo samples produced in carbon arc reactors [10], but here the chains were often arranged in large rings or bridges not seen in carbon arc soot. These rings were approximately circular, surrounded empty space, and manifested at several length scales (diameters from 40 to at least 300 nm). In the absence of an external magnetic field, the minimum energy configuration of a chain of magnetic spheres is a closed ring. The reactor product shown in Fig. 2 was pyrophoric and oxidized readily on first contact with air; this is typical of ultrafine powders that lack a passivating or protective layer. Higher magnification microscopy (not shown) revealed thin coatings on some particles, most likely due to the post-synthesis oxidation observed during powder collection.

Fig. 3 is a conventional TEM micrograph of carbon coated FeCo nanoparticles produced in the plasma torch reactor using the second synthesis protocol. Numerous particles with diameters < 20 nm can be seen embedded in a matrix of carbonaceous material. The unusual "stringy" contrast exhibited by the carbon matrix has appeared before in carbon arc soot produced using high-abundance Co and CoBSi starting materials [10]. Although further investigation into this morphology is clearly needed, Co is a well-known catalyst of single-walled carbon nanotubes [11]. In other fields of view traces of free graphitic carbon were seen as well as occasional nanoparticles encapsulated within graphitic cages. Although quantitative statistics were not compiled, the average particle size appears to be smaller in the coated sample compared to the bare FeCo sample, suggesting the carbon coating may have intervened in particle coalescence and coarsening pathways. No chains-of-spheres were seen in the coated sample.

X-ray powder diffraction (XRD) of the two samples revealed only FeCo alloy, fcc Co, Fe oxide, and carbon, although the possibility of Fe carbide could not be ruled out. The carbon-containing sample (Fig. 4, upper trace) contains predominantly FeCo alloy with small amounts of fcc Co and graphite. The sample produced using the carbon-less synthesis protocol shows only FeCo and Fe oxide. The peaks at 30.2°, 35.6°, 43.3°, 53.7°, 57.2°, and 62.9° match exactly the six most intense reflections of γ-Fe_2O_3. The small shoulder at 36.8° is consistent with the most intense peak of Fe_3O_4. The presence of oxide peaks in XRD is expected in the uncoated sample given the large, unprotected surface area and the visible evidence of vigorous oxidation observed during powder collection. Presumably, the high cobalt abundance (nominally 50 wt. %) prevents the remainder of the alloy powder from oxidizing readily.

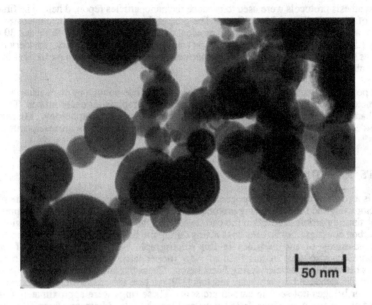

Fig. 2. Conventional TEM of bare (uncoated) FeCo nanoparticles collected from the RF plasma torch reactor wall. No carbon source was used during this synthesis.

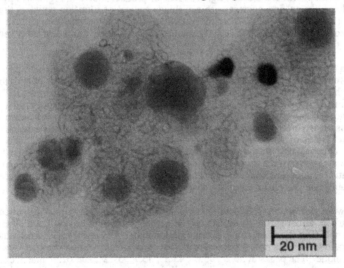

Fig. 3. Carbon coated FeCo nanoparticles produced in an RF plasma torch using metal powder feedstocks and acetylene as the carbon source. The processing conditions used to generate this product are described in the text.

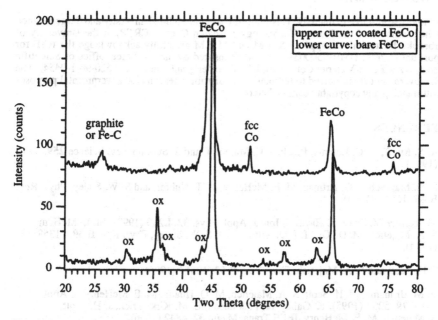

Fig. 4. X-ray powder diffraction of bare and carbon coated FeCo nanoparticles collected from the reactor wall. The lower trace (bare FeCo, no carbon) shows signs of the oxidation ("ox" indicates oxide peaks) expected from ultrafine particles lacking a protective carbon coating.

CONCLUSIONS

FeCo alloy nanoparticles were synthesized in an RF plasma torch reactor utilizing Ar as the plasma gas. Bare, uncoated particles were produced from a 50 wt. % Fe, 50 wt. % Co metal powder mixture. The resulting ultrafine powder oxidized upon contact with air, producing a mixture of FeCo alloy and γ-Fe_2O_3 as revealed by XRD. Conventional TEM of the particles showed chain-like agglomerates that were frequently arranged in a bridging, ring-like morphology. When a mixture of 70 wt. % Fe and 30 wt. % Co powders were processed in the plasma torch in the presence of acetylene as a gaseous source of carbon, coated nanoparticles (similar to those produced in carbon arc reactors) were generated in abundance. XRD of the carbon-containing powder detected predominantly FeCo alloy and small amounts of fcc Co and graphite. Conventional TEM of the soot revealed numerous particles below 50 nm in diameter embedded in a matrix of carbon.

These results establish RF plasma torch processing as a viable alternative to tungsten-arc, blown-arc, and other second generation techniques for producing carbon-encapsulated nanoparticles. The plasma torch process has been very well characterized, the synthesis parameters are well-behaved from a process control perspective, and the technique is easily scalable to industrial levels.

ACKNOWLEDGMENTS

The authors would like to thank N. T. Nuhfer of Carnegie Mellon University and Serge Gagnon and the members of the Plasma Technology Research Center (CRTP) at the University of Sherbrooke for technical support. S.A.M and M.E.M gratefully acknowledge the NSF for support under grant DMR-95000313. Effort sponsored by the Air Force Office of Scientific Research, Air Force Materiel Command, USAF, under grant number F49620-96-1-0454. The U. S. Government is authorized to reproduce and distribute reprints for governmental purposes notwithstanding any copyright notation thereon.

REFERENCES

1. R. S. Ruoff, D. C. Lorents, B. Chan, R. Malhotra, and S. Subramoney, Science **259**, 346 (1993).

2. S. A. Majetich, J. O. Artman, M. E. McHenry, N. T. Nuhfer, and S. W. Staley, Phys. Rev. B **48**, 16845 (1993).

3. M. Tomita, Y. Saito, T. Hayashi, Jpn. J. Appl. Phys. **32**, L280 (1993); M. E. McHenry, S. A. Majetich, M. De Graef, J. O. Artman, and S. W. Staley, Phys. Rev. B **49**, 11358 (1994).

4. J. H. Scott and S. A. Majetich, Phys. Rev. B **52**, 12564 (1995).

5. E. M. Brunsman, J. H. Scott, S. A. Majetich, M. Q. Huang, M. E. McHenry, J. Appl. Phys. **79**, 5293 (1995); K. Gallagher, F. Johnson, E. M. Kirkpatrick, J. H. Scott, S. Majetich, M. E. McHenry, IEEE Trans. Magn. **32**, 4842 (1996).

6. M. Ladouceur, G. Lalande, D. Guay, J. P. Dodelet, L. Dignard-Bailey, M. L. Trudeau, R. Schulz, J. Electrochem. Soc. **140**, 1974 (1993).

7. T. Hayashi, S. Hirono, M. Tomita, S. Umemura, Nature **381**, 772 (1996).

8. V. P. Dravid, J. J. Host, M. H. Teng, B. Elliot, J. H. Hwang, D. L. Johnson, T. O. Mason, J. R. Weertman, Nature **374**, 602 (1995); M. H. Teng, J. J. Host, J. H. Hwang, B. R. Elliot, J. R. Weertman, T. O. Mason, V. P. Dravid, D. L. Johnson, J. Mater. Res. **10**, 233 (1995).

9. M. I. Boulos, Pure and Appl. Chem. **57**, 1321 (1985).

10. J. H. Scott, PhD dissertation, Carnegie Mellon University, Department of Physics, 1996.

11. D. S. Bethune, C. H. Kiang, M. S. deVries, G. Gorman, R. Savoy, J. Vasquez, R. Byers, Nature **363**, 6430 (1993).

MAGNETIC PROPERTIES AND THERMAL STABILITY OF GRAPHITE ENCAPSULATED COBALT NANOCRYSTALS

J. J. HOST and V. P. DRAVID, Department of Materials Science and Engineering, Northwestern University, Evanston, IL 60208-3108

ABSTRACT

Small particles are of interest for magnetic study due to their range of magnetic properties with particle size and unusual magnetic properties. In this study we produced graphite encapsulated magnetic cobalt particles and measured their structural and magnetic properties both before and after annealing at 550° C for 6 hours. No large change in the structural properties (particle size, shape, and lattice parameter) was observed as a result of the annealing. There were, however, significant changes in the magnetic properties. The coercivity was found to decrease at all temperatures, especially at temperatures below 100° K. The possible reasons for this reduction in coercivity are examined, with the change in the shape of the particles appearing most likely.

INTRODUCTION

Nanocrystalline materials have been found to have interesting properties, including changes in magnetic phenomena[1]. In particular, it is known that ferromagnetic particles below a certain size are composed of a single magnetic domain, and can lose all coercivity (become superparamagnetic) below a critical size[2]. However, in attaining these small sizes, a large amount of surface area is produced. This leads to increased reactivity with the air, causing many small particles to partially or completely oxidize over time.

The discovery of graphite encapsulated nanocrystals[3] eliminated this problem. Due to the protective graphite sheets, metal particles encased in graphite are protected from the oxygen in the air, and can be used in applications where unencapsulated metal particles would rapidly oxidize. The fact that the graphite layers successfully protect these particles is shown by the fact that graphite encapsulated nanocrystals have been kept in strong acid for over a year with no degradation.

The magnetic properties are important for many of the possible applications of these materials. In these experiments we have examined the magnetic properties of the graphite encapsulated nanocrystals both before and after a high temperature anneal. This has allowed the study of both the thermal stability of the encapsulating graphite and the change in magnetic properties caused by the annealing cycle. It was found that the particles are ferromagnetic at all temperatures measured both before and after annealing for 6 hours at 550° C. Furthermore, the coercivity at all temperatures was reduced by annealing by as much as 100 Oe. Possible reasons for the observed reduction in coercivity are proposed.

EXPERIMENTAL

The particles were synthesized using the tungsten arc method, described elsewhere. Unencapsulated particles were removed by immersing the particles in an ultrasonic bath of concentrated nitric acid. Prior to magnetic tests, the particles were observed by Transmission Electron Microscopy (TEM) using an HF 2000 (Hitachi Corp. Tokyo, Japan). The lattice parameter and phase of the particles was measured by x-ray diffraction (wiyh an XDS 2000, Scintag Inc., Santa Clara CA) using Nelson-Riley[4] regression. The samples were approximately 1 mm thick, well above the penetration depth of about 10 µm for Cu Kα radiation. The x-ray diffraction apparatus was operated at 20 ma and 40 KV.

Mat. Res. Soc. Symp. Proc. Vol. 457 ° 1997 Materials Research Society

The particles were dispersed and immobilized in epoxy prior to magnetic measurement to minimize particle-particle interactions. Magnetic data (Magnetization vs. Field loops) were obtained using a SQUID (Superconducting Quantum Interference Device, MPMS Quantum Design Inc. San Diego, CA) at temperatures ranging from 5 to 300° K. Due to the sensitivity of the SQUID, the samples used were extremely small, with only about 1 mg of encapsulated Co powder. To test the stability of the graphite layers and observe any magnetic changes, the particles were annealed at 550° C for 6 hours in an evacuated tube with a piece of tantalum (to remove any oxygen present). Subsequent to annealing, XRD, TEM and SQUID studies were repeated, and the results were compared to those found prior to the annealing.

RESULTS/DISCUSSION

Structural Results (X-Ray Diffraction & TEM)

The annealing caused little detectable structural change. TEM measurements revealed that the particles appeared spherical both before and after being annealed, and that the particle size was close (withing statistical measurement error) to being the same. Similarly, x-ray diffraction did not reveal any measurable lattice parameter change. The lack of structural changes shows that the graphite sheets effectively protected the nanocrystals at 550° C. Similarly, the lack of a change in the lattice parameter showed that if any carbon was trapped in the particles, (which would have shown a lattice parameter change if dissolved carbon was present)[5], it was at a low concentration, resulting in no detectable lattice parameter change after annealing. Similarly, x-ray diffraction revealed that the particles were composed of FCC Cobalt both before and after the anneal. FCC is the high temperature phase for cobalt, normally only being present above 420°C. As has been advance previously, this was probably due the small particle nature of the material, as opposed to any "quenching" effects[8]. The lattice parameter and particle size data are summarized in Table I.

Table I
Structural Effects of Annealing

Property and Measurement Method	Before Annealing	After Annealing
Particle Size		
TEM (50-100 Particles Measured)	6 - 15 (nm)	6 - 17 (nm)
Size Estimated from Blocking Temp.	32.6 (nm)	35.8 (nm)
Lattice Parameter		
X-Ray Diffraction	3.54(2) Å	3.54(4) Å
(No Detectable Change)	Error of Measurement ≈0.004 Å	
Particle Phase		
X-Ray Diffraction	FCC	FCC

Magnetic Results (SQUID)

Magnetic results showed a decrease in the coercivity due to the annealing. Representative hysteresis curves taken at a high (300° K) and low (5° K) temperature both before and after the annealing are shown in Figure 1.

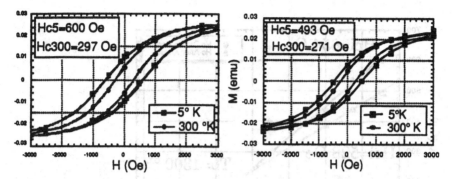

Figure 1. Hysteresis curves for graphite encapsulated cobalt nanocrystals at high and low temperatures both before and after annealing at 550° C for 6 hours. The coercivity is seen to decrease at both temperatures.

For superparamagnetic particles, the coercivity is expected to obey the formula[2]:

$$H_c = H_{ci}\left(1 - \left(\frac{T}{T_B}\right)^{\frac{1}{2}}\right) \qquad (1)$$

Which gives a linear dependence of the coercivity with $T^{1/2}$. Our coercivity data falls on a straight line both before and after annealing when plotted against $T^{1/2}$ as seen in Figure 2.

However, this does not necessarily show superparamagnetism, because the coercivity is expected to fall as the absolute temperature is raised[3]. The blocking temperature calculated from Figure 1 shows the temperature that the coercivity would vanish if the particles are superparamagnetic. The particle size can be calculated from this temperature, taking the crystalline anisotropy constant to be 2.7×10^6 (Ref. 10) by the following formula:

$$T_B = \frac{KV}{300k_B} \qquad (2)$$

This gives the size estimated from the blocking temperature given in Table I.

The coercivity decreased significantly after the anneal at all temperatures measured. This decrease ranged from over 100 Oe at the lowest temperature measured (5° K), to 26 Oe at the highest temperature measured (300° K).

Possible explanations for this decrease in coercivity include the removal of atomic defects or carbon atoms trapped in the particles as they formed, the rounding of the particles to minimize the surface energy, and an increase in the particle size. Though any of these factors could have played a role in the coercivity decrease measured, the concurrent structural studies indicate that the shape of the particles may have played a larger role than any particle size growth or the reduction of the carbon and defect contents.

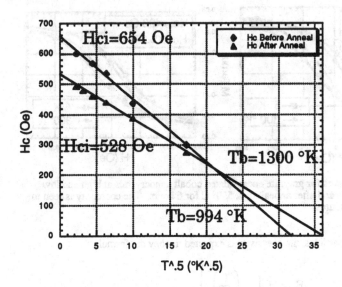

Figure 2. Coercivity as a function of temperature for the particles before and after annealing. The extrapolation of the coercivity to zero occurred at a higher temperature after annealing.

Carbon and Defect Contents

It is well known that the presence of solute atoms (such as carbon) and atomic defects (such as vacancies) can cause an increase in magnetic coercivity, which would account for the decrease in coercivity with their elimination by annealing. This would result in a decrease in the lattice parameter. However, the structural measurements do not support this possible explanation. As seen in Table I, the lattice parameter was not observed to contract. If carbon or vacancies were being "annealed" out, the lattice parameter should have changed. A lattice parameter change of 0.04 A is seen with the addition of only 1 wt % carbon in FCC iron[5]. Also, diffusion calculations show that the diffusion of carbon out of a small cobalt particle is extremely fast due to the short distance that these carbon atoms must diffuse to reach the surface (Figure 3). Because the arc is at a temperature of several thousand[6] degrees C, and a temperature of nearly 500° C was measured in the gas at a distance 4 inches from the arc, it is estimated that the particles are exposed to temperatures above 500° C for at least a tenth of a second as they leave the arc vicinity. As can be seen from Figure 2, this may have resulted in nearly all of the carbon diffusing from the particle. Similarly, atomic defects should also be greatly reduced during synthesis of the particles, as the recrystallization temperature of Co is 390°C, well below the 500° gas temperature mentioned earlier. For these reasons, it is expected that the concentration of both carbon atoms and atomic defects is expected to be low in the as-produced particles, and may not account for the decrease in coercivity with annealing.

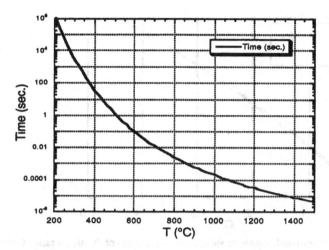

Figure 3. Diffusion of 99% of the carbon from a 10 nm diameter Co particle. At any temperature above 600 ° C, this takes less than one tenth of a second[7].

Shape Anisotropy

The coercivity can be strongly affected by the shape of the small particles, leading to an increase in the coercivity as the particles become oblong. Estimates of the change in coercivity based on calculated[2] demagnetizing factors give a large coercivity with only a slight departure from a spherical particle. This is shown in Figure 4, where a change in the aspect ration of only 1.1 to 1.09 results in a decrease in coercivity of nearly 100 Oe. As mentioned earlier, the particles were observed to be generally spherical both before and after the annealing cycle. However, small (and difficult to detect) departures from an aspect ratio of 1 could have caused an increase in the coercivity prior to the annealing, resulting in a decrease in the coercivity after the annealing due to the possible rounding of the particles. Further study, including a detailed measure of the aspect ratio of a large number of particles will be needed to investigate this possibility.

Particle Growth

Particles size can affect the coercivity of small particles, especially at small sizes, so changes in the coercivity could indicate a change in the particle size. However, as can be seen from Figure 4, particles below the single domain size (Ds) are expected to show an increase in coercivity with an increase in particle size. The Ds is for ferromagnetic particles is estimated to be between 10 and 50 nm[2], so our particles are probably single domain. Since the particle size could only be expected to increase with annealing, an increase in coercivity would be expected if particle size growth were occurring, the opposite of what we see. Also, no large particle growth was observed by TEM (see Table I).

CONCLUSION

The graphite encapsulated cobalt particles used in this study are ferromagnetic at temperatures below 300. Blocking temperature calculations show that the particles may be superparamagnetic at temperatures above about 1000° K, but the decrease in coercivity seen is

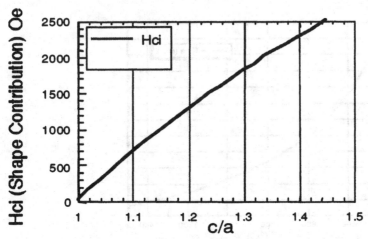

Figure 4.   Coercivity caused only by shape anisotropy as a function of the aspect ratio. Only a small departure of the aspect (c/a) ratio from one results in a significant increase in coercivity.

normal for any ferromagnetic material, so this does not necessarily show superparamagnetism. Annealing of the particles at 550° C resulted in very little structural change (particle size, shape and lattice parameter), but did result in a decrease in the magnetic coercivity. The decrease in coercivity seen in this study could have resulted from the removal (by diffusion) of carbon or crystalline defects from the particles during the annealing or by the rounding of the particles (thus reducing shape anisotropy) during the annealing. The rounding of the particles during annealing appears to be the most likely cause of the observed decrease in coercivity.

ACKNOWLEDGMENTS

The authors gratefully acknowledge the help of Mark Lowe and Mary Cheang in the initial production of the samples, M. Z. Lin in helping with use of the SQUID, and Kevin Peters who helped with the XRD. This work was supported by NSF NYI DMR-9357513.

REFERENCES

1.   G. C. Gangopadhyay, G. C. Hadjipanayis, C. M. Sorensen, and K. J. Klabunde, (Mater Res. Soc. Proc. 206, Boston, MA 1991) pg. 55.
2.   B. D. Cullity, Introduction to Magnetic Materials, Addison-Wesley Pub. Co. (1972)
3.   M. Tomita, Y. Saito, and T. Hayashi, Jpn. J. Appl. Phys.32 L280 (1993)
4.   Nelson, Riley. Proc. Phys. Soc. London, 57 pg. 160 (1945)
5.   W. F. Smith, Structure and Properties of Engineering Alloys McGraw-Hill Publishing Co. (1981)
6.   Connor, L. P. Editor. Welding Handbook American Welding Society (1991)
7.   Linde, D. R. Editor. CRC Handbook of Chemistry and Physics CRC Press (1991)
8.   J. J. Host, M. H. Teng, B. R. Elliott, J. H. Hwang, T. O. Mason, D. L. Johnson, and V. P. Dravid, "Graphite Encapsulated Nanocrystals Produced Using a Low Carbon:Metal Ratio" Submitted to Journal of Materials Research, 4/25/96
9.   R. M. Bozorth, Ferromagnetism D. Van Nostrand Co. (1951)
10.   W. Sucksmith, F.R.S. and J. E. Thompson, Proc. Roy. Soc. A, 254, pg. 362

MAGNETIC MICROSTRUCTURE OF A NANOCRYSTALLINE FERROMAGNET - MICROMAGNETIC MODEL AND SMALL-ANGLE NEUTRON SCATTERING

J. WEISSMÜLLER*, R.D. MCMICHAEL#, J. BARKER#, H.J. BROWN#, U. ERB‡, R.D. SHULL#
*Universität des Saarlandes, Saarbrücken, Germany; #National Institute of Standards and Technology, Gaithersburg, USA; ‡Queen's University, Kingston, Ontario, Canada

ABSTRACT

We report on a combined theoretical and experimental study of the magnetic microstructure of a single component, single phase, pore-free nanocrystalline ferromagnetic material. From the equations of micromagnetics we conclude that the magnetic microstructure is the convolution product of an anisotropy field microstructure and of a response function with a correlation length l_H that depends on the applied field H_a. We derive equations for small angle neutron scattering by such structures, and present experimental scattering data for electrodeposited nanocrystalline Ni, the first where for a wide range of H_a the dominant scattering contribution is from the purely magnetic microstructure, not from nuclear or magnetic contrast at pores or second phases. The variation of the scattering cross section with H_a is in excellent agreement with the theory, indicating that the underlying changes in the magnetic microstructure with H_a are not displacements of domain walls, but changes in l_H and hence in the magnetic response to an entirely stationary anisotropy field microstructure. At 20K the anisotropy fields are dominated by magnetocrystalline anisotropy, but at 300K the perturbation is from a much stronger interaction which maintains some moments aligned antiparallel to the field direction at H_a as high as 1.4MA/m (18kOe).

INTRODUCTION

In an idealized single phase, nanometer grain size polycrystalline ('*nanocrystalline*') ferromagnet subject to an external field, random jumps of the magnetic anisotropy field cause gradients of the magnetization at grain boundaries. Due to the magnetic exchange interaction there is excess energy in these gradients; therefore exchange interaction opposes the tendency of the magnetization to align itself with a low energy orientation in each grain. On the same grounds domain walls in single crystal ferromagnets are not atomically sharp but have a finite width w; for Ni $w \approx 100$nm. In a nanocrystalline ferromagnet with a grain size D of the order of 10nm a hypothetical *domain*, that is a region of homogeneous magnetization which is considerably larger than w, contains many grains with random orientations, and the expectation value for the net anisotropy in the domain decreases rapidly as D is reduced. The resulting decrease in coercivity is well explained by the random anisotropy model [1], originally derived for ferromagnetism in metallic glasses [2]. But since the net anisotropy is the very reason for the existence of domains, it is questionable whether domains and walls continue to exist and to represent the dominating elements of the magnetic microstructure in a nanocrystalline ferromagnet, or whether spatial variations of the magnetization of a different nature dominate.

Small-angle neutron scattering (SANS) experiments, which probe the magnetic microstructure on a range of length-scales including w and D, indicate the presence of spatial variations in the direction of the magnetization in nanocrystalline ferromagnets [3][4]. The central aim of our study is to explore the nature and origin of this deviation from the homogeneously magnetized state. We take advantage of the existence of a theoretical framework appropriate for describing the magnetic microstructure on the length scales of interest: the theory of micromagnetics [5]. Seeger and Kronmüller [6] have derived a solution of the equations of micromagnetics in the limit of high applied magnetic field based on which SANS due to dislocations in ferromagnetic single crystals can be understood [7]. In the present paper we apply the equations of micromagnetics to a macroscopically isotropic nanocrystalline ferromagnet and derive equations for the dependence of the differential SANS cross-section on applied magnetic field and on magnitude and direction of the scattering vector, which can be compared to experiment. Several characteristic length scales, which are suggested by the nuclear microstructure (grain size, sample size) and by various combinations of magnetic parameters (exchange constant, saturation magnetization, anisotropy energy, magnetic field) may be relevant for the magnetic microstructure [8][9]. Comparison of theory and experiment allows us to identify the dominant length scales.

SANS intensity arises from spatial variations of atomic density, composition, and magnetization. Therefore, investigations of the magnetic microstructure require that the variations in atomic density and in composition be minimized so that the signal from the magnetic microstructure is resolved. In investigations on nanocrystalline Fe-Si-B based alloys [4], superparamagnetic nanocomposites [10], and on single-component nanocrystalline ferromagnets [3][11] the strong scattering contribution from the a multi-phase and/or porous nature of the samples has so far precluded detailed conclusions on the magnetic microstructure of the idealized homogeneous nanocrystalline material. In the present study, we choose electrodeposited nanocrystalline Ni (n-Ni) [12] as a single-component, pore-free model system which is free of such complications.

THEORY

We consider the material to be homogeneous with respect to atomic density ρ_{atom}, saturation magnetization m_s, and exchange interaction A. The inhomogeneous nuclear microstructure enters the theory through $H_p(x)$, the (negative of the) derivative of a free energy density (free energy per unit volume) at x with respect to the magnetization vector $M(x)$ subject to $|M(x)| = m_s$, and hence a vector normal to M. H_p is due to magnetocrystalline and magnetoelastic anisotropy, has the dimension of a magnetic field and will be called the *anisotropy field*. We express $H_p(x)$ and $M(x)$ in terms of their Fourier transforms $h(q)$ and $m(q)$, respectively, and of the volumetric mean magnetization $<M>$:

$$H_p(x) = (2\pi)^{-3/2} \int\int\int_{-\infty}^{\infty} h(q) \exp(-iqx)\, d^3q \tag{1}$$

$$M(x) = <M> + m_s (2\pi)^{-3/2} \int\int\int_{-\infty}^{\infty} m(q) \exp(-iqx)\, d^3q \tag{2}$$

We use the Maxwell equations $\nabla\cdot(H+4\pi M)=0$ and $\nabla\times H=0$ to relate the magnetic field H to the sources in bulk, $\nabla\cdot M$. In the homogeneous material, the only other sources for H are the discontinuities in M at the (macroscopic) external surfaces, which induce a demagnetizing field $H_d = N_d<M>$ that varies slowly with position in the material and that we assume to be homogeneous. N_d is the demagnetizing factor. Therefore,

$$H(x) = H_a - H_d - 4\pi\, m_s (2\pi)^{-3/2} \int\int\int_{-\infty}^{\infty} (m(q)\cdot q)\, q \exp(-iqx) / q^2\, d^3q \tag{3}$$

with H_a the external applied field.

In the limit of high magnetic field, $H \gg H_p$, perturbations from the homogeneously magnetized state are small and the response of the magnetization to the applied and anisotropy fields satisfies the linearized Micromagnetics equation (compare Section 4.1 in [5]): for an orthonormal basis $\{e_x, e_y, e_z\}$ with H_a along e_z

$$(2A/m_s\, \{\nabla^2 M_x, \nabla^2 M_y, 0\} + m_s (H+H_p)) \times M = 0 \tag{4}$$

where for any vector f the scalars f_x, f_y, f_z and f are, respectively, the Cartesian coordinates of f relative to $\{e_x,e_y,e_z\}$ and the modulus of f. Due to linearity, (4) can be solved independently for each wavevector q. Without loss of generality, and motivated by SANS with the incoming neutron wavevector along e_x, and hence with the scattering vector in the plane containing e_y and e_z, we consider $q_x=0$, and denote by θ the angle between q and the applied field. The solution of Equation (4) in terms of (1)-(3) is then

$$m_x(q) = - h_x(q) / H_{eff};\quad m_y(q) = - h_y(q) / (H_{eff} + 4\pi\, m_s \sin^2\theta);\quad m_z(q) = 0 \tag{5}$$

The *effective field* H_{eff} depends on the *internal field* $H_i = H_a - H_d$ and on its *exchange length* l_H [8] by

$$H_{eff} = H_i (1 + l_H^2 q^2);\quad l_H = (2A / m_s H_i)^{1/2} \tag{6}$$

At high fields the Fourier transform of the magnetization is essentially the product of the Fourier coefficient of the anisotropy field and of the reciprocal of the effective field. Because of the convolution theorem, the product in reciprocal space corresponds in real space to a convolution with the Fourier transform of $1/H_{eff}$, which is an exponential with a characteristic length l_H. The central conclusion from the results (5) and (6) is therefore that *the magnetic microstructure is the convolution of the anisotropy field microstructure with an exponential response function with a characteristic length l_H that varies as the reciprocal root of the* *internal field.*

We restrict attention to elastic SANS with unpolarized neutrons for materials with isotropic nuclear and anisotropy field microstructures. From the expression for the scattering cross section of a single atom,

$$d\sigma = b^2_{nuclear} + b^2_{mag} (1 - (\epsilon \cdot Q)^2) \tag{7}$$

with $b_{nuclear}$ and b_{mag} the nuclear and magnetic scattering lengths and ϵ and Q unit vectors in the direction of the scattering vector k and of the magnetic moment, respectively [13], and from the Equation (5), we derive [14] an expression for the differential scattering cross section $d\sigma/d\Omega$ of the isotropic material:

$$d\sigma(k)/d\Omega = d\sigma_{res}(k)/d\Omega + S_H(k) R(k, H_i) \tag{8}$$

with $k=q$. The residual scattering $d\sigma_{res}/d\Omega$ is due to spin wave scattering and to nuclear and magnetic contrast from regions with reduced density, such as pores and grain boundaries, that are neglected in the micromagnetic model. In the high field limit, this term is independent of H_i. The *anisotropy field interference function* S_H depends only on the anisotropy field microstructure and on the modulus k of the scattering vector, but not on the applied field:

$$S_H(k) = 8\pi^3 b^2_{mag} \rho^2_{atom} |h(k)|^2 / (4\pi m_s)^2 / V \tag{9}$$

where V denotes the sample volume. The variation of the differential scattering cross section with the magnetic field is described by the *magnetic response function* R; for the present problem we find [15]

$$R(k, H_i) = (2\pi m_s / H_{eff})^2 (1 + (1+ 4\pi m_s \sin^2\theta / H_{eff})^{-2}) (1 + \cos^2\theta) \tag{10}$$

This equation implies a characteristic variation of the magnetic scattering cross section with the azimuthal angle θ (compare Figure 4 below), and with k and H_i. Often one is interested in an azimuthal (θ-) average of the scattering intensity, for which the response function is simply

$$R_\theta(k, H_i) = 6\pi^2 m_s^2 / H^2_{eff} \tag{11}$$

For given k and H_i, R can be computed with literature data; for Ni m_s= 692.3Am²/kg (55.09emu/g) at 298K and A= 8.6×10⁻¹²N (8.6×10⁻⁷ erg/cm) [2][16][17]. At each k the theory predicts that the azimuthal averaged total scattering cross section scales with R_θ according to Equation (8). In this relation there are therefore two free parameters at each k: $d\sigma_{res}/d\Omega$ and S_H; when $d\sigma/d\Omega$ is measured at two or more different magnetic fields then these parameters can be determined, for each k, from straight lines of best fit in a plot of $d\sigma/d\Omega$ versus R_θ.

Values for the exchange length l_H are readily computed: for magnetic fields of 1kA/m (13Oe) and 1000kA/m (13kOe), Equation (6) with the above values for m_s and A yields l_H=160nm and l_H=5nm, respectively.

EXPERIMENT

In this first report we present results from a single sample, 0.37mm thick electrodeposited n-Ni sheet with an area-weighted average grain size of 18(3)nm, determined from X-ray scattering data by the indirect deconvolution technique [18]. Archimedes immersion data indicate a mass density of 100.0(2)% of the

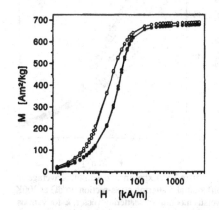

Figure 1 : Magnetization M on linear axis versus applied field H on logarithmic axis for n-Ni (●) and Ni reference sample (○) at 298K. Lines are guides to the eye.

Ni literature value, and hot extraction yields impurity levels below the resolution limits of 60 atomic ppm for H and 300 atomic ppm for N and O; no metallic impurities were detected by energy-dispersive analysis of X-ray fluorescence in a scanning electron microscope.

Magnetization data was recorded on a SQUID magnetometer calibrated with a reference of annealed coarse grained pure Ni of same mass, shape, and position as the sample. Figure 1 shows magnetization isotherms at 298K for sample and reference, which qualify n-Ni as a soft ferromagnet which can be brought close to saturation by moderate applied magnetic fields. By analysis of the approach to saturation in the magnetization isotherms we found a reduction in the saturation magnetization of n-Ni relative to the reference of 0.39(4)% at 5K and of 1.81(5)% at 298K. Coercivities were 1.4kA/m (18Oe) at 5K and 1.0kA/m (1.3Oe) at 298K.

SANS data was recorded at the 30m instrument on beamline 7 at the NIST Cold Neutron Research Facility, with a wavelength of 0.6nm and with the magnetic field applied horizontally and perpendicular to the (horizontal) neutron beam. Figure 2 displays azimuthal averaged differential scattering cross sections versus the modulus k of the scattering vector, recorded at 300K with different applied magnetic fields. The intensity is high for small fields; as the field is increased to the maximum experimental value of 1440kA/m (18.1kOe), the intensity decreases by more than two orders of magnitude. The intensity decrease is not a simple scaling, but the shape of the scattering curves changes: the wavevector at which the (negative) curvature on a log-log plot of $d\sigma/d\Omega$ versus k is at maximum increases systematically with increasing field. This indicates the existence of a characteristic length scale of the magnetic microstructure which decreases with increasing H_a.

Figure 3 compares the experimental data to scattering cross sections computed from Equation (8), with $d\sigma_{res}/d\Omega$ and S_H (also displayed) determined from fits to the experimental data. Demagnetizing fields for use in Equation (3) were computed from the experimental magnetization data, with estimates for N_d for spheroids with the same axis ratio as the samples. It is seen that the fit is in excellent agreement with experiment over the whole range of fields covered, that is 1.0kA/m (13Oe) to 1440kA/m (18.1kOe). Deviations are within

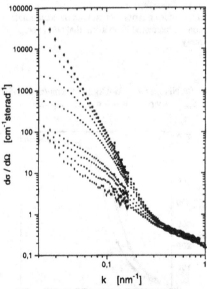

Figure 2 : Experimental azimuthal average differential scattering cross-section $d\sigma/d\Omega$ at 300K versus modulus of scattering vector, k, for various applied fields H_a. From top to bottom, values for H_a in kA/m (Oe in brackets): 1.0 (13), 41.0 (515), 81.6 (1025), 161 (2030), 462 (5810), 651 (8180), 995 (12500), 1440 (18100).

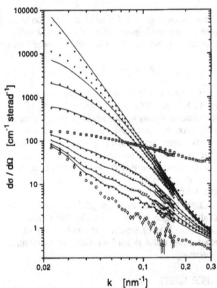

Figure 3 : Azimuthal average differential scattering cross-section $d\sigma/d\Omega$ at 300K versus modulus of scattering vector, k, for applied fields H_a as in Figure 2. Full symbols: experimental data. Lines: fit as discussed in text. (O) : residual scattering cross section $d\sigma_{res}/d\Omega$. (□) : anisotropy field interference function S_H.

error bars from uncertainty in the values for A and N_d.
SANS data recorded at 20K for fields up to 400kA/m
(5000Oe) show qualitatively similar scattering curves but
higher overall intensity, and fits of similar quality to those in
Figure 3 confirm the agreement between theory and experiment.

Figure 4 is a display of the aspect ratio (ratio of scattering
intensities perpendicular ($\theta=\pi/2$) and parallel ($\theta=0$) to the
direction of H_a) at $k=0.1\text{nm}^{-1}$ versus H_a at 20K and at 300K.
Contrary to theory (Equation (10) and dotted line in the figure), which predicts maximum intensity for $k\|H_a$ due to
small misalignments of the magnetic moments, the experimental intensity at 300K is highest for $k\perp H_a$. By Equation
(7) the observed
aspect ratio be shown to imply contrast from moments oriented nearly antiparallel to the field. However, the 20K data
approach the theory at higher fields.

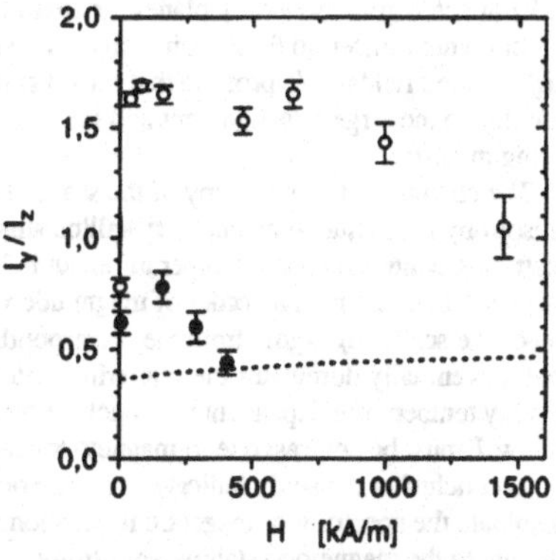

Figure 4 : Aspect ratio I_y/I_z (scattering
intensities perpendicular (I_y) and parallel (I_z)
to direction of applied magnetic field H) at
$k=0.1\text{nm}^{-1}$ versus H for n-Ni at 20K ($\bullet$) and at
300K (o). Dashed line: Theory (Equation
(10)). Experiment and theory are $\pm15°$ sector
averages in θ.

DISCUSSION AND CONCLUSIONS

We have derived a theory for the magnetic microstructure
of a nanocrystalline ferromagnet in the high field limit,
which is based on very general assumptions of homogeneity,
and in which the nuclear microstructure enters through the anisotropy fields. For isotropic microstructures we
derived expressions for the variation of SANS with magnetic field, scattering vector, and with the angle
between those quantities. Apart from the homogeneity conditions no assumptions are made on the geometry
of the microstructure; as a consequence the anisotropy field interference function $S(k)$ is perfectly arbitrary.
In the special case of random anisotropy, $S(k)$ is a constant. If the anisotropy field suffers random jumps
across the grain boundaries, then $S(k)$ is a weighted sum of the interference functions of the individual grains.

The observed small reduction, relative to the coarse-grained material, of m_s in nanocrystalline Ni and the
slightly enhanced temperature dependence indicate small, but measurable reductions in a mean magnetic
moment and in an effective spin-wave stiffness. The observed effect on m_s is much smaller than that reported
in early studies [19], but our result corrects later reports that magnetic moment and magnetic interactions at
grain boundaries are unchanged relative to bulk [20]. Together with the observation of negligible porosity
the smallness of the changes in m_s and in its temperature dependence validate the homogeneity assumptions
that underlie the theory, both with respect to the atomic density and with respect to the local magnetization
density and local exchange constant. Indeed, the azimuthal average SANS data are in good agreement with
the theory, in particular with respect to the overall variation of the intensity and with respect to the variation
of a characteristic length scale with the field; the agreement at small fields is even much better than would be
expected from the restrictive high field limit assumption underlying the theory. We conclude that the elements of magnetic microstructure that give rise to SANS are defects which obey the equations of micromagnetics, and that the changes in the magnetic microstructure with H_a, which give rise to changes in the
SANS signal are not displacements of domain walls, but result from changes in the correlation length and
hence in the magnetic response to an entirely stationary anisotropy field microstructure. In accordance with
our magnetization data, the good agreement between theory and SANS experiment indicates that the exchange coupling between neighboring grains, mediated by the matter at the grain boundaries, is strong.

While the anisotropy of the scattering pattern agrees with the theory at low temperatures and high fields, at
room temperature it does not indicate the predicted small misalignments of the magnetic moments. Instead,
the observed anisotropy suggests antiparallel orientation of moments at fields up to 1.8T. This cannot be due
to magnetocrystalline anisotropy, for which $H_p\approx 15\text{kA/m}$ (2000Oe) for Ni at 300K [21]; furthermore, measurements on cold worked Ni indicate that H_p from magnetoelastic anisotropy is also too weak to explain the
observation. We conclude that the observed misalignment of the moments at high fields originates from the
much stronger exchange interaction, as opposed to the weaker anisotropy fields. While the nature of these
defects is unclear, we might speculate on antiferromagnetic interactions due to modified atomic short-range

order at some grain boundary planes, or alternatively on 360° walls, that is topological defects in the spin system which suffer no force from homogeneous magnetic fields and can therefore persist in a material up to high applied fields. We propose that only a small volume fraction of the material contains such defects, but that due to the large misalignment and the resulting large scattering contrast their signal dominates the scattering intensity.

The change in the anisotropy of the scattering pattern at low temperatures is readily understood in terms of anisotropy fields due to magnetocrystalline anisotropy: these fields cause small misalignments and hence low scattering contrast at room temperature, but in Ni the magnitude of the magnetocrystalline anisotropy field increases by more than an order of magnitude when the temperature is reduced from 300K to 20K [21], and hence the scattering signal from the corresponding variation in M increases by two orders of magnitude, so that it eventually dominates the scattering pattern at low T. Since magnetoelastic anisotropy in Ni is only weakly temperature dependent we conclude from the evolution of the scattering with T that what is observed at low T must be contrast due to magnetocrystalline, not magnetoelastic anisotropy.

In conclusion the study indicates that micromagnetics defects are correctly modeled by the theory and dominate the spectrum of magnetic fluctuations in nanocrystalline Ni, but that only at low temperatures they are due to the magnetocrystalline anisotropy, which might be considered as the most obvious origin of anisotropy in nanocrystalline samples.

ACKNOWLEDGMENTS: This work was supported by Alexander von Humboldt-Foundation (J.W.) and by Deutsche Forschungsgemeinschaft (SFB 277).

REFERENCES

1:     Herzer, G., Mater. Sci. Engg. A133 (1991), 1-5.

2:     Harris, R., Plischke, M., and Zuckermann, M.J., Phys. Rev. Lett. 31 (1973), 160 - 162.

3:     Wagner, W., Wiedenmann, A., Petry, W., Geibel, A., and Gleiter, H., J. Mater. Res. 6 (1991), 2305. Löffler, J., Wagner, W., Van Swygenhofen, H., and Wiedenmann, A., Nanostruct. Mater. (in press).

4:     Kohlbrecher, J., Wiedenmann, A., and Wollenberger, W., Physica B213 - B214 (1995), 576- 581.

5:     Brown, W.F., 'Micromagnetics' (Wiley Interscience Publishers, New York 1963).

6:     Seeger, A., and Kronmüller, H., J. Phys. Chem. Sol. 12 (1960), 298-313.

7:     Göltz, G., Kronmüller, H., Seeger, A., Scheuer, H., and Schmatz, W., Phil. Mag. A54 (1986), 213.

8:     Kronmüller, H., in 'Moderne Probleme der Metallphysik', Vol. 2, edit. A. Seeger (Springer, Berlin 1966), 24 - 156.

9:     Holz, A., and Scherer, C., Phys. Rev. B50 (1994), 6209-6232.

10:    Childress, J.R., Chien, C.L., Rhyne, J.J., and Erwin, R.W., J. Magn. Magn. Mater. 104-107 (1992), 1585-1586.

11:    Moelle, C.H., Schriftenreihe Werkstoffwissenschaften 8 (Berlin, Köster 1996).

12:    Erb, U., Nanostruct. Mater. 6 (1995), 533 - 536.

13:    Halpern, O., and Johnson, M.H., Phys. Rev. 55 (1939), 898 - 923.

14:    Weissmüller, J., McMichael, R.D., Barker, J., Shull, R.D.; in preparation.

15:    The somewhat arbitrary terms $(4\pi m_s)^2$ in the definitions of S_H and R, that cancel in the product $S_H R$, reduce R to a dimensionless function, and S_H to a function with the units of a scattering cross section.

16:    Nosé, H., J. Phys. Soc. Japan 16 (1961), 2475-2481.

17:    Argyle, B.E., Charap, S.H., and Pugh, E.W., Phys. Rev. 132 (1963), 2051-2062.

18:    Weissmüller, J., in 'Nanomaterials: Synthesis, Properties, and Uses', editors A.S. Edelstein and R.C. Cammarata (Institute of Physics, Bristol, England, 1996), Chapter 10.

19:    Schaefer, H.E., Kisker, H., Kronmüller, H., and Würschum, R., NanoStruct. Mater. 1 (1992), 523.

20:    Kisker, H., Gessmann, T., Würschum, R., Kronmüller, H., and Schaefer, H.E., NanoStruct. Mater. 6 (1995), 925-928.

21:    Tebble, R.S., and Craik, D.J., 'Magnetic Materials' (Wiley, London 1969).

OPTICAL PROPERTIES OF METAL WIRE ARRAY COMPOSITES

LAURA LUO, T.E. HUBER*
Department of Applied Mathematics and Physics, Polytechnic University, Brooklyn, NY 11201

ABSTRACT

According to effective medium theories, electrically conducting composites consisting of parallel metal wires embedded in a transparent dielectric can propagate light in the direction of the wire length. We have prepared densely packed arrays (76% volume fraction) of 10-μm diameter indium wires by high pressure injection of glass microchannel plates. For wavelengths longer than 100 μm (k<100 cm^{-1}) the absorption of the wire array is almost three orders of magnitude smaller than that of an indium foil of equal thickness. The measured absorption increases as k$^{0.45\pm0.07}$ and can be accounted for by including magnetic dipole effects.

INTRODUCTION

The design of composite materials has flourished in the last few years.[1] Composites consisting of metals or semiconductors finely dispersed in a dielectric host are a dramatic example of how the electronic and optical properties are determined by the microstructure. Aspnes, Heller, and Porter,[2] on the basis of the Maxwell-Garnett (MG) model of the dielectric constant of composites, proposed that it is possible to prepare both optically transmissive and electrically conductive metallic films by appropriate design of the microstructure. High optical transmittance can be achieved when the metal units are isolated from each other in the direction of the photon electric field. The optical electric field can then induce a self-screening depolarization charge on the surface of the metal structures which prevents it from penetrating further into the metal. Elastic scattering is inhibited if the composite building units and spatial periodicity are smaller than the wavelength of light. An array of parallel metal cylinders would then be transparent to light propagating along the cylinders axis and of wavelength much larger than the cylinders diameter and separation. Conductivity is not impaired as long as the electron mean free path is much smaller than the wire diameter. Metal films of thickness greater than several hundred angstroms are optically opaque and the prospect of designing metal-insulator composites to be conductive while retaining polarization-independent optical transparency over micron-length scales is exciting. Also, this leads to a variety of possible applications in optoelectronic and photoelectrochemical technologies.[3] Moreover, there has been growing activity in the development of photonic band-gap materials structured in the direction perpendicular to the photon path.[4] While most of such effort has concentrated on dielectrics, there is recent interest in metallic-insulator structures for low frequency applications.[5]

There have not been experiments or materials which unambiguously test the aforementioned ideas. Heller, Aspnes, Porter, Sheng, and Vadimsky[3] have experimented with photoelectrochemically deposited Pt metal films. Because of the structural complexity, the films utilized are composed of small particles assembled in three levels of structure, and since the films are only partially conductive, the experiment is not conclusive. M.J. Tierney and C.R. Martin[6] have considered arrays of cylinders of diameter d small relative to the wavelength of light and of length l comparable to such wavelength. The metal particles were

Mat. Res. Soc. Symp. Proc. Vol. 457 © 1997 Materials Research Society

prepared by electrochemical deposition into the channels of a mesoporous membrane. In those experiments the cylinders are too short and their relative position is not well controlled. Recently, Foss, Hornyak, Stockert, and Martin[7] have studied arrays of gold cylinders of controlled radius and aspect ratio. Cylinders of diameter between 30 and 60 nm and length $l < 800$ nm were prepared. Although the optical absorption of such system follows qualitatively the MG model predictions, the system does not lend itself to a simple description at short wavelengths where the cylinder diameter $d \le \lambda$. The authors introduce a phenomenological modification of MG to account for dynamic depolarization effects. Aside from the fact that the conductivity of those arrays has not been demonstrated, the length of the cylinders, as well as their relative position within the membrane channels, has a considerable degree of randomness and a more clearly defined sample geometry is needed to test these ideas.

EXPERIMENT

In order to demonstrate the effect unambiguously it is important to use far-infrared (FIR) light. In this frequency range the electromagnetic field penetration depth δ is much smaller than the wavelength of light λ and the metal behaves as an almost perfect reflector. Here, we examine the FIR transmission of thick (90 μm) samples of 10-μm diameter In wire arrays. The wire arrays were prepared by filling the channels of a glass microchannel plate (MCP) with indium. The dense, 76% metal content, microstructure is highly reproducible. The composite is conducting as the In metal fills the full length of the channels. Our previous work has demonstrated the feasibility of injecting porous dielectrics of characteristic pore diameter as small as 5 nm with some selected low melting point materials (Se, Te, In) and alloys (Bi_2Te_3) by pressure-forcing their melts into the interconnected pores.[8] The MCP samples[9] were cleaned by subsequently washing in acetone, methanol, and distilled H_2O, and then dried at 150 °C for several hours; they were injected with the In melt at 400 °C and 60,000 psi. The composite samples were mechanically polished in the shape of a plate with the wires running perpendicular to the plate surface. Fig. 1 shows an electron micrograph of the wire array composite.

Figure 1. Top view of an hexagonal indium wire array prepared by high-pressure injection of the In melt into a glass microchannel plate. Indium crystallites can be seen in the 10-micron diameter wires.

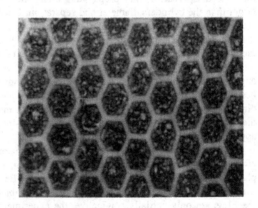

10 μm

Infrared absorption spectra were measured with a Bomem FTIR DA8 spectrophotometer.[10] A Hg lamp, Mylar film beam splitters, and DTGS detector were used. The instrument's beam divergence at the sample corresponds to a ratio of focal length to beam diameter of 4. An area of 3 mm^2 of sample was probed by the beam. Spectra were taken by averaging 8,000 scans. Figure 2 shows the transmission I/I_o of the empty MCP in the $k < 100$ cm^{-1} FIR region of transparency of the glass. The reference I_o is the intensity transmitted through an aperture of the same size as the sample under the same experimental conditions. The resolution is 1 cm^{-1}. The noise is typical of measurements in this frequency range, it increases at around 20 cm^{-1} as the lower limit of operation of the instrumental resources is approached. The empty MCP transmission spectrum shows a slow decrease with increasing frequency. This trend continues in the range between 100 and 200 cm^{-1} (not shown). Also shown in Fig. 2 is the transmission of the In-MCP composite (8000 scans). The transmission of the wire array also shows an overall decrease with increasing frequency and a much stronger frequency dependence.

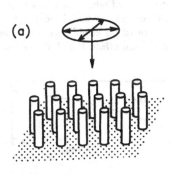
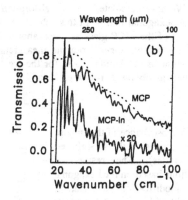

Figure 2. (a) Unpolarized FIR light impinges parallel to the channels. (b) Transmission of the microchannel plate (MCP) and of the 10-micron diameter In wire array composite (MCP-In). The dotted line is the result of a calculation based on Eq. 1 as described in the text. The thickness of the MCP is 100 μm and that of the composite is 91 μm.

DISCUSSION

Quasistatic effective medium theories of the dielectric constant of composites take into account the modification of the external electric field by the dipole fields of individual polarizable entities, in this case, the metal wires embedded in the insulating medium.[11,12] The structure shown in Fig. 1 can be approximated by a hexagonal lattice of very long cylinders. An approximate analytical solution which includes higher-order electric multipole terms and interparticle effects is available for a square lattice. For metal volume fraction $v_m < 0.78$, the dielectric constant of the composite is given by Rayleigh's expression:[13]

$$\frac{\varepsilon}{\varepsilon_i} = \frac{1 + \mathcal{P}v_m - a\mathcal{P}^2 v_m^4}{1 - \mathcal{P}v_m - a\mathcal{P}^2 v_m^4} \tag{1}$$

239

where $\mathcal{P} = (\varepsilon_m - \varepsilon_i)/(\varepsilon_m + \varepsilon_i)$ is 2π times the wire polarizability,[14] with ε_i and ε_m the dielectric constant of the insulator and metal, respectively, and $a = 0.306$ for a square lattice. For the hexagonal lattice, the coefficient a can be shown to be approximately $a = 0.13$.[15] For low metal volume fraction v_m the familiar MG result, $\varepsilon = \varepsilon_i (1 + \mathcal{P}v_m)/(1 - \mathcal{P}v_m)$ is obtained. The optical transmission is subsequently calculated from Eq. 1 in terms of the real and imaginary parts of the dielectric constant, both absorption and reflection losses are taken into account.[16]

The frequency dependence of the transmission of the empty MCP is due to the soda lime glass of which the MCP is made. Silica glasses show infrared absorption which varies as the square of the frequency and which is due to uncompensated charges distributed in the random medium.[17] To calculate the absorption of the empty MCP we use Eq. 1, v_m is the fractional volume of the empty channels in this case. The empty channels have an index of refraction equal to 1. For the glass we use an index of refraction $n_G = 2.6$ and absorption $\alpha_g = 0.085\ k^{1.97}$.[17] A good fit to the experimental data, shown in Fig. 2, is obtained for these parameters.

We have calculated the FIR absorption of the metal wire array composite using Eq. 1 for the hexagonal lattice. In this calculation we use the index of refraction and absorption coefficient of the MCP glass as given above. The optical constants of bulk In metal (ε_m) in the far-infrared are taken from the measurements of Golovashkin et al.[18] We adopt their modification of the Drude model for the weak anomalous skin effect. Fig. 3 shows the In absorbance as obtained from their data.

Figure 3. Absorbance of the In wire array composite in the far-infrared (open circles). The results of calculations based on the hexagonal lattice modification of Eq. 1 (Rayleigh) and on Eq. 2 (magnetic dipole) are shown as solid lines. The absorbance of a (thick) film of indium is also shown (dashed line).

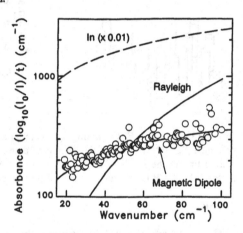

The composite absorbance as calculated from Eq. 1 is also shown in Fig. 3, its frequency dependence contrasts with the experimental results. Clearly, this simple approximation does not adequately represent the absorption of the composite, i.e., the calculated absorption is too small for $k < 50$ cm^{-1} and it has the wrong frequency dependence.

The optical properties of small spherical metal particles have been studied experimentally and theoretically for over twenty years because of their anomalous FIR absorption.[19] Sample preparation has been an important issue. For coated particles, the absorption is dominated by the enhancement of the electric dipole absorption due to the oxide coating. In that case, the absorption has a k^2 frequency dependence (like oxide glasses). It is believed that for bare particles clustering allows more closed paths for eddy currents to flow and enhances the absorption due to the magnetic dipole contribution. Although the matter of the absorption by small spherical metal particles is clouded by sample preparation issues, the fact is that in most experiments the absorption is enhanced.

We are not aware of a calculation of eddy current dissipation for cylinders. An estimate can be made as follows. In our case the penetration depth δ is much smaller than the cylinder diameter d and the main effect is the magnetic field exclusion from the wire due to induced eddy currents. Given a periodic magnetic field of amplitude B_0 and of direction perpendicular to the wire length, the eddy currents amplitude is approximately $I_0 = cdB_0/4$, where c is the speed of light. Absorption ensues because the current flow is dissipative. The current distribution is restricted to the periphery of the wire and shows an exponential damping with distance. The penetration length is $\delta = (c/k\sigma)^{1/2}/2\pi$, with σ the electrical conductivity of the metal. The average resistance per unit length of wire where eddy currents flow can be approximated by $1/(\sigma\delta d)$. If the average distance between wire centers is s, the power dissipated per unit volume is $c^2 d B_0^2 / 16 \sigma\delta s^2$. The absorption coefficient is given by the ratio of the power dissipated per unit volume to the electromagnetic field energy flow $cB_0^2/8\pi$. Therefore, since $s \sim d$, we find that the magnetic dipole (MD) contribution to the absorption is:

$$\alpha_{MD} = 2\pi^3 n_g v_m k(\delta/d) = \pi^2 n_g (v_m/d) (c/\sigma)^{1/2} k^{1/2} \qquad (2)$$

For $\sigma = 1.07 \times 10^{17}$ sec^{-1}, the conductivity of bulk indium, the skin depth at 30 cm^{-1} is $\delta = 154$ nm and we obtain $\alpha_{MD} = 57$ cm^{-1}. The frequency dependence predicted by Eq. 2, $k^{1/2}$, closely matches the experimental frequency dependence in the range investigated, which is $k^{0.47\pm0.07}$. A fit of Eq. 2 to the experimental data is shown in Fig. 3. Losses due to reflection have been neglected since they are much smaller than absorption losses. From this fit we obtain an effective surface conductivity of 1.4×10^{15} sec^{-1}. This is about two orders of magnitude less than the bulk conductivity. It is well known that actual eddy current losses in microwave resonators are always in excess to those calculated from bulk parameters due to surface roughness, oxide layer, and anomalous skin depth effects.[20] Such surface anomalies are likely to be more important at the higher frequencies employed here since the skin depth then becomes very small.

The FIR source used in these experiments is not well collimated. In fact, the ratio of the beam aperture to its focal length is approximately 0.25. The acceptance angle θ_A defined by the wires is approximately n_G $d/t = 0.26$, where d is the wire diameter and t is the wire length or sample thickness. Clearly, for angles of incidence $\theta \gg \theta_A$ the absorption increases since for large angles the photon electric field has a significant component along the wire length which is not screened. This would result in a somewhat larger absorption than that predicted by the simple approach described above. Although the propagation of electromagnetic waves perpendicular to the wire length and its polarization dependence is the subject of much recent work,[4] a systematic study of propagation for intermediate angles is still lacking.

CONCLUSIONS

It has been proposed that good optical transmission and high electrical conductivity can be simultaneously achieved in a metallic microstructure where there is electrical isolation along the direction of the photon electric field (i.e., the photon and current-driving electric field are perpendicular), as a self-screening charge is developed at the surface of the metal units by the local optical field. We have prepared a composite material which constitutes a physical model of this behavior. We find that the far-infrared transmission of a dense array of 10-μm diameter parallel wires along the wires' length is almost three orders of magnitude higher than that of a metal film of equal thickness. The composite absorption as calculated from quasistatic effective medium theories does not account for the experimental results. We propose that electromagnetic energy losses through dissipation by eddy currents, or magnetic dipole effects, can account for the frequency dependence of the measured absorption.

ACKNOWLEDGEMENTS

This work was supported by the U.S. Army Research Office through grant #DAAH04-95-1-0117. The authors thank S. Arnold and P. Sheng for valuable discussions concerning this work. Also, we gratefully acknowledge the assistance of D. Chacko with sample preparation and C. Huber for use of the FTIR.

* Also at the Graduate School of Arts and Sciences, Howard University, Washington, DC.

REFERENCES

1. An overview can be found in the Proc. 1st Int. Conf. Nanostructured Materials, Cancun, 1992, edited by M.J. Yacaman, T. Tsakalakos, and B.H. Kear, Nanostructured Materials 3 (1-6) (1993).
2. D.E. Aspnes, A. Heller, and J.D. Porter, J. Appl. Phys. 60, 3028 (1986).
3. A. Heller, D.E. Aspnes, J.D. Porter, T.T. Sheng, and R.G. Vadimsky, J. Phys. Chem. 89, 4444 (1985).
4. For a review, see Y. Yablonovitch, J. Opt. Soc. Am. B10, 283 (1993).
5. M.M. Sigalas, C.T. Chan, K.M. Ho, and C.M. Soukulis, Phys. Rev. B52, 11744 (1995).
6. M.J. Tierney and C.R. Martin, J. Phys. Chem. 93, 2878 (1989).
7. C.A. Foss, G.L. Hornyak, J.A. Stockert, and C.R. Martin, J. Phys. Chem. 98, 2963 (1994).
8. C.A. Huber, M. Sadoqi, T.E. Huber, and D. Chacko, Advanced Materials 7, 316 (1995).
9. Galileo Electro-Optics Corp., Galileo Park, Sturbridge, MA.
10. Bomem, Hartmann and Braun, Quebec, Canada.
11. D.E. Aspnes, Thin Solid Films 89, 249 (1982).
12. D. Stroud and F.P. Pan, Phys. Rev. B17, 1602 (1978).
13. L.K.H. van Beek in Progress in Dielectrics, Vol.7, edited by J.B. Birks (CRC Press, Cleveland, Ohio, 1967). Recent work on this expression includes N.A. Nicorovici and R.C. McPhedran, Phys. Rev. E54 1945 (1996).
14. L.D. Landau and E.M Lifshitz, Electrodynamics of Continuous Media, 2nd ed. (Pergamon, N.Y., 1966), p. 280.
15. T.E. Huber and L. Luo, to be published.
16. M. Born and E. Wolf, Principles of Optics (Pergamon, New York, 1959), Ch. 13.
17. W. Bagdade and R. Stolen, J. Phys. Chem. Solids 29, 2001 (1968).
18. A.I. Golovashkin, I.S. Levchenko, G.P Motulevitch, and A.A. Shuvin, JETP Letters 24, 1093 (1967).
19. D.B. Tanner, A. J. Sievers, and R.A. Buhrman, Phys. Rev. 11, 1330 (1975). T. Won Noh, S. Lee, Y. Song, and J.R. Gaines, Phys. Rev. B34, 2882 (1986).
20. For a review, see W.H. Hartwig, Proc. of the IEEE 61, 58 (1973).

THE FORMATION OF METAL/METAL-MATRIX NANOCOMPOSITES BY THE ULTRASONIC DISPERSION OF IMMISCIBLE LIQUID METALS

V. KEPPENS*, D. MANDRUS*, J. RANKIN**, L.A. BOATNER*
*Oak Ridge National Laboratory, Solid State Division, P.O. Box 2008, MS 6056, Oak Ridge TN 37831-6056, vk1@ornl.gov
**Brown University, Box D, 182 Hope St., Providence RI 02912

ABSTRACT

Ultrasonic energy has been used to disperse one liquid metallic component in a second immiscible liquid metal, thereby producing a metallic emulsion. Upon lowering the temperature of this emulsion below the melting point of the lowest-melting constituent, a metal/metal-matrix composite is formed. This composite consists of sub-micron-to-micron-sized particles of the minor metallic phase that are embedded in a matrix consisting of the major metallic phase. The zinc-bismuth case was used as a model system, and ultrasonic dispersion of a minor bismuth liquid phase was used to synthesize metal/metal-matrix composites. These materials were subsequently characterized using scanning electron microscopy and energy-dispersive x-ray analysis.

INTRODUCTION

The special properties that can be obtained by forming metal-matrix composites have previously been extensively documented. While much of the prior attention given to these materials has been focused on metal matrices reinforced with ceramic particles or fibers [1], the results reported for metal/metal-matrix composites show that the latter are no less interesting. For example, a new class of metal/metal matrix materials has been developed that exhibits extraordinary mechanical properties [2, 3, 4]. These materials are composed of a mixture of Cu plus 10-30% of a metal X that is immiscible with Cu. The mixture is severely deformed to produce a nanometer-scale microstructure of immiscible X filaments (or lamellae) within the Cu-matrix. Such processed composites have a strength that is substantially higher than those reported for any traditional Cu alloy.

In efforts to improve the mechanical properties – in particular the hardness – of materials, Singh *et al.* have dispersed approximately 20 wt.% Bi in Zn, using a melt-spinning technique [5]. This technique produced a metal/metal-matrix composite of nanosized Bi spheres entrained in a Zn matrix where the size of the Bi particles was controlled by adjusting the wheel speed used in the melt-spinning process. Hardness measurements performed on these materials showed that a decrease in the size of the Bi nanodispersoids leads to an increase in hardness.

In the present work, a new approach to the formation of bulk metal/metal-matrix composites is presented. High-intensity ultrasound has been used to disperse one metallic liquid in a second immiscible liquid metal thereby forming a metallic emulsion. When this emulsion is cooled, a metal/metal-matrix composite is formed consisting of minor-phase particles dispersed in the solidified major phase.

The basic idea of using ultrasound for mixing immiscible *liquids* is, of course, not new. In 1926, Wood and Loomis [6] reported that if two immiscible liquids such as oil and water are simultaneously subjected to ultrasonic radiation, an emulsion or colloidal suspension is formed as

243

a result of the forces acting at the interface between the liquids. A further extensive study of the mechanism of emulsification and coagulation by ultrasonic waves in water-oil and mercury-water/organic liquid systems was carried out by Bondy and Söllner [7,8].

A previous study of the influence of ultrasound on the production of unusual metallic mixtures was described by Schmidt and Ehret [9] and Schmidt and Roll [10]. Part of their work focused on the dispersion of 35 wt.% Pb in Al. By applying ultrasound with a frequency of 10 kHz to the melt, the Pb phase could be dispersed, forming spherical inclusions with a diameter of approximately 30 μm embedded in the Al matrix. However, the mixing of the components was incomplete, and a significant residue of Pb was found at the bottom of the crucible. More recently, a group in China has reported the use of ultrasound for the preparation of fine ceramic-particulate-reinforced metal-matrix composites [11] in which high-intensity ultrasound was used to disperse micrometer-size ceramic particles homogeneously in an aluminum matrix.

EXPERIMENTAL PROCEDURE

In order to investigate the application of ultrasound to the formation of metal/metal-matrix composites, the Zn-Bi case was selected as a model system, since both metals are relatively easy to handle based on their chemical reactivity and low melting points. Additionally, the wide miscibility gap in the Zn-Bi phase diagram (See Figure 1) made this system a particularly attractive candidate for the ultrasonic formation of metallic emulsions in varying concentrations.

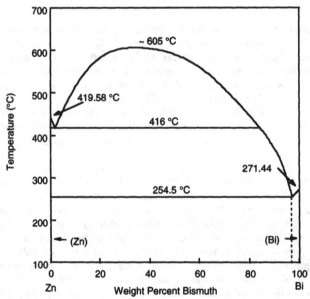

FIGURE 1: Equilibrium phase diagram of the Zn-Bi system showing the wide miscibility gap characteristic of this system [12].

The ultrasonic source used in the present experiments was a Misonix Sonicator Model W-385 that consists of a generator which feeds 20 kHz electrical energy to a transducer where it is transformed to mechanical vibrations. The ultrasonic energy is generated by a transducer that consists of a lead zirconate titanate piezoelectric driver. When subjected to an alternating applied voltage, this piezoelectric material expands and contracts at the 20 kHz driving frequency. The transducer is mechanically coupled to an acoustically resonant Ti-alloy horn assembly that vibrates in a longitudinal direction and transmits the high-frequency motion to the horn tip. A highly tapered Ti horn (termed a microtip) was used to achieve high-amplitude ultrasonic vibrations. A schematic representation of the ultrasonic processing system is shown in Figure 2.

The Zn-Bi composition selected for the present experiments consisted of 10 wt.% Bi. Once the two metals were weighed to achieve the appropriate proportions, they were then melted in a SiO_2 tube using a propane torch and were heated to approximately 650°C. To minimize oxide-formation, argon gas containing 4% H_2 was continuously sprayed over the surface. When the desired melt temperature was reached, the torch was turned off and the sonication process was initiated by immersing the vibrating microtip in the liquid. While the Ti-alloy used for the horn and tip represents the best horn material from the mechanical and acoustic point of view – combining outstanding acoustic properties with lightness, strength, abrasion resistance and chemical inertness – it tends to form an alloy with the Zn-phase. This interaction results in degradation of the microtip when it is immersed in the melt. Accordingly, in order to minimize the reaction of the Ti-tip with the sample, the sonication process had to be limited to short durations – typically 10-30 seconds. In the present experiments, the melt is first sonicated for 10-15 seconds while it begins cooling down at a typical cooling rate of 10°C per second. The solidification process is subsequently accelerated by spraying water on the outside of the SiO_2 tube, while maintaining the sonication conditions, until the solidification is complete. This results

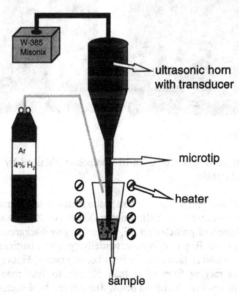

FIGURE 2: Schematic representation of the ultrasonic system.

in a total sonication time of 20-30 seconds. The resulting composite samples were then polished to obtain a flat surface suitable for examination in the scanning electron microscope.

EXPERIMENTAL RESULTS AND DISCUSSION

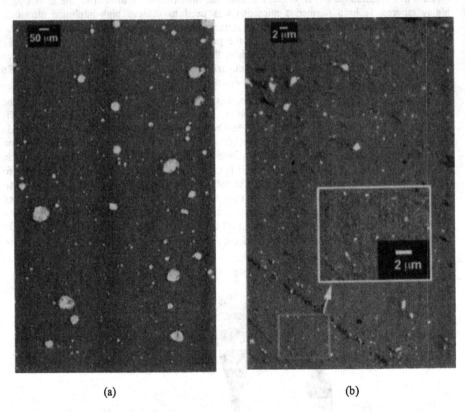

(a) (b)

FIGURE 3 (a) and (b): SEM image of a Zn-Bi composite obtained by sonication of the two molten immiscible liquid metals.

Figure 3(a) shows an SEM image of a composite metal/metal-matrix sample obtained after the sonication of a Zn plus 10 wt % Bi melt. Energy Dispersive X-ray analysis (EDX) confirms that the light colored dispersed particles are Bi, while the gray background consists of Zn. As evident in the micrograph, the Bi phase forms essentially spherical particles that are embedded in the Zn-matrix. This dispersion is, however, far from homogeneous: Figure 3(a) clearly shows Bi-particles with diameters ranging from more than 50 μm to less than 5 μm. Additionally, significantly smaller particles with diameters below 0.5 μm can be detected, as revealed in Figure 3(b). The presence of these sub-micron particles of Bi indicates that the application of high-intensity ultrasound can be a powerful tool for the formation of nanocomposite materials through the creation of metallic emulsions. However, it is clear from the observed microstructures that the major problem to be overcome in order to obtain uniform materials is achieving a marked improvement in the monodispersed size-distribution of the minor-phase particles.

In order to gain insight as to how metal/metal-matrix composites with a more-monodispersed minor phase can, in fact, be achieved by ultrasonic dispersion methods, a better understanding of the operative mechanisms responsible for the dispersion and the various parameters that control these mechanisms is needed. While previous investigations [6,7,8] have clearly shown that sonication of a liquid or melt significantly influences its behavior, there is, at present, no general consensus regarding the exact nature of the operative ultrasonic mechanism. According to Bondy and Söllner [7,8], the emulsification of immiscible liquids is due to the collapse of acoustic cavitation bubbles. Here "cavitation" refers to the formation, growth, and collapse of bubbles in liquids[13] that are initiated at nucleation sites where the tensile strength of the liquid is dramatically lowered -e.g., at small trapped gas bubbles. When sound passes through the liquid, these bubbles oscillate as a result of the rapid expansion and compression waves created by the sound field. As the bubble oscillates, it grows through several mechanisms and finally collapses catastrophically. When cavitation takes place near an interface, major changes in the nature of the bubble collapse occur. A markedly asymmetric collapse happens that generates a jet of liquid directed at the interface. This liquid jet may thus represent a mechanism for the injection of one liquid phase in the other. An other concept for the emulsification of immiscible liquids that does not involve cavitation has been presented by Suslick [14]. According to this model, ultrasonic compression and expansion effectively "stress" the liquid surfaces -eventually overcoming the cohesive forces that hold large droplets together. The larger droplets eventually burst into smaller droplets, and the liquids are thus emulsified.

Given the current lack of an accepted and appropriate model for the ultrasonic emulsification of liquids, an empirical investigation involving systematic variations of the ultrasonic parameters employed in the processing of a specific model system such as Zn-Bi is indicated. For a Zn-Bi sample with a fixed Bi concentration, three main parameters control the sonication process. These are: (1) the intensity of the applied ultrasound, (2) the sonication-time, and (3) the temperature of the melt as subjected to the ultrasound. Experiments are presently underway to explore the effects of variations of these important parameters in the case of the emulsification of Zn-Bi mixtures. In the case of experiments designed to explore increased time for ultrasonic processing, as noted earlier, the Ti-alloy used here for the ultrasonic horn tip reacts with Zn. Therefore, it has been necessary to keep the sonication time short and the temperature of the melt relatively low in order to minimize melt/tip interactions.. In an attempt to avoid this limitation, some preliminary experiments have been carried out with a stainless steel tip. The first results show no indications of a reaction between the tip and the melt, and this system will be utilized in future experiments carried out for the purpose of exploring the effects of variations in the parameters noted above on the emulsification of immiscible liquid metals.

CONCLUSION

The present investigations have shown that high-intensity ultrasound can be effectively used to disperse one liquid metallic component into another thereby forming metal/metal-matrix composites. In the case of the Zn-Bi system, the composite material consists of essentially spherical particles of Bi embedded in a Zn matrix. Dispersed nanophase Bi particles with diameters below 0.5 μm can be formed, however, the overall distribution of the size of the dispersed Bi phase is rather broad. Future investigations will, therefore, focus on new methods for more effectively controlling the size distribution of the minor metallic phase particles formed by ultrasonic dispersion techniques. Additionally, the resulting metal/metal-matrix Zn-Bi composite specimens will be characterized in terms of their physical, electronic, and mechanical properties and techniques for extending this general ultrasonic dispersal approach to higher melting point immiscible metals will be investigated.

ACKNOWLEDGMENTS

It is our pleasure to acknowledge the help of J. A. Kolopus, M. J. Gardner and D. W. Coffey. This work was supported in part by a grant from N.A.T.O and from the Fulbright Program, and by the Division of Materials Sciences, U.S. Department of Energy under contract DE-AC0596OR22464 with Lockheed Martin Energy Research, Inc.

REFERENCES

1. For a review, see I. A. Ibrahim, F. A. Mohamed, and E. J. Lavernia, J. Mater. Sci.
2. J. Bevk, J. P. Harbison, and J. L. Bell, Appl. Phys. 49, 6031 (1978).
3. D. Verhoeven, F. A. Schmidt, E. D. Gibson and W. A. Spitzig, J. Metals 38, 20 (1986).
4. D. Verhoeven, F. A. Schmidt, L. L. Jones, H. L. Downing, C. L. Trybus, E. D. Gibson, L. S. Chumbley, L. G. Fritzmeier, and G. D. Schnittgrund, J. Mater. Eng. 12, 127 (1990).
5. B. Singh, R. Goswami, and K, Chattopadhyay, Scripta Metall. Mater. 28, 1507 (1993).
6. W. Wood and A. L. Loomis, Phil. Mag. S.7. 4, 417 (1927).
7. C. Bondy and K. Söllner, Trans. Faraday Soc. 32, 556 (1936).
8. K. Söllner and C. Bondy, Trans. Faraday Soc. 32, 616 (1936).
9. G. Schmidt and L. Ehret, Zeitschr. Elektrochem. 43, 869 (1937).
10. G. Schmidt and A. Roll, Zeitschr. Elektrochem. 45, 769 (1939); 46, 653 (1939).
11. L. Ma, F. Chen, l. Shu, J. Mater. Sci. Lett. 14, 649 (1995).
12. Hansen, Constitution of Binary Alloys, 2nd ed. (McGraw-Hill Book Company, New York, 1958).
13. H. G. Flynn in Physical Acoustics, edited by W. P. Mason (Academic Press, New York, 1964), 1B, p. 57.
14. K. S. Suslick, Scientific American 260, 80 (1989).

ENERGETIC-PARTICLE SYNTHESIS OF NANOCOMPOSITE Al ALLOYS

D. M. FOLLSTAEDT, J. A. KNAPP, J. C. BARBOUR, S. M. MYERS and M. T. DUGGER
Sandia National Laboratories, Albuquerque, NM 87185-1056 (dmfolls@sandia.gov)

ABSTRACT

Ion implantation of O into Al and growth of Al(O) layers using electron-cyclotron resonance plasma and pulsed laser depositions produce composite alloys with a high density of nanometer-size oxide precipitates in an Al matrix. The precipitates impart high strength to the alloy and reduced adhesion during sliding contact, while electrical conductivity and ductility are retained. Implantation of N into Al produces similar microstructures and mechanical properties. The athermal energies of deposited atoms are a key factor in achieving these properties.

INTRODUCTION

Energetic atoms can often be used to form alloy layers with properties superior to those of alloys formed by purely thermal methods. Ion implantation can introduce essentially any species up to high concentrations (10's of atomic percent) in any substrate, independently of its solid solubility. Energetic particles can not only overcome such thermodynamic limitations, but can also alter the microstructure by displacing atoms in the alloy with their kinetic energy before coming to rest. Implantation can thus form surface alloys that are supersaturated solid solutions, amorphous phases, or densely precipitated layers. Similar processes occur during the deposition of alloy layers using isolated, energetic atoms.

Here we consider synthesis of precipitation-hardened Al(O) alloy layers by implantation of O^+ into Al and by two methods using athermal deposition of atoms: electron-cyclotron resonance (ECR) plasma deposition and pulsed-laser deposition (PLD). Each method produces a high density of nanometer-size oxide precipitates in an Al matrix. Such structures are rather ideal nanocomposites since the dispersed precipitates impart exceptional mechanical properties to the alloy layer while it retains key metallic properties of the matrix. We demonstrate the microstructure of O-implanted Al and discuss its strength and tribological properties in terms of precipitation hardening. The PLD and ECR alloy layers deposited on Si are shown to have similar microstructures, and electrical conductivity and ductility are shown to be metal-like for these alloys. Examination of N-implanted Al shows similar fine precipitates and enhanced mechanical properties. We discuss key features of alloy systems and energetic-particle synthesis that allow such desirable microstructures to be formed.

SYNTHESIS AND EVALUATION OF Al(O) ALLOY LAYERS

Ion implantation of O^+ into well-annealed Al with several energies from 25 to 200 keV was used to form alloys with a nearly constant composition extending ~0.5 μm deep [1]. The microstructures were evaluated with transmission electron microscopy (TEM), as seen in Fig. 1a for 17 at.% O [2]. The electron diffraction pattern demonstrates that when the fcc Al matrix is tilted a few degrees off a zone axis, there is a diffuse oxide reflection under each weakened Al reflection; detailed considerations indicate that the phase is γ-Al_2O_3. This phase (cubic spinel, $a_o = 0.790$ nm) precipitates instead of the equilibrium hexagonal phase (corundum), apparently because its atomic spacings closely match those of the Al matrix. Dark-field imaging with a diffuse reflection illuminates a high density of precipitates 1.5 - 3.5 nm in diameter that appear

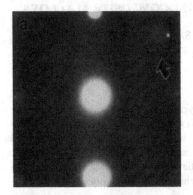
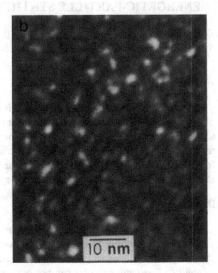

Figure 1. a) Electron diffraction pattern for O-implanted $Al_{83}O_{17}$ alloy with the Al matrix tilted off the [100] zone axis to show diffuse γ-Al_2O_3 reflections. b) Dark-field image of precipitates, obtained with a diffuse reflection.

10 nm

randomly dispersed and isolated from one another; see Fig. 1b. Their small size and lattice matching together imply that the precipitates are coherent with the Al matrix.

Several features of the Al-O alloy system appear responsible for the high density of small precipitates. These two elements react very exothermically with one another, giving O a very low solid solubility in Al, $<3\times10^{-8}$ at.% [3]. Further, each atom in the material is calculated to have been displaced many times, producing numerous point defects that provide nucleation sites for the oxide. The displacements and mobility of point defects in Al at room temperature probably also provide the mobility for O atoms to migrate and attach to precipitates.

The strengths of the Al(O) alloy layers are probed with ultra-low load "nanoindentation" to depths partially through the layer. Figure 2a shows the load response of Al implanted with 10 at.% O to indentation to 150 nm depth. Finite-element modeling was used to evaluate the mechanical properties of the alloy by using the known properties of the substrate and adjusting the properties of the layer to fit this experimental load versus depth response. The model divides the diamond tip, alloy layer and substrate into elements as in Fig. 2b, and their deformation during indentation is computed while maintaining contact with neighboring elements. The unloading portion of the curve is also shown in Fig. 2a; its initial slope at maximum depth is an elastic response that is used to evaluate Young's modulus. The loading portion of the curve depends on both Young's modulus and the yield stress. Finite-element modeling is essential to obtaining accurate properties for the layer because the observed response is due to the layer/substrate combination. For Al(O) layers on well-annealed Al, the response is softened by the substrate (yield stress = 0.041 GPa). The simulation in Fig. 2a was obtained with the commercial code ABAQUS [4] and fits the loading curve and unloading slope well with 2.4 ± 0.12 GPa for yield stress and 135 ± 7 GPa for Young's modulus of the alloy. Tests of implanted alloys with 5 - 20 at.% O give yield stresses of 1.4-2.9 GPa [1], much higher than for commercial Al alloys and like high-strength steels. The absolute error in evaluating these properties is judged to be < 20%.

The exceptional strength of O-implanted Al is due to the high density of hard precipitates. When the size and density are used in conventional dispersion hardening theory, good agreement

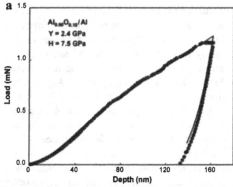
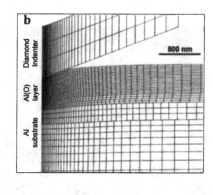

Figure 2. a) Nanoindentation load versus depth data (points) for Al implanted with 10 at.% O and simulation (curve) using a yield stress of 2.4 GPa. b) Mesh used for finite-element simulation in a).

is obtained with the magnitude and composition variation of the alloy strength [1]. Thus the mechanical properties are those of Al with a high density of particles blocking dislocation motion.

Implantation of O significantly improves the tribological properties of Al [1,5]. Whereas untreated Al shows "stick-slip" adhesion during pin-on-disk testing, implantation of as little as 5 at.% O eliminates it and changes the wear to a mild abrasive mode with a coefficient of friction of ~0.2. Benefits persist for several tens of cycles until the layer is worn through and stick-slip occurs. The reduced friction and wear are believed due to the high strength of the O-implanted surface layer since the strong material deforms less under the loaded pin and the contact area for adhesion is reduced correspondingly.

To form thicker Al(O) alloys, Al is evaporated onto a Si substrate with O_2^+ ions simultaneously incident from the ECR plasma source. These ions stream from the source with an energy of 30 eV, but the energy can be increased by biasing the specimen up to -300 V. The O atoms are

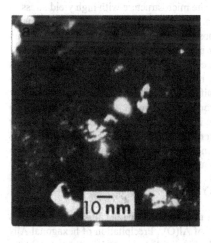

Figure 3. Dark-field images showing Al grains (larger white areas) and oxide precipitates (fine white spots) in a) ECR-synthesized $Al_{74}O_{26}$ and b) PLD-synthesized $Al_{69}O_{31}$ (greater magnification).

251

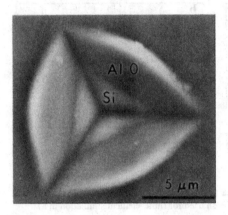

Figure 4. Secondary electron image of 1 μm -
deep indent of PLD Al$_{80}$O$_{20}$ alloy layer on Si.

calculated to penetrate the surface by 1.5-3.5 nm and to displace atoms in the growing layer. In these features, ECR synthesis is similar to high energy implantation, but the O$_2^+$ are implanted into a growing layer that can be micrometers thick. Dark-field imaging of an Al$_{74}$O$_{26}$ alloy layer shows fine grains of fcc Al (12 - 80 nm) with 1-2 nm oxide precipitates within the grains, as indicated in Fig. 3a. Indentation and modeling of ECR layers deposited near 100°C gives yield stresses of 0.62 - 1.3 GPa for 9 - 20 at.% O. This synthesis thus produces a microstructure similar to that for implantation and strengths that are high, although less than those of ion-implanted alloys of the same O concentration. The reduction in strength is not yet understood, but it is possible that deposited layers are not as fully dense as implanted layers.

Alloy layers of Al(O) can also be deposited with PLD by sequentially ablating Al and Al$_2$O$_3$ targets [6]. Composition is determined by the ratio of pulses per target, and the amount of material deposited per cycle is kept at ~1 monolayer to achieve a near-homogeneous deposition. The energies of atoms in the laser-ablated plasma plume are ~10 eV, which is insufficient to penetrate into the layer. However, the atoms reach the specimen with greater energies several hundred times k$_B$T (0.025 eV) and may thermally break surface bonds or directly displace surface atoms, which are less strongly bound. The dark-field image in Fig. 3b shows a microstructure of 5 - 25 nm fcc Al grains with ~1 nm oxides, similar to that with ECR but on a finer scale. The strengths of PLD alloys with 20 and 29 at.% O are 1.7 and 2.5 GPa, respectively [6], which are higher than for corresponding ECR alloys but below implanted alloys. Thus this relatively low energy deposition method also produces a nanocomposite microstructure with high yield stress.

The ECR and PLD layers allow us to examine two properties of the precipitated alloy layers. First, the electrical resistivities of ECR alloys deposited at 200°C with 27 at.% O and an evaporated pure Al layer were examined with a four-point probe. The alloy values of 65 - 89 μΩ-cm are significantly increased from that of the pure Al layer, 2.5 μΩ-cm, but are nonetheless metal-like and similar those of metal silicides. Electron scattering from the oxides apparently increases the resistivity, but the Al matrix is still significantly conductive. Second, the ductility of Al(O) is demonstrated in Fig. 4 by indentation through a 340 nm Al$_{80}$O$_{20}$ PLD layer and into the Si substrate to 1 μm total depth; the alloy/substrate interface can be seen. The alloy overlayer is smoothly displaced to the sides of the indent, but remains attached to the substrate and shows no evidence of fracture. Thus two key properties of metals are retained in our Al(O) with up to ~30 vol.% oxide, which is consistent with the continuous metal matrix and isolated precipitates.

EVALUATION OF N-IMPLANTED Al ALLOY LAYERS

We have recently examined N implantation of Al to determine whether high-strength alloys are produced and to compare alloy properties to those of Al(O). Precipitation of hexagonal AlN was expected, which could provide insight into the blocking of dislocations because the crystal

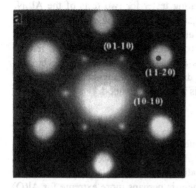
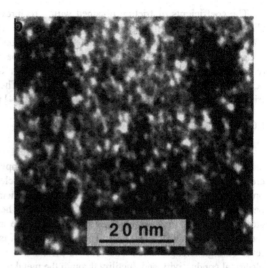

Figure 5. a) [111] diffraction pattern from Al implanted with 10 at.% N, showing weak AlN reflections (indexed), and b) dark-field image of AlN precipitates.

structure of AlN differs from that of cubic γ-Al_2O_3 and its bonds are more covalent. Annealed Al was implanted with 10 at.% N at room temperature and examined with TEM. Electron diffraction patterns show extra spots matching reflections of AlN, which align in the pattern of intense fcc Al reflections as in Fig. 5a. The precipitates align with the c-axis along a <111> matrix direction and with their {1-100} planes parallel to {2-20} Al planes. Dark-field imaging with these reflections shows precipitates ~2 nm in diameter as seen in Fig. 5b.

Nanoindentation was done on a specimen implanted with 5 at.% N to a depth of 0.5 μm, and the response curve is shown in Fig. 6. For loads $\leq$ 0.4 mN (depths $\leq$ 70 nm) the alloy layer appears quite strong; finite-element simulation using a yield stress of 2.15 GPa is seen to fit this portion of the curve well. However at higher loads, abrupt penetrations to greater depth are seen, termed "pop-ins" [7]. The origin of these features is not yet clear for N-implanted Al; they were not observed in O-implanted Al. Considering the initial part of the curve only, the $Al_{95}N_5$ alloy layer appears stronger than the corresponding $Al_{95}O_5$ alloy with yield stress = 1.4 GPa.

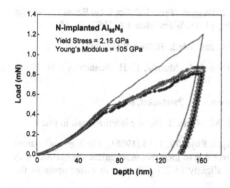

Figure 6. Nanoindentation of $Al_{95}N_5$ and fit to low-load portion using indicated alloy parameters.

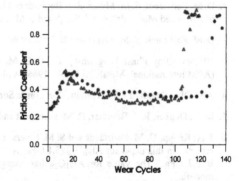

Figure 7. Friction coefficient versus wear cycles for N-implanted $Al_{95}N_5$.

The coefficients of friction obtained with a reciprocating tester for two tests of the $Al_{95}N_5$ alloy layer are shown in Fig. 7. The implantation has again eliminated stick-slip adhesion, and after a maximum at ~15 cycles the coefficient stabilizes at ~0.35 where it persists for over 100 cycles. Eventually, the implanted layer appears to be worn through and the adhesion returns (coefficient of friction approaches 1). This behavior is like that of O-implanted Al, but the benefits with only 5 at.% N appear to persist longer. The difference in steady-state coefficient of friction from earlier work with O implantation (~0.2) [4] is not yet understood.

CONCLUSIONS

By introducing insoluble species as individual energetic atoms into a metal matrix, nanocomposite structures can be formed with ion implantation. The numerous lattice defects created by the process provide a high density of nucleation sites, leading to nanometer-scale precipitate microstructures. This microstructural refinement is perhaps more extreme for Al(O) because of the high reactivity of the two elements. The ECR and PLD plasma syntheses using atoms with lower energies produce similar composite microstructures of 1-2 nm oxides within their small Al grains ($\lesssim$ 100 nm). The high density of isolated precipitates strengthens Al to the level of high-strength steels and reduces adhesion during dry sliding contact, while retaining electrical conductivity and ductility through the metal matrix. Implantation of N also produces a high density of fine precipitates in Al, and is very effective for strengthening and reducing adhesion during sliding contact. The indentation of N-implanted Al requires more study to understand "pop-in" abrupt yielding events.

ACKNOWLEDGEMENTS

The authors wish to thank G. A. Petersen for performing ion implantations, M. P. Moran for assistance with TEM examinations, and L. Sorroche for tribological testing. This work was supported by the United States Department of Energy under Contract DE-AC04-94Al85000, in part by its Office of Basic Energy Sciences and the Center of Excellence for Synthesis and Processing of Materials. Sandia is a multiprogram laboratory operated by Sandia Corporation, a Lockheed Martin Company, for the United States Department of Energy.

REFERENCES

1. D. M. Follstaedt, S. M. Myers, R. J. Bourcier and M. T. Dugger, "Proc. Intl. Conf. On Beam Processing of Advanced Materials", eds. J. Singh and S. M. Copley (TMS, Warrendale, PA, 1993), pp.507.

2. D. M. Follstaedt, S. M. Myers and R. J. Bourcier, Nucl. Inst. Meth. **B59/60**, 909 (1991).

3. "Binary Alloy Phase Diagrams", eds. T. B. Massalski, J. L. Murray, L. H. Bennett and H. Baker (ASM International, Metals Park, OH, 1986), Vol. 1, p. 143.

4. ABAQUS/Standard v.5.5, Hibbitt, Karlsson & Sorensen, Inc., Pawtucket, RI.

5. M. T. Dugger, R. J. Bourcier, D. M. Follstaedt and S. M. Myers, Tribology International, in press.

6. J. A. Knapp, D. M. Follstaedt and S. M. Myers, J. Appl. Phys. **79**, 1116 (1996). The high yield stress reported in this paper for $Al_{80}O_{20}$ (5.1 GPa) is in error due to incorrect indentation; the correct value is 1.7 GPa. The value for $Al_{71}O_{29}$ has changed slightly to 2.5 GPa due to refinements in the modeling.

7. A. B. Mann, J. B. Pethica, W. D. Nix and S. Tomiya, Mat. Res. Soc. Symp. Proc. **356**, 271 (1995).

MICROSTRUCTURAL VALUATION OF IRON-BASED COMPOSITE MATERIALS AS AN ECOMATERIAL

Norihiro Itsubo, Koumei Halada*, Kazumi Minagawa*,
and Ryoichi Yamamoto
Institute of Industrial Science, University of Tokyo, 7-22-1 Roppongi Minato-ku, Tokyo 106, Japan
*National Research Institute for Metals, 1-2-1, Sengen, Tsukuba, Ibaraki 305, Japan

ABSTRACT

One of an important method to realize is said that we should take recycle processes into consideration and select the material without the mixture of particular elements that make it difficult to recycle. Therefore, it is useful to control of microstructure for improvement.

From this point of view, we paid attention to "SCIFER (that is made from Kobe Steel Ltd.)" that has a recyclable formation (Fe-C-Si-Mn) and superior characteristic (tensile strength is 5000MPa). The grain size of this fiber is nano-size. In this study, we used this material and compounded it together with iron-matrix to make an iron-based composite for recycle and investigated the possibilities of realization. The difficulty of this study is to make this composite without injuring the fiber's microstructure. Therefore, we have adopted powder metallurgy which could fabricate composite at low temperature comparatively. Especially, Ultra Fine Particles (UFP) that would sinter at low temperature to bond the interface between fiber and matrix with keeping fiber's capacity. This method is useful to ascend the density of the matrix. Results are as follows.

(1) Utilization of UFP slurry made it possible to adhere UFP to the surface of fiber and seed powder. Still more, this procedure enabled it to make a thin film uniformly by selecting the condition of slurry density and procedure of dryness.

(2) Applying UFP to the surface of fiber and seed powder make it possible to get the bond between fiber and matrix. By the bond of interface, both fracture strength and energies have ascended remarkably due to pull out of fiber.

INTRODUCTION

Grovel problems around us (e.g., destruction of ozone layer, water and air pollution, acid rain and so on) become more and more serious[1][2][3]. We have borne the burden not only these environmental problems but limited resources, energy consumption, and population problems. Therefore, it is important to find of solutions to these problems[4][5]. "Eco-materials" is a keyword that was produced to realize sustainable developments with avoiding many grovel problems. When we design materials, it is necessary to consider both their capacities (e.g., tensile strength, elastic ratios) and environmental effects such as the quantity of waste. Therefore, we should select materials that are recyclable efficiently.

In this study, we took attention to SCIFER (that is made of Kobe Steel Ltd.) which has a recyclable composition (Fe-C-Si-Mn) comparatively. Still more this material has dual phases (ferrite and martensite) which enable to have superior characteristic. As a method of utilization of it, compounded it with iron matrix that has a same composition

Mat. Res. Soc. Symp. Proc. Vol. 457 ° 1997 Materials Research Society

for recycle and investigate a Fe-based metal matrix composite combined the property of recycle.

The most important problem to manufacture FRM is that how to bond the interface between fiber and matrix without destroying the fiber's strength by annealing. Especially, because the grain size of this fiber was made by cold drawing, the fiber was easy to be brittle by thermal exposure than the other fibers comparatively. To prevent interfacial reactions, it is available to coat the thin film on the surface of fiber in advance.

The aim of this study is to control the interface between fiber and matrix. To prevent from being brittle, using Ultra Fine Particles (<100nm) as a binder of them that sinter at low temperatures. To use the reinforcement effectively, it is important to make the films uniformly. Objective in this study is to establish the method which can apply UFP on the reinforcement uniformly and to investigate the effects of UFP to the iron-based composite.

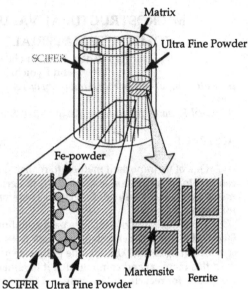

Fig.1 A schematic illustration of iron based composite, interface between matrix and fiber, and microstructure of fiber.

EXPERIMENTAL PROCEDURE

Establishment of the procedure to coat UFP on the surface of fiber

In order to make a coating film on the surface both of fiber and seed powder, it is important to pay attentions to scatter UFP around the surface uniformly. So we selected a slurry method that can use surface tension, and tried to make a coating film uniformly. Reinforcement is Fe-based fibers of 100É m diameter. Fig.2 reveals the coating process by the slurry method. In this procedure, we have attempted taking care of the two points; the density of UFP slurry and dryness of solvent that were considered to effect on the results.

Slurry was made of UFP (100nm) and ethanol that was used as a solvent. The conditions were

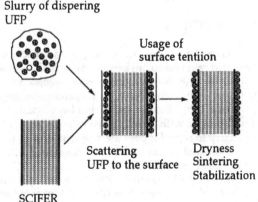

Fig.2 The schematic illustration of process applying UFP on the surface of fiber.

selected by macro observations; ethanol: UFP = 5:1, 10:1, 50:1, 100:1 (ratios of weight percentage) respectively.

The manufacturing method of slurry is stated as follows. Measured UFP and ethanol (purity 99.5%) have been admitted into a tube. After the mixture of fiber into the slurry, vibrations by ultra sonic wave have performed to break down aggregations of UFP and scatter them around the surface of reinforcements for three hours.

Investigation of the dry process was carried by three ways; (A) dried after putting up fiber in UFP slurry, (B) dried after pulling fibers out from UFP slurry and putting them horizontal, (C) applied slurry on the surface again after pulling fibers out from UFP slurry, these were performed taking the drying time and conditions of UFP in the slurry into account.

In the atmosphere of hydrogen these specimens were annealed at 673K for an hour in electric furnace to stabilize UFP on the surface of fiber. We used SEM and measured the thickness of coating film to investigate how the films have unevenness changing the density of slurry and process of dryness.

Applying UFP on the surface of iron powder

The method of applying UFP on the seed powder is required to keep UFP dispersion uniformly around the surface of the seed powder. Coating processes have been classified into the two types, namely, dry process and slurry process. In this study, we have selected the latter that can make a thin film easily with a solvent.

In this study, carbonyl iron powder (<30μm) was used to make matrix because that would sinter at low temperature comparatively. To make mixture powder between seed powder and UFP, we have mingled UFP with carbonyl powder at a ratio of 0.1-5 weight percentage. To make a coating film smoothly, these powders were mixed with ethanol (70ml) and agitated by a bowlmill method for three hours, and dried. Scanning electron microscopy (SEM) was used to characterize the surface of powder.

Manufacturing iron based composite using UFP coated.

To investigate interfacial bonding strength changing the thickness of coating films, we have selected three types of fibers: not applied UFP, 1μm applied UFP on the surface, and 10μm applied at previous procedures. Then, fibers and mixed powders were compounded respectively and green compacts were produced under a partial pressure of 392MPa for 10 seconds.

We have annealed these green compacts at 673K and 773K for an hour in the atmosphere of hydrogen to keep characters of reinforcements. Cooling rate is slow (10K/min) to prevent oxidation.

By these processes, we got iron based composites and checked the possibilities of this material by three points bending test and observations of the broken surface by scanning electron microscopy (SEM).

RESULTS and DISCUSSION

The investigation to apply coating films on the surface of fiber and seed powder
The effects of the density of slurry and dry processes on the thickness of coating films are shown in Fig.3.

In the case of process (B), there is scarcely any applied UFP on the surface of fiber, that results are independent of the density of slurry. This shows that the resistivity by the viscosity of slurry ascended by pulling out from slurry, and this force is greater than adhesive power on the surface.

In the case of (A) and (C), we have obtained rugged films in the condition of 5:1 because ultra sonic wave couldn't break UFP's cohesive force and disperse UFP uniformly around the surface. On the other hand, in dilute slurry, UFP slurries spread easily on the surface of fiber, because the viscosity is low. It is possible to control the thickness of films from 1 to 5mm.

According to this result, we found that UFP spread around the surface and adhered the whole of the surface. These reveal that utilization of ethanol as a solvent made it easy to stabilize UFP and build up a coating film by surface tension. Among the process, it should be take carefully that because UFP has very high cohesive energy, cause enlargement of powder.

The investigations of iron based composite using reinforcements applied UFP by three points bending tests and the observations the fractured surface are as follows.

Fig.4 shows a comparison of the fracture strength depend-

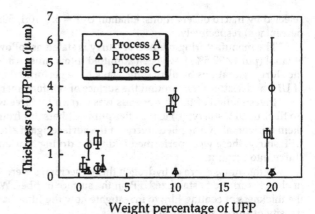

Fig.3 The relationships between the dryness process and the thickness of UFP films changing the weight percentages of UFP.

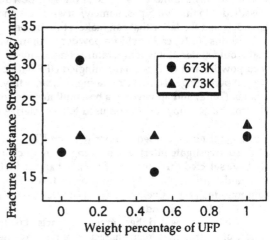

Fig.4 The relationship between weight percentages of UFP and bending strength.

ing on the weighting percentages of UFP by three bending tests. In the case of heating at 773K, the values are constant regardless of the quantity of UFP. In addition in these conditions, fibers have not been pulled out, and the fracture surfaces were smooth. This reveals that both heat and diffusion in the solid phase cause fiber to degrade.

On the other hand, in the case of 673K, the values were scattered and the sample of 0.1% UFP showed a maximum as seen in Fig.4. Load-displacement curve in this condition is compared with 1.0% applied in Fig.5. According to this figure, the case of 0.1% proved that stresses were propagated to fiber respectively, and indicate the raising both of maximum and fracture energy that was accompanied by the increase of fracture

N. Itsubo 4 of 6

working. As such condition, the pulled length was much longer than the other conditions. We have concluded that the release of stress and energy due to frictional force with pulling out have occurred after separation the interface between fiber and matrix.

CONCLUSION

Through the utilization of UFP to apply fiber that keeps the characteristic without annealing the high temperature, we have investigated the possibilities of realizing iron based composite. We can conclude the results as follows.

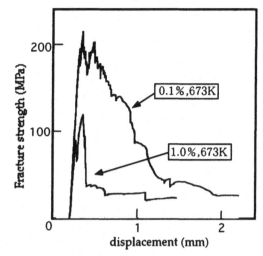

Fig.5 The relationships between fracture strength and displacement.

1. We have focused on the iron fiber that the grain size shows nano-size produced by heat treatment and cold hard working. It is necessary to sinter at low temperature in order to keep the characteristic of fiber.
2. Utilization of UFP as a binder makes it possible to bind between fiber and matrix. To keep the mechanical properties of fiber, UFP should be applied in manufacturing the iron based composite. In this study, it is possible to make films controlling the thickness by selecting density of slurry and dry method.
3. Applying UFP on the surface both of fiber and seed powder enable to get the bond of interface properly and rise fracture energy by the separation of interface and pulled out from matrix.

REFERENCES

1. H.spoel: Jounal of Metals, 42 4,(1990),38
2. UNEP: Environmental data report, 352pp., Basil Blackwell (1987)
3. Nuclear Instruments and Methods in Physics Reseach: A 308, (1991), p514
4. Y.Ando and R.Uyada: J.Crystal Growth 52, 178, (1981)
5. S.Iwata nd K.Hayakawa: Japan J. Appl. Phys., 20, 335 (1981)

NANOCRYSTALLINE SOLID SOLUTIONS OF Cu/Co AND OTHER NOVEL NANOMATERIALS

A.S. Edelstein,* V.G. Harris,* and D. Rolison,* J.H. Perepezko,** and D. Smith***

*Naval Research Laboratory, Washington, DC 20375
**Dept. of Materials Science, Univ. of Wisconsin, Madison-Madison, WI 53706
***Center for Solid State Science, Arizona State Univ., Tempe, AZ 85287

ABSTRACT

A brief review will be given of the preparation, synthesis and properties of the Cu/Co system. The special case of the synthesis of nanocrystalline $Cu_{.80}Co_{.20}$ by precipitation and reduction of hydroxides is discussed in more detail. It was found that the lattice constant of nanocrystalline $Cu_{.80}Co_{.20}$, determined from x-ray diffraction measurements, approximately fits Vegard's Law and the average nearest neighbor distance from both the Cu and Co atoms, determined from EXAFS measurements, is shifted from their bulk values. Samples given the minimum heat treatment needed to reduce the hydroxides contained Co-rich regions. Heat treatments cause the Co to segregate preferentially onto the surface of the Cu crystals. The presence of the Co delays the oxidation of the Cu surfaces.

INTRODUCTION AND BACKGROUND

The Cu/Co and Cu/Fe systems are similar in that they are simple, peritectic alloy systems with large miscibility gaps extending over nearly the entire concentration range. The similarity can be seen by comparing their phase diagrams (Fig. 1).[1] The positive enthalpies of formation, however, are different.[2] It is slightly more difficult to add Fe to Cu (ΔH=59 kJ/Mole) than Co to Cu (ΔH=44 kJ/Mole).

These systems are ideal systems for studying phase separation, i.e. spinodal decomposition in the non-stable region and nucleation and growth in the metastable region. For example, the phase separation of several different CuCo alloys annealed at 773 K has been studied theoretical.[3] More specifically Liu calculated the time dependence of the structure function S(k,T), which is the Fourier transform of the two-point correlation function and is directly proportional to the x-ray scattering intensity.[4] The calculation indicated that at 773 K, significant phase separation occurs in several minutes. High resolution microstructure analysis, using atom probe/field ion and transmission electron microscopy, have been made[5] of the decomposition of $Cu_{.90}Co_{.10}$. The composition profiles indicate a spinodal type decomposition. In addition, pure Co particles precipitate by heterogeneous nucleation at the grain boundaries. The exotherms due to phase separation measured by differential scanning calorimeter measurents[6] on nanocrystalline solid solutions of $Cu_{.50}Fe_{.50}$ and $Cu_{.50}Co_{.50}$ prepared by mechanical attrition are very similar. Vickers hardness measurements were also made on these samples. It was concluded that the strength of the solutions depended on both solid solution hardening and grain boundary hardening. Some care must be exercised in preparing samples to study phase separation because often, the as-prepared material is already a phase-separated nanomaterial. This is especially likely the further one is away from the solid solubility limits.

261

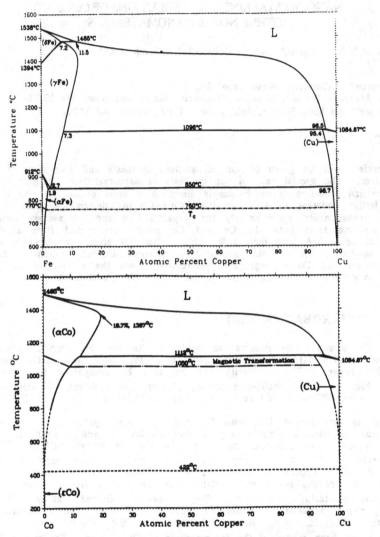

Figure 1 - Phase diagrams for Cu/Fe and Cu/Co.

A variety of techniques have been employed to prepare both Cu/Fe and Cu/Co in the form of nanocrystalline materials. For example, nanocrystalline Cu/Fe has been prepared by rapid quenching,[7-9] and vapor quenching using thermal evaporation[10], sputtering,[11] and using reverse micelles.[12] Nanocrystalline Cu/Co has been prepared by mechanical alloying,[6,13] rapid quenching,[7,9] electrodeposition,[14] and sputtering.[15,16] The latter technique has been used to prepare supersaturated solid solutions.[17,18] These approaches for producing nanocrystalline materials involve either rapid quenching, decreasing the crystallite size by high energy collisions, or limiting the size by forming an oxide shell.[12] If the crystallite size is sufficiently small, the surface energy contribution to the free energy becomes large enough that the free energy of the system is reduced by forming a solid solution.[19]

Monolayer films of Cu and Co on a Ru(001) surface have also been employed[20] to investigate surface alloy formation by interdiffusion across a linear surface. Several interesting features of this study are that, (1) monolayer films of Cu and Co are completely miscible and (2) the interface remains sharp even after a substantial number of atoms have crossed the interface. The concentration profiles are very different from those obtained by Fickian interdiffusion in which the profiles can be fitted to error functions. The unusual concentration profiles are explained[20] by a combination of adatom surface diffusion and exchange of adatoms with atoms in the monolayer.

Besides using these materials as model systems for studying nucleation and precipitation kinetics, there has been increased interest in the Cu/Co system and related systems when it was found that suitable heat treated films of $Cu_{.80}Co_{.20}$ exhibited a large magnetoresistance.[17,18] Because the magnetoresistance is so much larger than that seen previously in metallic systems such as permalloy, it was given the name giant magnetoreistance, GMR. The relationship between the GMR and the microstructure has been investigated[21] in sputter deposited Cu/Fe and Cu/Co films. There is need for materials with large magnetoresistance for use in read heads in magnetic storage devices. GMR films composed of several layers of ferromagnets with substantially different coercivities have been studied for use in spin valves[22] and as a new type of random access memory.[23] Decreasing the size of the devices is important. Using electrodeposition,[24] multilayered arrays of CuCo have been electrodeposited into the cylindrical pores of polycarbonate membranes. The quality of these devices is very dependent on the microstructure. Thus, it is interesting that researchers have been able to perform scanning friction force microscopy on CuCo ribbons.[25] The devices are fabricated in the form of multilayers since particulate samples require too large of fields to show a large change in resistivity.

In this paper a chemical approach is described for making nanocrystalline particles of Cu/Co and Cu/Fe. Some aspects concerning the structure and magnetism of these nanocrystals was presented previously.[26,27] We use a chemical method in which we first precipitate hydroxides and then reduce the oxides to the metallic forms. The products are very different for the two systems. For $Cu_{.80}Fe_{.20}$, even with the minimum heat treatment necessary to reduce the hydroxides, the nanocrystalline material is clearly phase-separated. For $Cu_{.80}Co_{.20}$, though it is not a single supersaturated solid solution, the Cu-rich and Co-rich regions are mixed on a much finer scale. A polyol process has also been employed to chemically synthesize nanocrystalline Cu/Co.[28]

METHOD OF PREPARATION

Preparation of $Cu_{.80}Co_{.20}$

Nanocrystalline $Cu_{.80}Co_{.20}$ was prepared starting from a solution of either the chloride or nitrate salts of cobalt and copper in distilled deionized water (DDW). This solution was sprayed into a stirring solution of NaOH. The NaOH solution was prepared using an excess (5%) of sodium hydroxide in DDW. After the addition of the metal salt solution, the resulting hydroxide-metal solution was stirred for an additional hour. Then the hydroxide-metal solution was filtered, washed and dried.

The dried powder was passed through a -325 mesh sieve.   The resulting hydroxide precipitates were then heated in flowing hydrogen gas (25 ml/min) for one hour at temperatures between 215 and 650 °C.

Preparation of Cu.80Fe.20

Nanocrystalline Cu.80Fe.20 was prepared using a procedure similar to that used for Cu.80Co.20, except iron (II) chloride and copper (II) chloride hexahydrate were used to make the metal solution.

CHARACTERIZATION

The methods that were used for characterizing the samples   included x-ray diffraction, extended x-ray absorption fine structure (EXAFS), x-ray photoelectron spectroscopy (XPS), nuclear magnetic resonance (NMR), transmission electron microscopy and magnetization measurements.   The following discussion presents some of the major features of these investigations.

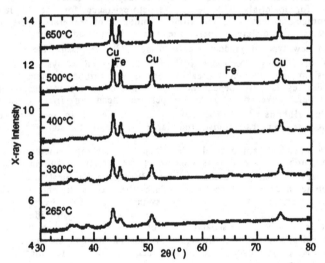

Figure 2 -   X-ray diffraction spectra from a series of $Cu_{0.80}Fe_{0.20}$ samples heated in flowing hydrogen gas for 1 hour at the specified temperatures.

X-ray Diffraction

The samples were examined by x-ray diffraction measurements by making θ-2θ scans using Cu K_α radiation in a conventional x-ray diffractometer.   Figures 2 and 3 show a comparison of the scans taken from Cu.80Co.20 and Cu.80Fe.20 after different heat treatments for one hour at the specified temperatures.   Also shown is a scan on an as-prepared Cu.80Co.20 hydroxide sample.   In the case of Cu.80Fe.20, the ratio of the intensity of the Fe to Cu diffraction peaks at each temperature is approximately what one would expect for the phase-separated elements based on the composition and the scattering factors. For $Cu_{.80}Fe_{.20}$ with increasing heat treatment temperature, the diffraction peaks of both the Cu and Fe increase in height and become narrower as the crystallites grow, but the ratio of their intensities remains approximately constant. The d-spacings of the Fe diffraction peaks correspond to the usual bcc phase of bulk Fe.

The x-ray spectra of annealed samples of $Cu_{.80}Co_{.20}$, shown in Fig. 3, are very different from that observed for the $Cu_{.80}Fe_{.20}$ samples. At the lowest heat treatment temperatures, 265 °C and 333 °C, there is only a single set of peaks corresponding to an fcc phase. This result does not necessarily imply that there is a single phase solid solution. Michaelson[29] has studied Cu/Co multilayer samples and has found that, for approximately equal thickness Cu and Co layer and modulation wavelength less than 10 nm, the x-ray spectra are the same as those of a solid solution. Thus, the x-ray spectra implies that there is either a single phase present, or that the structure is very fine grained.

In either case, it is clear that the hydroxides of the two systems behave very differently when they are reduced at low temperatures in flowing hydrogen gas. We believe that the Cu and Fe do not form mixed precursors, i.e. the Fe and Cu form

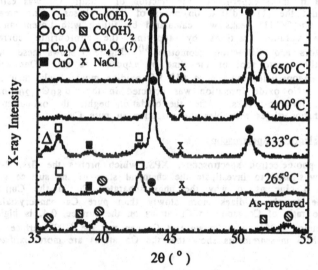

Figure 3- X-ray diffraction pattern from a series of $Cu_{.80}Co_{.20}$ samples heat treated for 1 hour in flowing hydrogen gas at the specified temperatures.

separate hydroxide and oxide crystals. Thus, upon reduction separate Fe and Cu crystallites are formed. Further, the decrease in surface energy of forming coherent fcc interfaces appears insufficient to overcome the tendency of Fe to crystallize into a bcc structure. The situation appears to be different in the case of Cu/Co. For this system, the data is most easily interpreted by supposing that either the hydroxide or oxide precursors contain a mixture of Cu and Co. Even though the x-ray spectrum consists of peaks consistent with a single structure, after the precursor is reduced at a low temperature, the product is either (1) a homogeneous supersaturated solid solution, (2) a supersaturated solid solution with a nonuniform composition, or (3) it consists of a very fine-grained nanostructure of Cu-rich and Co-rich phases. In any case, the Co atoms reside in fcc structures. Apparently, the presence of the Cu forces the Co to have an fcc structure and not the usual hexagonal, close-packed, room temperature equilibrium structure. An examination of ternary phase diagrams for the metals and oxygen supports our suggestion about the Cu/Fe and Cu/Co precursor phases being different in that there are more possible mixed precursors for Cu and Co than there are for Cu and Fe.

Values for the lattice parameter a were obtained from five x-ray diffraction peaks in our θ-2θ plots. These values were plotted versus $\cos\theta\cot\theta$. The values of a for $Cu_{1-x}Co_x$ were obtained by extrapolating linear fits of this data to the y axis. Values for a at x=0 and 1.0 were obtained from a sample heat treated at 600 °C for one hour to insure phase separation. Values for a at 0.2 were obtained from samples heat treated at 215 °C for one hour The values obtained for a are 0.36128, 0.36061 nm, and 0.35456 nm at x=0.0, 0.2, and 1.0, respectively. These values are compared with Vegard's law in Fig. 4. The error for the lattice parameter a was taken to be the errors in the linear fits discussed above. One sees that the a value for $Cu_{.80}Co_{.20}$ is in approximate agreement with Vegard's law but deviates toward larger values of a. This result does not allow one to select between the three possibilities listed above.

The oxidation of Cu-Co and Cu nanocrystals as a function of time of exposure was investigated by x-ray diffraction measurements.[27] Since only fcc Cu and Cu_2O were detected in these experiments, the fraction of metallic Cu was estimated from the areas under the (111) diffraction peaks of these phases. The area under the $Cu(111)$ and $Cu_2O(111)$ peaks was calculated by fitting with Lorentzian curves and converted to volume fractions by applying the appropriate corrections for structure factors and diffraction geometries.[30] Figure 5 shows these estimates as a function of the square root of the time of exposure to air. One sees that the $Cu_{.80}Co_{.20}$ nanoparticles reduced at 650 °C oxidize more slowly than the pure Cu nanoparticles. No oxide formation was detected in the $Cu_{.80}Co_{.20}$ nanoparticles after exposure for two weeks. After the oxidation begins, the oxidation rate in the $Cu_{.80}Co_{.20}$ nanoparticles resembles the oxidation in pure Cu.

X-ray Photoelectron Spectroscopy

X-ray photoelectron spectroscopy, XPS, which probes the first 2-5 nm from the surface, was used to investigate the chemical state of the surface atoms. This data provide a clue as to why the phase-separated Cu in the $Cu_{0.80}Co_{0.20}$ nanocrystalline sample oxidizes more slowly than pure Cu nanocrystals. It was found that the ratio of Co atoms to Cu atoms on the surface, 0.4, is higher than the bulk stoichiometry, 0.25. It is known that Co can wet the (100) surface of Cu.[31,32] Further, the XPS measurements show that the Co atoms are more oxidized than the Cu atoms.

Magnetization

Magnetization measurements were performed on a $Cu_{.80}Co_{.20}$ sample that had been heat treated at 215 C with a SQUID magnetometer. Figure 6 shows a plot of the saturation magnetization as a function of temperature. The temperature dependence may be due to coupling between ferromagnetic regions. Such coupling was inferred in ferromagnetic resonance measurements[33] on sputtered films. The saturation magnetization of bulk Co at room temperature is 161 emu/gm. Magnetizationmeasurements[34] on $Cu_{1-x}Co_x$ samples, prepared by sputtering,

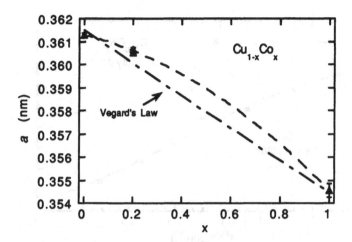

Figure 4 - Lattice constant a in nm versus Co concentration in $Cu_{1-x}Co_x$.

indicate that a homogeneous supersaturated solid solution of $Cu_{.80}Co_{.20}$ should not have any ferromagnetic component. The saturation magnetization[35] of as-prepared, ball-milled samples of $Cu_{.80}Co_{.20}$ also is much smaller than the values shown in Fig. 6. Thus, despite the x-ray data, it is clear that our $Cu_{.80}Co_{.20}$ sample heated for one hour at 215 °C is not a homogeneous solid solution. On the other hand, since the saturation magnetization per gram of Co is smaller than that of bulk fcc Co (161 emu/gm), some of the Co is in some form other than pure Co. Some of the Co is either in a supersaturated solid solution or is in the form of an oxide.

EXAFS

Extended x-ray absorption fine structure, EXAFS, measurements were performed to study the evolution of local structure around both Co and Cu atoms with heat treatment temperature. The x-ray absorption spectra above the K absorption edges of Cu and Co of the as-prepared hydroxide sample and heat treated $Cu_{.80}Co_{.20}$ samples were measured at the National Synchrotron Light Source. Data were analyzed using standard procedures[36,37] resulting in Fourier transforms of the EXAFS data. Figure 7 shows the Fourier transformed (FT) Co and Cu EXAFS data from samples annealed at temperatures ranging from 265-650 °C. The Fourier peaks represent atom shells where the position of the peak centroid corresponds to the bond distance (without electron phase shift correction) and the amplitude of the peaks is determined by the coordination and the atomic disorder (both thermal and structural) of the atoms contributing to the shell. These data are useful for comparing the local environment around the absorbing atoms to one another as a function of temperature or in comparison to known structures.

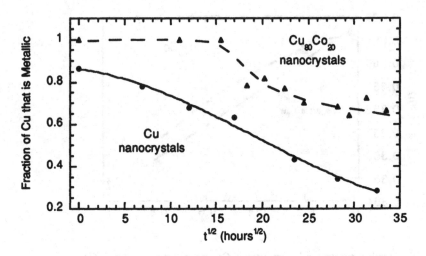

Figure 5 - Comparison of the oxidation of nanocrystalline $Cu_{0.80}Co_{0.20}$ and nanocrystalline Cu. Both samples were reduced at 650 $^{\circ}$C for one hour.

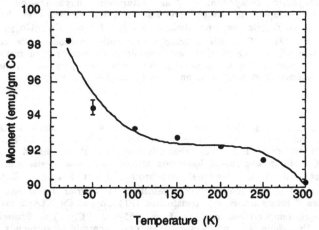

Figure 6 - The magnetic moment per gm Co in a sample of $Cu_{.80}Co_{.20}$ heat treated 215 $^{\circ}$C for one hour measured in a field of 30 kOe as a function of temperature.

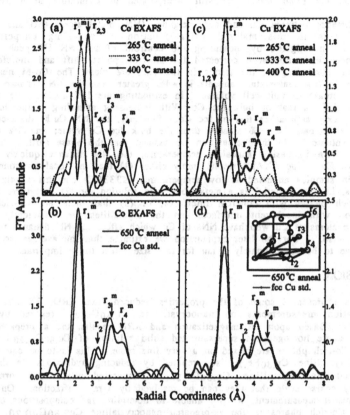

Figure 7 - Fourier transformed (FT) EXAFS data versus radial distance from the absorbing atom. The peaks superscripted 'o' and 'm' correspond to neighbors in an oxide and close-packed metal phases, respectively. The subscript denotes the atomic shell number relative to the absorbing atom. Cu foil data are included in panels (b) and (d) for comparison purposes. (Data analyzied using a k-range of 2.5-12.5Å^{-1} and k^2-weighting.)

In each panel some of the Fourier features are labeled with a super- and subscripted r. The subscript denotes the number of the coordination shell around the absorber, e.g. 1 corresponding to the nearest neighbor shell. The superscript, either an 'o' or 'm', denotes a neighbor in an oxide phase or a metallic phase, respectively.

In Fig. 7, the evolution of the Co and Cu environment indicates that the atoms initially reside in hydroxide phases, either mixed metal - or single metal - hydroxides, and are reduced after heat treatments to an fcc metal phase. However, in the Co EXAFS data one sees that the Co atoms are reduced at lower temperature than the Cu atoms. This is best seen in panel (a), where after an anneal at 333°C the Co appears nearly completely reduced to an fcc metal, whereas in panel (c) the Cu

269

atoms after the same anneal are still incorporated in a mixture of hydroxide/oxide phases.

The distance between Cu and near neighbor (NN) metal atoms (M) and Co and its NN M atoms in this material as a function of annealing temperature were qualitatively investigated by measuring the centroid of the NN FT peak. Although the distances have not been corrected for electron phase shift and therefore do not reflect true bond lengths, the trends in the data are clear. The Co-M near neighbor distance after low temperature anneals is ~4% greater than the NN distance in fcc Co; a 10.6% increase in unit cell volume. In estimating the NN distance we assume that the phase shift is that of bulk fcc Co. With increasing annealing temperature the NN distance quickly approaches that of fcc Co. Similarly, the NN Cu-M distance after low temperature anneals is ~4% smaller than the bulk fcc Cu value; an 11% decrease in unit cell volume. In this calculation we assume that the phase shift is that of bulk fcc Cu. With increasing annealing temperature, the NN distance quickly approaches that of fcc Cu. The EXAFS results imply either (1) that the local environment of Co and Cu in samples annealed at low temperatures ($\leq 333^\circ$C) contains a mixture of the two atoms, i.e. the material is a metastable solid solution, or (2) that there are Cu-rich and Co-rich phases with coherent interfaces, or (3) that there are Cu-rich and Co-rich region with incoherent interfaces but the crystallites are sufficiently small that the surface atoms of Cu who have NNs of Co which affect the NN distance trends of Cu and visa versa. For the later explanation we estimate that the average crystallite size would have to be approximately 2 nm for this size effect to be important.

DISCUSSION

We have presented some of the properties deduced from XRD, EXAFS, XPS, and magnetization measurements on nanocrystalline $Cu_{.80}Co_{.20}$ prepared by reducing hydroxides. Based upon the magnetization and XRD results, the as-prepared material is not a single homogeneous supersaturated solid solution of $Cu_{.80}Co_{.20}$ but the Cu-rich and Co-rich phases are mixed on a very fine scale. This is to be contrasted with the nanocrystalline $Cu_{.80}Fe_{.20}$ also prepared by reducing hydroxides. In the case of as-prepared nanocrystalline $Cu_{.80}Fe_{.20}$, the separated phases are formed on a sufficiently coarse scale that they can be resolved by x-ray diffraction. On the basis of the present measurement, we are unable to determine the compositions of the Cu-rich and Co-rich phases in the as-prepared nanocrystalline $Cu_{0.80}Co_{0.20}$. From the magnetization measurements, we can conclude that not all the Co is in the form of pure metallic Co.

One interesting and potentially useful property of our $Cu_{.80}Co_{.20}$ nanocrystals is that the Co delays the oxidation of the Cu.[27] This occurs because the Co forms preferentially on the surface of the Co. The Co atoms near the surface oxidize before the Cu atoms near the surface oxidize. It appears that the oxidation rate of the Cu atoms only becomes large after most the Co atoms or Co surface atoms have oxidized.

ACKNOWLEDGMENT

The financial support of the Office of Naval Research and useful discussions with H. Aaronson are gratefully acknowledged.

REFERENCES

(1) The phase diagrams were taken from *Binary Alloy Phase Diagrams*, 2nd Ed., ed. T.B. Massalski, ASM Int.. Materials Park, OH, 1990, Vol.2.

(2) Turchanin, M. A. *Russian Metallurgy* 1995, *5*, 9.

(3) Liu, J.-M. *J. of Mater. Sc.* 1996, *31*, 2807.

(4) Langer, J. S.; Bar-on, M.; Miller, H. D. *Phys. Rev. A* 1975, *11*, 1417.

(5)   Busch, R.; Gärtner, F.; Borchers, C.; Hassen, P.; Bormann, R. *Acta Mater.* **1996**, *44*, 2567.

(6)   Shen, T. D.; Koch, C. C. *Acta Mater.* **1996**, *44*, 753.

(7)   Nakagawa, Y. *Acta Metal.* **1958**, *6*, 704.

(8)   W. Klement, J. *Trans. Metall. Soc. AIME* **1965**, *233*, 1180.

(9)   Elder, S. P.; Munitz, A.; Abbaschian, G. J. *Mat. Science Forum* **1989**, *50*, 137.

(10)   Kneller, E. F. *J. Appl. Phys.* **1964**, *35*, 2210.

(11)   Sumiyama, K.; Yoshitake, T.; Nakmura, Y. *J. Phys. Soc. of Japan* **1984**, *53*, 3160.

(12)   Tanori, J.; Duxin, N.; Petit, C.; Lisiecki, I.; Veillet, P.; Pileni, M. P. *Colloid Polym. Sci.* **1995**, *273*, 886.

(13)   Gentre, C.; Oehring, M.; Bormann, R. *Phys. Rev. B.* **1993**, *48*, 13244.

(14)   Blythe, H. J.; Fedosyuk, V. M. *J. Magn. Magn. Mater.* **1996**, *155*, 352.

(15)   Kneller, E. *J. Appl. Phys.* **1962**, *33*, 1353.

(16)   Childress, J. L.; Chien, C. L. *Phys. Rev. B* **1991**, *43*, 8089.

(17)   Berkowitz, A. W.; Mitchell, J. R.; Carey, M. R.; Young, A. P.; Zhang, S.; Spada, F. E.; Parker, F. T.; Hutten, A.; Thomas, G. *Phys. Rev. Lett.* **1992**, *68*, 3745.

(18)   Xiao, J. Q.; Jiang, J. S.; Chien, C. L. *Phys. Rev. Lett.* **1992**, *68*, 3749.

(19)   Yavari, A. E.; Desré, P. J.; Benameur, T. *Phys. Rev. Lett.* **1992**, *68*, 2235.

(20)   Schmid, A. K.; Hamilton, J. C.; Bartelt, N. C.; Hwang, R. Q. *Phys. Rev. Lett.* **1996**, *77*, 2977.

(21)   Takanash, K.; Park, J.; Sugawara, T.; Hono, K.; Goto, A.; Yasuoka, H.; Fujimori, F. *Thin Solid Films* **1996**, *275*, 106.

(22)   Sakakima, H.; Irie, Y.; Kawawake, Y.; Satomi, M. *J. Magn. Magn. Mater.* **1996**, *156*, 405.

(23)   Wang, Z.; Nakamura, Y. *J. Magn. Mag. Mat.* **1996**, *155*, 161.

(24)   Nagodawithana, K.; Liu, K.; Searson, P. C.; Chien, C. L. In *Proceedings of the Symposium on Nanostructured Materials in Electrochemistry*; Searson, P. C., Meyer, G. J., Eds.; The Electrochemical Society, Inc.: Pennington, New Jersey, USA, 1995; pp 237.

(25)   Correla, A.; Garcia, N.; Massanell, J.; Costa-Krämer, J. L. *Appl. Phys. Lett.* **1996**, *68*, 340.

(26)   Harris, V. G.; Kaatz, F. H.; Browning, V.; Gillespie, D. J.; Everett, R. K.; Ervin, A. M.; Elam, W. T.; Edelstein, A. S. *J. Appl. Phys.* **1994**, *75*, 6610.

(27)   Kaatz, F. H.; Harris, V. G.; Kurihara, L.; Rolison, D. R.; Edelstein, A. S. *Appl. Phys. Lett.* **1995**, *67*, 3807.

(28)   Chow, G. M.; Kurihara, L. K.; Kemner, K. M.; Schoen, P. E.; Elam, W. T.; Ervin, A.; Keller, S.; Zhang, Y. D.; Budnick, J.; Ambrose, T. *J. Mater. Res.* **1995**, *10*, 1546.

(29)   Michaelsen, C. *Phil. Mag. A* **1995**, *72*, 813.

(30)   *International Tables for X-ray Crystallography*; The Kynoch Press: Birmingham, UK, 1968; Vol. 3.

(31)   Miguel, J. J. d.; Cebollada, A.; Gallego, J. M.; Miranda, R.; Schneider, C. M.; Schuster, P.; Kirschner, J. *J. Magn. Magn. Mat.* **1991**, *93*, 1.

(32)   Pescia, D.; Zampieri, G.; Stampanoni, M.; Bona, G. L.; Willis, R. L.; Meir, F. *Phys. Rev. Lett.* **1987**, *58*, 933.

(33)   Bai, V. S.; Bhagat, S. M.; Kishman, R.; Seddat, M. *J. Magn. Magn. Mater.* **1995**, *147*, 97.

(34)   Childress, J. L.; Chien, C. L. *J. Appl. Phys.* **1991**, *70*, 5885.

(35)   Ueda, Y.; Ikeda, S.; Chikazawa, S. *Jpn. J. Appl. Phys.* **1996**, *35*, 3414.

(36)   Sayers, D. E.; Bunker, B. A. In *X-ray Absorption: Basic Principles of EXAFS, SEXAFS, and XANES*; Koningsberger, D. C., Prins, R., Eds.; Wiley: New York, 1988.

(37)   *Report on the International Workshops on Standards and Criteria in XAFS*; Hasnain, S. S., Ed.: Howard, Chichest, UK, 1991, pp 751.

MICRO-TENSILE TESTING OF NANOCRYSTALLINE Al/Zr ALLOYS

M. Legros*, K. J. Hemker*, D. A. LaVan*, W. N. Sharpe, Jr.*, M. N. Rittner**, and J. R. Weertman**;
*Dept. of Mechanical Eng., Johns Hopkins University, Baltimore, MD 21218-2686,
**Dept. Of Materials Science and Eng., Northwestern University, Evanston, IL 60208-3108.

ABSTRACT

A novel micro testing machine has been used to perform tensile tests on nanocrystalline Al/Zr microsamples with grain sizes ranging from 10 to 250 nm. The problems associated with testing such small specimens (200μm x 200μm in the gage section) were overcome by using a contact-free interferometric strain gage (ISDG) and alignment and low friction loading were assured by use of a linear air bearing. The postulated relationship between yield stress and hardness was investigated and will be discussed. The effect of the microstructure and the grain size of the compacts on their mechanical behaviour are also analysed.

INTRODUCTION

By reducing the free path of dislocation motion in a crystalline material, one can expect to increase its resistance to deformation and therefore enhance its mechanical properties such as hardness and yield strength [1,2]. Grain boundaries offer formidable obstacles to dislocation motion and the effect of grain sizes in mechanical strength is generally described by the Hall-Petch (HP) relationship [3-7], which has been evidenced in many materials and especially metals [8-12]. The hardness of nanocrystalline metals has been widely investigated this past ten years and it has been suggested that hardness follows the HP relationship down to a critical grain size [4]. When a grain size of several nanometers is reached, the hardness of the nanocrstalline material, as compared to the coarse grain material, can be magnified by a factor as high as 6 to 10 [4]. A parallel increase in yield stress has been projected from these unusually high hardness values. Unfortunately, nanocrystalline metals are generally produced in very small quantities and for this reason very few compressive tests and even fewer tensile tests have been performed to date [9,13-17].

The recent development of a novel microsample testing machine has greatly facilitated the mechanical testing of very small specimens [18]. Dog-bone tensile specimens with a gage section of approximately 250 μm x 250 μm and an effective gage length of 1.8 mm can be pulled in tension using a load frame that applies loads on the order of 20 pounds and measures strain using a non-contact interferometric strain gage (ISDG). This test set-up assures proper alignment and low friction loading of the specimen by use of a linear air bearing. The ISDG strain measurement device is a critical component of this testing because it provides a means in which displacement can be measured directly on the sample surface without actually touching the sample and interfering with the testing. The resolution of this system is approximately 0.5 MPa and 10 μstrain, which make it an ideal technique for surpassing the geometrical hurdles associated with the tensile testing of very small nanocrystalline samples.

The purpose of this study was to take nanocrystalline Al and Al-Zr specimens that have been prepared at Argonne National Laboratory using inert gas condensation (IGC) and uniaxial compression and to perform microsample tensile tests on these materials. This project has been further expanded to include a comparison of the results of these tensile tests with microhardness measurements and TEM microstructural observations of the same nanocrystalline materials. Here, emphasis will be placed on stress-strain curves obtained on compacts with grain sizes ranging from 10 to 250 nm, and a comparison of these results with tensile tests made previously from somewhat larger pieces of the same material at Northwestern University [17].

273

EXPERIMENTAL

Material processing

The alloys tested in this study were prepared at Northwestern University. Precursor powders of Al and Zr were produced using the Inert Gas Condensation (IGC) technique [11, 16]. After evaporation by electron-beam heating and condensation on a liquid nitrogen cold finger, the nanocrystalline mixed metals were scraped off of the cold finger and collected in powder form. These powders were transported, under high vacuum, to a heatable die that was employed for compaction. The powders were uniaxially pressed under 1.4 GPa at 100 °C [16]. The resulting disks were 9mm in diameter and 100 to 800 μm thick. Their average relative density was determined to be always greater than 93%.

Since pure nanocrystalline Al was found to exhibit grain growth, even at room temperature, small amounts of Zr were added to the Al in an attempt to stabilize the grain size of the alloys and keep it in the nanometer regime [16]. The Zr was added during evaporation by periodically moving the electron beam from the Al to the Zr crucible. Chemical analysis and grain size determination of the Al-Zr alloys were performed using both x-ray diffraction [16] and electron microscopy (TEM). The average oxygen and Zr contents are reported along with the corresponding grain sizes in Table 1. Oxygen was found to be mainly located at the free surface of the pellet, especially in cases where the average O content was beyond 4 -5 wt.%. Room temperature stability of the as-produced Al-Zr samples was verified by comparing the TEM microstructure taken a few weeks after processing with that conducted more than one year after fabrication. There was no visible change in the microstructure. Moreover, *in-situ* TEM heating experiments revealed that no significant grain growth takes place at temperatures below 400°C [16] in the case of samples containing several % or more of Zr.

Preparation of microsample tensile specimens

The as pressed pellets were initially sectioned to produce a small but conventional tensile specimens (4mm x 7mm) that were tested using a conventional MTS machine at Northwestern University. The results of these tests, which were conducted with small strain gages glued to the flat faceof the gauge section, have been published elsewhere[16, 17]. Dog-bone shaped microsample tensile specimens (Fig. 1) have been punched out of the left-over wings of the nanocrystalline pellets. The punching of these delicate specimens is greatly facilitated by the use of a specially machined graphite electrode on a plunger EDM that is equiped

Fig. 1 : SEM picture of a micro tensile specimen. The gauge section is 200 x 200 μm.

with a Micro Fin power controller. The punched specimens are mechanically polished to a mirror finish and a final thickness of ~ 200 μm. Once polished, two small reflective markers, microhardness indents, are placed on the nanocrystalline specimens using a Vickers microhardness indenter. These indents serve as reflective markers for the interferometric strain displacement gage (ISDG) [20].

The microsample tensile machine

The microsample testing machine provides a method in which micro scale samples can be tested in both tension and compression [21, 24]. The testing machine consist of the basic components found in a typical testing frame, but have been scaled down to handle the unique demands associated with micron scale testing. The microsample load frame is actuated by a low speed screw drive and employs an air bearing that maintains alignment and reduces friction so that loads on the order of .001 lb can be measured. The load is measured directly with a miniature load cell. The dog-bone shape of the specimens allows the ends of the specimen to fit into matching wedge-shaped grips and the specimen seats itself into the grip when pulled in tension. The complete description of this machine can be found in [19-21].

Table 1: Average chemical composition, density and grain size of the nanocrystalline alloys used in this study.

Sample	Zr wt. %	O wt. %	% Density	GS (nm)
A	0	2	98	250
B	7	2	97	10
C	7	2	97	70
D	15	2	95	20
E	31	1	94	15

The principle of the ISDG consists of measuring the relative displacement of two reflective features, microhardness indents, on the specimen. Shining a laser on the specimen leads to diffraction of the coherent beam and results in a fringe pattern. The relative displacement of these fringes can be measure using a photodiode array and related to the strain in the specimen. The relative displacement of the fringes (Δm) is related to the strain of the specimen (ε) by the relation :

$$\varepsilon = \lambda \, \Delta m / d_0 \, \sin\alpha_0 \qquad (1)$$

where α_0 is the angle between the incident beam and the indent facets, d_0 is the initial spacing between indents and λ the wavelength of the laser [20, 21]. By averaging the relative displacement on two diode arrays, resolution on the order of microstrain can be achieved. Specimen bending can be accounted for by measuring the strain on both sides of the sample [22].

<u>Microstructural characterization</u>

Electron microscopy (TEM and EDX) and x-ray diffraction (XRD) have been used to analyse the chemical composition and underlying micro(nano)structure of the specimens tested in this study. TEM samples have been prepared by either electropolishing or use of a tripod polisher and subsequent ion-milling.

RESULTS AND DISCUSSION

The chemical composition and average grain size of the different samples tested are reported in Table 1. These values were obtained by using EDS on the as-pressed samples, and are reported in greater detail in [16].

Systematic arrays of microhardness indents were used to measure variations in hardness for each of the nanocrystalline alloys. Hardness values for all alloys are reported in Table 2. A dual distribution of hardness values was evidenced in the higher Zr containing alloys. This variation in hardness was not random; instead the overall specimens could be divided into harder and softer regions. Fig. 2 is the plot of hardness expressed in GPa as a function of $d^{-1/2}$, where d is the average grain size. As can be seen, the average values of hardness for the different alloys seems to follows a Hall-Petch relationship down to sample D. However, the measured values of hardness were found to have a bi-modal distribution at smaller grain sizes, and the evidence of a HP behavior appears to be much less conclusive than is suggested by the averaged data.

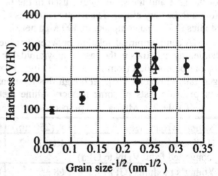

Fig. 2 : Hardness as a function of $d^{-1/2}$ for samples A-E. Triangles represent average values of H for spec. D, E.

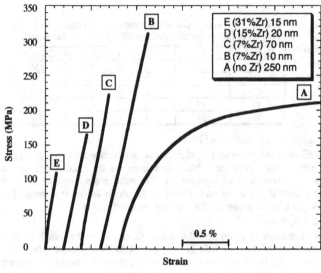

Fig. 3 : Tensile stress-strain curves of nanocrystalline alloys A-E.

less conclusive than is suggested by the averaged data.

The microsample stress - strain curves for alloys A - E are shown in Fig. 3. In these tests, the strain has been measured directly on the specimens and Young's modulus can be taken directly from the elastic portion of these curve. The modulus values that have been calculated from each of the curves are given in Table 2 and can be seen to be very close to the reference value for the Young's modulus of bulk aluminum (69 GPa) [27]. By contrast, the stresses realized in these tests are significantly higher than the generally referenced value of the yield strength of bulk aluminum (σ_y=20 MPa). The stress-strain curve for nanocrystalline aluminium (A) exhibits nonlinear permanent deformation and is suggestive of dislocation motion and plasticity. By comparison, the Al-Zr alloys remained very linear and fractured with very little or no plasticity. This lack of plasticity, even at very high stresses, had been taken as a strong indication that dislocation motion is effectively inhibited or blocked in these nanocrystalline alloys. This brittle behaviour is expected when the size of the grains of a previously ductile metal is reduced to the nanometer scale [4, 23]. It is however, interesting to note that the fracture strength decreases with increasing Zr content, as seen on Fig. 5, which emphasizes the role of the flaws in this type of alloy [28]. Material from the same disks as used for samples A and B had been tested previously at Northwestern. The slightly higher values reported here, especially in the case of the more ductile sample (A) have been attributed to a miscalibration of the compliance of the glue and the strain gauge used in the previous study (but the maximum strains have been found to be very close). The fracture stresses of sample B are in decent agreement between the two studies (250 Mpa compared to 300 Mpa here).

A semi-empirical relation between hardness and yield stress has been established for non hardening materials in the case of a pyramidal indenter [26]. This relation states that the yield stress σ_y can be obtained from hardness (H) by converting from VHN to GPa and dividing by a factor generally close to 3 [26]. Although this relationship between H and σ_y has been found to hold in compression tests of some nanocrystalline metals [13], it has not been validated in the small

Table 2 : Mechanical properties collected from tensile tests and Vickers indentations.

Sample (Grain size)	A (250 nm)	B (10 nm)	C (70 nm)	D (20 nm)	E (15 nm)
Hardness (converted from VHN to GPa)	1.04 ±.12	2.40 ±.20	1.25 ±.20	2.50 ±.15* 1.40 ± .10*	2.60 ±.20* 1.5 ± .15*
Young's modulus (GPa)	68 ±5	69 ±3	72 ±5	69 ±4	71 ±6
Max strain (%)	2.4	0.6	0.3	0.25	0.13
Yield Stress(MPa)	75				
Fracture stress (MPa)	200	315	220	165	110

* The average values of hardness for samples D and E are respectively 2.20 GPa (±.60) and 2.40 GPa (±.60)

number of tensile tests that are available in the literature [14-16]. In the present study, the ratio of hardness (1 GPa) and tensile yield strength (75 M Pa) for the nanocrystalline Al alloy (A) has been measured to be 13. The fact that this ratio is not 3 suggests that this empirical relationship may not be appropriate for all nanocrystalline materials and may be explained, at least in part, by the fact that alloy A can be seen to undergo significant strain hardening on the stress-strain curve in Fig. 3. The fact that alloys B-E fractured before yielding precluded comparing them with their hardness values, and served to highlight the importance of fracture in nanocrystalline materials.

Microstructural observations

TEM observations have been used to characterize the underlying micro(nano)structure of the alloys tested in this study. Fig. 4 is a TEM micrograph of alloy B. This alloy appeared to be the most homogeneous of the Al-Zr alloys when observed with an optical microscope and SEM, but TEM observations like the one shown in this figure

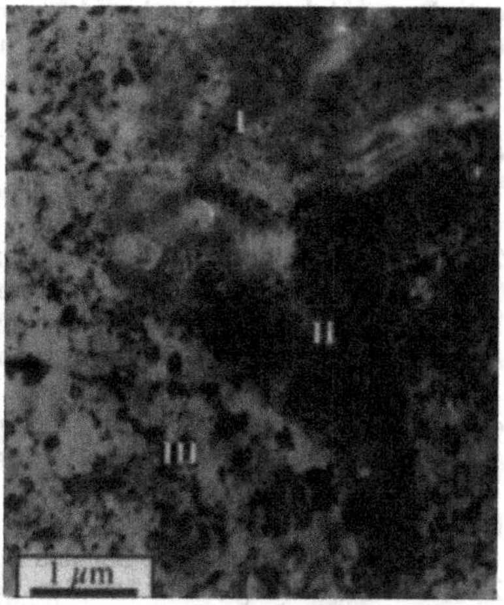

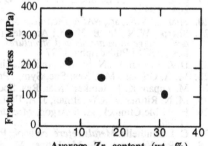

Fig. 4 : Multi modal nanostructure of sample B. The number1 I, II and III are related to zones of different average grain sizes.

indicate signifcant variations in the microstructure. Three different regions (marked I, II and III) can be seen in this micrograph, and grain sizes in these regions range from about 5nm (I) to more than 100nm (III). This variation in grain size has been directly related to the local concentration of Zr; the the larger the concentration of Zr the smaller the grain size. In addition, two oxides Al_2O_3 and ZrO_2 and one intermetallic Al_3Zr were found to coexist with the main Al phase. But, by using dark field imaging the volume fraction of these phases compared to Al was only estimated to be on the order of a few percent. The size and shape of the grains of these second and third phases were generally found to scale the grain size of the Al phase. These observation are consistent with those reported previously on the same material.

The relationships between the underlying micro(nano)structure and the various mechanical properties that have been measured in this study can be reasoned as follows. The bimodal distribution of hardness that has been measured in alloys D and E (and to a lesser extent C) may be caused by the spacial distribution of grain sizes; areas with smaller grain sizes are harder while those with bigger grain sizes are softer. This is supported by the fact that areas with the similar hardness values generally coincided with areas that exhibited a constant optical contrast, while changes in optical contrast (relating to changes in local average grain size) generally related to changes is hardness. By constrast, changes and variations in grain size did not have an effect on the elastic modulus that was measured in the microsample tensile tests. Although small, the cross-section of the microsample specimens contained more than one million grains and small changes in the size and distribution of these grains would not be expected to affect the measurement of Young's modulus. The effect of microstructure on fracture strength can perhaps best be divided into two parts. Increasing the Zr content in the alloys not

Fig. 5 : Fracture stress as a function of Zr content for samples B-E.

only decreased the grain size but also increased the heterogeneity of the material. Of the two, the effect of having a heterogeneous microstructure appears to be more influencial. As can be seen in Fig. 5, the material fails faster at higher levels of Zr concentration.

CONCLUSIONS
1. The addition of Zr to nanocrystalline Al reduced grain growth but also led to a more heterogeneous micro(nano)structure.
2. Young's modulus of the alloys tested in this study were not affected by variations in the grain size or the heterogeneity of the micro(nano)structure.
3. Like microhardness, the yield strength of nanocrystalline Al is significantly higher than for bulk Al. But the empirical yield strength to hardness ratio of 1/3 was not observed in this alloy. The measured ratio in this study was close to 1/14.
4. Yielding was suppressed in the nanocrystalline Al-Zr alloys and fracture appeared to be related to inhomogeneities in the microstructure more than to the average grain size.

REFERENCES
1. J. R. Weertman , Mat. Sci. and Eng., A166, 161-167, (1993)
2. R. W. Armstrong , Mat. Res. Soc. Synp. Proc., 362, 9-17, (1995)
3. J. C. M. Li, Y. T. Chou , Metall. Trans., 1, 1145-1158, (1970)
4. R. W. Siegel, G. E. Fougere , Mat. Res. Soc. Symp. Proc., 362, 219-229, (1995)
5. T. R. Smith, R. W. Armstrong, P. M. Hazzledine, R. A. Masumura, C. S. Pande , Mat. Res. Soc. Symp. Proc., 362, 31-37, (1995)
6. A. H.Cholski, A. Rosen, J. Karch,H. Gleiter , Scripta Met., 23, 1679 (1989)
7. J. Lian, B. Baudelet , Nanostructured Materials, 2, 415-419, (1993)
8. G. E. Fougere, J. R. Weertman, R. W. Siegel , Nanostructured Materials, 5, 2, 127 (1995)
9. A. Kumpmann, B. Günther, H. D. Kunze , in *Mechanical Properties and Deformation Behavior of Materials Having Ultra-Fine Microstructures*, M. Nastasi et als. eds, 241-254, Kluwer Academic Publishers, (1993)
10. L. Wong, D. Ostrander, U. Erb, G. Palumbo K. T., Aust , in Nanophases and Nanocrystalline Structures, R. D. Schull and J. M. Sanchez eds, 85-93, The Minerals, Metals & Materials Society, (1994)
11. B. Günther, A. Baalmann, H. Weiss , Mat. Res. Soc. Symp. Proc.,195, 611, (1990)
12. C. Suryanarayana, F. H. Froes, Metall. Trans., 23A, 1071-1081, (1992)
13. R. Suryanarayanan, Ph. D. thesis, Washington University at St Louis (1996)
14. G. W. Nieman, J. R. Weertman, R. W. Siegel, Scripta Metall. 23, 2013 (1989)
15. G. W. Nieman, J. R. Weertman, R. W. Siegel, J. Mater. Res. 6, 1012 (1991)
16. M. N. Rittner, Ph. D. thesis, Northwestern University (1996)
17. M. N Rittner, J. R. Weertman, J. A Eastman, K. B. Yoder, D. S. Stone , to appear in Mat. Sci. and Eng.
18. W. N. Sharpe, R.O. Fowler, ASTM STP 1204, Small Specimen Test Techniques Applied to Nuclear Reactor Vessel Thermal Annealing and Plant Life Extension, (1993), 386-401.
19. C. C. Koch, T. D. Shen, T. Malow, O Spaldon , Mat. Res. Soc. Symp. Proc., 362, 253-264, (1995)
20. Sharpe, W.N., Jr., *NASA Technical Memorandum* , 101638, (1989).
21. Sharpe, W.N., Jr., B. Yuan ; Accepted in *Nontraditional Methods Of Sensing Stress, Strain and Damage in Materials and Structures*, SATM STP 1318, G. F. Lucas and D. A. Stubbs, Eds., ASTM Philadelphia 1997
22. D. A. LaVan, W. N. Sharpe , to appear in SEM proceedings ...(1997)
23. M. A. Otooni , Mat. Res. Soc. Symp. Proc., 362, 45-49, (1995)
24. M. Zupan, K. J. Hemker , Mat. Res. Soc. Symp. Proc., symposium Z, this conf., (1996)
25. M. N. Rittner, J. R.Weetman, J. A. Eastman, Acta Mater., 44, n°. 4, 1271-1286, (1996)
26. F. A. Mc Clintock, A. S Argon, *Mechanical behaviour of materials*, Addison-Wesley Pub. (1966)
27. C. J. Smithells, *Metals reference book*, Butterworths, London, (1976)
28. J. R. Weertman, N. M. Rittner M. and C. Youngdahl, in Mechanical Properties and Deformation Behavior of Materials Having Ultra-Fine Microstructures, M. Nastasi et als. eds, 241-254, Kluwer Academic Publishers, (1993)

CHARACTERIZATION OF CONSOLIDATED RAPIDLY SOLIDIFIED Cu-Nb RIBBONS

F. Ebrahimi and M. L.C. Henne
Materials Science and Engineering Department
University of Florida, Gainesville, FL 32611

ABSTRACT

Copper-niobium ribbons produced by melt-spinning were compacted by swaging and consolidated using HIPping. Final processing to obtain *in-situ* composites was done by swaging. The strength of the composite is discussed in terms of the composition and morphology of the niobium phase as evaluated using electron microscopy techniques.

INTRODUCTION

The need for high strength and high conductivity (electrical and/or thermal) materials has led to the development of copper-based *in-situ* composites [1]. The high melting point of Nb in conjunction with its low solubility make the Cu-Nb composites to be resistant to coarsening at elevated temperatures and hence good candidates for use at high temperatures as well as applications that require high strength and high electrical conductivity. Conventionally, the *in-situ* composites are produced by heavy deformation of as cast alloys, which contain a large dendritic second phase. Because of deformation, the second phase particles become elongated into a filament shape. The strength of these composites depends on the interfilamentary spacing of the second phase through a Hall-Petch type relationship [2]. The major drawback to these *in-situ* composites is the large amount of processing involved to produce the fine microstructure required to provide the high strength.

The objective of this project has been to produce an *in-situ* composite utilizing rapidly solidified Cu-Nb ribbons. The fine structure of the rapidly solidified alloy is expected to reduce significantly the amount of deformation needed to produce a nano-scale interphase spacing. In a previous study [3] we have analyzed the microstructure of Cu-Nb ribbons produced by melt spinning. The as-melt spun ribbon showed a normal distribution of Nb particles with an average spacing of approximately 82 nm. Upon heat treatment at 900°C for three hours the particles grew with the largest ones (~300nm) at the copper grain boundaries. Simultaneously very fine particles were precipitated, indicating that the as-melt spun copper was supersaturated with Nb. In this study the results of mechanical and microstructural characterization of a consolidated cu-7wt%Nb composite is presented and discussed.

EXPERIMENTAL PROCEDURES

Rapidly solidified ribbons were produced in a chill block melt spinning system [4]. The ribbons were compacted in a pure copper tube of 1.9 cm diameter and then the tube was swaged to a final diameter of 1.27 cm. The sample was HIPped at 800°C under 207 MPa pressure for 4 hours. The copper can and the tubing were machined off of the sample before final swaging. The sample with a diameter of 1.03 cm was swaged to a wire of 0.254 cm in diameter, which resulted in a true strain of 2.8. The

279

microstructure and the composition of the sample were characterized in the HIPped and in the wire form using analytical electron microscopy techniques.sec^{-1}. Thin foil specimens for TEM (transmission electron microscopy) were prepared using the jet electropolishing technique. Five tensile specimens were cut from the wire sample and were tested at a strain rate of 1.7×10^{-4}

RESULTS AND DISCUSSION

Figure 1a shows an SEM (scanning electron microscope) back scattered image of the sample in the as-HIPped condition. The individual ribbons are distinguishable in this micrograph and the microstructure appears inhomogeneous. This inhomogeneity arises from the variation in the microstructure of the ribbons across their thickness [3]. During melt spinning the areas close to the wheel surface cooled faster and thus a finer microstructure was obtained. In Figure 1a the apparently lighter areas are regions with larger niobium particles. The average composition of the sample was measured to be 92.1%Cu - 7.20% Nb - 0.67% Si using microprobe analysis. The quartz tube in which the Cu-Nb alloy was melted during melt spinning is believed to be the source of silicon. Although the interior of the tube was coated with zirconia, because of the high melting temperature of the alloy the coating did not completely prevent the dissolution of silicon.

Except for a low density of round pores the compaction and HIPping processes resulted in a complete dense material. However, some of the interfaces between ribbons were not bonded perfectly and appeared dark as marked by arrows in Figure 1a. Occasionally regions with a much coarser microstructure were found. Figure 1b presents an example of such microstructure. These regions are believed to be pieces of the master alloy which did not melt completely and were ejected with the melt during melt spinning. As shown in Figure 1b these remains are very brittle and cracked during the final swaging process.

 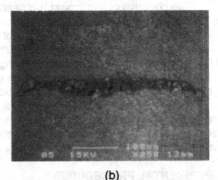

(a) (b)

Figure 1. SEM images showing (a) the microstructure of the as-HIPped material and (b) the microcracking of a piece of the master alloy at a strain of 0.43.

Figure 2 presents a TEM micrograph of the as-HIPped microstructure. The niobium particles showed a wide size distribution similar to the heat treated specimens reported previously [3].The largest particles were in the order of 200-300 nm and they developed by the coarsening of the niobium particles which had formed during melt spinning in the ribbons. A distribution of much finer particles was also observed. These particles precipitated from the supersaturated copper matrix during the HIPping process.

The tensile specimens made from the wire sample showed an average ultimate tensile strength of 514 Mpa (459 to 542 MPa) and an average maximum uniform strain and total elongation of 2.5% and 6.7%, respectively. Figure 3 presents an SEM fractograph of the fracture surface. The fracture occurred by microvoid coalescence mechanism. The voids were submicron in size and were possibly initiated at the niobium particles. Delamination was detected at two locations on the fracture surface as can be seen in Figure 3: at the ribbon boundaries (marked as B) and at the regions with the coarse microstructure (marked as C). The delamination process is believed to contribute to the scatter observed in the tensile results. Occasionally, the pores that formed upon the HIPping were also detected on the fracture surface (marked as A).

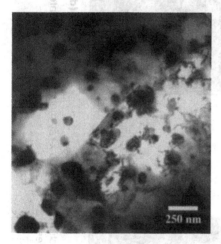

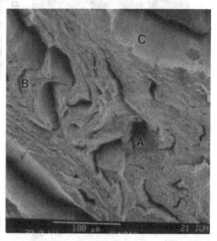

Figure 2. TEM micrograph of the as-HIPped material.

Figure 3. SEM fractograph of a HIPped and swaged tensile sample.

At the effective true strain of 3 the ultimate tensile strength of copper and niobium are 370 and 420 MPa, respectively [2]. Considering that the volume fraction of niobium is close to its weight percent in copper, the predicted strength of the composite produced in this study based on the rule of mixtures is 374 MPa, which is much lower than the observed strength of 514 MPa. Assuming that the niobium particles deform to fibers upon swaging, for an average particle diameter of $d_0 = 250$ nm in the HIPped microstructure and at the true strain of $\eta = 2.8$ the fiber diameter, $d = d_0 e^{-\eta/2}$, and the length, $l = 2/3 d_0 e^{\eta}$, would be 62 nm and 677 nm, respectively. The dislocation mean free path in the copper matrix can be estimated from $\overline{\lambda} = (1-f)d/f$ to be 823 nm, where f is the volume fraction of niobium. Extrapolated from the results for conventionally processed *in-situ* composites [2] the strength of the composite produced in this study is estimated to be around 1000 MPa rather than the observed strength of 514 MPa.

TEM analysis of the composite after the final swaging revealed that not all of the niobium particles were elongated into fibers. Figure 4a shows TEM micrograph of part

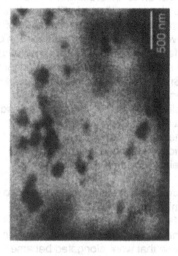

(a) Bright Field Image

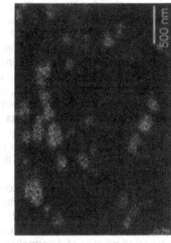

(b) Cu x-ray

(c) Nb x-ray

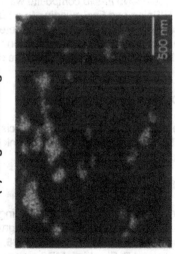

(d) Si x-ray

Figure 4. Scanning transmission electron microscope x-ray mapping of consolidated wire showing the incorporation of Si in the undeformed Nb particles.

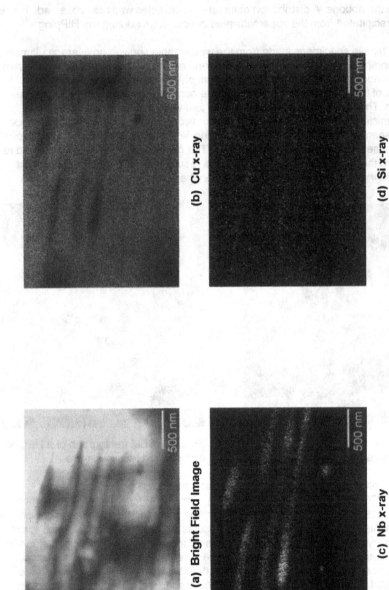

(a) Bright Field Image

(b) Cu x-ray

(c) Nb x-ray

(d) Si x-ray

Figure 5. Scanning transmission electron microscope x-ray mapping of consolidated wire showing the absence of Si in elongated Nb particles.

283

of the composite which contained non-deformed spherical particles. Figure 5a presents micrographs of a region of the microstructure which did show elongated fiber-like niobium phase. X-ray mapping was conducted in an effort to understand why the two shapes were observed. The Cu, Nb, and Si maps of the areas shown in Figures 4a and 5a are presented in b, C, and d of each Figure, respectively. A high concentration of Si was found in the spherical niobium particles, however, no measurable accumulation of Si was detected in the elongated particles. This difference suggests that solid solution strengthening of niobium makes the particles more resistant to deformation and hence to elongation. The inhomogeneous distribution of Si may be attributed to the possible variation in the silicon content of the ribbons produced at different runs. It was also noticed that consistently the larger particles were elongated but the smaller ones had remained spherical. This observation suggests that the plastic constraint around the larger particles is higher and hence they carry higher stresses than the smaller particles do. Another possibility is that smaller particles contained more silicon. The results of this study is inconclusive regarding the relative contribution of these two mechanisms to the observed size effect.

The lower strength of the composite produced in this study when compared to the estimated strength, based on the results for a microstructure consisting of continuous fibers in copper, is associated with the fact that not all of the niobium particles became elongated. Furthermore, due to the small initial size of the particles combined with the small amount of deformation, those that were elongated became like short fibers rather than continuous fibers. The results of this study suggests that the morphology of the second phase play an important role in the level of strength achievable in *in-situ* composites.

SUMMARY

A dense Cu-7%Nb *in-situ* composite was produced by melt spinning, consolidation, HIPping and final swaging to a true strain of 2.8. The consolidation process was successful in producing a completely dense material. However, the microstructure obtained after final swaging consisted of short fibers and spherical niobium particles, which did not strengthen the copper as effectively as continuous fibers do.

ACKNOWLEDGMENTS

Authors wish to acknowledge the financial support of National High Magnetic Laboratory in Tallahassee, Florida and to thank Pratt and Whitney for HIPping the Cu-Nb specimens.

REFERENCES

1. J. D. Verhoeven, W. A. Spitzig, H. L. Downing, C. L. Trybus, L. L. Jones, H. L. Downing, L. G. Fritzemeier, and G. D. Schnittgrund, J. Mater. Eng. **12**, p. 127 (1990).
2. W. A. Spitzig and P. D. Kroz, Acta metall. **36**, p. 1709 (1988).
3. M. L. C. Henne and F. Ebrahimi, MRS proceedings **362**, p. 211 (1995).
4. M. L. C. Henne, The Development of an *In-situ* Composite Using a Rapidly Solidified Material, Master Thesis, 1996, University of Florida.

THE COMPOSITION EFFECT ON THE NANOCRYSTALLIZATION OF FINEMET AMORPHOUS ALLOYS

J.Zhu, T.Pradell*, N.Clavaguera** and M.T.Clavaguera-Mora
Grup de Física de Materials I, Dept. de Física, Universitat Autònoma de Barcelona 08193 Bellaterra, Spain
* ESAB, Universitat Politécnica de Catalunya, Urgell 187, 08036-Barcelona, Spain
** Física de l'Estat Sòlid, Dept. ECM . Facultat de Física, Universitat de Barcelona, Diagonal 647, 08028 Barcelona, Spain

ABSTRACT

Differential Scanning Calorimetry (DSC), X-ray Diffraction (XRD), Neutron Diffraction (ND) and Mössbauer Spectroscopy (MS) were used to study the nanocrystallization process of $Fe_{73.5}Cu_1Nb_3Si_{22.5-x}B_x$ ($x=5$, 7, 8, 9 and 12) amorphous alloys. Both the temperature range and the activation energy of Fe(Si) phase precipitation from the amorphous martrix increase with the initial B composition. The initial Si composition influences the mechanism of the nanocrystallization: for the Si rich samples, the beginning of nucleation and growth processes is interface controlled, for the B rich samples it is diffusion controlled. Secondary crystallization from the remaining amorphous is mainly Fe_3B and Fe_2B, the ratio of Fe_3B/Fe_2B being dependent on the initial composition too.

INTRODUCTION

Many analysis are reported about the the Cu and Nb addition effect on the crystallization of the FINEMET amorphous materials[1,2,3]. The amount of Cu is very small, only 1at.%, it decreases the percipitation temperature of the Fe(Si) phase because it is antisoluted in iron and increases the Fe(Si) nucleation rate. On the other hand, by Nb addition, the Fe(Si) phase crystallization temperature range is getting extended. Most of the Nb atoms are piled at the grain boundary [1], and hinder the Fe(Si) grain growth. Therefore, the Fe(Si) grain can get the nanometre scale[3,4,5].

As well as the Cu and Nb addition, the influence of Si and B composition has been analysed[6]. For a fixed Cu and Nb content, such as 1at.%Cu and 3at.%Nb, the ratio of Si/B composition changes the crystallization process sensitively. The Si and B composition influence not only the thermal behaviour during crystallization, but also the crystalline phase structure and composition. With the aim to understand further the Si and B composition effect for the FINEMET amorphous crystallization process, the DSC, XRD, ND and MS techniques were used in this research work.

EXPERIMENTAL

The $Fe_{73.5}Cu_1Nb_3Si_{22.5-x}B_x$ ($x=5$, 7, 8, 9 and 12) FINEMET amorphous ribbon were produced by planar flow casting method, with 15mm width and about $20\mu m$ thickness. The heat treatment and calorimetric experiments were carried out by DSC on a Perkin Elmer DSC-7. The XRD measurements were performed using the Bragg-Bretano $\theta/2\theta$

power diffractometer Siemens D-500 instrument at room temperature. The ND experiments were taken in the ISIS Rutherford Appleton Laboratory on a "Liquids and Amorphous Differactometer" by "*in situ*" measurements of the diffraction patterns during the heat treatment. The Mössbauer spectra were measured at room temperature using a ^{57}Co in Rh source with calibration of α-Fe foil, and fitted with a histogram magnetic hyperfine-field distribution by the Brand [7] and Hesse-Rubartsch method[8]. The sub-spectra of the nanocrystalline Fe(Si) phase were fitted by superimposing sextets of Lorentzian lines corresponding to the different Fe neighbourhoods in Fe(Si) phase, in these case the Gaussian-shaped hyperfine distribution was related with the remaining amorphous phase.

RESULTS

Using the DSC continuous scan regime, all the thermal phenomena display in the DSC curves. Fig.1 shows the DSC curves with heating rate 10°C /min. The Curie temperature (T_C) of the amorphous phase shows as a small peak at $\sim 320°C$. Relaxation processes appear in the temperature range from 350°C till the Fe(Si) crystallization. Then the Fe(Si) phase precipitates in the first main exothermal peak. One remark is that the peak position and the onset temperature are shifted to high temperature with increasing of B composition. At the same time, the shape of the peak also changes, for the sample $x=5$, the peak is sharp and narrow, the peak area is 100 ± 5 J/g; but for the sample $x=12$, the peak becomes broad and stump and its area is only 45 ± 3 J/g. By the Kissinger and the multiple scan methods, the average activation energy of Fe(Si) phase crystallization was evaluated, it increases from 370 ± 5 to 433 ± 5 kJ/mol when the B composition increases from $x=5$ to $x=12$.

From the DSC curves, another important difference is the second main exothermal peak, which corresponds to the Fe(B) phase formation: the onset temperature and the

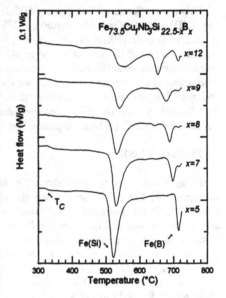

Figure 1: The DSC curves of FINEMET samples $x=5$, 7, 8, 9 and 12, at heating rate 10°C/min.

peak position are shifted to low temperature with increasing of B composition. As will be presented afterwards, the Fe(B) phases obtained after the second main exothermal peak are different for each sample. Finally, a small exothermal peak related most probably to the recrystallization is observed at about 710°C .

Based on the analysis of DSC isothermal experiments, the JMAE exponent, n, has been calculated. Fig.2 presents the JMAE exponent as a function of the transformed fraction for two samples $x=5$ and $x=8$ obtained at several annealing temperatures. The exponent n decreases to zero with the transformed fraction. But the value of n in the beginning of crystallization is different for the different samples. This means that the controlling mechanisms during the crystallization process are different. For the sample $x=5$ it is an interface controlled grain growth with a increasing nucleation rate at the beginning of crystallization[9], then the nucleation rate saturates and diffusion limited growth dominates. For the sample $x=8$, the beginning of crystallization is diffusion controlled growth with increasing nucleation rate[9], and then diffusion limited growth till the end of the process.

In Fig.3 we present the lattice parameters of nanocrystalline Fe(Si) phase obtained by XRD in samples $x=5$, 7, 9 and 12 (shown as squares). The two lines with the cross and the circle symbols are the values for bcc structure and DO_3 superlattice as reported in the references[6, 11-20], respectively. For the samples $x=5$, 7 and 9, the data are on the line corresponding to DO_3 superlattice structure, with a Si content decreasing from 21 ± 1 to 15 ± 1 at.%. For the sample $x=12$, the datum is on the intersection between both bcc and DO_3 lines. In this case the Si content in the nanocrystals is about 10 at.% which is too low to see the superlattice diffration peaks in this sample by XRD. These results show that the initial Si composition directly influence the Fe(Si) nanocrystalline phase Si content.

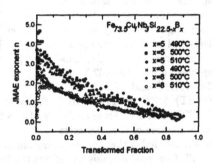

Figure 2: The JMAE exponent,n, of Fe(Si) crystallization process as a function of transformed fraction, annealing the samples $x=5$ and $x=8$ at 490, 500 and 510°C.

The ND results show the structure evolution during crystallization. In Fig.4 we present the "*in situ*" measured diffraction pattern during heating sample $x=12$ up to 720°C at 10°C /min. The diffraction peaks of the Fe(Si) phase appear at 510°C and the peaks of the Fe(B) phase appear at 640°C , in agreement with the exothermal peaks observed in the DSC curves (see Fig.1).

Fig.5 shows the comparison of the ND patterns obtained from each sample after heating up to 720°C . The composition effect becomes more evident. The intensity of the diffraction peaks of DO_3 superlattice structure decreases with the increasing of B composition. The Fe_3B phase (both orthorhombic and tetragonal are possible) decreases but the Fe_2B phase (tetragonal structute) increases with the increasing of B composition. Therefore, not only the Si content in Fe(Si) phase, but also the ratio of Fe_3B/Fe_2B phases depends on the initial Si/B ratio.

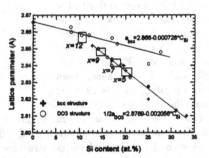

Figure 3: The lattice parameter of Fe(Si) phase obtained by XRD in the samples $x=5$, 7, 8 and 12 as a function of silicon content.

The Mössbauer spectra were obtained for two series of pre-annealed samples $x=5$ (annealing at 490°C for different time) and $x=12$ (annealing for 1 hour at different temperature), respectively. The Mössbauer spectra show five subspectra corresponding to the different near-neighbour (NN) sites in DO_3 and a remain amorphous subspectrum. The five sub-spectra are: 8NN(32.5 T), 7NN(31.8T and 30.6T), 6NN (28.8T), 5NN(24.5T) and 4NN

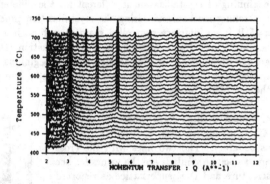

Figure 4: "*in situ*" measured ND patterns of sample $x=12$ up to 720°C with heating rate 10°C/min.

(19.8T)[11,21-23]. From analysis of the area of the subspectra it was concluded that the remaining amorphous phase was more rich in Fe in the sample $x=12$. Consequently, this fact affects the Fe(B) phase crystallization temperature and the ratio of Fe_3B/Fe_2B phases.

When the percentage of each NN site normalized to the nanocrystalline Fe(Si) phase, the changes of the relative intensity for each NN site indicate the kinetic path of the nanocrystals SRO parameter during the crystallization. Considering the sample $x=5$, the relative intensity of NN sites changes during annealing at 490°C as shown in Fig.6(a). For instance, when the annealing time is 10 or 30 min, the Si content in the Fe(Si) phase corresponds to around 20 at.%, as obtained by comparison with the theoretical calculation[23]; but when the annealing time is long enough, the Si

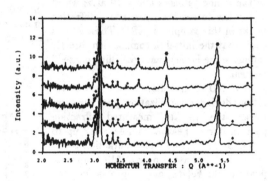

Figure 5: The ND patterns of the samples heated up to 720°C . Bragg peaks: ∗ bcc Fe(Si), ⋆ DO_3 superlattice structure, ♦ tetragonal or orthohombic Fe_3B, ● orthohombic Fe_3B, ○ tetragonal Fe_2B.

content in Fe(Si) increases and reaches a stable value of 21 at.%[21]. Also show in the Fig.6(a) is sample $x=5$ pre-annealed at 510°C for 1 hour. It is fully nanocrystallized and the Si content in the Fe(Si) phase is about 22 at.% [22].

In Fig.6(b) the results of sample $x=12$ are plotted as a function of annealing temperature. After annealing at 510°C , the Si content of the nanocrystalline Fe(Si) phase is close to 20 at.% (the ideal cross point of 4NN and 5NN [23]); after annealing at 550°C , the sequense of the relative intensity of 5NN, 6NN and 4NN seems close to the ideal cross point of 5NN and 6NN with Si content of 15.5 at.%[23].

Therefore, the kinetic path of the SRO during the crystallization is also strongly related with the initial composition. The Si content of the Fe(Si) phase evolves differently according to the initial Si composition. For the sample $x=5$ it increases from 20 to 22 at.%, whereas for sample $x=12$ it decreases from 20 to 15.5 at.%. The final Si content obtained by MS also agrees with the XRD experimental results (see Fig.3).

For both samples $x=5$ and $x=12$, at the beginning of crystallization, the Fe(Si) phase moreless have the same Si content 20at.%, it means that the Fe(Si) phase is easily nucleating from the amorphous with an Fe80Si20 enviroment. In sample $x=5$, the initial ratio of Fe/Si is 73.5/17.5, which is close to 80/20, so the atoms do not need to diffuse when the nucleation process occurs. However, in the sample $x=12$, the initial ratio of Fe/Si is 73.5/10.5, which is far from 80/20, so atoms need to diffuse when the nuclei grow. Both the activation energy and the onset tempeatuure re higher than that in sample $x=5$. This explanation is in good agreement with the value of the n exponent already given indicative of the different mechanism controlling the beginning of crystallization.

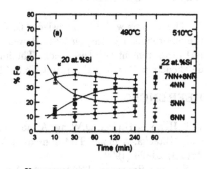

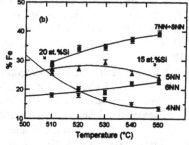

Figure 6: The relative intensity of nearneighbour sites of Fe(Si) phase for sampes $x=5$(a) and $x=12$ (b).

CONCLUSIONS

Though the DSC, XRD, ND and MS studies of the the nanocrystallization process of $Fe_{73.5}Cu_1Nb_3Si_{22.5-x}B_x$ ($x=5$, 7, 8, 9 and 12) amorphous alloys, we conclude that both thermal behaviour and structure of the crystallized phases are strongly related with the initial Si/B composition ratio.

The temperature onset, interval and the activation energy of Fe(Si) nanocrystallization increase with the increasing of initial B composition. On the other hand, the appearnce of DO_3 ordered Fe(Si) phase decreases with the increasing of the initial B composition. The mechanism of the nanocrystallization change as well with the initail Si composition. For the Si rich samples, the beginning of nucleation and growth process are interface controlled, but for the B rich samples they are diffusion limited. The SRO parameters of Fe(Si) follow different paths during crystallization.

The secondary crystalline phases formed from the remaining amorphous are mainly Fe_3B and Fe_2B, the transformation temperature range and the ratio of Fe_3B/Fe_2B are strongly depending on the initial composition too.

ACKNOWLEDMENTS

The authors wish to acknowledge Dr.G.Herzer, Vacuumschmelze GmbH, for providing the amorphous ribbons; Dr. W.S.Howells, ISIS Rutherford Appleron Laboratory, for the ND measurement and Discussion. The finacial support of the 'Comision Interministerial de Ciencia y Tecnologia' (project No.MAT96-0769) is greatefully acknowledged.

REFRENCES

1. K.Hono, K.Hiraga, Q.Wang, A.Inoue and T.Sakurai, Surface Science **266** 285 (1992)
2. N.Kataoka, A.Inoue, T.Masumoto, Y.Yoshizawa and K.Yamauchi, Jpn.J.Appl. Phys., **28(10)** L1820 (1989)
3. F.M.van.Bouwelen, J.Sietsma and A.van.dan.Beukel, J.Non-Cryst.Sol **156-158** (1993)
4. A.R.Yavari and O.Drbohlav, Mater. Trans., JIM, **36(7)** 896 (1995)
5. M.Müller, N.Mattern and L.Illgen, J.Mag.Mag.Mat., **112** 263 (1992)
6. M.Müller, N.Mattern and L.Illgen, Magnetically Soft Nanocrystalline Fe, Bd.82 (1991) H.12, 895
7. R.A.Brand, 1990 NORMOS Program version 1990
8. J.Hesse and A.Rubartsch, J.Phys.E:Dci.Instrum, **7** 526 (1974)
9. J.Málek, thermachimica Acta. **267** 61 (1995)
10. N.Clavaguere, J.Non-crys.Solids, **162** 40 (1993)
11. G.Rixecker, P.Schaaf and U.Gonser, J.Phys.:Condens.Matter, **4** 10295 (1992)
12. Y.Ueda, S.Ikeda and K.Minami, Mater.Sci.Eng., **A179/A189** 992 (1994)
13. U.Köster, U.Schünemamm, M.Blank-Bewerdorff, S.Bauer, M.Sutton and G. B. Stephenson, Mater. Sci. Eng., **A133** 61 (1991)
14. W.B.Pearson, A Handbook of Lattice spacings and Structures of Metals and Alloys Pergamon Press. Oxford 1964
15. R.M Bo Zorth, Ferromagnetism, Van Northerland New York,1951
16. K.Khalaff and J.Schubert, J.Less-Common Met., **35** 341 (1974)
17. A.A.Waniewska, M.Gutowski, M.Kuzminski and E.Dynowska, Nanophase Materials, edited by G.C.Hdjipanayis and R.W.Siegel, p.721
18. P.R.Swann, L.Granä and B.Lethins, Met.Sci., **9** 90 (1975)
19. J.Bigot, N.Lecaude, J.C.Perron, C.Milan, C.Remiarinjaona and J.F.Rialland, J.Mag. Mag.Mat., **133** 299 (1994)
20. E.Gaffet, J.Alloys and Comp., **194** 339 (1993)
21. T.Pradell, N.Clavaguera, J.Zhu, and M.T.Mora, J.Phys.:Condens.Matter, **7** 4129 (1995)
22. T.Pradell, N.Clavaguera, J.Zhu, and M.T.Mora, Conference Proc. of ICAMA-95, Rimini, Italy, 1995, p409
23. G.Rixecker, P.Schaaf and U.Gonser, phys.sol.sta.(a), **139** 309 (1993)

Synthesis, Characterization and Mechanical Properties of Nanocrystalline NiAl

M. S. Choudry[1,2] , J. A. Eastman[2] , R. J. DiMelfi[2] and M. Dollar[1]

[1]Department of Mechanical, Materials and Aerospace Engineering, Illinois Institute of Technology, Chicago, IL 60616

[2]Argonne National Laboratory, Argonne, IL 60439

ABSTRACT

Nanocrystalline NiAl has been produced from pre-cast alloys using an electron beam inert gas condensation system. In-situ compaction was carried out at 100 to 300°C under vacuum conditions. Energy dispersive spectroscopy was used to determine chemical composition and homogeneity. Average grain sizes in the range of 4 to 10 nm were found from TEM dark field analyses. A compression-cage fixture was designed to perform disk bend tests. These tests revealed substantial room temperature ductility in nanocrystalline NiAl, while coarse grained NiAl showed no measurable room temperature ductility.

INTRODUCTION

Intermetallic alloys are of interest for a variety of potential applications because of their high strength-to-weight and stiffness-to-weight ratios and because they often possess excellent elevated temperature properties. The fundamental limitation in the commercial utilization of these materials invariably is their inherent ambient temperature brittleness, which adversely affects material handling and fabricability.

Schulson and Barker [1] reported an improvement in tensile ductility at 400°C in ordered NiAl as a result of grain size refinement to sizes as small as 20 μm. While Schulson's and Barker's results demonstrated that grain size refinement can increase ductility, these results did not give any information about the room temperature deformation behavior of NiAl. A number of recent studies have begun investigating the possibility of improving room temperature ductility of intermetallic alloys through further size refinement to the nanocrystalline regime. In this respect, investigators [2,3] produced n-NiAl through different processing routes and concluded that some evidence of room temperature ductility is present in these materials. In case of [2], NiAl was ball milled to obtain grain sizes in the μm to nm regime. They used miniaturized disk bend testing and observed that materials that had low carbon content showed some evidence of ductility while materials with higher carbon content showed no ductility. In addition to carbon, their materials also contained substantial amounts of other impurities and questions remain regarding the possible effects of these impurities, as well as the broad grain size distribution of their materials on the observed properties. On the other hand, the investigation by Haubold and co-workers [3] was done on cleaner nanocrystalline NiAl produced by the inert gas condensation (IGC) method, but their mechanical properties evaluation was limited to microhardness tests done on a single sample annealed at various temperatures. A more comprehensive study of mechanical behavior of nanocrystalline NiAl with low impurity levels is clearly needed and motivates the present study.

In the present investigation, the ambient temperature deformation behavior of n-NiAl was investigated using biaxial disk bend testing (BDBT). Disk bend testing offers possible advantages over tensile testing in characterizing nanocrystalline samples, particularly since thin disk-shaped samples can be tested in as-produced form. This method eliminates the need to machine dogbone-shaped samples and subsequently attach grips and strain gauges to typically small samples produced by the IGC method. Previous mechanical tests on nanocrystalline NiAl [2] used the miniaturized disk bend testing technique with 3 mm diameter specimens. In the present study, biaxial disk bend testing is used with 9 mm diameter disks produced directly from our synthesis facility without any post production sample machining. This avoids the potential introduction of stresses in the specimen. The load was applied through a flat punch, as seen in Fig. 1, which also eliminated the uncertainties associated in determining the exact contact area between the specimen and the punch. Knowing the contact area accurately is important in calculating the exact value of yield stress from the measured displacements.

EXPERIMENTAL

Nanocrystalline NiAl was produced by the inert gas condensation process using an electron beam evaporation system [4]. Nickel and NiAl were evaporated from separate crucibles with dwell times chosen to yield the desired 50:50 composition. A Ni source was required to compensate for the larger vapor pressure of Al compared to Ni. The powder produced was transported under vacuum to an adjoining chamber where 9 mm diameter disks were compacted at 1.4 GPa. High vacuum compactions at temperatures of 100 - 300°C were performed, resulting in sample densities that varied from 70 - 90% of theoretical. Chemical analysis was done by energy dispersive spectroscopy (EDS) and average grain size was determined by dark field transmission electron microscopy (TEM).

BDBT was performed to characterize the mechanical behavior of both nanocrystalline and coarse grained NiAl samples. All samples were mechanically polished prior to testing to produce a $0.05 \mu m$ surface finish. A compression cage fixture was designed and built for these measurements. A schematic representation of the BDBT apparatus is shown in Fig. 1. The 9 mm diameter specimens were freely supported on a 7 mm diameter ring and load was applied to the center of the disk through a flat punch having a 1.15 mm diameter. The BDBT load-displacement data were analyzed following the procedures outlined in [5].

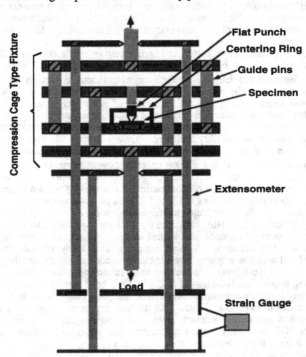

Figure 1. Schematic drawing of a cross section of the BDBT apparatus.

RESULTS AND DISCUSSION

Chemical analysis results for two nanocrystalline specimens and one coarse grained (CG) NiAl specimen are shown in Table I. From each specimen, an EDS/SEM analysis was done on three random locations from a relatively large area. EDS/TEM analysis was done in a nanoprobe mode with a nominal probe size of ~10 nm, thus producing information on a much more local scale. RBS scans covering an area of approximately one mm^2 were used to obtain macroscopic compositions. Clearly the small scatter in the composition obtained by all three methods indicates that the samples are very homogenous in composition on a microscopic as well as macroscopic level. Also, the good agreement between all three methods is further evidence of sample homogeneity. This behavior differs from that of nanocrystalline Al-Zr [6] produced in the same system, where large variations in the composition were obtained within the same sample. It is possible that in the case of Al-Zr, the limited diffusion rate of Zr in Al during evaporation and/or consolidation prevents homogenization. Faster diffusion of Ni and Al in NiAl than of Zr in Al is believed to result in a more uniform composition in the present case.

Table I Chemical composition, density, and mechanical properties of two nanocrystalline NiAl and one coarse grained (CG) NiAl sample.

Sample #	EDS(SEM) at.% Ni.	EDS(TEM) at.% Ni.	RBS at.% Ni.	Density % Theor.	σ_y MPa	Fracture Mode
NiAl-1	48.9±0.3	48.1±1.3	48.6±0.7	84.67	91.32	Ductile
NiAl-2	48.6±0.4	48.8±0.9	48.1±0.5	90.05	134.67	Ductile
CG- NiAl	50.3±0.1	-	-	100	-	Brittle

Figs. 2(a) and (b) show a dark field TEM image of n-NiAl-1 and a histogram of the grain size distribution. The grain size distribution is quite narrow and uniform throughout the specimen with an average size of ~6 nm in this particular case.

Before disk bend testing nanocrystalline NiAl specimens, the performance of the BDBT apparatus was evaluated by testing standard coarse grained stainless steel and aluminum samples. These materials have very well characterized mechanical properties. Load-displacement curves obtained were similar to load-displacement curves obtained by other investigators on similarly ductile materials when subjected to disk bend testing [7]. Also, yield stress and modulus values obtained were within ±5% of the literature values for the same materials tested by more conventional tensile test methods.

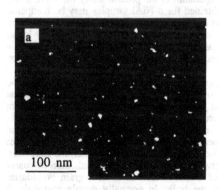
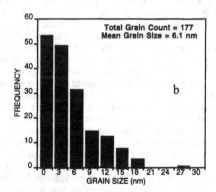

Figure 2. (a) Dark field TEM image of nanocrystalline NiAl-1, and (b) histogram showing the grain size distribution for the same specimen.

A representative room temperature load-displacement curve for n-NiAl-1 and coarse grained NiAl sample strained at 1.4×10^{-6} sec^{-1} is shown in Fig. 3. After correction for the expected initial non linear region [7], the yield stresses of nanocrystalline samples were calculated from the total loads at yielding (taking into account the sample thicknesses according to the procedures described in [5]). Yield stress values are given in Table I. Coarse-grained NiAl samples fractured prior to yielding and the stresses at fracture were typically approximately 75 MPa, well below the yield stresses of nanocrystalline samples.

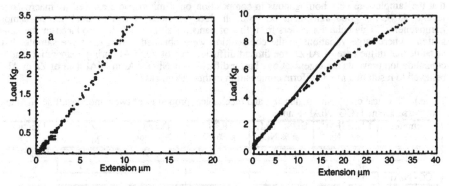

Figure 3. Load-displacement curves obtained at room temperature for (a) coarse grained NiAl (sample failed prior to yielding), and (b) nanocrystalline NiAl-1.

It has been shown by Li et al. [7] that for disk bend tests plastic yielding occurs at the onset of the deviation from linearity in the load-displacement curve and that the yield strength obtained by disk-bend testing is comparable to the tensile yield strength. A significant and most important observation in the present study is that a deviation from linearity is observed in the load-displacement curve for nanocrystalline NiAl, indicating that NiAl in nanocrystalline form exhibits yield behavior and thus measurable room temperature ductility. A deviation from linearity was not observed in eight CG-NiAl samples tested under similar conditions. Strains-to-failure of 0.1 - 0.2% and 0.016 - 0.06% were observed at room temperature in the present study for n-NiAl and CG-NiAl, respectively. The values of total strain obtained for n-NiAl samples may be hampered by the residual porosity since stress concentrations could develop at the voids resulting premature fracture. Therefore, the measured strain-to-failure values in the present case are most likely lower limits. Further improvements in ductility may be obtained if denser specimens can be obtained in the future. The failure mode of n-NiAl was determined to be ductile from postmortem examination of broken pieces using SEM, as seen in Fig. 4.

The apparent improvement in the room temperature ductility in n-NiAl can be understood by examining the potential deformation mechanisms in nanocrystalline materials. For materials that are normally ductile at ambient temperature in coarse grained form (e.g. fcc metals) a significant reduction in ductility is observed when grain sizes are reduced to the nanometer range [6,8]. In these fcc metals, deformation occurs by dislocation generation and motion. It has been pointed out by numerous authors (e.g., [9]) that dislocation generation is increasingly difficult as grain sizes decrease, leading to a decrease in dislocation-based ductility in nanocrystalline materials. Likewise, diffusional mechanisms are expected to be more active in nanocrystalline materials than in coarse grained materials due to the large volume fraction of atoms located in or near grain boundaries [10], which display far higher diffusion rates than the bulk. In normally ductile materials, it appears that any increase in ductility afforded by diffusional mechanisms is insignificant compared to the competing loss of ductility due to the hindrance of dislocation motion [11]. In contrast to the behavior of materials that are normally ductile in coarse grained form, the present studies indicate that in the case of a normally brittle material such as NiAl where dislocation-based ductility is limited even for coarse-grained material, deformation of nanocrystalline samples is likely affected to a measurable extent by enhancements of diffusional mechanisms (such as grain boundary sliding and Coble creep) that accompany grain size refinements.

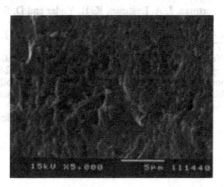

Figure 4. SEM micrograph showing a ductile-appearing fracture surface of n-NiAl-1.

Another aspect of the observed improved ductility to consider is the test method itself. Since the stress state during a disk bend test is complex and involves both tensile and compressive stresses, it is possible that this may have an effect on the measured ductility of nanocrystalline materials. For example, significantly larger ductilities are seen for nanocrystalline Cu tested in compression [12] than in tension [8]. However, this cannot be the only reason for the measured enhanced ductility in case of n-NiAl, since coarse grained specimens did not show any evidence of ductility when also tested by BDBT under similar conditions.

CONCLUSIONS

The conclusions drawn from this study are:
1) Ordered single phase nanocrystalline NiAl can be successfully synthesized using an electron beam inert gas condensation system. The chemical composition and grain size distribution is quite uniform on both microscopic and macroscopic scales.
2) The substantial room temperature ductility observed in nanocrystalline NiAl samples suggests that diffusional mechanisms such as grain boundary sliding controlled by grain boundary diffusion are contributing substantially to deformation.

ACKNOWLEDGMENTS

This research was supported by AFOSR Grant# FY9620-92-J, and by U.S. Department of Energy BES-DMS Contract# W-31-109-ENG-38. The authors gratefully acknowledge the help of Dr. John Kramer for Instron operation and experimental setup, Bernie Kestle for TEM specimen preparation, Pete Baldo for performing the RBS experiment, and Dr. Stanslaw Dymek for TEM operation.

REFERENCES

1. E. M. Schulson and D. R. Barker, Scripta Met. 17, 519 (1983)
2. T. R. Smith, Nanostruct. Mater. 5, 337 (1995)

3. T. Haubold, R. Bohn, R. Birringer and H. Gleiter, Mater. Sci. and Eng. **A153**, 679 (1992)
4. J. A. Eastman, L. J. Thompson and D. J. Marshall, Nanostruct. Mater. **2**, 377 (1993)
5. S. Timoshenko and S. Woinowsky-Krieger, Theory of Plates and Shells, 2nd ed., McGraw-Hill, New York, NY, (1959)
6. M. N. Rittner, T. R. Weertman, J. A. Eastman, K. B. Yoder and D. S. Stone, Accepted for publication in Materials Science and Engineering A, (1996)
7. H. Li, F. C. Chen and A. J. Ardell, Metall. Trans. **A22**, 2061 (1991)
8. P. G. Sanders, Ph.D. Thesis, Northwestern University, (1996)
9. G.W. Nieman, J. R. Weertman and R.W. Siegel, J. Mater. Res. **6**, 1012 (1991)
10. G. Palumbo, S. J. Thrope and K. T. Aust, Scripta Metall. et Mater. **24**, 1347 (1990)
11. J. A. Eastman, M. Choudry, M. N. Rittner, C. J. Youngdahl, M. Dollar, J. R. Weertman, R. J. DiMelfi and L. J. Thompson, Submitted to the proceedings of the annual meeting of the Minerals, Metals, and Materials Society, Orlando (1997)
12. C. Y. Youngdahl, J. R. Weertman and J. A. Eastman, Submitted to Scripta Met. (1996)

STRUCTURAL EVOLUTION OF Fe RICH Fe-Al ALLOYS DURING BALL MILLING AND SUBSEQUENT HEAT TREATMENT

H.G. JIANG, R.J. PEREZ, M.L. LAU, E.J. LAVERNIA
Department of Chemical Engineering and Materials Science
University of California, Irvine. Irvine, CA 92697-2575, jhang@eng.uci.edu

ABSTRACT

X-ray diffraction (XRD) and differential scanning calorimetry (DSC) have been utilized to investigate the structural evolution of Fe rich Fe-Al alloys during ball milling. It is found that b.c.c. solid solutions can be formed either through ball milling alone or through ball milling together with heat treatment. Thermal diagrams of the milled Fe-Al powders reveal exothermic peaks corresponding to the formation of α-Fe(Al) solid solution (in both Fe-4wt.%Al and Fe-10wt.%Al) and the formation of FeAl intermetallic compound (in Fe-10wt.%Al). The transformation kinetics of α–Fe(Al) solid solution in Fe-4wt.%Al were found to follow the Johnson-Mehl-Avrami equation.

INTRODUCTION

The synthesis of Fe-Al alloys by high energy ball milling has been described in a number of studies, the majority of which were directed toward the intermetallic compositions of Fe_3Al and FeAl [1-4], and accordingly only a few studies have been conducted on Fe rich Fe-Al alloys [5-7]. Recently, however, Rawers et al. [5] and Perez et al. [6, 7] reported that the addition of relatively small quantities of Al (less than 10wt.%) to Fe during ball milling can enhance the thermal stability of the as-milled nanocrystalline Fe-Al alloys. This would hinder the grain growth of these powders during subsequent thermomechanical consolidation, thereby aiding in the fabrication of dense nanocrystalline solids. The physical origin behind this enhancement of thermal stability is speculated to be due to the formation of iron aluminum oxides and/or solid solution of Al in Fe [5, 7]. In order to obtain deeper insight into these promising stabilization mechanisms, careful study of the dissolution of 0-10wt.% Al in Fe during ball milling is required.

Recent evidence has suggested that the formation of solid solutions using ball milling may, in certain cases, be extremely slow. For example, cryogenic ball milling of Fe-3wt.%Al elemental powders for 25 hrs has been shown to produce negligible increase in the Fe lattice parameter, indicating an absence of dissolved Al atoms in the lattice [7]. The present study represents an effort to evaluate the kinetics of the mechanically induced dissolution of the Fe-Al system at conventional (i.e., ambient) ball milling temperatures using SPEX milling.

EXPERIMENTAL

Blended Fe (99.9% pure, 100-200 mesh) and Al (99.97% pure, 40 mesh) powders were prepared in Fe-4wt.%Al and Fe-10wt.%Al compositions for ball milling in a SPEX model 8000D shaker mill. The powders and stainless steel milling balls (in a 1:4 mass ratio) were sealed within the stainless steel vials under a protective argon atmosphere. XRD measurements were carried out in a Siemens D5000 diffractometer equipped with a graphite monochromator using Cu Kα and Mo Kα radiation. Thermal analysis of powder samples taken during the ball milling process was performed using a Perkin-Elmer DSC 7 system. In order to protect the powder samples from oxidation during DSC, they were covered by flowing argon.

RESULTS

Figures 1 and 2 show the XRD patterns of the Fe-4wt.%Al and Fe-10wt.%Al blended elemental powders as a function of milling time. From Figure 1, it is apparent that diffraction peaks corresponding to Al were not visible in the blended Fe-4wt.%Al powders, due to the small

297

Mat. Res. Soc. Symp. Proc. Vol. 457 ° 1997 Materials Research Society

Al content. In Fe-10wt.%Al powder, Al peaks which may be clearly discerned in the blended powders (Figure 2) were found to disappear with 40 minutes of milling. The corresponding shift in α-Fe peak position, however, was small relative to that expected for a solid solution of 10wt.%Al in α-Fe [8]. Broadening of the XRD peaks became apparent with increasing ball milling time, indicating the refinement of grains of the milled Fe-4wt.%Al and Fe-10wt.%Al powders.

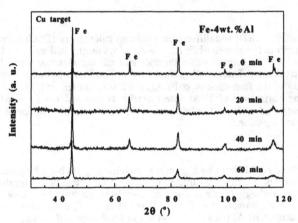

Figure 1. XRD patterns of Fe-4wt.%Al powders milled for 0, 20, 40 and 60 min.

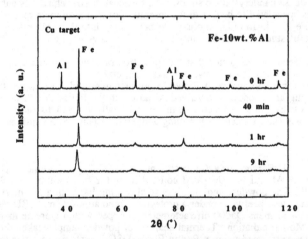

Figure 2. XRD patterns of Fe-10wt.%Al powders milled for 0, 40, 60 and 540 min.

Figures 3 and 4 show the DSC thermal diagrams of the Fe-4wt.%Al and Fe-10wt.%Al powders during the initial period of ball milling. There was one exothermic peak in the milled Fe-4wt.%Al powders and two exothermic peaks in Fe-10wt.%Al powders. All exothermic peaks disappeared after prolonged ball milling (about 60 min for Fe-4wt.%Al and 120 min for Fe-10wt.%Al). In order to determine the nature of the exothermic peaks, additional Fe-10wt.%Al samples milled for 10 min were heated in the DSC to three predetermined points: (a) prior to the first peak, (b) after the first peak and (c) following the second peak. The samples were then cooled at the maximum rate obtainable (approximately 320°C/min), and analyzed by XRD, as shown in Figure 5. The figure clearly shows that Al diffraction peaks were present prior to the first peak, but were no longer part of the spectra taken immediately afterward. This suggests that the first peak represents the heat released upon the formation of α-Fe(Al) solid solution. In order to confirm the importance of the addition of Al in producing the peak, pure Fe powders were prepared using identical milling conditions. DSC analysis of these powders revealed no exothermic reactions. For the sample passing the second exothermic peak (Figure 5 C), a set of new XRD peaks emerged, which were indexed as FeAl intermetallic compound, rather than Fe₃Al or Fe(Al) solid solution. This observation can be hardly interpreted by the equilibrium phase diagram, which determines a single α-Fe(Al) solid solution phase in this composition. Even the concept of metastable phase boundary extension [2] which successfully explained the coexistence of Fe₃Al and α-Fe(Al) solid solution in this composition range fails to justify this because FeAl phase is not in the vicinity of α-Fe(Al) solid solution in the Fe-Al phase diagram. Further studies regarding the mechanism of the formation of the FeAl phase is under way.

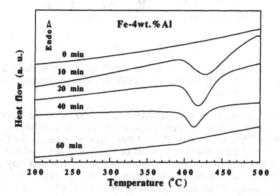

Figure 3. DSC diagrams of the Fe-4wt.%Al milled for different times.

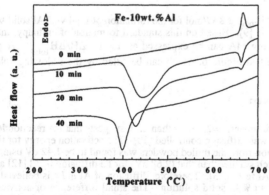

Figure 4. DSC diagrams of the Fe-10wt.%Al milled for different times.

299

In order to study the kinetics of the formation of solid solution during ball milling, the ball milled Fe-4wt.%Al, which has single exothermic peak in the thermal diagram, was

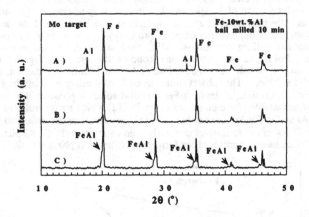

Figure 5. XRD patterns of Fe-10wt.%Al powder milled for 10 min and annealed prior to (sample A) and after (sample B) the first exothermic peak and after the second exothermic peak (sample C) in DSC diagram.

employed for XRD and DSC analysis. It is reasonable to assume here that the formation of solid solution consists of both a mechanically induced component (during ball milling) and a thermally induced component (during DSC heating), with the latter being reflected in the DSC diagrams (As seen in Figures 3 and 4). If the thermal component of enthalpy of solid solution formation is known, the mechanical component of enthalpy of formation of solid solution may be calculated provided that the overall standard enthalpy of formation (ΔH_{total}) is available. In the present experiment, the standard enthalpy of formation of solid solution Fe_xAl_{1-x} (x atomic content) at 293 K can be approximated by [9]

$$\Delta H_{total} = (-38000 + 28000)x(1-x) \tag{1}$$

which yielded a standard ΔH_{total}=3.8 kJ/mol for the formation of Fe-4wt.%Al solid solution from elemental powders at 293 K [9]. Based on this standard formation of enthalpy, the fraction of solid solution formed during MA can be expressed as $f = 1 - \Delta H/\Delta H_{total}$. A plot of f vs. milling time t is illustrated in Figure 6, which can be fitted by the following Johnson-Mehl-Avrami type equation:

$$f = 1 - \exp(-0.0003\, t^{2.3}) \tag{2}$$

The value of the kinetic parameter, n=2.3 (less than 2.5), suggests that the reaction leading to the formation a solid solution was diffusion controlled [10]. The activation energy for the formation of Fe-4wt.%Al solid solution from the milled powders was found to be 1.42 eV using Kissinger's analysis [11], which is much smaller than that (2.68 eV) for Fe diffusion in Al [12] and relatively close to that (1.95 eV) of Al in Fe [13]. Therefore, diffusion of Al in Fe is believed to dominate the formation of the Fe-4wt.%Al solid solution. The small difference of activation energies between 1.42 and 1.95 eV might be due to the influence of various defects introduced during ball milling. However, in a system possessing a positive mixing enthalpy, the mechanism governing the formation of supersaturated solid solution during ball milling may be different from those

300

possessing negative mixing enthalpy. It was recently observed by Bellon et al. [14] that pure shear deformation leads to the formation of homogeneous supersaturated solid solution while diffusion opposes the mixing process during ball milling. As such, it is reasonable to speculate that the role played by diffusion in the formation of solid solution during MA depends on the thermodynamic driving force for solid solution formation. The more energetically favorable the formation of a solid solution is, the larger a role diffusion plays.

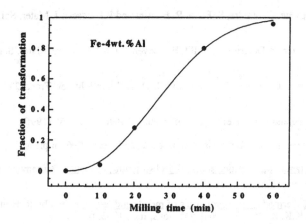

Figure 6. Formation of Fe-4wt.%Al solid solution as a function of milling time.

The role played by mechanical mixing in promoting solid solution formation can be reflected by the shift of exothermic peaks to lower temperatures with increasing milling time (Figures 3 and 4). Extensive ball milling can remarkably minimize the layer-to-layer thickness (the distance between the center of neighboring layers) between Fe and Al, and then accelerate the formation of solid solution. This enhancement of reaction due to smaller layer to layer distance has also been found in multilayer films exhibiting an obvious shift of exothermic peaks to lower temperature (in the thermal diagrams) with decreasing the unit bilayer thickness [15, 16]. In addition, the refining of grain size to the nanoscale during mechanical milling introduces a large percentage of grain and interphase boundaries and various defects, all of which could promote the accelerated formation of solid solution through a diffusion mechanism.

CONCLUSIONS

XRD and DSC can be employed to quantitatively study the kinetics of the solid solution formation during ball milling. DSC analysis of the milled Fe-Al powders yields exothermic peaks corresponding to the formation of solid solution (in both Fe-4wt.%Al and Fe-10wt.%Al) and the formation of an FeAl intermetallic compound (in Fe-10wt.%Al). The transformation kinetics of α-Fe(Al) solid solution in Fe-4wt.%Al were found to follow the Johnson-Mehl-Avrami equation. Assessment of the kinetic parameter, n, indicates that a diffusion controlled mechanism (Al in Fe) is responsible for the formation of Fe-4wt.%Al solid solution.

ACKNOWLEDGMENTS

The financial support of the Office of Naval Research (Grants No. N00014-94-1-0017 and No. 00014-93-1-1072) is gratefully acknowledged.

REFERENCES

1. S. Enzo, R. Frattini, R. Gupta, P.P. Macri, G. Principi, L. Schiffini and G. Scipione, Acta Mater. **44**, p. 3,105 (1996).

2. D.G. Morris and S. Gunther, Acta Mater. **44**, p. 2,847 (1996).

3. E. Bonetti, G. Scipione, G. Valdre, S. Enzo, R. Frattini and P.P. Macri, J. Mater. Sci. **30**, p. 2,220 (1995).

4. K. Wolski, G. Le Caer, P. Delcroix, R. Fillit, F. Thevenot and J. Le Coze, Mater. Sci. Eng. **A207**, p. 97 (1996).

5. J. Rawers, G. Slavens, D. Govier, C. Dogan and R. Doan, Metall. Mater. Trans. **27A**, p. 3,126 (1996).

6. R.J. Perez, B. Huang and E.J. Lavernia, NanoStructured Mater. **7**, p. 565 (1996).

7. R.J. Perez, H.G. Jiang and E.J. Lavernia, NanoStructured Mater. 1996, in press.

8. W.B. Pearson in Handbook of lattice spacings and structures of metals, Pergamon Press, London, Vol.2, 1967, p. 560.

9. Selected values of thermodynamic properties of metals and alloys, edited by R. Hultgren, R.L. Orr, P.D. Anderson, K.K. Kelley, John Wiley & Sons, Inc., 1963, p. 415.

10. A.S. Shaikh and G.M. Vest, J. Am. Ceram. Soc. **69**, 682 (1986).

11. H.E. Kissinger, Anal. Chem. **29**, 1,702 (1957).

12. G.M. Hood, Phil. Mag. **21**, 305 (1970).

13. A. Vignes, J. Philibert, N. Badia, Diffusion Data **3**, 269 (1969).

14. P. Bellon and R.B. Averbach, Phys. Rev. Lett. **74**, 1,819 (1995).

15. L.A. Clevenger, C.V. Thompson and R.C. Cammarata, Appl. Phys. Lett. **52**, 795 (1988).

16. H.G. Jiang, J.Y. Dai, H.Y. Tong, B.Z. Ding, Z.Q. Hu and Q.H. Song, J. Appl. Phys. **74**, 6,165 (1993).

MICROSTRUCTURE, MECHANICAL PROPERTIES AND WEAR RESISTANCE OF WC/Co NANOCOMPOSITES

KANG JIA AND TRAUGOTT E. FISCHER
Department of Materials Science and Engineering, Stevens Institute of Technology, Hoboken, NJ 07030

ABSTRACT

The microstructure, mechanical properties, abrasion and wear resistance of WC-Co nanocomposites synthesized by the spray conversion technique by McCandlish, Kear and Kim have been investigated. The binder phase of WC-Co nanocomposites is enriched in W and C, compared to conventional cermets. Small amorphous regions exist in the binder despite the slow cooling after liquid phase sintering. Few dislocations are found in the WC grains. The increased WC content and the amorphous regions modify (i.e. strengthen) the binder phase of the composites. Vickers indentation measurements show a hardness of the nanocomposites reaching 2310 kg/mm². While the toughness of conventional cermets decreases with increasing hardness, the toughness does not decrease further as the WC grain size decreases from 0.7 to 0.07 μm but remains constant at 8 MPam$^{1/2}$. Scratches caused by a diamond indenter are small, commensurate with their hardness. These scratches are ductile, devoid of the grain fracture that is observed with conventional materials. The abrasions resistance of nanocomposites is about double that of conventional materials, although their hardness is larger by 23% only. This is due to the lack of WC grain fragmentation and removal which takes place in conventional cermets. Sliding wear resistance of WC/Co is proportional to their hardness; no additional benefit of nanostructure is obtained. This results from the very small size of adhesive wear events in even large WC grains.

INTRODUCTION

WC-Co nanocomposite powders [1] have been synthesized by a novel thermochemical synthesis method, called Spray Conversion Processing (SCP)[2]. These have been sintered by the RTW company into nanostructure WC-Co composites with WC grain size of about 70 nm. The present work investigated the microsctructures of the nanocomposites and of conventional cermets and their effect on the hardness, the toughness measured by Vickers indentation, the resistance to abrasion and sliding wear of the materials.

Small WC grains may not only change their own properties but also those of the binder phase [1]. For a given binder concentration, a smaller average WC grain size means a smaller average thickness of the binder phase. The cobalt binder has fair amounts of tungsten and carbon in solid solution. During cooling after sintering, these two elements precipitate on neighboring carbide grains; as a result there is a gradient of tungsten concentration in the binder phase. High contents of tungsten in the binder phase give improved performance, for instance in milling operations. This has been exploited in many commercial grades by means of heat-treatment or by decreasing the carbon content. It is known that a high concentration of tungsten and carbon atoms in cobalt can increase its martensite transforming temperature (FCC to HCP) to about 750 °C from 417 °C. This avoids the formation and expansion of the brittle hexagonal close-packed cobalt at low temperature, increases the a-Co content and improves the transverse rupture strength and toughness of the cemented carbides.

303

Mat. Res. Soc. Symp. Proc. Vol. 457 © 1997 Materials Research Society

The present investigation has shown that, indeed, reduction of the WC grain size to the nanometer range influences the structure and composition of the binder phase with attendant effects on the hardness, toughness, abrasion and sliding wear resistance of the materials. While the resistance to both forms of wear increases with the hardness of the samples, departures from this law depend on the WC grain size in ways that are different for the two wear forms, showing that the wear mechanisms are different and respond to different material properties. The details of the investigation will be published elsewhere, we present here an overview of the results.

THE SAMPLES

Conventional carbides, in which the average WC grain size ranges from 0.7 to 2.5 µm and the cobalt content from 6 to 20 wt.%, were obtained from the Federal Carbide Company (FC) and the RTW company. The nano-structured carbides with carbide grain size of 0.07 µm and cobalt contents from 7 to 15 wt.% were sintered by the RTW Company (RTW) from powders synthesized by Nanodyne through the Spray Conversion technique [2]. In all figures NA indicates nanostructured samples, FC and RTW conventional samples, and the numbers indicate the weight percent cobalt content. Details of the sample preparation and measurement techniques are published elsewhere [2,3].

MICROSTRUCTURE

The nanophase cemented carbide has a fine structure with smaller carbide grain size and binder thickness than the conventional material. Not much difference is observed in the microstructure and composition of WC phase of conventional and nano-structured WC-Co composites. The dislocation density in the WC crystals of the nano-structured samples is lower than the conventional ones, and no inclusions are observed in either.

There is a significant difference in the binder phase of conventional and nano-structured composites, as shown in Figure 1. An amorphous phase is observed in the binder phase of the nano-structured samples. The tungsten content detected within the amorphous regions is much higher than in the crystalline binder phase. It is not clear why the amorphous phase exists and how much of it there is in the binder of nano-structured samples.

Nano-structured composites have a higher tungsten content in the binder phase than conventional ones. This not only increases the volume fraction of the binder phase in the nano-structured composite but also raises the ratio of FCC/HCP of the cobalt. The small size of the WC grains in the nano-structured composite promotes the solubility of WC in both liquid and solid cobalt, which stabilizes the FCC phase of cobalt during cooling from sintering.

HARDNESS AND TOUGHNESS

An unambiguous relationship exists between hardness and the mean free path in conventional and nano-structured composites separately: hardness increases with decreasing binder mean free path. The ascending rate of the hardness with decreasing binder mean free path in nano-structured samples is much faster than that in the conventional ones (Figure 2). The high hardness of nano-structured cemented carbides results not only from the ultra fine microstructure, but also from the alloy-strengthening of the binder phase.

The bulk fracture toughness is related to crack propagation through the phases of the material. Palmqvist indentation toughness confirms that toughness decreases with increasing hardness in

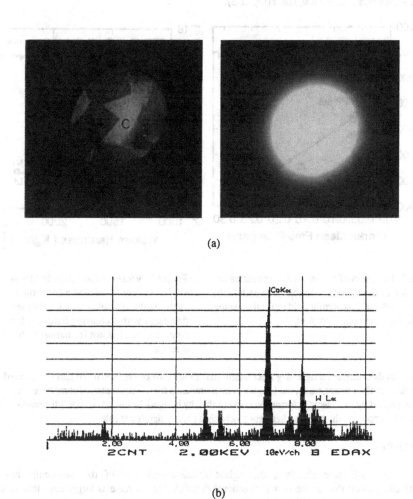

(a)

(b)

Figure 1. a) Left: Transmission electron micrograph of WC/Co namocomposite.
Rignt: Electron diffraction of binder area C
b) EDS spectrum of area C of a): the composition is 43 wt% W,
57 wt% Co

305

conventional composites but the increase of hardness in the nano-structured composites does not decrease their bulk fracture toughness (Figure 3).

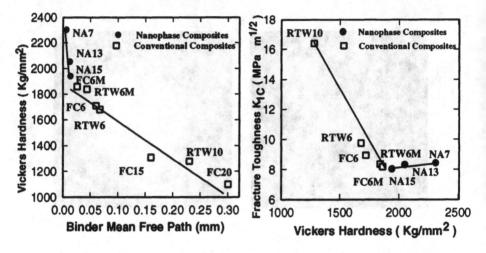

Figure 2: Hardness of the WC/Co composites as a function of the dislocation mean free path in the Co binder phase. The symbols identify the samples, the numbers the cobalt content.

Figure 3: Vickers indentation hardness of WC/Co composites as a function of their hardness. Note that the toughness decreases with increasing hardness for conventional, but not for nanostructured cermets.

This implies that different toughening mechanisms may exist in the conventional and nano-structured composites. In the present study, the toughness measurements for the conventional composites are better explained by the plastic deformation concept than by bridging ligaments, whereas for nano-structured composite, the bridging ligament mechanism plays a significant role.

ABRASION

Against all three abrasives, the highest abrasion resistance of the nanocomposites is approximately twice that of the best conventional material. This increase is larger than that of the hardness alone which is about 26% (Figure 4).

The resistance of WC/Co composites to abrasion by diamond increases with increasing hardness of the material and decreasing WC grain size. The latter contributes more to the gain in abrasion resistance of nanocomposites than hardness does.

SLIDING WEAR

The sliding wear of the conventional and nanostructured WC-Co composites can, in first approximation, is inversely proportional to the macroscopic hardness of the samples (Figure 5); it can be expressed by a Blok-Archard equation with wear coefficient $k = 6.9 \times 10^{-6}$.

Departures from the simple hardness dependence that is expressed by the solid line depend in a somewhat complex manner on the cobalt content, the grain size and the hardness of the composite. Smaller carbide grain sizes tend to lower wear resistance, in spite of an increase in hardness.

The wear of the cermets can also be expressed as increasing with the cobalt content. The wear of nanostructured materials, at equal cobalt content, is only 60% of that of the conventional cermet. The most effective technique to increase sliding wear resistance of conventional cermets is to reduce the cobalt

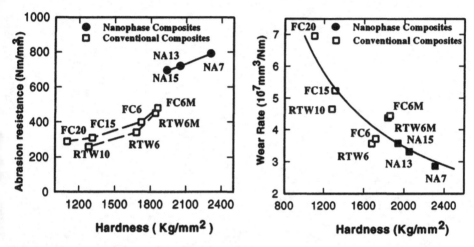

Figure 4: Abrasion resistance of WC/Co composites as a function of their hardness. There is an additional contribution to wear provided by the decrease in WC grain size.

Figure 5: Sliding wear rate of WC/Co composites as a function of their hardness.

content and increase the grain size. In nanocomposites, by contrast, the reduction in grain size does not decrease the wear resistance. Their higher wear resistance is commensurate with their hardness.

ACKNOWLEDGMENTS

Support for this works by the Office of Naval Research under Contract # N00014-91-J-1661 is gratefully acknowledged. The authors also thank Professor Bernard Kear of Rutgers University and Dr. Larry McCandlish of Nanodyne Inc. for providing them with the samples, and Professor Bernard Gallois for valuable discussions.

REFERENCES

1. R. Birringer, Materials Science and Engineering, A117, 1989, 33-43

2. L. E. McCandlish, B. H. Kear and B. K. Kim, Materials Science and Technology, 6, 1990, 953-960

3. K. Jia and T. E. Fischer, Wear 200 (in press)

4. K. Jia and T. E. Fischer, Wear of Materials 1997, San Diego Cal. April 21

SYNTHESIS OF NANOCOMPOSITE THIN FILM Ti/Al MULTILAYERS AND Ti-ALUMINIDES

R. BANERJEE, X. D. ZHANG, S. A. DREGIA AND H. L. FRASER
Department of Materials Science and Engineering, The Ohio State University, Columbus, OH 43210

Abstract

Nanocomposite Ti/Al multilayered thin films have been deposited by magnetron sputtering. These multilayers exhibit interesting structural transitions on reducing the layer thickness of both Ti and Al. Ti transforms from its bulk stable hcp structure to fcc and Al transforms from fcc to hcp. The effect of ratio of Ti layer thickness to Al layer thickness on the structural transitions has been investigated for a constant bilayer periodicity of 10 nm by considering three different multilayers: 7.5 nm Ti / 2.5 nm Al, 5 nm Ti / 5 nm Al and 2.5 nm Ti / 7.5 nm Al. The experimental results have been qualitatively explained on the basis of a thermodynamic model. Preliminary experimental results of interfacial reactions in Ti/Al bilayers resulting in the formation of Ti-aluminides are also presented in the paper.

Introduction

Laminated intermetallic nanocomposites based on Ti-aluminides are potentially important for use as high temperature structural thin film coatings in aerospace applications [1]. These laminated composites are expected to have attractive mechanical properties. The ability to tailor the microstructure of such thin films on a nanoscale makes magnetron sputtering a suitable technique for processing these multilayers. Recently, a series of intriguing structural transitions were reported in Ti/Al multilayered thin films in which the Ti:Al thickness ratio was fixed at 1:1 [2,3,4,5]. At room temperature, bulk Ti and Al have *hcp* and *fcc* structures respectively. On reducing the bilayer thickness (also known as the compositionally modulated wavelength, CMW) to ~ 10 nm, an *fcc* Ti / *fcc* Al structure was observed in these multilayers. Further reduction of the CMW to ~ 5 nm resulted in an *hcp* Ti / *hcp* Al structure. A plausible explanation for these structural transitions was also presented by the authors [5] based on a model initially proposed by Redfield and Zangwill [6]. Subsequently, a different model has been proposed to account for these structural transitions [7]. In the present paper the influence of Ti:Al ratio on these structural transitions in Ti/Al multilayers will be investigated and interpreted on the basis of the more recent model [7,8]. Furthermore, preliminary experimental results of intermetallic compound formation at Ti/Al interfaces in annealed Ti/Al bilayers will also be discussed in this paper.

Experimental Procedure

The thin films have been deposited using a custom designed UHV magnetron sputtering system. The base pressure was ~ 1 x 10^{-9} torr prior to sputtering and the argon pressure was 2 x 10^{-3} torr during the sputtering process. The Ti/Al multilayered thin films have been deposited using pure Ti and Al targets, which were 3" in diameter and 0.25" thick. Oxidized Si (100) wafers with a 200 nm oxide layer on the surface were used as the substrate for deposition. Targets consisted of pure Ti and Al, with the sputtering power being 200 W DC for Ti and 160 W DC for Al. The substrate temperature during deposition did not exceed 323 K. Three different multilayers were deposited, each with a constant value of CMW=10nm, and each with a total multilayer thickness of 350nm, i.e. with 35 bilayers. The three multilayers had individual layer thicknesses of 7.5 nm Ti / 2.5 nm, 5 nm Ti / 5 nm Al and 2.5 nm Ti / 7.5 nm Al, respectively. These multilayered thin films were characterized by cross-section transmission electron microscopy (TEM) using a Philips CM200 microscope operating at 200 kV and by high resolution electron microscopy (HREM) using a Hitachi H 9000 NAR microscope operating at 300 kV. The details of the cross-section sample preparation procedure are discussed elsewhere [9]. The Ti/Al bilayered thin films have been deposited under the same conditions as the multilayers. These consisted of a 750 nm thick Ti layer and a 750 nm Al layer, and the bilayered films were deposited on oxidized Si(100) wafers. Sections of the wafer were annealed at 748 K under a protective Ar atmosphere for 180 minutes. Cross-section TEM specimens have been prepared from the annealed bilayers and studied in a Philips CM200 TEM operating at 200 kV.

Results and Discussion

Ti/Al Multilayers

A cross-section selected area diffraction (SAD) pattern recorded from the 7.5 nm Ti / 2.5 nm Al multilayer, shown in Fig.1(a), consists of reflections corresponding to the $<2\bar{1}\bar{1}0>$ and $<10\bar{1}0>$ zone axes for *hcp* Ti and the $<1\bar{1}0>$ and $<1\bar{2}1>$ zone axes for *fcc* Al. This indicates that the Ti and Al layers grow with a {0002} and {111} texture respectively. A cross-section HREM image (Fig.1(b)) from the same multilayer revealed an *hcp* ABABAB... stacking sequence of close-packed planes in the Ti layers and an *fcc* ABCABC... stacking sequence of close-packed planes in the Al layers. Both Ti and Al adopt their bulk stable structures of *hcp* and *fcc* respectively in this multilayer. In the 5 nm Ti / 5 nm Al multilayer, both Ti and Al exhibit an *fcc* stacking sequence as shown in Fig.2. The Ti layers undergo a transition but the Al layers retain their bulk structure in this multilayer. Finally, in the 2.5 nm Ti / 7.5 nm Al multilayer, both Ti and Al adopt *fcc* stacking sequences. The absence of the {0$\bar{1}$10} type reflections in the selected area diffraction pattern from this multilayer (Fig.3(a)) indicates the transition of *hcp* Ti to *fcc* Ti. The HREM image in Fig.3(b) also confirms this observation.

Since the various transitions from stable to metastable structures occur in thin films as the layer thicknesses decrease, i.e. as the interfacial area-to-volume ratio increases, it seems reasonable to introduce a model based on changes in interfacial energies as the driving force for transformation [7,8]. Thus, the net free energy change in any transformation can be represented as a combination of the bulk free energy, strain energy and interfacial energy contributions as represented below :

$$\Delta G = \Delta G_V . V + \Delta G_S . V + \gamma . A \tag{1}$$

where ΔG is the net free energy change, ΔG_V, the strain-free volume free energy change for the bulk, ΔG_S, the strain energy contribution and γ, the interfacial energy. V and A represent the volume and surface area, respectively. For each layer in a multilayered structure of Ti/Al, an expression for the net free energy change can be written by normalizing per unit cross-sectional area as follows :

$$\Delta G / A = \Delta G' . d + \gamma \tag{2}$$

where $\Delta G'$ is the volume free energy change for *hcp-fcc* type transitions in either Ti or Al, and d

310

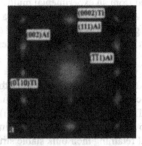 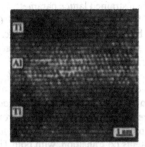

Fig.1 (a) A cross section SAD pattern from the 7.5/2.5 multilayer. The spots have been indexed on the basis of the <2$\bar{1}\bar{1}$0> and <10$\bar{1}$0> zone axes for hcp Ti and <1$\bar{1}$0> and <1$\bar{2}$1> zone axes for fcc Al. (b) A high resolution image from the same multilayer showing the hcp stacking in the Ti layers and fcc stacking in the Al layer.

Fig.2. A high resolution image showing the fcc stacking sequence in both Ti and Al layers of the 5/5 multilayer.

Fig.3.(a) A SAD pattern from 2.5/7.5 multilayer. The absence of {0~110}hcp type reflection suggests that Ti has transformed from hcp to fcc. (b) A high resolution micrograph from the same multilayer confirming the fcc stacking in both layers

is the thickness of a layer in the multilayer. Therefore, the net free energy change for unit repeat distance in a Ti/Al multilayer (which consists of one layer of Ti and one layer of Al) can be expressed as given below :

$$\Delta G / A = \Delta G'(Ti).f_{Ti}.\lambda + \Delta G'(Al).f_{Al}.\lambda + 2\Delta\gamma \qquad (3)$$

where $\Delta G'(Ti)$ and $\Delta G'(Al)$ represent the free energy difference between the hcp and fcc forms of Ti and Al respectively. f_{Ti} and f_{Al} represent the volume fraction (i.e. thickness) of Ti and Al in a bilayer of the multilayer and λ is the bilayer period (CMW). Equation (3) represents the functional dependence of $\Delta G/A$ on the CMW. A schematic representation of $\Delta G/A$ versus λ is shown in Fig. 4, in which is has been assumed that $\Delta\gamma(hcp\ Ti/\ hcp\ Al) < \Delta\gamma(fcc\ Ti/\ fcc\ Al) < \Delta\gamma(hcp\ Ti/fcc\ Al) < 0$, and that $\Delta G'(Ti) < \Delta G'(Al)$. These latter values of $\Delta G'$ for both Ti and Al have been assumed to be constant and independent of the CMW. Also $f_{Ti} = f_{Al} = 0.5$ for this case. Under these assumptions, the schematic diagram in Fig. 4 shows three distinct phase stability regimes, i.e. hcp Ti / fcc Al at large CMW values, fcc Ti / fcc Al at intermediate values of CMW and hcp Ti / hcp Al at small values of CMW, which are in concert with the experimentally determined changes observed with $f_{Ti} = f_{Al}$[3]. Thus, the structural changes in the Ti and Al layers as a function of layer thickness can be explained using this formulation provided the hierarchy of $\Delta G'$ and γ values described above are correct. Although calculations of $\Delta G'(Ti)$ and $\Delta G'(Al)$ using bulk lattice parameters do not agree with one of these assumptions [10], more recent calculations show that when strained lattice parameters recorded from an experimental multilayers with a CMW value of 5.2 nm are used, in fact $\Delta G'(Al) > \Delta G'(Ti)$ [11]. Also, using a supercell calculation, it has been shown that $\Delta\gamma(hcp\ Ti/fcc\ Al)$ is greater than either $\Delta\gamma(hcp\ Ti/hcp\ Al)$ or $\Delta\gamma(fcc\ Ti/fcc\ Al)$, although at present calculation reveals little difference between the energies of the latter two interfaces [11].

The results or trends are for equal thicknesses of Ti and Al. The effect of changing the ratio of layer thicknesses of Ti and Al will now be considered. According to the model (i.e. equation (3)), the slope of the line representing the transition from hcp Ti to fcc Ti is given by ($\Delta G'(Ti).f_{Ti}$) and that for the line representing the transition from fcc Al to hcp Al by ($\Delta G'(Al).f_{Al}$). Since the h/f line (Fig. 4) represents no transition, with both Ti and Al retaining their bulk stable structures of hcp and fcc respectively, its slope remains zero irrespective of the Ti:Al ratio. The slope of the line f/f increases on changing the Ti:Al ratio from 1:1 to 3:1 because the fraction of Ti increases in the multilayers, whereas by the same reasoning the slope of this line would decrease when the ratio is changed to 1:3. The slope of the line h/h decreases when the Ti:Al ratio changes to 3:1 and increases as this ratio is changed to 1:3. This has been schematically represented in Fig. 5. Since the point of intersection of the two transition lines (h/h and f/f) shifts to a higher value of λ and the intersection of the line f/f with the untransformed line (h/f) shifts to a lower value of λ as the Ti:Al ratio changes from 1:1 to 3:1, the phase stability region for fcc Ti / fcc Al would tend to be decreased whereas the phase stability region for hcp Ti / hcp Al would be increased (refer to Fig. 5). Conversely, changing the ratio of Ti:Al to 1:3 has the opposite effect of increasing the fcc Ti / fcc Al stability region and reducing the hcp Ti / hcp Al stability region. It is now possible to compare the trends in phase stabilities exhibited by the experimental results with the trends which emerge from the model, depicted in Fig. 5. Consider first the multilayer samples with a Ti:Al ratio of 3:1 and thickness of the individual layers being 7.5nm and 2.5nm for Ti and Al, respectively. Experimentally it has been revealed that the Ti layers adopt the hcp structure whereas the Al layer adopts the fcc structure. From the reasoning given above and consideration of Fig. 5, it is clear that for a layer ratio of 3:1, the extent of the fcc Ti / fcc Al region would be decreased, and so the structures adopted in the multilayers, i.e. the stable ones, are consistent with this trend. Interestingly, at this thickness of Al in a sample with a Ti:Al ratio of 1:1, the Al layer would have the hcp structure. This result cannot be explained on the basis of the model of Redfield and Zangwill [6]. For the multilayer with Ti:Al ratio of 1:3, experimentally both the Ti and Al layers adopt the fcc structure. The model would suggest a trend for increased stability of the fcc Ti / fcc Al region, and so the experimental observations are consistent with this. Interestingly, at a layer thickness of 2.5nm in a multilayer with a Ti:Al ratio of 1:1, the Ti layer would have adopted the hcp structure, a result which again cannot be explained on the basis of the model of Redfield and Zangwill [6].

Annealed Ti/Al bilayers

After annealing bilayers at 748 K for 180 minutes, an interfacial reaction between the Ti and Al layers resulted in a reaction zone which consisted predominantly of grains of Al_3Ti. Ti and Al were present in addition to the reaction products which means that complete reaction had not taken place. Fig. 6(a) is a bright field micrograph from the cross-section specimen which shows grains of Al_3Ti adjacent to the grains of Al. Fig. 6. (b) shows a dark field micrograph of an Al_3Ti grain. Close examination of the interface between Ti and Al_3Ti revealed grains of a second phase. Fig. 7(a) shows two particles of this phase lying at the interface between Al_3Ti and Ti. Microdiffraction patterns from this phase, shown in Figs. 7(b) and (c) can be indexed as the [110] and [111] zone axes diffraction patterns from γ-TiAl. Energy dispersive spectroscopy (EDS) analysis

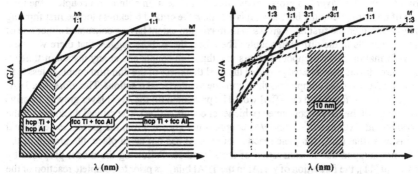

Fig.4. A schematic plot of ΔG/A versus the bilayer (l). The f/f line represents only Ti transforming from hcp to fcc. Similarly, the h/h line represents Al transforming. The h/f line represents the bulk structures for both Ti and Al with no transitions occuring in either.

Fig.5. A schematic plot representing the effect of changing the Ti:Al ratio on the boundaries of different phase stability regions. Three different Ti:Al ratios, 1:3, 1:1 and 3:1 have been considered. CMW value of 10 nm is expected to lie in the hashed region.

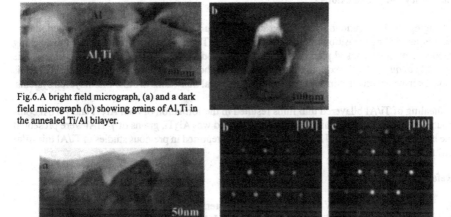

Fig.6.A bright field micrograph, (a) and a dark field micrograph (b) showing grains of Al₃Ti in the annealed Ti/Al bilayer.

Fig.7.(a) A bright field micrograph showing grains of γ-TIAl at the interface between Al₃Ti and Ti.(b) [101] and (c) [110] zone axes diffraction patterns from the γ-TiAl grains.

of this phase revealed an average composition of Ti-54 at % Al which is in the composition phase field of γ-TiAl in the Ti-Al binary phase diagram [12].

The formation of γ-TiAl in Ti/Al thin film diffusion couples has not been reported previously. It should be noted that research on compound formation in Ti/Al bilayers has been carried out for more than 20 years [13]; but there has never been any report of formation of intermetallic compounds other than Al₃Ti in such bilayers. A possible explanation for the absence of intermetallic compounds other than Al₃Ti was suggested by Gosele and Tu [14], although their model refers to planar morphologies rather than the more equiaxed transformation products observed in the present study. They introduced the concept of missing compounds in thin film diffusion couples as compared with bulk diffusion couples. A critical thickness is defined where a transition occurs from interfacial reaction to diffusion control of the rate of reaction. A second reaction product layer will not form until the first layer has grown to its critical thickness. In a thin-film couple if the total thickness of the film (including both layers) is less than the critical thickness for the first forming compound layer, then the first layer grows to its maximum extent having consumed either one of the parent materials in the diffusion couple. Therefore, the authors suggest that there will be a sequential formation of the compound layers in a thin film couple and at any instant of time no more than three layers can co-exist provided the total thickness of the thin film couple is less than the critical thickness of the compounds that form due to interfacial reactions. Therefore, in a Ti/Al diffusion couple, the formation of γ-TiAl would be prevented till the Al_3Ti layer grew to its critical thickness or all the Al was consumed (whichever occurs earlier). The critical thickness for the Al_3Ti compound layer in a thin film Ti/Al couple is not known. From the present experimental results it is clear that since complete reaction of pure Al has not taken place, the growth of the Al_3Ti has not stopped due to exhaustion of the Al source. Following the model proposed by Gosele and Tu [11], the formation of γ-TiAl in the Ti/Al bilayers prior to complete reaction of the Al layer to form Al_3Ti implies that the critical thickness of the Al_3Ti layer is less than ~ 1000 nm (the maximum thickness to which the Al_3Ti layer can grow if it consumes the 750 nm thick Al layer completely).

Summary and Conclusions

Changing the Ti:Al ratio in sputter deposited Ti/Al multilayers with the same bilayer periodicity, causes interesting transitions in the structure of the Ti and Al layers, in terms of the stacking sequence of close-packed planes. Using a simple model based on the interfacial energies,, it is possible to qualitatively explain these transitions. It should be noted that the applicability of this model can be extended to investigate possible structural transitions in other multilayered systems.

Annealing of Ti/Al bilayered thin films resulted in the formation of both Al_3Ti and γ-TiAl compounds. Though the predominant compound formed was Al_3Ti, grains of γ-TiAl were present at the interface between Ti and Al_3Ti, which were not reported in previous studies of Ti/Al thin-film diffusion couples.

References

1.   K. S. Chan and Y. W. Kim, Acta Metall. Mater., 43(2), 439 (1995).
2.   R. Ahuja and H. L. Fraser, J. Elec. Mater., 23(10), 1027 (1994).
3.   R. Ahuja and H. L. Fraser, JOM,
4.   R. Ahuja and H. L. Fraser, Mat. Res. Soc. Symp. Proc. 317, 479 (1994).
5.   R. Banerjee, R. Ahuja and H. L. Fraser, Phys. Rev. Lett. 76(20), 3778 (1996).
6.   A. C. Redfield and A. M. Zangwill, Phys. Rev., B34(2), 1378 (1986).
7.   S. A. Dregia, R. Banerjee and H. L. Fraser, submitted to Scripta met. & mat.
8.   R. Banerjee, X. D. Zhang, M. Asta, A. A. Quong, S. A. Dregia and H. L. Fraser, submitted to Phys. Rev. Lett.
9.   R. Ahuja, Ph. D. Thesis, The Ohio State University, 1994.
10. M. Asta and D. de Fontaine, J. Mater. Res., 8(10), 2554 (1993)
11. M. Asta and A. A. Quong, private communication
12. C. McCullough, J. J. Valenicia, C. G. Levi and R. Mehrabian, Acta Metall. Mater., 37, 1321(1989).
13. E. G. Colgan, Mat. Sci. Rep., 5,1 (1990).
14. U. Gosele and K. N. Tu, J. Appl. Phys., 53, 3252 (1982).

ATOMISTIC STUDY OF CRACK PROPAGATION AND DISLOCATION EMISSION IN CU-NI MULTILAYERS

Jeff Clinedinst and Diana Farkas
Department of Materials Science and Engineering, Virginia Tech, Blacksburg VA 24061

ABSTRACT

We present atomistic simulations of the crack tip configuration in multilayered Cu-Ni materials. The simulations were carried out using molecular statics and EAM potentials. The atomistic structure of the interface was studied first for a totally coherent structure. Cracks were simulated near a Griffith condition in different possible configurations of the crack plane and front with respect to the axis of the layers. Results show that interface effects predominately control the mechanical behavior of the system studied.

INTRODUCTION

Nano-composite multilayers have become the focus of a great deal of research recently. The ability ot precisely control the composition, microstructure and interface properties make them very attractive for magnetic and electrical applications. Properties such as strength, toughness, and oxidation resistance can be enhanced by varying the repeat length, layer thicknesses, and the interface structure [1], which is thought to be an important factor controlling the mechanical behavior of such systems [2]. A recent study by Thompson et. al [3] has suggested that misfit stresses at the interfaces, arising from different lattice parameters, contribute significantly to crack propagation and dislocation emission in metallic multilayers.

Processing techniques such as physical vapor deposition, sputtering, and ion-beam assisted deposition have been used to make multilayers [4]. A large number of materials can be used to make multilayers and they usually consist of a hard or brittle material in combination with a ductile one. Common systems include ceramics with a polymer and metals with intermetallics or other metals [5].

Difficulties are encountered with multilayer systems when mechanical testing and characterization becomes necessary [5], and computer simulation techniques are becoming increasingly more attractive. Studies by Kalonji et. al [2] and Yamamoto et. al [6] have implemented EAM (embedded atom method) potentials with lattice statics and molecular dynamics (MD) modeling, respectively, to successively estimate elastic constants of several metal/metal multilayered systems. In a more general study, Anderson et. al [7] employed a crack-dislocation model to examine the ductile-to-brittle behavior of generic multilayers.

The present study uses EAM potentials and molecular statics simulations to first calculate the surface and interfacial energies and then to study crack propagation and dislocation emission in a Cu/Ni multilayer system. This paper presents some results obtained for several crystallographic orientations.

SIMULATION TECHNIQUES

The many body interatomic potentials used for pure copper and nickel were developed by A.F. Voter [8] and Voter et. al [9] respectively, and were used for all calculations and simulations in the presented results. Both potentials were modified to an effective pair scheme by setting the first derivative of the embedding function at the perfect lattice electronic density equal to zero. In

315

addition, the perfect lattice electronic densities were normalized and set equal to 0.34 for both copper and nickel. Finally, a mixed Cu/Ni pair potential was created by averaging the pure copper and nickel pairs. This average is expected to describe the experimental thermodynamics of the system which is nearly ideal. The potentials used gave lattice parameters of 0.3615 nm and 0.352 nm for copper and nickel, respectively, resulting in a lattice misfit of 2.6%.

As previously stated, interfacial effects are primarily responsible for the observed mechanical behavior of multilayer systems, and thus it is very important to understand the energetics of the interface. The surface energies of the bulk copper and nickel were calculated by way of an energy minimization technique. The interfacial energy was determined by an extrapolation technique described by Farkas et al. [10] whereby the long range, elastic misfit energy can be effectively isolated from the inelastic local interface energy. The Cu/Ni interface in the laminates was simulated as completely coherent using an average lattice constant for the directions parallel to the interface, while a minimum energy distance was used for the direction perpendicular to the interface. In all simulations the interface was oriented along (010).

Once these energies were determined, initial supercells of forty-eight planes were used to construct, by repetition, the simulation blocks. The cracks were introduced in the simulation block according to the anisotropic elasticity solution given by Sih and Liebowitz [11]. The elastic constants used for this solution were average values for pure copper and nickel. The regions far away from the crack tip were kept fixed at the positions given by elasticity, whereas periodic boundary conditions were used along the crack front. Three crystallographic orientations were modeled for the multilayers, as well as for both pure copper and nickel.

RESULTS AND DISCUSSION

The calculated values of surface energy for the {100} and {110} planes, for bulk copper and nickel, are given in Table I. In both metals the {110} planes have the highest surface energy. The values for copper are considerably lower than the corresponding values for nickel. Table I also shows values of the Griffith stress intensity and the interface energy calculated from the surface energies and the elastic constants for copper and nickel given by Ackland et al. [12] and Simmons et al. [13], respectively.

Table I Surface Energies and Griffith Stress Intensities for Copper and Nickel

Surface	Surface Energies, mJ/m^2		Interface Energy, mJ/m^2	K Griffith, MPa*m$^{1/2}$	
	Cu	Ni		Cu	Ni
{100}	1328	1760	24	0.584	0.880
{110}	1472	1968	---	0.670	1.034

In order to study the mechanical behavior of the multilayered structures, it was first essential to understand the behavior of the bulk copper and nickel. Two orientations for both metals were simulated slightly above the Griffith stress and are presented as Figure 1. The vertical line in these, and all subsequent figures, denotes the original position of the crack tip. As can be seen in Figures 1a and 1c, both copper and nickel, respectively, exhibited brittle cleavage in the (010)[001] orientation. This is expected since in this configuration there are no available slip systems for dislocation glide. In the (101)[10$\bar{1}$] orientation with a <110>{111} type slip system, both copper and nickel emitted two 1/6<121> Shockley partials, and are shown in Figures 1b and 1d, respectively. The crack tip blunting corresponds to a total dislocation of 1/3[101], which is the sum of the two partials; 1/6[121] + 1/6[1$\bar{2}$1]. This blunting and closure of the crack, particularly in the pure copper (Fig. 1b), after emitting dislocations is evidence of ductility in the metals.

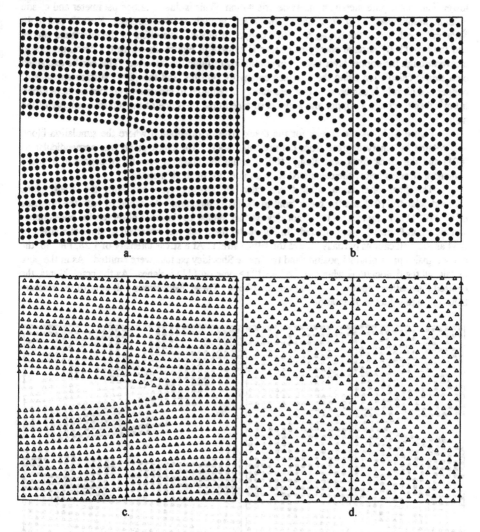

a.　　　　　　　　　b.

c.　　　　　　　　　d.

Figure 1: Cracks in the (010)[001] and (101)[10$\bar{1}$] orientations for copper, Figs. a. and b., and nickel, Figs. c. and d, respectively.

With an understanding of the behavior of the bulk metals, simulations of three orientations of the multilayered structures were studied. In all of the following figures copper is represented by black circles and nickel by triangles.

The first orientation was with the crack plane perpendicular to the [010] direction and the crack front along [001]. This orientation is referred to as delamination since the crack lies within the interface. The crack was observed to propagate, without dislocation emission, at stresses lower than in the pure metals in the same orientation. This is due to lattice parameter and elastic moduli mismatches. Figure 2 shows the crack growth at two stresses, K=0.32 and 0.48 MPa*m$^{1/2}$, both below the $K_{Griffith}$ for either copper or nickel. The effect of the interfacial energy and the low {100} surface energy of copper is evident in Figure 2b where the crack prefers to cleave through the copper layer instead of continuing through the interface. This can be explained by a simple calculation using the interface energy of 24mJ/m^2 and the {100} surface energies of copper and nickel. It can be seen that less energy is required to create two copper surfaces (2656 mJ/m^2) than to create a copper and a nickel surface in the presence of the interface energy (3064 mJ/m^2).

Figure 3 shows the results for the crack arrester orientation where the simulation block was rotated by ninety degrees from the previous block. The crack plane was perpendicular to [001] and the crack front was along [100]. Again the crack is seen to propagate in a brittle manner at stresses below the calculated $K_{Griffith}$ values for the bulk metals.

The final orientation studied was also an arrester type but with the crack plane perpendicular to [101] and the crack front in the [10$\bar{1}$] direction. As expected with the presence of a <110>{111} slip system two Shockley partials similar to the ones observed for pure copper and nickel, in Figure 1, were emitted. In Figure 4a, the crack was observed to close at a stress of 0.64 as the dislocations partially relieve the lattice strain. At a stress intensity of 1.28, Fig. 4b, the crack remains in its original position and two more Shockley partials were emitted. As in the pure metals, all the dislocations where of the 1/6<121> type on {111} planes. As the crack blunts, the emitted dislocations appear to stay at the interfaces. These dislocations will control the fracture behavior in this ductile orientation.

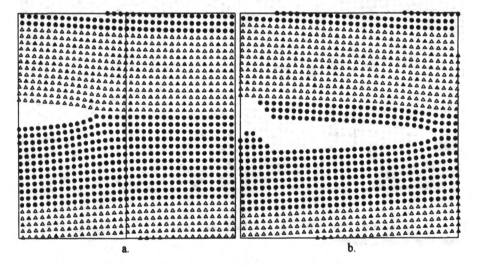

a.   b.

Figure 2: Crack growth in the delamination orientation, (010)[001], at stresses of a). 0.32 and b). 0.48 MPa*m$^{1/2}$.

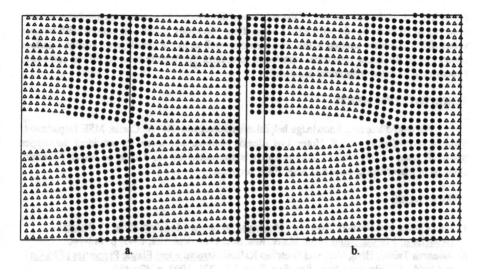

Figure 3: Crack growth in the arrester orientation, (001)[100], at stresses of a). 0.448 and b). 0.48 MPa*m$^{1/2}$.

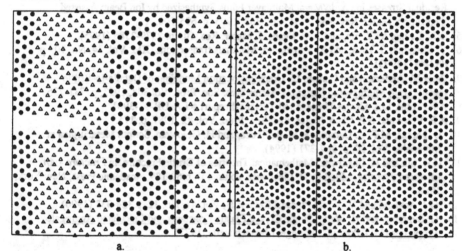

Figure 4: Dislocation emission in the (101)[10$\bar{1}$] orientation and stresses of a). 0.64 and b) 1.28 MPa*m$^{1/2}$.

CONCLUSIONS

In this paper we presented molecular statics simulations of crack tip configurations for copper, nickel, and Cu/Ni multilayered structures. Three orientations were studied for each structure; two were variations of the cube edge directions and the third was oriented so as to provide a <110>{111} slip system. It was found that the pure metal and the multilayers exhibited brittle fracture in the orientations of the cube edge directions. In contrast, ductile behavior was observed in all cases for the (101)[10$\bar{1}$] orientation. Shockley partial dislocations with burgers vector 1/6<121> were emitted in the pure metals as well as in the multilayers. Finally, interface effects due to the lattice mismatch were found to control the fracture behavior of the composite structures in all three orientations examined.

ACKNOWLEDGEMENTS

We would like to acknowledge helpful discussions with Dr. W. Curtin, MSE Department, Virginia Tech. Thanks to A.F. Voter, Los Alamos National laboratory, for providing the copper potentials used in this study. This research is supported by the Office of Naval Research, Division of Materials Sciences and NSF, FAW program.

REFERENCES

1. James Belak and Davis B. Boercker, Molecular Dynamics Modeling of the Mechanical Behavior of Metallic Multilayers, Mater. Res. Soc. Proc. Vol. 308, 1993, p. 743-746.
2. Ademola Taiwo, Hong Yan, and Gretchen Kalonji, Structure and Elastic Properties of Ni/Cu and Ni/Au Multilayers, Mater. Res. Soc. Proc. Vol. 291, 1993, p. 479-484.
3. D.S. Lashmore and R. Thompson, J. Mater. Res. 7, p. 2379-2386 (1992).
4. C.E. Kalnas, L.J. Parfitt, M.G. Goldiner, G.S. Was and J.W. Jones, Mechanical Properties and Residual Stresses in Oxide/Metal Multilayer Films Synthesized by Ion Beam Assisted Deposition, Mater. Res. Soc. Proc. Vol. 308, 1993, p. 695-700.
5. D.K. Leung, M.Y. He, and A.G. Evans, J. Mater. Res. 10, p. 2379-2386 (1995).
6. Y. Sasajima, S. Taya, and R. Yamamoto, Materials transactions, JIM, Vol. 34, No. 10, p. 882-887 (1993).
7. Peter M. Anderson and Canhao Li, Crack-dislocation Modeling of Ductile-to-Brittle transitions in Multilayered Materials, Mater. Res. Soc. Proc. Vol.308, 1993, p. 731-736.
8. A.F. Voter, Los Alamos Unclassified Technical report # LA-UR-93-3901.
9. A.F. Voter and S.P. Chen, Mater. Res. Soc. Proc. Vol. 82, 1987, p. 175.
10. D. Farkas, M.F. de Campos, R.M. de Souza, and H. Goldstein, Scripta Metallurgica et Materialia, Vol. 30, p. 367-371 (1994).
11. G.C. Sih and H. Liebowitz, Mathematical Theories of Brittle Fracture, in *Fracture - An Advanced Treatise*, edited by H. Liebowitz, volume II, p.69-189, Academic Press, New York, 1968.
12. G.J. Ackland and V. Vitek, Physical Review B, Vol. 41, No. 12, p.324-333 (1990)
13. G. Simmons and H. Wang, Single Crystal Elastic Constants and Calculated Aggregate Properties (MIT Press, Cambridge, Massachusetts, 1977).

Part V

Oxide, Non–Oxide, and
Oxide–Metal Nanocomposites

POLYMERIZABLE COMPLEX SYNTHESIS OF NANOCOMPOSITE BaTi$_4$O$_9$/RuO$_2$ PHOTO-CATALYTIC MATERIALS

Masato KAKIHANA
Materials and Structures Laboratory, Tokyo Institute of Technology, Nagatsuta 4259, Midori-ku, Yokohama 226, Japan, kakihan1@rlem.titech.ac.jp

ABSTRACT

The "polymerizable complex (PC)" technique, a kind of gel technologies, is based on formation of a polyester resin precursor in which various metal ions can be uniformly distributed keeping their initial stoichiometric ratio. The approach allows for the synthesis of multicomponent oxides at reduced temperatures. Feasibility of the PC method is demonstrated for the synthesis of BaTi$_4$O$_9$ at 700-900 °C. BaTi$_4$O$_9$ was subsequently converted to nanocomposite materials by modifying its surface with ultrafine particles of RuO$_2$, and they were used as photocatalysts for decomposition of water into H$_2$ and O$_2$ under irradiation of light from a high-pressure Hg lamp operated at 100 W. High-resolution transmission electron microscopic observations indicates uniform dispersion of spherical RuO$_2$ particles of ~2 nm in diameter on the host BaTi$_4$O$_9$ surface. The nanocomposite BaTi$_4$O$_9$/RuO$_2$ (1 wt % Ru relative to BaTi$_4$O$_9$) material prepared by the PC method at 800 °C showed a photo-catalytic activity ~3 times higher than that of a material prepared by the conventional ceramic technique at 1100 °C.

INTRODUCTION

Fine powders of semiconducting oxides with deposited metal and/or metal oxide particles have been widely used as heterogeneous photocatalysts for innumerable chemical reactions. Among the many photocatalytic reactions, splitting of water assisted by light has become one of the most active areas in heterogeneous photocatalysis, since it can be a promising chemical route for energy renewal and energy storage. The photocatalytic splitting of water on TiO$_2$ electrodes discovered by Fujishima and Honda in 1972 [1] is a prototypic example of this concern, and there exists a vast body of literature describing the potential application of TiO$_2$-based photocatalysts for water decomposition. In recent years, a new type of photocatalysts for water decomposition have emerged, which include *multi*-component oxides combined with metal oxides such as Na$_2$Ti$_6$O$_{13}$/RuO$_2$ [2, 3] and BaTi$_4$O$_9$/RuO$_2$ [4, 5] and ion-exchangeable layered compounds such as K$_4$Nb$_6$O$_{17}$ [6-10], KCa$_2$Nb$_3$O$_{10}$ [11] and K$_2$La$_2$Ti$_3$O$_{10}$ [12]. It is also known in the latter systems that their photocatalytic activity for water decomposition was greatly enhanced when the host compounds were combined with either Ni or Pt [6-10].

One of the serious problems in studying these new composite materials as photocatalysts is the great difficulty in preparing the host compounds in their pure form at relatively low temperatures, viz. 700-900 °C. For instance, BaTi$_4$O$_9$ has been synthesized by the conventional ceramic route, which involves mixing BaCO$_3$ and TiO$_2$, followed by repeated cycles of grinding and firing at high temperatures (1000-1300 °C) [13-16]. Alternatively, attempts to synthesize BaTi$_4$O$_9$ at lower temperatures have been carried out by sol-gel techniques using metal-alkoxides, but without success [17, 18]. BaTi$_4$O$_9$ phase did not crystallize from gels derived from barium and titanium alkoxide precursors until ~900 °C, below which BaTi$_5$O$_{11}$ rather than BaTi$_4$O$_9$ was a major phase formed. With a further increase in firing temperature, Ba$_4$Ti$_{13}$O$_{30}$ as well as BaTi$_4$O$_9$ showed up with strong reduction of BaTi$_5$O$_{11}$. Firing the gels at 1100 °C [17] or 1300 °C [18] was required to obtain phase pure BaTi$_4$O$_9$. This rather unexpected result may arise from the differential hydrolysis and condensation rates of Ba and Ti alkoxides, wherein the latter is much more readily hydrolyzed than the former even by small amount of water, thus making it difficult to prepare homogeneously mixed Ba and Ti oxides of the required stoichiometry. (A similar type of problem will be indeed demonstrated in this paper; see Fig.3.) The situation in the synthesis of the ion-exchangeable layered compounds is more or less the same as that in BaTi$_4$O$_9$, in which processing temperatures

323

higher than 1000 °C are usually required. Not only the poor uniformity of the product but also the large grain growth owing to the high-temperature heat treatment is a principal obstacle to the application of these compounds as hosts for the photocatalytic composite materials with higher activities. It is, therefore, necessary to explore another solution route, that enables the synthesis of these host compounds in their pure form at reduced temperatures (700-900 °C).

The principal aim of this paper is to demonstrate the feasibility of a simple polymerizable complex (PC) route, known as the Pechini method [19], for the synthesis of photocatalysts. BaTi$_4$O$_9$ and its composites combined with RuO$_2$ have been chosen as the focus of the present work, and the PC technique was successfully applied to the synthesis of pure BaTi$_4$O$_9$ at 700-900 °C [20]. The PC method is based on polymerization between citric acid (CA) and ethylene glycol (EG) in the presence of soluble metal salts with required amounts to form a polyester-type resin, inside which metal-CA complex species remain soluble to sustain the scale of mixing of different metals almost molecularly homogeneous [21-27]. An important aspect of this technique in the synthesis of BaTi$_4$O$_9$ is that the individual metal CA complexes can be immobilized in a rigid polyester network while preserving the initial stoichiometric ratio of Ba and Ti upon polymerization (see a schematic picture shown in Fig.1). Immobilization of the metal complexes in a rigid polymer

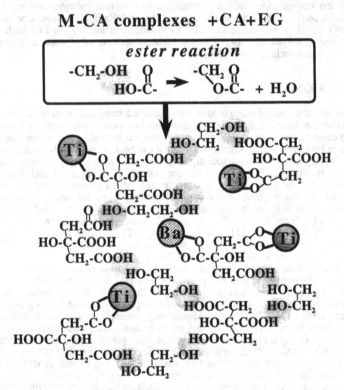

Fig. 1 Concept of polymerized complex (PC) method (schematically drawn). The fundamental reaction is esterification between citric acid (CA) and ethylene glycol (EG). This reaction occurs in sequence, leading to a polyester resin, inside which Ba and Ti are uniformly distributed keeping their initial stoichiometry (Ba/Ti=1/4).

can largely reduce a marked tendency for Ti-species to be preferentially hydrolyzed, that might occur in the alkoxide-based sol-gel technique. This remarkable character of the **PC** route allows us to synthesize $BaTi_4O_9$ at reduced temperatures without passing through formation of other barium-titanates with Ba/Ti ratios deviated from the desired Ba/Ti=1/4 stoichiometry. The surface of $BaTi_4O_9$ fine particles thus prepared should be modified by RuO_2 to form nanocomposite photocatalytic materials. The photocatalytic activity for the decomposition of water in the $BaTi_4O_9$/RuO_2 material prepared *via* the **PC** route is compared with that in a sample prepared by the conventional ceramic technique in order to demonstrate the potential advantage of the **PC** method in fabricating photocatalysts with higher activities.

EXPERIMENT

Preparation of Host $BaTi_4O_9$

Figure 2 is a flow chart outlining the procedures followed to prepare $BaTi_4O_9$ by the **PC** method. Below described is the process for preparing ~4 g of $BaTi_4O_9$. Titanium tetraisopropoxide $(Ti[OCH(CH_3)_2]_4 : Ti(OiPr)_4)$ and barium carbonate $(BaCO_3)$ were chosen as sources of titanium and barium, respectively. Ethylene glycol $(HOCH_2CH_2OH : EG)$ was used as a solvent at the initial stage of processing, while anhydrous citric acid $(HOOCCH_2C(OH)(COOH)CH_2COOH:$ CA) was used as a complexing agent to stabilize Ba and Ti ions against water evolved during the polyesterification between EG and CA at the later stage of processing. A 40 mmol of $Ti(OiPr)_4$ (11.4 g) was first dissolved into 1.6 mol of EG (99.2 g ~ 92 ml). A large excess of CA (0.4 mol=76.8 g) relative to titanium (40 mmol) was added with continuous stirring to the $Ti(OiPr)_4$/EG solution to convert $Ti(OiPr)_4$ to stable Ti-CA complexes. After achieving complete dissolution of CA, 10 mmol of $BaCO_3$ (1.97g) was added and the mixture was magnetically stirred for 1 h to drive off carbon dioxides evolved in the decomposition reaction of barium carbonate to produce a transparent solution of Ba- and Ti-CA complexes. All the procedures described above were carried out at ~50 °C. The clear solution thus prepared, while stirred with a magnetic stirrer, was heated at ~130 °C to accelerate esterification reactions between CA and EG and to remove excess solvents. The prolonged heating at ~130 °C produced viscous, bubbly mass that formed a brown transparent glassy resin upon cooling. No visible formation of precipitation or turbidity has been observed during the polymerization. Charring the resin at 350 °C for 2 h in a box furnace resulted in a black solid mass, which was lightly ground into a powder with a Tefron rod. The powder thus obtained is referred to as the "powder precursor" hereinafter. The powder precursor was heat-treated in air for 2 h at temperatures between 650 °C and 900 °C.

It is worthwhile to mention here that the **PC** technique above described involves *no* step of pH control in contrast to the case of the so-called amorphous

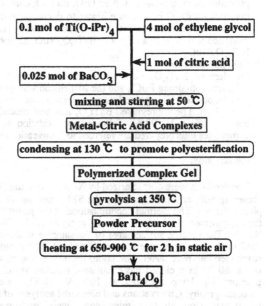

Fig. 2 Flow chart for the polymerizable complex (**PC**) procedure used to prepare $BaTi_4O_9$.

citrate/water gel process wherein pH of solution is usually adjusted to 3-6 by addition of ammonia to avoid any precipitation during gellation [21]. It is however anticipated, even when CA is used as a complexing agent, that addition of ammonia to a solution comprising of Ba and Ti may promote a preferential hydrolysis of titanium giving rise to inhomogeneity of the resulting gel. In the present PC technique utilizing CA and EG, addition of ammonia to the Ba/Ti=1/4 solution was also carried out on purpose to demonstrate how badly ammonia affects the phase evolution of $BaTi_4O_9$. In this additional experiment, approximately 2 ml of 17 mol/l aqueous ammonia solution (~4 times the molar quantity of titanium) was added to the original Ba/Ti=1/4 CA/EG solution to adjust its pH to ~3. Despite that the rest of procedures was the same as those depicted in Fig.2, the addition of ammonia to the original solution significantly hindered the polymerization between CA and EG, rendering the system non-viscous until most of EG and water were removed.

For the purpose of comparison, $BaTi_4O_9$ was also prepared by the conventional solid state reaction at 1100 °C for 10 h using an intimate mixture of $BaCO_3$ and TiO_2 achieved by mechanical grinding for 2 h.

Preparation of Nanocomposites $BaTi_4O_9/RuO_2$

Photocatalytic composite materials are typically made by impregnation of a powder of the host compound with an aqueous solution of a metal salt, followed by appropriate heat-treatments. This conventional way has been indeed applied to prepare $BaTi_4O_9/RuO_2$ (refereed to "*conventional route*") [4, 5, 20]. In this work, to improve the dispersion of RuO_2 on the surface of $BaTi_4O_9$ a kind of gel coatings was employed as well (refereed to "*PC route*"). Details of the preparation of $BaTi_4O_9/RuO_2$ (1 wt% of Ru relative to $BaTi_4O_9$) in these two different ways are written below:

i) *PC route*.
Powders of $BaTi_4O_9$ *via* the **PC** method were suspended into CA/EG=1/4 solutions containing $RuCl_3$ (CA/$BaTi_4O_9$=1/0.4). The suspension was stirred at ~130 °C to promote polyesterification between CA and EG. $BaTi_4O_9$ coated with a resin comprising of Ru was heat-treated at ~450 °C in a mantle heater to decompose the polymeric resin. The resulting brown powder was heated at 500 °C for 2 h under flowing H_2/N_2 gas (H_2 2% + N_2 98%), followed by oxidation at 475 °C in air for 7 h. $BaTi_4O_9/RuO_2$ composites thus prepared are refereed to "PC-$BaTi_4O_9/RuO_2$".

ii) *Conventional route*.
Powders of $BaTi_4O_9$ *via* the conventional ceramic method were suspended into aqueous solutions containing $RuCl_3$ and the suspension was stirred for 4 h at ~70 °C until most of water was evaporated. This procedure was repeated once more. The resulting mass was dried at 100 °C for 12 h. The impregnated $BaTi_4O_9$ was heat treated at 500 °C for 2 h under flowing H_2/N_2 gas (H_2 2% + N_2 98%), followed by oxidation at 475 °C in air for 7 h. $BaTi_4O_9/RuO_2$ composites thus prepared are refereed to "conventional-$BaTi_4O_9/RuO_2$".

Characterization

The products were characterized by X-ray diffraction (XRD) using CuKα radiation and Raman scattering with an excitation using the 514.5 nm line of an Ar laser to identify various possible phases formed. The thermal decomposition of the powder precursor was investigated by means of thermogravimetry-differential thermal analysis (TG-DTA) in static air with a heating rate of 10 °C/min. The specific surface area of the samples was measured by the conventional three-points BET method using nitrogen gas as absorbent. The powdered $BaTi_4O_9/RuO_2$ photocatalysts were suspended into pure water. The photo-decomposition of water by $BaTi_4O_9/RuO_2$ was then carried out at 60 °C in a closed gas-circulation reaction vessel under irradiation of light from a high-pressure Hg lamp operated at 100 W. H_2/O_2 gases evolved were analyzed by a gas chromatography. Observations and elemental analyses of RuO_2 nanoparticles on $BaTi_4O_9$ surfaces by transmission electron microscopy were performed with a JEOL JEM-2010 electron transmission microscope operated at 200 kV.

RESULTS AND DISCUSSION

<u>Effects of addition of ammonia to Ba/Ti=1/4 CA/EG solutions on Phase Evolution of BaTi$_4$O$_9$</u>

One of the common problems in the synthesis of BaTi$_4$O$_9$ by wet chemical routes is the difficulty in controlling the rapid hydrolysis of Ti-alkoxides (chemicals most frequently used as sources of Ti) compared with barium species. For instance, in the previously reported sol-gel synthesis of BaTi$_4$O$_9$ [18], gels derived from Ba and Ti alkoxide precursors have produced strongly multiphase samples even after the heat-treatment at 1200 °C. The preferential hydrolysis of Ti-alkoxides can form Ti-rich clusters, which destroys the cation composition of the original solution. This result tells us that Ti-alkoxides should be modified by certain organic compounds such as acetic acid to control the degree of hydrolysis and subsequent polycondensation reactions [21, 28-30]. A similar event has been observed in the present PC synthesis of BaTi$_4$O$_9$, when the pH of the original solution was adjusted to 3 by addition of ammonia. An interesting comparison can then be made as to phase evolution of BaTi$_4$O$_9$ between the original Ba/Ti=1/4 CA/EG solution (pH~0.1) and the one modified by ammonia (pH~3). Figure 3 shows XRD patterns of the powder precursor (derived from the *ammonia* containing solution) calcined in air at different temperatures

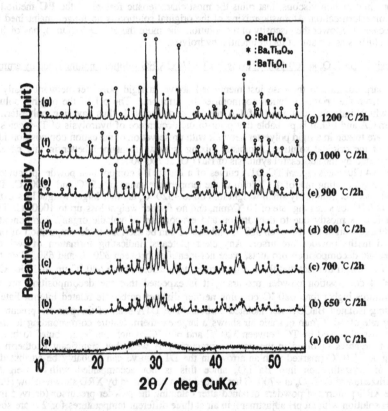

Fig.3 X-ray diffraction patterns of the Ba/Ti=1/4 composition powder precursor (derived from an ammonia containing solution (pH=3)) calcined in static air for 2 h at 600 °C (a), 650 °C (b), 700 °C (c), 800 °C (d), 900 °C (e), 1000 °C (f), and 1200 °C (g).

for 2 h. An immediate conclusion drawn from Fig.3 is that the process of phase evolution is quite complicated: $BaTi_5O_{11}$ rather than the desired $BaTi_4O_9$ formed as a main phase between 650 °C and 800 °C, $BaTi_4O_9$ showed up only after calcination at 900 °C accompanied with strong reduction of $BaTi_5O_{11}$ but with evolution of a new phase $Ba_4Ti_{13}O_{30}$, and no phase pure $BaTi_4O_9$ could be obtained even after calcination at 1200 °C. Formation of $BaTi_5O_{11}$ in preference to $BaTi_4O_9$ at temperatures below 800 °C may imply that a -Ti-O-Ti- network is built up around a barium ion thus creating clusters locally rich in Ti with respect to Ba. An important concern is that addition of ammonia plays a role not only to promote the preferential hydrolysis of $Ti(OiPr)_4$ but also to inhibit ester reactions between CA and EG. The inhibition of ester reactions by ammonia comes from an increased pH of the solution, that significantly enhances the dissociation of protons from carboxylic acid (-COOH) groups in CA according to the following type of reaction:

$$CA\text{-}(COOH)_3 + n\,OH^- \rightarrow CA\text{-}(COOH)_{3-n}\,(COO^-)_n + n\,H_2O \quad (1).$$

Ester reactions occur only between carboxylic acid (-COOH) groups of CA and hydroxyl (-OH) groups of EG (see Fig.1), while the dissociated carboxylate (-COO⁻) groups do not participate in the esterification. Thus an increased fraction of COO⁻ in CA due to an increase in pH renders the solution fluid or non-viscous, that ruins the most characteristic feature of the **PC** method based upon polyesterification. Atomistic mixing of the original solution is no longer maintained in this case because the lower the viscosity of the solution, the more the chance $Ti(OiPr)_4$ has of forming Ti-rich clusters as a result of its preferential hydrolysis.

Synthesis of $BaTi_4O_2$ at 700-900 °C using Ba/Ti=1/4 CA/EG solutions nonmodified by ammonia

In sharp contrast to the case just mentioned above, a rigid polyester network easily forms starting from the original solution nonmodified by ammonia. The pH of the original solution is ~0.1, which indicates that most of CA remain undissociated, thus favoring polyesterification between CA and EG. It is probable that prior to the occurrence of hydrolysis of Ti-species, metal species are frozen in a rigid polyester network with preservation of the cation composition identical to that of the original solution. Atomistic mixing is now maintained better in this case, as is demonstrated by the successful synthesis of $BaTi_4O_9$ at 700-900 °C.

Figure 4 illustrates typical TG-DTA curves of a Ba/Ti=1/4 composition powder precursor fired in air using a heating rate of 10 °C/min in the temperature range of 25 °C and 1000 °C. The TG curve shows a continuously small weight loss up to ~320 °C, another larger weight loss extending up to ~680 °C for a heating rate of 10 °C/min, and no further weight loss up to 1000 °C. The first weight loss is mostly due to dehydration and evaporation of volatile organic components. The second large weight loss between 320 °C and 680 °C can be ascribed to decomposition of organics involved in the powder precursor. Any clear plateaus, indicating formation of well-defined intermediate decomposition products, were not identified between 320 °C and 680 °C on the TG curve. In view of the fact that there is no further weight loss above 680 °C up to 1000 °C in the Ba/Ti=1/4 composition powder precursor, it is expected that the decomposition product at temperatures higher than 680 °C contains neither distinct carbonate related intermediate phase (including isolated $BaCO_3$) nor isolated carbons. The DTA scan of the powder precursor at a heating rate of 10 °C/min in static air shows a large exotherm feature corresponding to the large weight loss observed by TG between 320 °C and 680 °C, which can be attributed to burnout of most of organics involved in the powder precursor. A weak but significant exotherm feature starting at ~730 °C (marked with an arrow) in the DTA curve can probably be attributed to the onset of crystallization into $BaTi_4O_9$ since this is not accompanied with a weight loss. Crystallization of $BaTi_4O_9$ at ~700 °C has indeed been confirmed by XRD shown below (Fig.5).

The XRD patterns of powders obtained after calcining the powder precursor (derived from the low pH solution without pH adjustment) in air at three different temperatures for 2 h are shown in Fig.5 in 2θ range of 10-55°. Powder precursors heat treated above 650 °C were all white, which was usually indicative of complete burnout of the residual carbon consistent with the TG-DTA

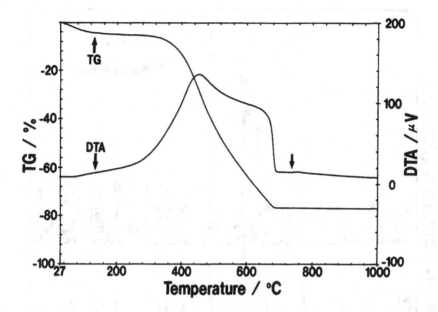

Fig.4 TG-DTA curves of the Ba/Ti=1/4 composition precursor (derived from the original solution nonmodified by ammonia) in air with a heating rate 10 °C/min.

analysis (Fig.4). The powder precursor fired at 650 °C was primarily amorphous in structure, as shown by the broad continuum in the XRD in Fig.5 (a). Drastic crystallization has occurred during the heat treatment of the powder precursor in air at 700 °C for 2 h (Fig.5(b)). The width of the principal lines somewhat sharpens at 800 °C (Fig.5(c)) but the overall shape of the pattern remains unchanged. Note however that the broad continuum in Fig.5(a) can be slightly recognized in the XRD pattern (Fig.5(b)) of the sample heat-treated at 700 °C. (This residual amorphous component appears to affect the photocatalytic activity of $BaTi_4O_9/RuO_2$ in a negative way as shown later.) All the well-defined peaks in the XRD patterns of Figs.5(b) and (c) exhibited a pure orthorhombic phase of $BaTi_4O_9$ in good agreement with the diffraction pattern observed for this compound by Phule et al. [17]. It should be stressed here that the impurity phases $BaTi_5O_{11}$, $Ba_4Ti_{13}O_{30}$ and $BaTi_2O_5$, which are most frequently formed as by-products during the synthesis of $BaTi_4O_9$, were not detected by XRD. Another important aspect worthwhile to mention is that no reflections from $BaCO_3$ and TiO_2 were observed as distinct intermediate phases prior to the formation of $BaTi_4O_9$ during the thermal decomposition of the powder precursor at 650 °C. Since it is known that well-crystalline $BaCO_3$ and TiO_2 are obtained at 500 °C from similar thermal decomposition of powder precursors containing only Ba or Ti [31, 32], no detection of $BaCO_3$ and TiO_2 from the Ba/Ti=1/4 composition precursor should be considered as an implication of almost perfect mixing of the constituent cations in the powder precursor. This would in turn indicate that $BaTi_4O_9$ does not form through a solid-state reaction between isolated $BaCO_3$ and TiO_2 fine particles but forms directly from the XRD amorphous **PC** precursor above 700 °C without significant segregation of the individual metals. The temperature (700 °C) at which $BaTi_4O_9$ starts to crystallize *via* the **PC** method is so far the lowest processing temperature for this compound.

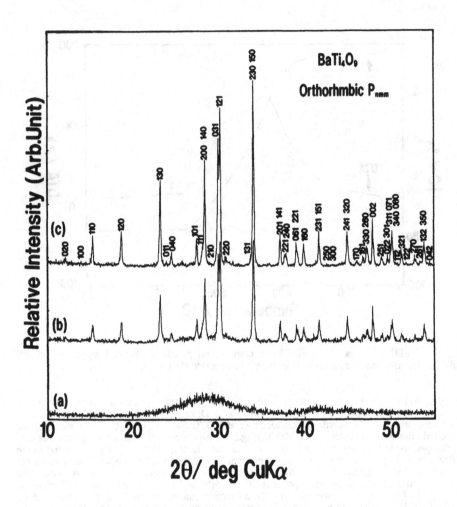

Fig.5 X-ray diffraction patterns of the Ba/Ti=1/4 composition powder precursor (derived from the original solution nonmodified by ammonia) calcined in static air for 2 h at 650 °C (a), 700 °C (b) and 800 °C (c).

The purity of the $BaTi_4O_9$ sample was also checked by Raman spectroscopy, a technique which is capable of detecting impurities consisting of very small crystallites not easily identified with the XRD technique because of their diffuse reflections. The Raman pattern of a powder obtained after calcining the powder precursor in air at 700 °C for 2 h is shown in Fig.6 in a frequency range of 60-1200 cm^{-1}, where all the lines well coincide in both position and relative intensities with those of the orthorhombic $BaTi_4O_9$. The Raman spectrum also shows no evidence for the presence of $BaCO_3$ because of the complete absence of the strongest Raman peak at 1059 cm^{-1} characteristic of $BaCO_3$.

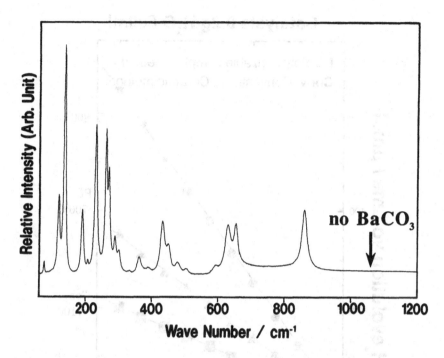

Fig.6 A Raman spectrum of the Ba/Ti=1/4 composition powder precursor (derived from the original solution nonmodified by ammonia) calcined in static air for 2 h at 700 °C.

Photocatalytic activities of $BaTi_4O_9/RuO_2$

Figure 7 shows H_2 gas evolution volumes with illumination time for $BaTi_4O_9/RuO_2$ samples prepared at 700 °C or 800 °C by the PC route, and their photocatalytic activity is compared with that of another sample prepared at 1100 °C by the conventional ceramic route. The corresponding O_2 gas evolution volumes were confirmed to be equal to half the volumes of H_2 gas for all the samples tested; i.e. H_2O is stoichiometrically decomposed into H_2 and $(1/2)O_2$. As expected from the low surface area (< 1 m^2/g) of the conventional-$BaTi_4O_9/RuO_2$, the PC-$BaTi_4O_9/RuO_2$ samples with higher surface areas (8-20 m^2/g) have higher photo-catalytic activities compared to those for the conventional-$BaTi_4O_9/RuO_2$. Quite interestingly, however, the PC-$BaTi_4O_9/RuO_2$ sample prepared at 700 °C showed photocatalytic activities much less than expected from its relatively large surface area (~20 m^2/g). The second PC-$BaTi_4O_9/RuO_2$ sample prepared at 800 °C showed ~2 times higher photocatalytic activities compared to those for the first PC-$BaTi_4O_9/RuO_2$ sample prepared at 700 °C despite its lower surface area (~8 m^2/g). One possible explanation for this contradiction is that the host $BaTi_4O_9$ compound prepared at 700 °C contains a lot of defects such as amorphous components (as indeed revealed by XRD; see Fig.5 and text), which may act as recombination centers for photo-induced electrons and holes thus reducing the net photocatalytic activity significantly. The unexpected larger photocatalytic activity in PC-$BaTi_4O_9/RuO_2$ prepared at 800 °C can be explained by taking into account the balance of two conflicting factors that affect the overall photocatalytic activity. When the processing temperature for $BaTi_4O_9$ is increased from 700 °C to 800 °C, the expected increment of the photocatalytic activity owing to the decrease in the

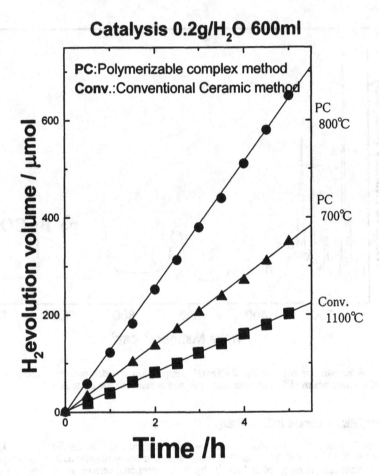

Fig.7 Photoassisted water decomposition on RuO_2-impregnated $BaTi_4O_9$. The host $BaTi_4O_9$ powders were prepared by the **PC** method at 700 °C (▲) and 800 °C (●) or by the conventional ceramic route at 1100 °C (■). The amount of RuO_2 is the same for all the $BaTi_4O_9$ powders (1 wt% of Ru relative to $BaTi_4O_9$). The weight of each catalyst used for test is 0.2 g.

fraction of amorphous components may excel the other expected decrement of the photocatalytic activity owing to the decrease in the surface area. For genuine crystals of $BaTi_4O_9$ practically free from any defects, the photocatalytic activity should then be primarily governed by their surface area. This was indeed confirmed by the observation of the strong reduction in the photocatalytic activity with the increase in the preparation temperature from 800 °C to 900 °C: i.e. the decrement of the photocatalytic activity is almost consistent with what is expected from the decrease in the surface area. Thus the remarkable acceleration of the photocatalytic reaction in the **PC**-$BaTi_4O_9$/RuO_2 sample prepared at 800 °C can be considered as a consequence of utilizing such a sample well optimized with respect to the balance between the crystallinity and the surface area.

<u>High-resolution electron microscopic image of BaTi$_4$O$_9$/RuO$_2$</u>

The BaTi$_4$O$_9$ compound is characterized by the presence of the pentagonal-prism tunnel structure, which is supposed to serve as a prevention against aggregation of RuO$_2$ particles [5]. Since the photocatalytic process can be influenced not only by the particle size of the host BaTi$_4$O$_9$ but also by how evenly and finely RuO$_2$ particles are distributed over the surface of BaTi$_4$O$_9$, a high-resolution electron microscopic image for one of the PC-BaTi$_4$O$_9$/RuO$_2$ samples was taken as a photograph (Fig.8). A number of dark spots of ~2 nm in diameter are homogeneously distributed on the surface of BaTi$_4$O$_9$, and from the energy analysis of the characteristic X-ray peak they were confirmed to be composed of Ru elements. It can be concluded that the high photocatalytic activity of the nanocomposite PC-BaTi$_4$O$_9$/RuO$_2$ material is at least partly attributable to the uniform distributions of RuO$_2$ nanoparticles on BaTi$_4$O$_9$ with relatively large surface areas.

CONCLUSION

A pure BaTi$_4$O$_9$ has been successfully synthesized by heat-treating the PC powder precursors in air at reduced temperatures (700-900 °C) for 2 h. The success in lowering the processing temperature for BaTi$_4$O$_9$ down to 700 °C implies that the molecular-level mixing of Ba and Ti seems to be likely throughout the PC process. It has been shown that the PC route is a promising methodology for preparing BaTi$_4$O$_9$ with relatively large surface areas that can be subsequently used as highly active nanocomposite BaTi$_4$O$_9$/RuO$_2$ photocatalysts for water decomposition.

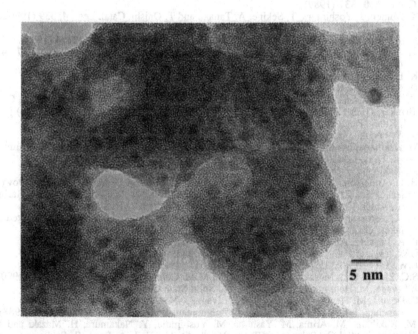

5 nm

Fig.8 A high-resolution transmission electron microscopic image of PC-BaTi$_4$O$_9$/RuO$_2$ (1 wt% Ru relative to BaTi$_4$O$_9$) prepared at 700 °C. A lot of dark spots correspond to RuO$_2$ particles.

ACKNOWLEDGMENT

Supports from Iketani Science and Technology Foundation and Japan Society for the Promotion of Science are gratefully acknowledged. I am grateful towards M. Tada (TIT), Mr. Y. Yamashita, Dr. M. Arima, Dr. K. Yoshida (Riken Corporation), and Prof. T. Sato (Tohoku Univ.) for their very active participation in the work presented here. Thanks are also due to Prof. K. Domen (TIT), Dr. M. Yashima and Prof. M. Yoshimura (TIT) for their valuable discussions. The assistance of Mr. K. Ibe (Electron Optics Division, JEOL Ltd.) with TEM observations is greatly appreciated.

REFERENCES

1. A. Fujishima and K. Honda, Nature, **37**, 238 (1972).
2. Y. Inoue, T. Kubokawa and K. Sato, J. Chem. Soc., Chem. Commun., 1298 (1990).
3. Y. Inoue, T. Kubokawa and K. Sato, J. Phys. Chem., **95**, 4095 (1991).
4. Y. Inoue, T. Niiyama, Y. Asai and K. Sato, J. Chem. Soc., Chem. Commun., 579 (1992).
5. Y. Inoue, Y. Asai and K. Sato, J. Chem. Soc., Faraday Trans., **90**, 797 (1994).
6. K. Domen, A. Kudo, A. Shinozuka, A. Tanaka, K. Maruya and T. Onishi, J. Chem. Soc., Chem. Commun., 356 (1986).
7. K. Domen, A. Kudo, M.Shibata, A. Tanaka, K. Maruya and T. Onishi, J. Chem. Soc., Chem. Commun., 1706 (1986).
8. A. Kudo, A. Tanaka, K. Domen, K. Maruya, A. Aika and T. Onishi, J. Catal. **111**, 67 (1988).
9. K. Sayama, A. Tanaka, K. Domen, K. Maruya and T. Onishi, Cat. Lett., **4**, 217 (1990).
10. A. Kudo, K. Sayama, A. Tanaka, K. Asakura, K. Domen, K. Maruya and T. Onishi, J. Catal., **120**, 337 (1989).
11. K. Domen, J. Yoshimura, T. Sekine, A. Tanaka and T. Onishi, Catal. Lett., **4**, 339 (1990).
12. K. Domen, private communication.
13. S. G. Mhaisalkar, W. E. Lee and D. W. Readey, J. Am. Ceram. Soc. **72**, 2154 (1989).
14. H. M. O'Bryan, Jr., J. Thomson, Jr. and J. K. Ploudre, J. Am. Ceram. Soc., **57**, 450 (1974).
15. K. Lukaszewicz, Proc. Chem., **31**, 1111 (1957).
16. D. E. Rase and R. Roy, J. Am. Ceram. Soc., **38**, 102 (1955).
17. P. P. Phule and S. H. Risbud, Better Ceramics through Chemistry III, edited by C. J. Brinker, D. E. Clark and D. R. Ulrich [Mater. Res. Soc. Symp. Proc. **121**, 275 (1988).].
18. J. J. Ritter, R. S. Roth and J. E. Blendell, J. Am. Ceram. Soc., **69**, 155 (1986).
19. M. P. Pechini, U. S. Patent No.3, **330**, 697 (1967).
20. M. Kakihana, M. Arima, T. Sato, K. Yoshida, Y. Yamashita, M. Yashima and M.Yoshimura, Appl. Phys. Lett., **69**, 2053 (1996).
21. M. Kakihana, J. Sol-Gel Sci. Tech., **5**, 7 (1996).
22. H. U. Anderson, M. J. Pennel and J. P. Guha in "Advances in Ceramics" : Ceramic Powder Science, Vol. 21, edited by Messing, G.L., Mazdiyasni, K.S., McCauley, J.W. and Harber, R.A., Amer. Ceram. Soc., Westerville, OH, p.91 (1987).
23. N. G. Eror and H. U. Anderson in "Better Ceramics Through Chemistry II" edited by Brinker, C.J., Clark, D.E. and Ulrich, D.R., Mater. Res. Soc. Proc. **73**, 571 (1986).
24. P. A. Lessing, Amer. Ceram. Soc. Bull. **168**, 1002 (1989).
25. L.W. Tai and P.A. Lessing, J. Mater. Res. **7**, 502 (1992).
26. L.W. Tai and P.A. Lessing, J. Mater. Res. **7**, 511 (1992).
27. S.C. Zhang, G.L. Messing, W. Huebner and M.M. Coleman, J.Mater. Res. **5**, 1806 (1990).
28. J. Livage, M. Henry and C. Sanchez, Prog. Solid State Chem., **18**, 259 (1988).
29. S. Doeuff, M. Henry, C. Sanchez and J. Livage, J. Non-Cryst. Solids, **89**, 206 (1987).
30. C. Sanchez, J. Livage, M. Henry and F. Babonneau, J. Non-Cryst. Solids, **100**, 65 (1988).
31. M. Kakihana, M. Arima, M. Yashima, M. Yoshimura, Y. Nakamura, H. Mazaki and H. Yasuoka, in "Sol-Gel Science and Technology" edited by E. J. A. Pope, S. Sakka and L. C. Klein, Ceramic transactions, Amer. Ceram. Soc., Westerville, OH, **55**, 65 (1995).
32. M. Arima, M. Kakihana, Y. Nakamura, M.Yashima and M.Yoshimura, J. Am. Ceram. Soc. **78** (1996) in press.

HEAT TREATMENT OF NANOCRYSTALLINE Al$_2$O$_3$-ZrO$_2$

BRIDGET M. SMYSER*, JANE F. CONNELLY*, RICHARD D. SISSON, JR.*, VIRGIL PROVENZANO**
*Worcester Polytechnic Institute, Department of Materials Engineering, Worcester, MA 01609
**Naval Research Laboratory, Code 6320, Washington, DC 20375-5343

ABSTRACT

The effects of grain size on the phase transformations in nanocrystalline ZrO$_2$-Al$_2$O$_3$ have been experimentally investigated. Compositions from 10 to 50 vol% Al$_2$O$_3$ in ZrO$_2$ were obtained as a hydroxide gel. The powders were then calcined at 600 °C for 17 hours and heat treated at 1100 °C for 24 and 120 hours and at 1200 °C for 2 hours. The phase distribution and grain size were determined using x-ray diffraction and transmission electron microscopy. The initial grain size after calcining was 8-17 nm. It was determined that the critical ZrO$_2$ grain size to avoid the tetragonal to monoclinic phase transformation on cooling from 1100 °C was between 17 and 25 nm. Samples containing 50% Al$_2$O$_3$ maintained a grain size below the critical size for all times and temperatures. The 30% Al$_2$O$_3$ samples showed the same behavior in all but one heat treatment. The remainder of the samples showed significant grain growth and at least partial transformation to the monoclinic phase.

INTRODUCTION

In recent years nanocrystalline materials have been the subject of a great deal of research. Nanocrystalline materials have grain sizes typically less than 100 nm, and in some materials the grain size can be as low as 5-10 nm.[1] Because these materials have up to 50 vol% grain boundaries, they often have radically different properties than conventional materials. Properties altered by grain size include lowered thermal conductivity, increased hardness, and even low temperature ductility in some ceramics. [2-4]

ZrO$_2$ undergoes a series of transformations upon cooling. The cubic-to-tetragonal transformation occurs at 2370°C and the tetragonal-to-monoclinic transformation occurs at 1170 °C. [5] This second transformation is martensitic in nature and causes a 3-5 vol% increase, which can lead to cracking in the material. [6] To eliminate this volume change the high temperature phases are typically stabilized by dopants such as Y$_2$O$_3$ and CeO$_2$. [7] The resulting partially stabilized ZrO$_2$ (PSZ) is widely used for thermal barrier coatings and for ZrO$_2$ toughened Al$_2$O$_3$ ceramics.

The tetragonal phase of ZrO$_2$ can also be stabilized by reducing the grain size below a critical amount. Grains become too small to nucleate the monoclinic phase, and thus the tetragonal phase can exist at room temperature without the addition of stabilizing dopants. Garvie developed the following equation for critical grain size in pure unconstrained ZrO$_2$ based on surface free energy considerations, which is valid below T$_b$:

$$r_c = \frac{-3(\Delta \sigma)}{q(1 - T/T_b)} \tag{1}$$

where r$_c$ is the critical grain size, $\Delta\sigma$ is the difference in surface free energy between the high temperature phase and the low temperature phase, q is the heat of transformation per unit volume of an infinite crystal, and T$_b$ is the transformation temperature of an infinite crystal. [8] Figure 1 shows the variation of the critical grain size with temperature up to T$_b$. In order to be able to

335

Mat. Res. Soc. Symp. Proc. Vol. 457 ® 1997 Materials Research Society operate at a low temperature or thermally cycle the material without allowing the tetragonal-to-monoclinic transformation, the grain size must be maintained at ~20nm.

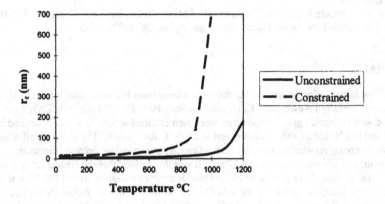

Figure 1: Theoretical Critical Grain Size in Tetragonal ZrO$_2$

The critical grain size increases with operating temperature or if constrained by a matrix.[6, 9, 10] For a zirconia particle constrained by a matrix, the equation for critical grain size was given by Lange as:

$$r_c = \frac{6(\gamma_m - g_s\gamma_t)}{[|\Delta G^c| - \Delta U_{se}]}$$ (2)

where γ_m and γ_t are the interfacial surface energies in the monoclinic and tetragonal states, g_s is the ratio of the interfacial surface areas of the two phases, ΔG^c is the free energy of the system and ΔU_{se} is the strain energy associated with the transformed particle.[6] The critical grain size decreases after thermal cycling, and heavily depends on the processing. In the literature the reported critical grain size ranges from 15-30 nm, because of differences in measurement and processing. [6, 11-13]

Maintaining the critical grain size can be accomplished by taking advantage of the properties of 'nearly immiscible systems'. In these systems, of which the ZrO$_2$-Al$_2$O$_3$ system is an example, two solid phases with little or no mutual solid solubility can control each other's grain growth. This can be accomplished by interpenetrating phases acting as diffusion barriers, or by small grains of one phase pinning the grains of the other phase. [14, 15] This second phenomena has been observed in ZrO$_2$ toughened Al$_2$O$_3$ systems with low concentrations of ZrO$_2$[14]. It is hoped that small Al$_2$O$_3$ grains in low concentrations can act to pin ZrO$_2$ grains in high ZrO$_2$ - Al$_2$O$_3$ systems. The goal of the current research was to determine the optimum combination of Al$_2$O$_3$ and ZrO$_2$ needed to maintain grain size at high temperature and thus retain the tetragonal phase. In addition, the research sought to verify measurements of critical grain size and attempt to determine the mechanism for grain size control.

EXPERIMENT

Colloidal dispersions of 20% ZrO$_2$ and 20% Al$_2$O$_3$ in water were obtained from Alfa/Aesar. ZrO$_2$ gels containing 10,20, 30 and 50 vol% Al$_2$O$_3$ were prepared from the colloidal dispersions and calcined at 600 °C for 17 hours in air. Samples were subsequently heat treated at 1100 °C for 24 and 120 hours and at 1200 °C for 2 hours. X-ray diffraction (XRD) was used to determine the percentage of each phase present. Transmission electron microscopy (TEM) and selected area diffraction (SAD) were used to determine the grain size and phase analysis and to examine the microstructure.

RESULTS

Figure 2 shows the percent of ZrO_2 in the tetragonal phase in each sample after heat treatment. In all of the calcined samples, the ZrO_2 was 100% tetragonal. The Al_2O_3 was found to be either γ-AlOOH (boehmite) or γ-Al_2O_3, a cubic transition form of Al_2O_3. After 24 hours at 1100 °C, the 10% Al_2O_3 samples contained 100 % monoclinic ZrO_2, and the 20% samples contained less than 60% tetragonal. The 30% and 50% samples maintained 90% or greater tetragonal ZrO_2 even after 120 hours at temperature. The Al_2O_3 transformed from γ-Al_2O_3 to varying amounts of θ- and α-Al_2O_3.

The evolution of the XRD spectra of the 10% Al_2O_3 sample can be seen in figures 3 and 4. In the region from 25-40 °2θ, the tetragonal (111) peak can be seen in figure 3. In figure 4, this peak has disappeared and has been replaced by the monoclinic (111) and (1̄11) peaks. In contrast, figure 5 shows the XRD spectra of the 50% Al_2O_3 sample after 120 hours, showing that the tetragonal phase has been retained. The Al_2O_3 phases often proved difficult to identify. This is due to the fact that the highest intensity peaks coincided for several different phases of Al_2O_3, and in some cases are hidden by nearby ZrO_2 peaks. There is therefore some uncertainty in the Al_2O_3 phase identification.

Figure 6 shows the results of the grain size measurements performed on TEM micrographs. In the calcined samples, all of the compositions showed a grain size of less than 20 nm. After 24 hours, the 10% and 20% samples had grown the most, and had exceeded the critical grain size for tetragonal ZrO_2. Most of the samples with a high percentage of tetragonal ZrO_2 maintained a grain size of 15-25 nm. However, a few samples had grains as large as 28 nm while maintaining the tetragonal phase. This agrees with the literature values of 15-30 nm as the critical grain size for tetragonal ZrO_2. Dark field TEM images were used to determine the size of the Al_2O_3 grains in several samples. The Al_2O_3 grain size varied widely, from 10-50 nm, with the majority of the grains in the 20-50 nm range. XRD measurements of the grain size proved to be difficult, as the sample quantities were limited, leading to low intensity data and a low degree of confidence in the results.

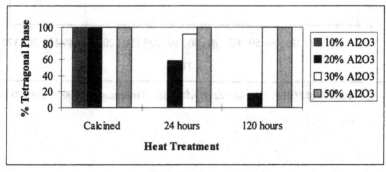

Figure 2: Tetragonal ZrO_2 Evolution During Heat Treatment

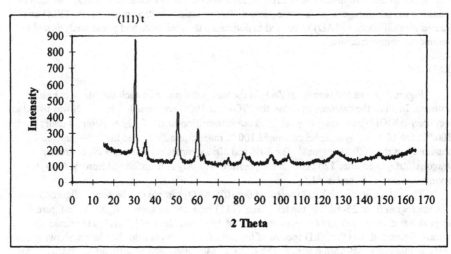

Figure 3: XRD Pattern of Calcined 10% Al$_2$O$_3$-90% ZrO$_2$

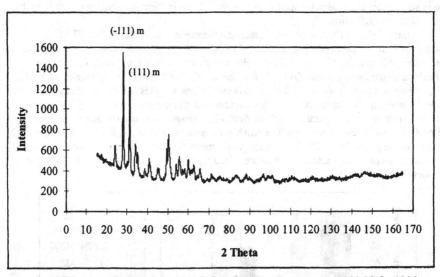

Figure 4: XRD Pattern of 10% Al$_2$O$_3$-90% ZrO$_2$ After Heat Treatment at 1100 °C for 120 hours

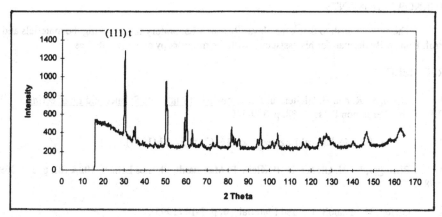

Figure 5: XRD Pattern of 50% Al_2O_3-50% ZrO_2 After Heat Treatment at 1100 °C for 120 hours

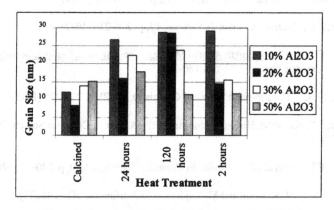

Figure 6: Average ZrO_2 Grain Size Measured by TEM

CONCLUSIONS

In general the 30% and 50% samples maintained desired grain size and phase distribution. The 20% samples did show some retention of the tetragonal phase, which implies that 20% Al2O3 might be enough for full stabilization if the mixing were improved. The critical grain size for tetragonal ZrO_2 was confirmed at 15-30 nm. The Al_2O_3 was found to be in transition phases such as γ-and θ-Al_2O_3 more frequently than the stable α-Al_2O_3 phase. The Al_2O_3 grains were found to be larger than the ZrO_2 grains. This rules out the possibility of a pinning mechanism for the grain size control, as the Al_2O_3 grains would need to be smaller than the ZrO_2 grains for pinning to occur. EDX studies are planned in order to conclusively identify the means of grain size control. In addition, attempts will be made to reduce the Al_2O_3 grain size, and to observe its effect on the grain size and phase evolution in the ZrO_2.

ACKNOWLEDGMENTS

The authors wish to thank the Naval Research Laboratory for providing the materials and Prof. Ronald Biederman for his assistance with the microscopy and XRD studies.

REFERENCES

1.	Birringer, R. and H. Gleiter, in Encyclopedia of Materials Science and Engineering, R.W. Cahn, Pergamon Press, 1988, p. 339-349.

2.	Ashley, S., Mechanical Engineering, 116, (2) p. 52-57 (1994).

3.	Birringer, R., H. Gleiter, H.-P. Klein, P. Marquardt, Physics Letters, 102A,(8), p. 365-369 (1984).

4.	Gleiter, H., NanoStructured Materials, 6, p. 3-14 (1995).

5.	Yoshimura, M., Ceramic Bulletin, 67,(12), p. 1950-1955 (1988).

6.	Lange, F.F., Journal of Materials Science, 17, p. 225-234 (1982).

7.	Curtis, C.E. . in Forty-Eighth Annual Meeting, The American Ceramic Society, (Buffalo, NY, American Ceramic Society,1946), p. 180-196.

8.	Garvie, R.C., The Journal of Physical Chemistry, 82,(2), p. 218-224 (1978).

9.	Lange, F.F. and M.M. Hirlinger, Journal of the American Ceramics Society, 70,(11), p. 827-830 (1987).

10.	Lange, F.F., Journal of the American Ceramics Society, 69,(3), p. 240-42 (1986).

11.	Aita, C.R., C.M. Scanlan, and M. Gajdardziska-Josifovska, JOM, 46,(10), p. 40-42 (1994).

12.	Chatterjee, A., S.K. Pradhan, A. Datta, M. De, D. Chakravorty, Journal of Materials Research, 9,(2), p. 263-265 (1994).

13.	Garvie, R.C. and M.F. Goss, Journal of Materials Science, 21, p. 1253-1257 (1986).

14.	Duclos, R. and J. Crampon, Scripta Metallurgica et Materialia, 24, p. 1825-1830 (1990).

15.	French, J.D., M.P. Harmer, H.M. Chan and G.A. Miller, Journal of the American Ceramics Society, 73,(8), p. 2508-10 (1990).

FERROELECTRIC LEAD ZIRCONATE TITANATE NANOCOMPOSITES FOR THICK-FILM APPLICATIONS

Kui Yao[*], Weiguang Zhu[*] and Xi Yao[**]
[*]School of EEE, Nanyang Technological University, Singapore 639798;
[**]Electronic Materials Research Laboratory, Xi'an Jiaotong University, Xi'an 710049, China.

ABSTRACT

In this paper, we report our experimental results of lead zirconate titanate (PZT) thick-films prepared from the sol-gel derived nanocomposites. Transparent Pb-Zr-Ti-B-Si gels were synthesized from various metal organic precursors. PZT grains with the size of one hundred nanometers were homogeneously grown from the gels in the annealing process. Then PZT thick-films were prepared using the gel-derived glass-ceramic powders with average particle size below 300 nm. Surficial morphology, dielectric and ferroelectric properties of the films were studied. The characteristics of such films are attributed to the nano-sized composite structure of the gel-derived PZT glass-ceramic precursors.

INTRODUCTION

PZT-based thick-films have shown great potential as piezoelectric sensors and actuators [1-3]. The glass phase is one of the important constituents as binding phase in the films. Conventionally, the thick-film pastes are prepared by mixing the powders of ferroelectric phase and glass phase using a ball-mill process [1,4]. The agglomeration of fine ferroelectric powders limits the improvement of the homogeneity of the mixture. The glass-ceramic technique, however, results in finer crystallites and the glass phases and thus could be useful for fabricating PZT-based thick-film or multilayer packages with improved structure and properties.

Although PZT ceramics have been used widely for many years, PZT glass-ceramics have not yet been well developed. Only in recent years has any work been done on glass-ceramics with a PZT phase [5]. Usually, ferroelectric glass-ceramics are prepared by a melting/quenching process. Understandably, fine PZT glass-ceramics with a substantial amount of pure PZT crystalline phase are difficult to fabricate by conventional melting, because of the high volatility of Pb and the high melting temperature of Zr. The authors have reported the preparation for ferroelectric glass-ceramics using the sol-gel method [6-7]. In this paper, the PZT thick-films fabricated by the sol-gel derived glass-ceramic have been studied.

EXPERIMENTAL

Pb-, Zr-, Ti-, B-, and Si- containing transparent gels were synthesized from metal alkoxides or organic metal salts using the sol-gel technique. Lead acetate ($Pb(CH_3COO)_2 \cdot 3H_2O$), titanium butoxide ($Ti(OCH_2CH_2CH_2CH_3)_4$), zirconium acetyl-acetonate ($Zr(CH_3COCHCOCH_3)_4$), tripropyl borate ($B(OCH_2CH_2CH_3)_3$), and ethyl silicate ($Si(OCH_2CH_3)_4$) precursor materials were dissolved into the common solvent 2-ethoxyethanol, and the solutions turned into gels in the air. In the present study, the atomic ratio of Pb:Zr:Ti:B:Si was 1.1:0.5:0.5:0.2:0.2.

After aging and drying, the gels were pyrolysed at 650°C. The perovskite PZT crystallites as small as 100 nm were formed in the firing process. The fired gels were then crushed and ground by the ball milling with ethanol to obtain fine powders. Due to the nature of the glass-ceramics, the thick-film pastes were prepared directly by blending the powders with an organic vehicle without adding any other glass frit. PZT based thick-films were then screen printed on alumina substrates with Pt/Au bottom electrodes and sintered at 1100°C for 10 minutes. Au top electrodes were printed and fired for electric characterization.

The particle sizes of the gel-derived powders were measured using a BI-90 particle size analyser. The PZT films were printed using an AMI 251 screen printer and their cross-sectional profiles were examined using a RTH 112/1550 surgraphic recorder. The surfaces of the samples were observed using a Cambridge Stereoscan 360 scanning electron microscope (SEM). The dielectric properties were characterized using a HP 4284A precision LCR meter and the leakage currents were measured using a HP 4156A precision semiconductor parameter analyzer. The ferroelectric hysteresis loops of the films were measured under virtual ground model using a RT66 ferroelectric test system.

RESULTS AND DISCUSSION

It has been found that the crystallization of Pb-Zr-Ti-B-Si (PZTBS) gel is controlled by the nucleus sites of the entire matrix and the perovskite PZT grains as small as 100 nm have formed during the pyrolysis process. Research on the crystallization of the PZT phase have revealed that the small tetragonal ZrO_2 precipitates may be the original crystal nuclei of the lead zirconate titanate crystalline phase in PZTBS [8]. Figure 1 shows the measured particle sizes of the gel-derived powders after ball milling. The volume median diameter is around 260 nm, and so the powders are obviously smaller than the conventional ceramic powders used for thick-film pastes. The powders are aggregated by many grains bonded with glass phase together. In fact, the PZT grain sizes are even smaller than the particle sizes shown in Fig. 1.

The cross-sectional profile from a sintered PZT film is shown in Fig. 2. The thickness of the film is uniform and about 50 μm, and its surface is smooth. The adhesion of the films to the electrodes is strong. The surficial micrographic picture in Fig. 3 shows a densified structure and the homogeneous mixture of the PZT grains with the glass phase, although a few pores exist in the picture. Such a surficial structure is much denser than that of the thick-films prepared using conventional ceramic and glass powders with similar composition. The fineness and

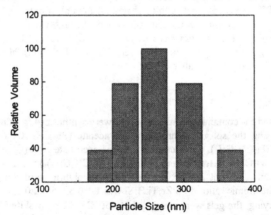

FIG. 1. The particle size distribution for the sol-gel derived PZTBS powders annealed at 650°C and then ball milled.

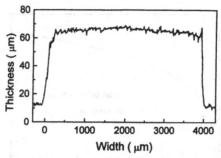

FIG.2. Cross-sectional profile from a PZTBS glass-ceramic thick-film
sintered at 1100°C for 10 min.

homogeneity of the gel-derived precursor composed of 100 nm sized PZT grains and the glass phase are believed to contribute to the resulted good quality of the films. In the firing process for the films, the PZT grains in the glass-ceramic have developed greatly and the average surficial grain size is over 1 μm for the glass-ceramic derived thick-films annealed at 1100°C for 10 min.

The leakage current of the film measured with increasing voltage varying from 1 to 100 V, at 1 V step increments, is shown in Fig. 4. The leakage current increases from around 1 to 100 pA with the applied voltage increasing from 1 to 100 V. The calculated resistivity, based on the measured leakage current and the dimension of the film, is also given in the same figure and the value is around 7×10^{12} ohm-cm. This resistivity is as high as that of bulk ceramics and is suitable to produce practical ferroelectric devices.

FIG.3. Surficial micrographic picture of the PZTBS glass-ceramic thick-film
annealed at 1100°C for 10 min.

343

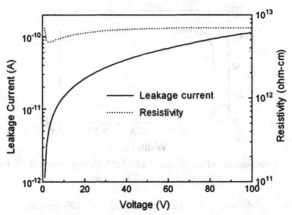

FIG. 4. Leakage current and resistivity under increasing voltage for the PZTBS glass-ceramic thick-film annealed at 1100°C for 10 min.

The dielectric loss is below 1% in the range of 100 Hz - 100 kHz, and increased to 1.3% at 1 MHz, as shown in Fig. 5. This figure also indicates that the dielectric constant of the film is about 90, a value much lower than the bulk ceramic samples. This phenomenon is believed to be attributed to the existence of the glass gradients with low dielectric constant in the present glass-ceramic thick-films. Since both the Ti and Zr elements are precipitated from the amorphous gel to form the PZT grains, the possible deviation of actual Zr/Ti ratio from the morphotropic boundary may also contribute to the decrease in the dielectric constant.

The polarization-electric field hysteresis loops, which are the confirming evidence for ferroelectricity, are given in Fig. 6 for the PZT films, with alternating fields varying between 150 and 200 kV/cm. With allowance for resistance compensation, a remanent polarization of 1

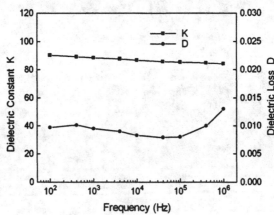

FIG.5. Dielectric constant and loss at different frequencies for PZTBS glass-ceramic thick-film annealed at 1100°C for 10 min.

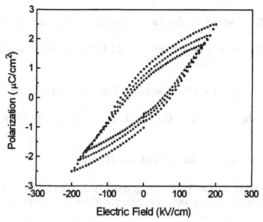

FIG. 6. The polarization - electric field hysteresis loops of the PZTBS
glass-ceramic thick-film annealed at 1100°C for 10 min.

$\mu C/cm^2$ and a coercive field of 75 kV/cm have been observed under a 200 kV/cm peak driving field. Since the leakage current is only 40 nA under such a high electric field, the remanent polarization and the coercive field remain almost the same value even in the case without resistance compensation. The remanent polarization is about one order of magnitude less than the PZT thin ceramic films and the same order of magnitude as the reported value for conventional PZT thick-films with glass added as the binding phase [4,9-10]. The coercive field for our samples is higher compared to PZT ceramic samples. The reason is believed to be the existence of the glass phase in the thick-films. For the glass-ceramic, the possible deviation of Zr/Ti ratio from the morphotropic phase boundary can also contribute to the smaller remanent polarization. Stronger ferroelectricity is expected if the PZT concentration in the glass-ceramic is further improved and/or the Zr/Ti ratio is better controlled.

CONCLUSIONS

The ferroelectric PZT thick-films have been prepared from gel-derived composite precursors of one hundred nanometer sized PZT grains and glass phase. The fineness and homogeneity of the precursor used for the pastes result in denser films with high resistivity and low dielectric loss. So the gel-derived glass-ceramics are gifted with some advantages for thick-film applications over the conventional process. The results also show that the dielectric constant and remanent polarization of the glass-ceramic films are lower than the ceramic films. Further improvement of the ferroelectric properties can be expected in the future research.

ACKNOWLEDGMENTS

We gratefully acknowledge the assistance of Ng Yong Chiang during our experiments.

REFERENCES

1. B. Morten, G. De Cicco and M. Prudenziati, Sensors and Actuators A **31**, 153 (1992).

2. G. De Cicco, B. Morten and M. Prudenziati, IEEE Trans. Ultrasonics, Ferroelectrics and Frequency Control **43**, 73 (1996).

3. H. Moilanen and S. Leppavuori, Microelectronics International **37**, 28 (1995).

4. H. D. Chen, K. R. Udayakumar, L. E. Cross, J. J. Bernstein and L. C. Niles, J. Appl. Phys. **77**, 3349 (1995).

5. B. Houng and M. J. Haun, Ferroelectrics **154**, 107 (1994).

6. K. Yao, L. Y. Zhang and X. Yao, in Electroceramics IV **2**. (The European Ceram. Soc., Proc. 4th Inter. Conf. Electronic Ceram. and Appl., Archen, Germany, 1994), pp. 1303.

7. K. Yao, L. Y. Zhang and X. Yao, Chinese Science Bulletin **40**, 694 (1995).

8. K. Yao, L. Y. Zhang, X. Yao and W. Zhu, J. Am. Ceram. Soc., Revised.

9. K. R. Udayakumar, P. J. Schuele. J. Chen, S. B. Krupanidhi and L. E.Cross, J. Appl. Phys. **77**, 3981 (1995).

10. B. Morten, G. De Cicco, A. Gandolfi and C. Tonelli, Hybrid Circuits **28**, 25 (1992).

RAPID CONSOLIDATION OF NANOPHASE Al₂O₃ AND AN Al₂O₃/Al₂TiO₅ COMPOSITE

David A. West, Rajiv S. Mishra and Amiya K. Mukherjee
Department of Chemical Engineering and Materials Science
The University of California, Davis
Davis, CA 95616

ABSTRACT

A rapid consolidation technique has been utilized in producing single phase Al_2O_3 in less than 10 minutes at 1400 °C resulting in a grain size less than 500 nm. TiO_2 has been added in hopes of obtaining Al_2O_3/Al_2TiO_5 nanocomposites in sintering times less than 30 minutes. The sintering process involves resistance heating of a graphite die containing the powder at heating rates of about 10 °C/s. The resistance heating step is preceded by a preparatory step consisting of DC voltage pulses applied across a prepressed powder compact. The retention of the nanostructure is attributed to the rapid heating rate although the possible effect of the DC pulses are also discussed. An Al_2O_3/Al_2TiO_5 composite has been produced during a short anneal immediately following sintering of an Al_2O_3/TiO_2 nanocomposite. Substantial grain growth has been observed to occur during the transformation taking the composite to the microcrystalline regime.

INTRODUCTION

Nanocrystalline powders are used in processing metal and ceramic parts in hopes of maintaining the nano-sized structure during the consolidation step. Unique properties have been observed in consolidated nanostructured materials (n-materials) but there is much that is not understood about the details of the structure/property relationships. Therefore in addition to the interest of basic science there is significant interest in producing n-materials for engineering applications based on the observed improvements in mechanical strength, toughness and formability.

The ability to produce ceramics with a stable nanostructure holds promise for producing ceramic materials capable of superplastic forming processes. Ceramic materials such as zirconia and titania have been successfully and routinely sintered to near theoretical density while maintaining the nanostructure. Nanophase alumina has been more difficult to produce by conventional sintering and hot-pressing presumably due to the high and anisotropic grain boundary energy making grain growth difficult to overcome. However, extensive powder processing techniques have been utilized to facilitate rapid and low temperature consolidation thus resulting in ultra-fine structures [1,2].

In order to stabilize the nanostructure in fully consolidated Al_2O_3, additives and secondary phases such as MgO and TiO_2 are required. TiO_2 enhances the sintering of alumina, and above 1280 °C forms Al_2TiO_5 in Al_2O_3. Al_2TiO_5 is of interest because a substantial amount of an Al_2TiO_5 secondary phase in Al_2O_3 has been observed to increase the high temperature ductility of an Al_2O_3/ZrO_2 composite by minimizing the grain growth during deformation [3]. In addition, the large thermal expansion mismatch between Al_2O_3

Mat. Res. Soc. Symp. Proc. Vol. 457 © 1997 Materials Research Society

and Al_2TiO_5 introduces residual stresses in the Al_2O_3 matrix near the interphase boundaries and it is suggested that this may influence the toughness of the composite [4-6].

It is the purpose of this study to apply a technique that has been successful in consolidating Al_2O_3 composites in very short times maintaining the nanostructure without the need for extensive powder preprocessing [7-13] to forming an Al_2O_3/Al_2TiO_5 nanocomposite.

EXPERIMENTAL PROCEDURES

Powders

The starting powders used in this study are described in Table I.

Table I: Starting Powders

	Manufacturer	phase	particle size	surface area
Al_2O_3	Baikowski	100% α	30 nm	50 m^2/g
Al_2O_3 + 250 ppm MgO	Baikowski	100% α	30 nm	50 m^2/g
TiO_2	Nanophase Technologies	80% anatase 20% rutile	32 nm	48 m^2/g

The composite powders were mixed by ultrasonic agitation in a methanol solution. For the single phase materials the powders were placed directly into the graphite dies for pressing without any solution processing.

The 70 vol% Al_2O_3 / 30 vol% Al_2TiO_5 composite was produced by adding 12.5 wt% TiO_2 to the Al_2O_3 powder taking into account the volume increase associated with the $Al_2O_3 + TiO_2 \rightarrow Al_2TiO_5$ reaction.

Sintering

A schematic of the sintering process is illustrated in Figure 1. The powders are initially briefly pressed at room temperature under 10 MPa in a graphite die with an inner diameter equal to 2 cm. DC pulses (750 A) of 90 ms duration are then applied across the die for a 60 s period. Immediately following this preparatory procedure, a constant current (1500 to 2000 A) is then applied heating the die at rates of up to 10 °C/s. After sintering, the dies were removed from the chamber and subsequently cooled to room temperature. Final specimen thicknesses were about 3 mm.

Characterization

Densities were measured using the Archimedes method with methanol being used as the immersing medium. Relative densities were calculated using the following theoretical densities: Al_2O_3 : 3.987 g/cm^3, TiO_2 : 4.245 g/cm^3, Al_2TiO_5 : 3.70 g/cm^3. Particle sizes of powders and dislocation generation were verified by TEM micrographs and grain sizes of densified compacts were determined by SEM. The phase analysis for the composites was characterized by XRD.

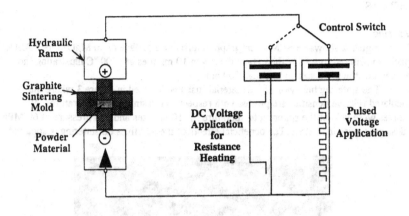

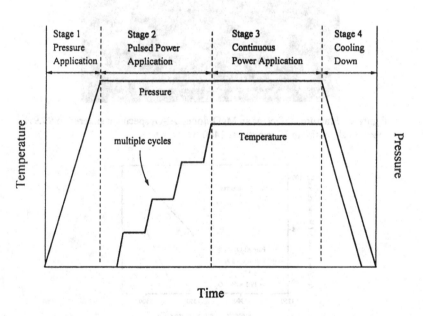

Figure 1. Schematic of the sintering process.

RESULTS

Sintering
    Figure 2 shows an SEM micrograph of a fracture surface of an MgO-doped Al$_2$O$_3$ specimen sintered to 99.9% theoretical density in 10 minutes at 1400 °C illustrating the retention of the grain size to less than 500 nm.
    The sintering behavior of the materials used is depicted in Figure 3 where the calculated relative densities are plotted with respect to the sintering temperature. All samples were held at the sintering temperature for 10 minutes and at a pressure of 60 MPa unless otherwise indicated. The beneficial effects of the additives on sintering is apparent.

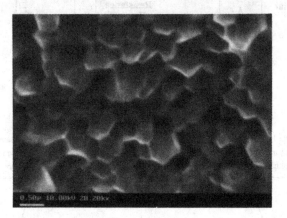

Figure 2. Fracture surface of an MgO-doped Al$_2$O$_3$ specimen sintered to 99.9% theoretical density in 10 minutes at 1400 °C.

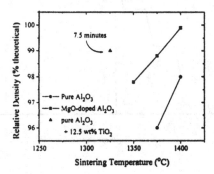

Figure 3. Theoretical density versus sintering temperature plot for pure Al$_2$O$_3$ and with MgO and TiO$_2$ additives.

Formation of Al₂TiO₅ secondary phase

Formation of Al_2TiO_5 begins after the specimen nears theoretical density [14]. Figure 4a shows the diffraction pattern of a composite sample sintered at 1400 °C obtaining full density after five minutes and the only phases present are the initial Al_2O_3 and TiO_2. In Figure 4b can be seen the pattern for a specimen held further at 1400 °C for 25 minutes showing a full transformation of the Al_2TiO_5 phase. The transformation was reported by Mishra et al. to take place during similar processing conditions within 10 minutes for a γ-Al_2O_3 + 2 vol% TiO_2 composite and the ultra-fine grain size was maintained [13]. Figure 5a is a fracture surface from an Al_2O_3/TiO_2 nanocomposite where the grain size can be seen to have been maintained during sintering. This is to be contrasted with the fracture surface shown in Figure 5b of a composite sintered to near full density then subsequently held for 30 minutes during which Al_2TiO_5 is formed. It is apparent from the comparison of these two specimens that substantial grain growth has taken place during the transformation. This grain growth will have to be overcome in order to obtain the nanocomposite.

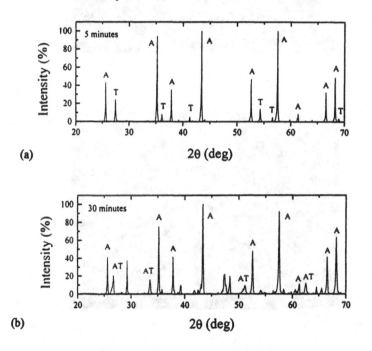

(a)

(b)

Figure 4. Diffraction patterns for (a) an Al_2O_3/TiO_2 specimen sintered to 99.8% theoretical density in 5 minutes at 1400 °C and (b) an a similar specimen held further at 1400 °C for 30 minutes.

(a)

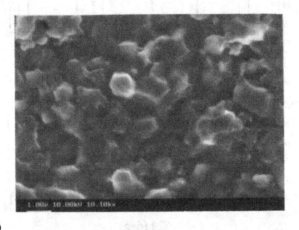

(b)

Figure 5. Fracture surfaces for (a) an Al$_2$O$_3$/TiO$_2$ nanocomposite sintered to 99.0% theoretical density in 7.5 minutes at 1325 °C and (b) an Al$_2$O$_3$/Al$_2$TiO$_5$ composite held at 1400 °C for 30 minutes.

DISCUSSION

The ability for consolidating nanophase ceramics using this technique is apparent. However, the exact mechanisms involved in the rapid-sintering process are still yet to be determined unambiguously. Notwithstanding, the beneficial effects of the pulsing step have been considered to be related to particle surface cleaning due either to concentrated current density at particle-particle contacts [7] or charge buildup and subsequent dielectric breakdown occurring between neighboring particles [13]. In the latter case, the sinterability of a given material will therefore be influenced by the materials dielectric constant and strength.

The difficulty in determining the mechanisms involved in this process is due partly to some difficulty in controlling processing parameters accurately and consistently as well as for lack of precise analytical techniques for determining the character of the dielectric breakdown if in fact it is occurring. Further analysis is underway to explore the influence of the pulsing cycle on the consolidation of specimens by way of control of the processing parameters and for characterization of the effect of the pulsed current.

Aside from the possible influence of the DC pulses on consolidation, the retention of the ultra-fine grain size in this study could also be attributed to the pressure application combined with the rapid heating rate. The influence of pressure on sintering is understood to be beneficial to inhibiting grain growth during sintering. The influence of heating rate can be understood in terms of the following competing processes during densification: particle coarsening (surface diffusion) and densification (lattice and grain boundary diffusion). The activation enthalpy of surface diffusion usually being lower than that for lattice (or grain boundary) diffusion, allows particle coarsening to dominate at lower temperatures. Therefore rapid heating rates promote densification and minimize coarsening.

The inability for obtaining the Al_2O_3/Al_2TiO_5 nanocomposite using this technique is not yet understood. Figure 6 is a TEM micrograph showing a γ-Al_2O_3/Al_2TiO_5 composite where the Al_2TiO_5 secondary phase is seen to form. The dislocation generation is apparent indicating the residual stresses produced during cooling due to the thermal expansion mismatch. These residual stresses likely will influence crack propagation through the Al_2O_3 matrix. Further mechanical testing and modeling should provide detailed information on this effect on the toughness of the composite.

CONCLUSIONS

Nanophase Al_2O_3 and an Al_2O_3/TiO_2 nanocomposite have been obtained by a resistance heating processes involving heating rates up to 10 °C/s. The rapid heating rate reduces the time spent at lower temperatures where coarsening dominates and allows densification to dominate at the higher temperatures.

Al_2TiO_5 is observed to form in the Al_2O_3/TiO_2 nanocomposite during a short annealing step after densification is complete. The time required for the transformation to occur is large enough to allow significant coarsening to occur taking the Al_2O_3/Al_2TiO_5 composite out of the nanocrystalline regime. Further analysis into inhibiting the grain

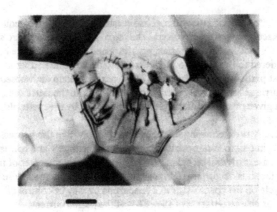

Figure 6. TEM micrograph displaying dislocation generation at the particle interfaces. (bar = 333 nm)

growth during the transformation will be required in order to obtain an Al_2O_3/Al_2TiO_5 nanocomposite.

ACKNOWLEDGMENTS

This work was supported by the ceramics division of National Science Foundation under grant NSF DMR-9314825. The authors wish to thank Professor K. Yamazaki for the use of his laboratory facilities.

REFERENCES

1. Masato Kumagai and Gary L. Messing, J. Am. Ceram. Soc. **68** [9], p. 500 (1985).

2. T. Yeh and M. D. Sacks, J. Am. Ceram. Soc. **71** [10], p. 841 (1988).

3. John Pilling and James Payne, Scripta Met. **32** [7], p. 1091 (1995).

4. P. F. Becher, Annu. Rev. Mater. Sci. **20**, p. 179 (1990).

5. G. C. Wei and P. F. Becher, J. Am. Ceram. Soc. **67**, p. 571 (1984).

6. A. V. Virkar and D. L. Johnson, J. Am. Ceram. Soc. **60**, p. 514 (1977).

7. R. Groza, S. H. Risbud and K. Yamasaki, J. Mater. Res. **7**, p. 2643 (1992).

8. J. Hensley Jr., C. H. Shan, S. H. Risbud and J. R. Groza, PM in <u>Aerospace, Defence and Demanding Applications</u>, p. 309, Metal Powder Industries Federation, Princeton, N. J. (1993).

9. J. Groza, Scripta Met. **30**, p. 53 (1994).

10. S. H. Risbud, J. R. Groza and M. J. Kim, Philos. Mag. B. **69**, p. 525 (1994).

11. S. H. Risbud, and C. H. Shan, Materials Letters, **20**, p. 149 (1994).

12. R. S. Mishra, J. A. Schneider, J. F. Shackelford, A. K. Mukherjee, Nanostructured Materials, **5** [5], p. 525 (1995).

13. R. S. Mishra, A. K. Mukherjee, K. Yamasaki, K. Shoda, J. Materials. Res. **11** [5], p. 1144 (1996).

14. B. Freudenberg and A. Mocellin, J. Am. Ceram. Soc. **70** [1], p 33 (1987).

SYNTHESIS OF OXIDE–COATED METAL CLUSTERS

ROBERT A. CRANE, JONATHAN T. MATTHEWS and RONALD P. ANDRES
School of Chemical Engineering, Purdue University, West Lafayette, IN 47906

ABSTRACT

"Fish-eye" particles consisting of metal clusters (Ag, Cu) a few nanometers in diameter encapsulated within a thin layer (~1 nm) of silica are produced using aerosol synthesis procedures. We present a method for predicting stable "fish-eye" nanostructures and describe synthesis techniques for producing significant quantities of silica-encapsulated metal nanoparticles.

For many metal/oxide pairs, gas phase formation of oxide encapsulated metal particles is thermodynamically favorable. Using known surface free energies and binary phase diagrams, it is possible to predict whether SiO_2-encapsulated metal clusters will form in the gas phase. Two conditions which must be satisfied are: 1) that the surface free energy of the metal is higher than that of Si; and 2) that the metal composition in the particle is greater than the eutectic composition in the metal/Si phase diagram. $Ag-SiO_2$ and $Cu-SiO_2$ are two examples of systems which readily form "fish-eye" structures.

Two types of gas phase cluster sources are used at Purdue for producing encapsulated metal nanoparticles. The Multiple Expansion Cluster Source (MECS) is a well established apparatus which produces small quantities (~50 mg/hr) of very uniform materials using resistive heating for evaporation. The new Arc Cluster Evaporation Source (ACES) offers much higher production rates (>1 g/hr) using DC arc evaporation. These two cluster sources make possible the study of a unique class of materials.

INTRODUCTION

Metal nanoparticles have numerous potential applications in electronics, bioengineering, and catalysis. The difficulty is preventing agglomeration when the particles are consolidated at high concentration. One method to prevent agglomeration is to coat the metal nanoparticles with a thin layer of silica.

Gas phase methods allow synthesis of metal clusters with nanometer-scale diameters and a narrow size distribution [1]. Two types of sources are used at Purdue: a Multiple Expansion Cluster Source (MECS) and an Arc Cluster Evaporation Source (ACES). Using these sources, it is also possible to produce bi-component (metal–silicon) clusters in the gas phase [2]. Upon exposure to ambient oxygen, the silicon in such bi-component clusters is readily oxidized to SiO_2.

In this paper, a qualitative thermodynamic method for prediction of structures of metal–silicon–oxygen particles formed in the gas phase is presented, and gas phase methods for producing silica-encapsulated metal nanoparticles are described in detail. Results are presented for synthesis of silica-encapsulated copper and silver nanoparticles.

Mat. Res. Soc. Symp. Proc. Vol. 457 © 1997 Materials Research Society

PREDICTION OF STRUCTURES OF METAL–SILICON–OXYGEN CLUSTERS FORMED IN THE GAS PHASE

The most stable structure of nanoparticles containing metal, silicon, and oxygen can be predicted from thermodynamic considerations. Although ternary phase diagrams do not exist for all metal–silicon–oxygen systems, it is possible to predict stable tie lines from the energies of formation of the binary compounds. If the Gibbs free energy of reaction for the formation of metal and SiO_2 from a metal silicide and oxygen is negative, then there exists a stable tie line between the metal and silicon dioxide. These are the most stable phases for most metal–silicon–oxygen systems. The energetically most favorable structure will be concentric spheres with the component having the higher surface free energy at the core.

Consider a bi-component particle of metal and silicon. In the liquid phase, each of the coinage metals are completely miscible with silicon. Upon cooling, phase segregation occurs. In the silver–silicon system, for example, the two components are completely immiscible at temperatures below the eutectic point. Thus, a two phase particle will form upon cooling.

The surface free energies of the coinage metals are much higher than that of silicon. However, the initial structure of a bi-component particle will depend upon the component that solidifies first. If the composition of the mixture is richer in metal than the composition at the eutectic point, the metal will solidify first, within a liquid pool of silicon. As the temperature is lowered further, the silicon solidifies around the silver core. Upon exposure to ambient oxygen, the silicon is oxidized to silica, and a "fish-eye" particle is formed.

Consider now the case of a composition which is on the silicon-rich side of the eutectic point. Upon cooling, silicon condenses first. However, the metal will not solidify around the silicon since it has a higher surface free energy. Significant thermal annealing is then required to drive this particle to the desired encapsulated structure. Thus, for gas phase synthesis of "fish-eye" metal–SiO_2 particles: 1) the surface free energy of the metal should be greater than that of Si; and 2) the metal composition in the particle should be greater than the eutectic composition in the metal/Si phase diagram.

PRINCIPLES OF GAS PHASE SYNTHESIS METHODS

The methods used at Purdue for nanoparticle synthesis involve rapid thermal quenching using a cold, inert gas. Two types of sources are used. Although they are physically very different, they are conceptually similar.

Multiple Expansion Cluster Source (MECS)

In the MECS, metal and silicon are evaporated from individual graphite crucibles enclosed in a resistively heated oven and the vapor is carried out of the oven by an inert gas, typically helium. Figure 1 shows a schematic representation of the MECS. The oven is heated above the melting temperatures of the cluster materials at a pressure of typically 100 torr. The hot gas from the oven is expanded through a sonic orifice into a quench zone held at typically 20–30 torr, where cold inert gas (helium) is introduced. The quench gas serves to drop the temperature and dilute the stream, preventing particle–particle coalescence. The aerosol stream containing metal–Si clusters passes through a high temperature region to anneal the particles [3].

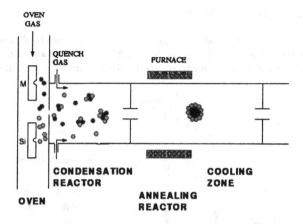

Figure 1: Multiple Expansion Cluster Source (MECS)

The cluster aerosol can be expanded into a vacuum chamber held at 10^{-5} torr to expose TEM grids to the clusters, or the clusters can be contacted with a fine mist of surfactant spray and the particles collected as a colloidal suspension [1].

The chief advantage of the MECS is its ability to control particle size on the nanometer scale. This is achieved primarily by varying the quench to oven gas ratio. Increasing the quench flow rate produces smaller particles. The MECS is capable of producing very narrow particle size distributions of clusters a few nanometers in diameter. Figure 2 is a bright-field transmission electron micrograph of silica-encapsulated copper particles produced in the MECS and sampled from the gas phase. These particles were produced at a rate of approximately 50 mg/hr. The diffraction rings shown in the insert in Figure 2 give lattice spacings which match exactly those of *fcc* copper, indicating that the silica layer acts as an oxidation barrier at room temperature. Uncoated copper clusters are always found to be cuprous oxide.

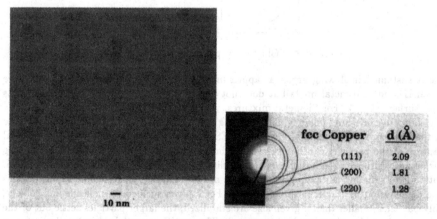

Figure 2: Copper–Silica Clusters Produced in MECS

Arc Cluster Evaporation Source (ACES)

The ACES, shown schematically in Figure 3, uses a high-current distributed DC arc to evaporate material from a carbon crucible, which serves as the anode of the arc. This is a variation of a cluster source reported previously by Mahoney and Andres [4]. The

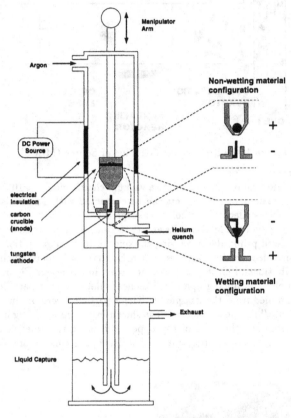

Figure 3: Arc Cluster Evaporation Source (ACES)

arc is sustained in flowing argon at approximately atmospheric pressure. When a pure metal is used, the metal melts but does not flow through the crucible orifice due to its high surface energy. For Si–metal mixtures, which wet carbon, it is necessary to modify the source by reversing polarity and inserting a tungsten rod to nearly plug the aperture in the bottom of the crucible. Since the ACES operates at atmospheric pressure, colloidal capture of the particles can be achieved by bubbling the gas through a liquid containing surfactant. Although the initial particle size distribution produced in ACES is quite broad, with clusters diameters ranging from 1 to 50 nm in diameter, it is possible to narrow this size distribution by fractional precipitation. For a given surfactant molecule and solvent, only particles smaller than a given size are stabilized; the larger particles settle out of the solution. Centrifugation speeds this process. The resulting particle size distributions can be very narrow. This is best examplified by gold clusters which are stabilized by dode-canethiol in mesitylene. Figure 4a displays a micrograph of the gold clusters as-prepared. Figure 4b displays a micrograph of the same colloidal dispersion after concentration and centrifugation to remove all but the smallest clusters.

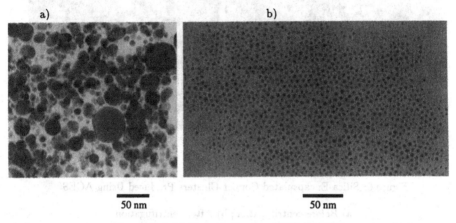

a)   b)

50 nm   50 nm

Figure 4: Gold Clusters Produced Using ACES

Before (a) and after (b) centrifugation.
(The gold clusters are stabilized using dodecanethiol as surfactant.)

Size-separation techniques also apply to silica-coated clusters. Figure 5 displays mi-crographs of ultrafine silver–silicon clusters produced in ACES. The clusters have been stabilized by octyltrimethoxysilane dissolved in mesitylene. It is difficult to discern the SiO_2 coating on such small particles. These particles were formed at an evaporation rate exceeding 1 g/hr. Figure 6 shows micrographs of copper–silica clusters produced using the

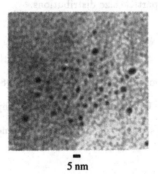

5 nm

Figure 5: Silica-Encapsulated Silver Clusters Produced Using ACES
(These clusters all have diameters smaller than 5 nm. The clusters are stabilized using octyltrimethoxysilane as surfactant.)

ACES, before and after size selection. The silica overlayer is apparent in the micrograph of these larger clusters.

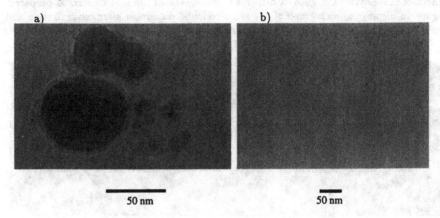

a)   b)

50 nm   50 nm

Figure 6: Silica-Encapsulated Copper Clusters Produced Using ACES

a) Before centrifugation; b) After centrifugation
(The clusters are stabilized using octyltrimethoxysilane as surfactant.)

CONCLUSIONS

Silica-encapsulated metal nanoparticles can be synthesized by gas phase processes. The Multiple Expansion Cluster Source produces nanoparticles with very narrow particle size distributions and good control of mean particle diameter. The Arc Cluster Evaporation Source produces a broader size distribution of particles, but at rates exceeding 1 g/hr. Clusters captured in the liquid phase as sterically stabilized colloidal particles can be size-selected to yield very narrow particle size distributions.

REFERENCES

1. L.-C. Chao and R. P. Andres, Journal of Colloid and Interface Science **165**, 290 (1994).

2. A. Patil and R. Andres, Journal of Physical Chemistry **98**, 9247 (1994).

3. A. Patil, D. Paithankar, N. Otsuka, and R. Andres, Z. Phys. D **26**, 135 (1993).

4. W. Mahoney and R. Andres, Materials Science and Engineering **A204**, 160 (1995).

Embedded Gold Clusters: Growth in Glass and Optical Absorption Spectra

P.G.N. Rao*, R.H. Doremus, Ph.D.**
Materials Science & Engineering Dept., Rensselaer Polytechnic Institute, Materials Research Center, Troy, NY 12180-3590, *raop2@rpi.edu, **doremr@rpi.edu, contact author

ABSTRACT

Optical absorption of nanosized particles in glass was measured. The plasma absorption band was completely spread out for particles below about 1.5 nm in diameter, as also has been found by gold particles in water and a polymer. The growth kinetics suggested growth of spherical particles controlled by diffusion of gold from the matrix to the particles, and a constant number of growing particles, giving a narrow size distribution of particles. For particles below about one nm in diameter (31 gold atoms), the optical absorption was proportional to λ^{-4}, as expected if the absorption results from free electrons in the particles.

INTRODUCTION

Optical absorption of metallic clusters can demonstrate the transition from atoms to bulk metal. The kinetics of growth of the particles help to determine their size. Small gold particles about 10 nm in diameter have a high non-linear optical absorption [1] ($x^3 = 2(10)^{-7}$), suggesting application in devices [2,3]. There are reviews of optical absorption measurements and theories for small metallic particles in Refs. [4-8]. Recent measurements on embedded silver particles are in Refs. [9-11], and of gold particles in Refs. [12, 13].

The optical absorption spectra of small particles of metals with optical properties involving free electrons show an absorption band in the visible range caused by collective oscillations of the free electrons. When the particles are smaller than the electron mean free path, this plasma absorption band broadens [4, 12-16]. For the smallest gold particles, the plasma absorption band is completely spread out [12, 13, 15].

In the present work, we grew small gold particles in glass and measured their optical absorption spectra. The particles were nucleated with UV light and grown at a higher temperature, so they had a narrow size distribution.

THEORETICAL BACKGROUND

Equations for Particle Growth

It is assumed that all particles are nucleated at time t=0, are spherical, and grow to the same size by diffusion of gold from the matrix to the particles. Then the volume fraction W of total precipitation of many particles as a function of time t is [17, 18]:

$$W = 1 - e^{-kt^{\frac{3}{2}}} \tag{1}$$

in which k is a growth coefficient:

$$k = \frac{8\pi N D^{\frac{3}{2}}}{3}\left(\frac{2C^{\frac{1}{2}}}{\rho}\right) \tag{2}$$

363

in which N is the number of particles per unit volume, D is the diffusion coefficient of gold dissolved in the glass, and ρ is the density of gold in concentration units.

<u>Equations for Optical Absorption</u>

The optical absorption, α, of a collection of spherical particles much smaller than the wavelength of light, λ, is given by the Mie equation:

$$\alpha = \frac{18\pi NV n^3 \varepsilon_2 / \lambda}{\left(\varepsilon_1 + 2n^2\right)^2 + \varepsilon_2^2} \tag{3}$$

where N is the number of particles per unit volume, V is the volume of a particle, n is the refractive index of the embedding medium, and ε_1 and ε_2 are the real and imaginary parts of the complex dielectric constant, ε, of the particles:

$$\varepsilon = \varepsilon_1 - i\varepsilon_2 \tag{4}$$

The optical properties of gold used here are those measured by Otter [20] and Thèye [21]; the optical properties of Otter measured for gold, silver, and copper were judged to be the most reliable for these metals [15, 22].

In the Drude free electron model, the dielectric constant is [10, 23]

$$\varepsilon = \varepsilon_1 - i\varepsilon_2 = \varepsilon_i - \frac{w_p^2}{w^2} - \frac{iw_p^2\gamma}{w^3} \tag{5}$$

$$w_p^2 = \frac{N_e e^2}{\varepsilon_0 m} \tag{6}$$

$$\gamma = \frac{v_F}{l} \tag{7}$$

In these equations, w_p = the plasma frequency, N_e = the number of free electrons per unit volume, ε = the electronic charge, ε_0 = the dielectric constant of free space, and m = the electronic mass, possibly an effective mass. ε_i is the frequency-independent part of ε; ε_i is often assumed to be one, but in gold, it equals about 10. Further γ is a damping coefficient that in the Drude theory is given by Eq. 7; v_F is the electron velocity at the Fermi energy and l is the mean free path of the electrons. The optical conductivity is

$$\sigma = \frac{w_p^2 l}{v_F} \tag{8}$$

If the optical absorption is plotted as a function of W, the result is a Lorentzian band with width γ at half-maximum. Experimentally, the width of this plasma absorption band becomes greater for particles smaller than about 40 nm in diameter because the mean free path of the electrons in bulk gold is greater than the particles diameter. Thus the optical conductivity of the metal in the particles is lower than the bulk conductivity, giving a higher ε_2 value for the particles.

An equation for the effective mean free path, l_e, of the electrons in the particles as a function of particle radius, R, is [14]:

$$\frac{1}{l_e} = \frac{1}{R} + \frac{1}{l_0} \qquad (9)$$

in which l_0 is the mean free path in bulk gold, which is about 40 nm.

EXPERIMENTAL METHODS

The base glass composition contained 63.9 wt.% SiO_2, 9.6 wt.% B_2O_3, and 26.5 wt.% K_2O. This composition was chosen because it dissolved a larger amount of gold than many soda-lime and aluminosilicate compositions. Added to the base composition was 0.2 wt.% of gold as $HAuCl_3 \cdot H_2O$. Twenty gram batches of glass were melted in alumina crucibles at 1400 °C to 1500 °C, and then quenched rapidly by pouring onto a metal plate. Strains were annealed by holding the glass for a few minutes at 400 °C and cooling slowly. Samples were then cut, and polished to give a smooth surface.

The glass was irradiated with a high intensity mercury UV lamp for 15 to 20 min. to nucleate particles. Then samples were heated in hydrogen at 550 °C or 600 °C to grow the particles. The hydrogen served to reduce the gold ions to atoms in the glass.

Optical absorption spectra were measured with a spectrophotometer (Perkin-Elmer 330). The samples were about 2 mm thick. Crushed pieces of glass containing gold particles were examined in the transmission electron microscope (Philips CM-12).

EXPERIMENTAL RESULTS

A sample of glass was heated for four hours at 550 °C to remove radiation damage and then at 600 °C for 4, 8, 12, 16, 20, and 24 hrs.. It is estimated that the growth is twice as fast at 600 °C as at 550 °C based on an activation energy of about 84 kJ/mole [24]. Thus two hours was added to the times at 600 °C to account for the heating at 550 °C. The volume fraction of gold particles was determined from the optical absorption at 0.44 μm wavelength, and is shown as a function of time in Fig. 1. The experimental data fit Eq. 1 with 3/2 as the exponent for time. This result is consistent with the diffusion-controlled growth of spheres [17] of uniform size; thus it suggests a narrow size distribution of gold particles.

The optical absorption spectra for samples heated at 600 °C for different times are shown in Fig. 2. After 20 hours, the spectrum shows a shoulder near 0.54 μm from the plasma absorption band; at earlier times, the spectra show no plasma absorption band.

Crushed samples of the glass heated for 20 hrs. were examined in the TEM. The glass showed intensity variations on the scale of about one nm, and it was not possible to identify gold particles of this size. In vitreous silica, dark-field electron microscopy revealed coherently scattering regions about one nm in size [25]; the authors concluded that their results agreed with a random network model for the silica. In earlier work, gold particles above about 2 nm in diameter were easily identified in a silicate glass [26]. We conclude that in the 20 hr. sample, there were no gold particles larger than about 2 nm in diameter.

DISCUSSIONS

The sizes of the gold particles in our specimens can be estimated from the shapes of the optical absorption bands and the kinetics of particle growth. For gold particles about 5 nm in diameter in glass, the plasma absorption band has a maximum at a wavelength of about 0.525 μm. The ratio of the height of this absorption maximum to the absorption at 0.44 μm decreases linearly with particle size as the size decreases, for sols with uniform particle size [4, 15]. Thus, this ratio can be used to estimate particle sizes.

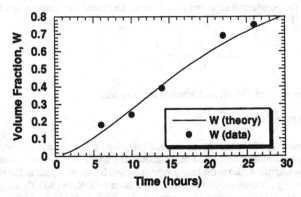

Figure 1: Volume fraction of gold in particles as a function of time. Points are from optical absorption at 0.44 μm. Line is from Eq. 1 with n = 3/2.

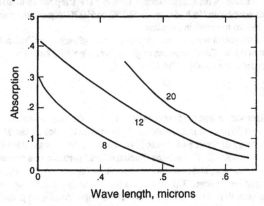

Figure 2: Measured optical absorption spectra of small gold particles in glass. Samples irradiated in UV light and heated at 600 °C for time in hours on curves.

From measured particle sizes and absorption ratios, we estimate that the particles in the 20 hr. sample in Fig. 2 were about 1.3 nm, in diameter.

The particle sizes in the samples at shorter heating times can be calculated from the kinetics of the particle growth. The volume of the particles grew proportional to the 3/2 power of time, or the radius to the square root of time.

Then the 12 hr. sample has particles $\sqrt{\dfrac{12}{20}} \cdot 1.3 = 1.0$ nm in diameter (31 atoms) and the 8 hr.

sample has particles $\sqrt{\dfrac{8}{20}} \cdot 1.3 = 0.82$ nm in diameter (17 atoms).

The absorption for the 12 hr. sample is plotted against the reciprocal of the wavelength to the fourth power in Fig. 3. A straight line results from wavelengths longer than about 0.45 μm. Other samples, one of which is also plotted in Fig. 3, showed this dependence. These samples were estimated to contain particles about 1 nm (31 gold atoms) in diameter. Also in the figure are absorption data measured

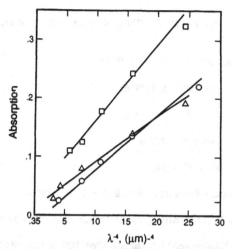

Figure 3: Optical absorption of small gold particles as a function of λ^{-4}. Present work in glass: (O) 12 hr. at 600 C, (□) 68 hr. at 600 C (another melt), absorption displaced for clarity. (Δ) Ref. 12, in water. In all samples, the particle diameter was about one nm (31 gold atoms).

by Duff, Baiker, and Edwards for gold particles in water [12]. The mean particle size was measured to be about one nm in diameter in the electron microscope.

The λ^{-4} dependence of absorption in these samples results from the free electron character of the electrons in them. Eq. 9 shows that for a particle with $R \ll l_0$, the effective mean free path, l_e, is equal to the radius. The small value of l_e leads to a large damping constant, γ (Eq. 7) and also of ε_2 (Eq. 5). Thus the value of ε_2^2 in the denominator of Eq. 3 becomes much larger than $(\varepsilon_1 + 2n^2)^2$, so for these small clusters:

$$\alpha = \frac{18\pi NV n^3}{\varepsilon_2 \lambda} \qquad (10)$$

Since ε_2 is proportional to λ^3 (Eq. 5), α is proportional to λ^{-4}.

These results show that the optical absorption of these gold particles embedded in glass is dominated by the free electrons in the gold, even though interband transitions cause absorption at wavelengths below about 0.5 µm in bulk gold. The lack of a plasma absorption band in these particles in glass also occurs for particles of similar sizes in water [12] and an organic polymer [13].

ACKNOWLEDGMENTS

Mr. Raymond Dove, Materials Science and Engineering, Rensselaer Polytechnic Institute, and Dr. Lu-Min Wang, Earth and Planetary Science, University of New Mexico, performed the electron microscope measurements.

REFERENCES

1. J. Tanahashi, et al., J. Appl. Phys. 79, p. 1244 (1996).

2. A. Leitner, Mol. Phys. 70, p. 179 (1990).

367

3. R.F. Haglund, L. Yang, R.H. Magruder, J.E. Wittig, K. Beiker, and R.A. Zuhr, Opt. Lett. **18**, p. 373 (1993).

4. U. Kreibig and L. Genzel, Surf. Sci. **156**, p. 678 (1985).

5. W.P. Halperin, Rev. Mod. Phys. **58**, p. 533 (1986).

6. W.A. de Heer, Rev. Mod. Phys. **65**, p. 611 (1993).

7. M. Brack, Rev. Mod. Phys. **65**, p. 677 (1993).

8. R.A. Broglia, Cont. Phys. **35**, p. 95 (1994).

9. S. Fedrego, W. Harbich, and J. Buttet, Phys. Rev. **B47**, p. 10,706 (1993).

10. H. Hövel, S. Fritz, A.H. Hilger, U. Kreibig, and M. Vollmer, ibid **B48**, p. 18,178 (1993).

11. K.P. Charlé, L. König, J. Rabin, and W. Schulze, Z. Phys. **D35**, p. 159 (1996).

12. D.G. Duff, A. Baiker, and P.P. Edwards, Langmuir **9**, p. 2301 (1993).

13. R. Lamber, S. Wetjen, G. Schulz-Ekloff, and A. Baalmann, J. Phys. Chem. **99**, p. 13,834 (1995).

14. W.J. Doyle, Phys. Rev. **111**, p. 1067 (1958).

15. R.H. Doremus, J. Chem. Phys. **40**, p. 2389 (1964); **42**, p. 414 (1965).

16. U. Kreibig, J. Phys. F: Metal Phys. **4**, p. 999 (1974).

17. R.H. Doremus, Rates of Phase Transformations, Academic Press, Orlando, FL, 1985, p. 22.

18. P. Rao and R.H. Doremus, J. Noncryst. Solids **203**, p. 202 (1996).

19. G. Mie, Ann. Phys. **25**, p. 377 (1908).

20. M. Otter, Z. Physik **161**, p. 163 (1961).

21. M.L. Thèye, Phys. Rev. **B2**, p. 3060 (1970).

22. R.H. Doremus, S.C. Kao, and R. Garcia, Appl. Opt. **31**, p. 5773 (1992).

23. N.F. Mott and H. Jones, "The Theory of the Properties of Metals and Alloys", Dover, p. 105ff (1958).

24. R.H. Doremus, in Symposium on Nucleation and Crystallization in Glasses and Melts., Am. Cer. Soc., Columbus, OH, 1967, p. 19.

25. P. Chaudhari, J.F. Gracyk, and L.R. Herd, Phys. Stat. Sol **51b**, p. 801 (1972); Phys. Rev. Lett. **29**, p. 425 (1972).

26. R.H. Doremus and A.M. Turkalo, J. Phys. Chem. Glasses **13**, p. 14 (1972).

LOWERED DIFFUSIVITY IN TiO_2 WITH A NANOPHASE DISPERSION OF SiO_2

GARY M. CROSBIE
Ford Research Laboratory, Ford Motor Company, MD 3182 SRL Bldg., P.O. Box 2053,
20000 Rotunda Dr., Dearborn, MI 48121-2053, gcrosbie@ford.com

ABSTRACT

In materials that are subject to environmental reactions, one way to increase durability
is to decrease the chemical interdiffusion rate. In metal systems, mass transport by diffusion
is enhanced -- if any change -- with finer grained microstructures, because of the greater
number of high diffusion interface and boundary paths. In non-metals, increased transport is
not necessarily the case, as other mechanisms may control the overall diffusivity.
C. Wagner [1] showed the theoretical basis of transport in certain nanocomposite cases based
on space charge layers at interfaces.

In an oxide-in-oxide nanocomposite, one can use the tools of bulk analysis to study the
near-interface effects on transport. The immiscible pair, TiO_2-SiO_2, was used as a model
system. Nanophase powders (50 and 200 m^2/g, respectively) from flame hydrolysis were
dispersed with high shear, then freeze-dried and hot-pressed to near theoretical density.
Transmission electron microscopy was used to show < 10 nm particles in the composites
produced at the time of diffusivity measurement. The rate of change of four-point electrical
conductivity in response to a change in oxygen partial pressure was used to estimate chemical
diffusivity. The re-equilibration differences between TiO_2 diffusivity for bars of about 10
times different square cross-section areas confirmed the interpretation of the conductance rate
changes to be the result of bulk interdiffusion.

The interdiffusion rate in the nanoscale dispersoid composites was more than 5 times
lower than for same SiO_2-fraction composites after coarsening which were like TiO_2 with no
second-phase. Results are in qualitative agreement with models for impurity segregation to
interfaces and for space charge layers near the dispersed nanoparticles.

INTRODUCTION

Nanocomposite materials are allowing analogies that have served elsewhere in
materials science to be broken. Transport properties of compounds with point defects is now
being recognized as one such area. Liang [2], Novotny and J. B. Wagner [3], and Maier [4]
have shown that adding nanoscale insulating Al_2O_3 to salts can increase the ionic conductivity
in solid electrolytes used in batteries. The effects can be attributed to space charge layers.
In metals, mass transport is typically enhanced -- if any effect -- with a finer grained
microstructures. A decrease in mass transport would be atypical.

For oxidation case, C. Wagner [1] showed a theoretical basis for a nanocomposite
transport case being lowered with extended interface effects, in a electronic semiconductor
due to space charges around dispersed particles. For oxide scales, the work was extended in
an approximate theory to mixed ionic/electronic non-stoichiometric compounds by Crosbie
[5]. In this report, we summarize and update the subsequent experimental work [6] which
was directed to show that mass transport can be reduced by the addition of a submicron
second-phase dispersion. Also, we compare the results to nonstoichiometric transport models.
Use of nanoscale structure to alter transport is of growing interest.

EXPERIMENT

Composites from 0 to 10 volume percent SiO_2 in TiO_2 were prepared by interdispersion and hot-pressing of ultrafine commercially-available powders. These highly pure, high surface area precursor powders had been produced by flame hydrolysis of the tetrachorides, $TiCl_4$ and $SiCl_4$. To create the intimate mixture, first, the minor constituent (SiO_2) was added to deionized water and dispersed with high shear and ultrasonic agitation. Then the TiO_2 was added and dispersed in the same two ways to produce a slurry. To minimize aggregation in water removal, the slurry was quick-frozen (in an acetone-dry ice bath) as a shell layer inside a lyophylizing bottle and then the ice content was sublimated to vacuum. The resulting low density composite powder was then compressed in loosely fitting graphite dies that allowed outgassing of adsorbed water (as shown by IR) in time-temperature holds under vacuum at 200 and 650°C before a final ramp (under an axial pressure of 24 MPa with a hold at 1200°C for 1 h. The square cross-section bars were then machined for four point d.c. conductivity measurements at temperatures from 1100 to 1200°C.

Diffusivity values were derived from the 90 and 95% re-equilibration points of electrical conductance response to a step-change in oxygen partial pressure. As with many other non-stoichiometric oxides, the electrical conductivity of TiO_2 is dependent on the partial pressure of oxygen. In Fig. 1, data for the baseline samples of this work (TiO_2 with no SiO_2 added) show the expected trend [7] of lower conductivity at higher oxygen partial pressure. By proper flow design [6], one can make the step change to a higher oxygen partial pressure quickly, without disturbing the overall gas flow rate. A sample re-equilibration curve is plotted on a logarithmic time-base in Fig. 2 for such an oxidation step.

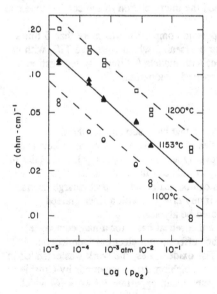

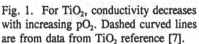

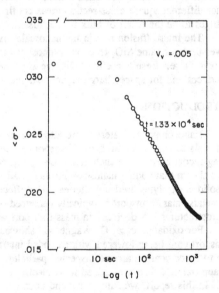

Fig. 1. For TiO_2, conductivity decreases with increasing pO_2. Dashed curved lines are from data from TiO_2 reference [7].

Fig. 2. During re-equilibration after a step-change to higher pO_2, the bar conductance decreases.

By comparing the times for percentage changes of overall bar conductance (averaged over the non-uniform conductivities in axial aligned surfaces within the square section bars) to a finite element model, one can obtain a value for diffusivity. This diffusivity is the chemical diffusivity, which has contributions from mass transport and an accelerating term for the gradient during the transient, which can be large for such small non-stoichiometries.

To confirm that the measurements were of transport through the bulk, paired samples (with ~10× different cross-section areas) were run simultaneously in a dual four-probe conductance rig. With similar diffusivity calculated from conductance re-equilibration values for each pair, we were able to confirm the re-equilibration rate-limiting was based on transport through the bulk, rather than any surface-reaction or gas-change-rate limitation.

RESULTS

The chemical diffusivity measurements are shown in Fig. 3 for samples held for days at 1153°C. The plotted data for re-equilibration runs (each 1500 sec in duration) after step changes from 0.02 atm to 0.42 atm pO_2. For 1153°C hold times less than 10^5 sec, each sample with the SiO_2 nanoscale addition shows markedly slower re-equilibration rates than for the baseline TiO_2 without SiO2. The largest initial decrements in diffusivity are found for the largest volume fraction additions. The lowest diffusivity values are partly affected by the limit imposed by the re-equilibration measurement run time and the cross-sectional area.

The electrical conductivity measurements over the same days of holding at 1153°C are shown in Fig. 4. These conductivities are the based on the conductance values at the end of

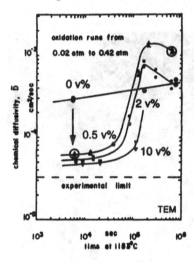

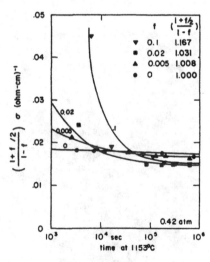

Fig. 3. Nanoscale SiO_2 added to TiO_2 causes a decrease (arrow) in the interdiffusion coefficient, relative to TiO_2 alone. The largest drops are for the largest vol.%'s. After 10^5 sec at 1153°C, the decreases are diminished.

Fig. 4. Conversely, the electrical conductivity is increased with a nanoscale SiO_2 addition. Asymptotic values at long times are lower than for TiO_2 alone, even with the classical two-phase factor [8] applied here.

each re-equilibration (oxidizing and reducing) run. Even though a factor is applied to each value to offset the series-parallel blocking by a dispersed spherical insulating phase, allof the first day nanocomposite conductivities are higher than those of TiO$_2$ with no SiO$_2$ added.

DISCUSSION

The co-forming of a TiO$_2$ matrix with a nanoscale dispersion of SiO$_2$ causes a decrease (Fig. 3) of more than a factor of 5 in the chemical diffusivity and lesser increases in electrical conductivity (Fig. 4). Over time, these effects diminish, with the samples held for a week at 1153°C becoming more like the baseline TiO$_2$ material produced by the same processing route.

To interpret the source of the initial effects and subsequent changes (over days holding time), models based on particle size were investigated. In the transmission electron micrograph, shown in Fig. 5, particles finer than 5 nm are visible above the electron noise after more than 2 hr at 1153°C. However, after 7 days, the few remaining fine particles have coarsened (Fig. 6) and a coarse SiO$_2$ phase as large as 10 μm can be seen optically. During this time, grain size coarsening also was observed.

At temperatures below 1550°C, the equilibrium phases present in the TiO$_2$ matrix - SiO$_2$ dispersion system consist solely of the binary oxide end members, rutile and the temperature-dependent SiO$_2$ phases. Aliovalent doping of the rutile by any dissolved SiO$_2$ is not expected to affect defect chemistry or electrical or mass transport, since both the Ti and Si cations have the same nominal valence. This feature helps simplify interpretation.

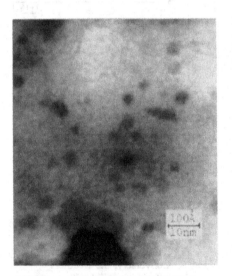

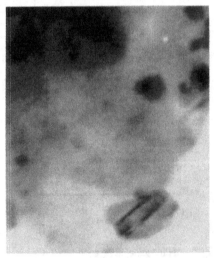

Fig. 5. Structure after 8.3×10^3 sec at 1153°C showing particles finer than 5 nm (transmission electron micrograph).

Fig. 6. After 6.4×10^5 sec at 1153°C, the dispersed particles (imaged by TEM) have coarsened (and SiO2 is visible with an optical microscope).

For the relatively low volume fractions, f, added here, the classical theory predicts the effect on transport in a two-phase composite with a spherical, nonconducting second phase [8] to be a decrease of transport somewhat greater than the volume fraction:

$$\sigma_d = (1 - f) \, \sigma_o \, / \, (1 + 0.5 \, f) \tag{1}$$

The measured conductivity values do not fit this model, even qualitatively. Importantly, the classical model does not point to any effect of coarsening particles. The lowering of diffusivity by more than 5 times with a 0.5% addition is quantitatively at variance with this model prediction (of less than 1%). However, the classical theory may explain small differences (from baseline) for the coarsened samples. Since the SiO_2 phase appears to partially wet the TiO_2 in such instances, the decreases in conductivity (vs. baseline at long times) are somewhat larger than for the classical correction used in Fig. 4.

For near-particle space charge layers or other internal interface effects to be detected, the matrix point defect concentration needs to be low, even for nanoscale particles. In TiO_2, the point defect chemistry is such that there are relatively small native defect concentrations in an accessible range of oxygen partial pressures, near one atmosphere oxygen (as shown in Fig. 1). Although the oxygen pressure dependence (fitted slope of -0.199 in Fig. 1 at 1153°C) is not a sure basis to identify ionic point defect type, TiO_2 is regarded as having both titanium interstitials and oxygen vacancies [9]. For the model discussion which follows, it is the sign of the ionic defect that matters in the models, not the particular species. In both types (and combinations of the two), the ionic defect carries a positive charge. At least two models qualitatively agree with the results are based on point defects charges. In both, the opposite effects on conductivity and diffusivity are linked to the opposite signs of electrons and the ionic point defects. Segregation of impurities and space charge are now described:

Within the nanoscale composites, segregation of impurities to interfaces at particle surfaces or grain boundaries can effectively make the remaining matrix more pure. For instance, if Fe^{3+} or Al^{3+} impurities (present in these samples at 20 to 30 ppmw) segregate to dispersoid surfaces, the matrix electroneutrality condition would shift in a way to produce more electrons and fewer positively charged ionic species. After dialysis of some, but not all of the impurity governing the electroneutrality, the matrix conductivity σ_o is estimated [6] as:

$$\sigma_o = (1 + 3 \, f \, r^{-1} \, \Gamma_o \, d \, / \, [\, A \,]_{total}) \, \sigma_b \tag{2}$$

In this estimate, r is the particle radius; Γ_o is the fraction of coverage of a layer of thickness, d; and σ_b is the bulk conductivity of the matrix alone when doped to the same total impurity level, $[\, A \,]_{total}$. For TiO_2 diffusivity, the plus sign in Eq. 2 is reversed, because the ionic defects are of a positive charge. These estimated changes are qualitatively consistent with the observations of transport properties. Once the dispersoid has coarsened or recrystallized, the impurities would return to the matrix bringing the transport behavior close to that of the baseline material, as observed.

The matrix electroneutrality can also be shifted by a space charge in the immediate vicinity of dispersed particles which take on a colloid-like charge. A positive charge on the dispersed particles would lead to more electrons and fewer titanium interstitials or oxygen vacancies in the matrix near (on a Debye length scale) to the particles. An approximate solution has been given by Crosbie [5] based on an electronic defect only model by C. Wagner [1]. With a positive particle charge (based on tighter binding of O^{2-} to TiO_2 than SiO_2), the electrical conductivity of the matrix is increased and the diffusivity decreased, as observed. An estimate of the effect on transport properties for the TiO_2 case with SiO_2 is:

$$\sigma_o = (1 + 0.83 \text{ g } |z| \text{ f } L_D^2 / r^2) \sigma_b \qquad (3)$$

where the new terms are: g, a factor (about 20 in strong adsorption) which depends on surface potential and slightly on the geometry of the dispersion; z, the charge relative to the lattice of the ionic defect; and L_D, the Debye length, which depends on the bulk defect concentrations as they are controlled by partial pressure of oxygen and impurities.

As with the impurity segregation model, the plus sign in Eq. 3 is reversed for the diffusivity case, because the ionic defects are of a positive charge. The estimated changes are qualitatively consistent with the observations of transport properties of TiO_2-SiO_2.

Assuming a $t^{-1/3}$ particle growth, one can compare the segregation model to the space charge model, because they have different particle size dependencies. For impurity segregation, the conductivity increment should follow a $t^{-1/3}$ function of time. For space charge layers the increment should follow a $t^{-2/3}$ function. Experimental fits of the 0.5% and 2% data range from time exponents of -0.43 to -0.64. These values are intermediate between the two model values (of -0.33 and -0.67). So one model is not supported over the other. Also to be noted, interface-related effects such as space charge and impurity segregation can occur at grain boundaries, although not likely to be as large.

CONCLUSION

An example of decreased mass transport with the addition of a nanoscale dispersion was established experimentally. The chemical diffusivity (interdiffusion) in TiO_2 is decreased by the addition of a nanoscale dispersion of SiO_2. Models based on space charge layers and impurity segregation qualitatively can explain the decreased diffusivity, the complementary increase in electrical conductivity (with the addition of a fine insulator), and the relaxation (as the dispersed phase coarsens) over time back towards the transport properties of TiO_2 without any dispersion.

ACKNOWLEDGMENTS

The author is grateful for advice and suggestions of J. B. Wagner, Jr., D. L. Johnson, S. T. Gonczy, and C. Wagner; and for support of this work by the National Science Foundation through a graduate fellowship and the Northwestern University Materials Research Center.

REFERENCES

1. C. Wagner, J. Phys. Chem. Solids 33, 1051 (1972).
2. C. C. Liang, J. Electrochem. Soc. 120, 1289 (1973).
3. J. Nowotny and J. B. Wagner, Oxid. of Metals 15, 169, 1981.
4. J. Maier, Physica Status Solidi B 123, K89 (1984); J. Phys. Chem. Solids 46, 309 (1985); Mater. Res. Bull. 21, 909 (1986).
5. G. M. Crosbie, Corros. Sci. 17, 913 (1977).
6. G. M. Crosbie, J. Solid State Chem. 25, 367 (1978).
7. R. N. Blumenthal, J. Coburn, J. Baukus, and W. M. Hirthe, J. Phys. Chem. Solids 27, 643 (1966).
8. R. E. Meredith and C. W. Tobias, Advan. Electrochem. Electrochem. Eng. 2, 15 (1962).
9. D. C. Sayle, C. R. A. Catlow, M.-A. Perrin, and P. Nortier, J. Phys. Chem. Solids 56, 799 (1995).

PREPARATION OF CERAMIC NANOCOMPOSITE
WITH PEROVSKITE DISPERSOID

T. Nagai*, H. J. Hwang**, M. Sando**, and K. Niihara+
*Synergy Ceramics Lab., Fine Ceramics Research Association
1-1 Hirate-cho, Kita-ku, Nagoya 462, Japan, tnagai@nirin.go.jp
**National Industrial Research Institute of Nagoya, AIST / MITI
1-1 Hirate-cho, Kita-ku, Nagoya 462, Japan
+The Institute of Scientific and Industrial Research, Osaka University
8-1 Mihogaoka, Ibaraki, Osaka 567, Japan

ABSTRACT

In order to introduce ferroelectricity into structural ceramics, we synthesized novel ceramic composites containing submicron sized perovskite-type ferroelectrics (ceramic / perovskite-type ferroelectrics nanocomposites). Although perovskite compounds easily react with other ceramics, $MgO/BaTiO_3$ nanocomposites were successfully fabricated using conventional sintering technique. The microstructure of nanocomposites and the phase stability of the dispersed $BaTiO_3$ particulate are discussed.

INTRODUCTION

Several ceramic-based nanocomposites such as Al_2O_3/SiC or Al_2O_3/Si_3N_4 system [1 - 4] have been developed in the field of engineering ceramics, and their improved mechanical properties have been reported. Recently, nanocomposites incorporated with nano-sized soft and weak materials such as metals and BN into Al_2O_3 and Si_3N_4 ceramics have been found to give enhanced mechanical strength, also [5 - 8]. Furthermore, this type of nanocomposites exhibit new functional properties, for example ferromagnetic property or improved corrosion resistance to molten metal.

From the view point of introducing new function to ceramic-based nanocomposites, ceramic/perovskite-type ferroelectrics system is promising. Incorporating ferroelectric perovskite into structural ceramics allows us to utilize not only ferroelectricity but also piezoelectricity, ferroelasticity and so on. In the field of electronic ceramics, polymer matrix or glass matrix composites containing ferroelectric phase have been studied [9 - 14]. However, there have been no studies on ceramic/ferroelectrics composites prepared by conventional sintering technique.

In order to fabricate ceramic nanocomposites with perovskite dispersoids (= particle dispersion in composites), phase compatibility between the matrix and the dispersoid is important because perovskite compounds easily react with various oxide ceramics and change into non-ferroelectric phase. In the present study, we have chosen MgO as a ceramic matrix, and succeeded in preparing MgO/BaTiO$_3$ nanocomposites. Microstructure and phase stability of the composites investigated by SEM, X-ray diffractometry and Raman spectroscopy will be discussed.

EXPERIMENT

MgO/BaTiO$_3$ composites were prepared using a conventional powder metallurgical method. The starting materials were MgO (Ube Chemical Industries Ltd, particle size 100 nm) and BaTiO$_3$ powder (Sakai Chemical Industry Co. Ltd.). Average particle sizes of BaTiO$_3$ powder were chosen as 300 nm and 1 μ m. Both BaTiO$_3$ powders had tetragonal structure. The amount of BaTiO$_3$ powder was varied from 5 to 20 vol%. MgO and BaTiO$_3$ were weighed and mixed in a nylon mill for 16 hours with ZrO$_2$ balls. n-Butyl alcohol was used as a medium of the ball-milling. The dried powder was sieved through 110 mesh screen and calcined at 800 ℃ for 30 minutes in air. The powder was hot-pressed at 1350 ℃ under a pressure of 40 MPa for 1 hour in nitrogen atmosphere. Sintered specimens were ground to thickness of 2 mm, and annealed at 1300 ℃ for 8 hours in air.

X-ray diffraction analysis was carried out at room temperature using a diffractometer RIGAKU RU-200B. Cu-K$_a$ radiation was used as the X-ray source.

Raman spectra were measured using a JASCO-NR1100 spectrometer using 514.5 nm line of an Ar-ion laser operating at 100 mW. Measurement were performed at a 90° geometry with instrumental resolution of 1 cm^{-1}.

Microstructure details of the samples were observed using a scanning electron microscope (SEM, JEOL JSM-6320FK).

RESULTS AND DISCUSSION

X-ray diffraction patterns of as-hot-pressed MgO/BaTiO$_3$ composites are shown in Fig. 1. All of the peaks in the profiles are assigned as MgO or BaTiO$_3$, and no additional phase is detected. Based on the X-ray diffraction analysis, it is concluded that BaTiO$_3$ does not react with MgO during the sintering process.

Some peaks corresponding to hexagonal BaTiO$_3$, which is high temperature phase of BaTiO$_3$, appear in the X-ray profile taken from the composite prepared from BaTiO$_3$ powder with particle size of 300 nm, as shown in Fig. 1 (b). Moreover, both of the as-hot-pressed composites prepared from the 300 nm and 1 μ m BaTiO$_3$ powders exhibited light blue color. It suggests that the BaTiO$_3$ dispersoid was reduced during the sintering process.

Therefore, we performed an annealing process to oxidize BaTiO$_3$ dispersoids. After annealing in air, the color of the composite turned white suggesting that the specimen was successfully oxidized. The X-ray diffraction pattern of the re-oxidized composite did not show hexagonal BaTiO$_3$.

Typical SEM images of the MgO/BaTiO$_3$ composites are shown in Fig. 2. The grain size of the MgO matrix is between 1 and 4 μ m in both specimens, and it is smaller than those of monolithic MgO sintered under the same conditions; monolithic MgO consists of grains between 5 to 10 μ m in size.

As shown in Fig. 2 (b), BaTiO$_3$ particles in the composite prepared from the 300 nm BaTiO$_3$ powder are located both within the MgO grains and at the triple points of the grain boundaries. The

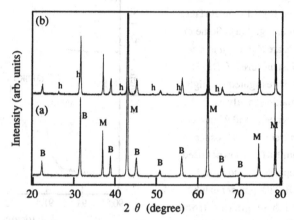

Fig. 1. X-ray diffraction patterns of MgO/BaTiO$_3$ composites. The average particle size of starting BaTiO$_3$ powder is (a) 1 μ m and (b) 300 nm. M is MgO, B is BaTiO$_3$, and h is hexagonal BaTiO$_3$.

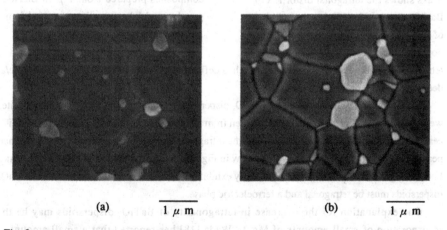

Fig. 2. SEM micrographs of MgO/BaTiO$_3$ composites prepared from (a) 1 μ m and (b) 300 nm BaTiO$_3$ powder.

377

particles of the intragranular BaTiO$_3$ are about 100 to 300 nm in size. It is concluded that nanocomposite microstructure was successfully realized in MgO/BaTiO$_3$ system.

The most of the BaTiO$_3$ dispersoids are located at the grain boundaries in the composite prepared from 1 μ m BaTiO$_3$ powder, as shown in Fig. 2 (a). Some of BaTiO$_3$ dispersoids are about 1 μ m size.

The crystal structure of BaTiO$_3$ dispersoids were investigated by high-angle X-ray diffraction analysis in the range of 2 θ between 90.5 and 93 degree. The profiles taken from the composites as well as the 300 nm BaTiO$_3$ powder used to prepare the composite are shown in Fig. 3. The profile of BaTiO$_3$ powder exhibits split peaks, which originates from tetragonal ferroelectric phase of BaTiO$_3$ [15]. In the MgO/BaTiO$_3$ composites, the shoulder on the lower-angle side of the peaks shows the tetragonal distortion of BaTiO$_3$ dispersoids [16], but the splitting of the peaks is not evident like the BaTiO$_3$ powder. This result indicates that the tetragonality of the BaTiO$_3$ dispersoid, which is defined as the ratio of the lattice parameter, c/a, decreased in both of the composites.

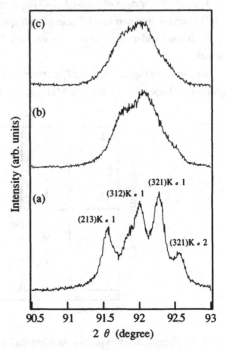

Fig. 3. High angle X-ray diffraction patterns of (a) 300 nm BaTiO$_3$ powder, (b) MgO/BaTiO$_3$ composites prepared from 1 μ m BaTiO$_3$ powder, and (c) composites prepared from 300 nm BaTiO$_3$ powder.

To confirm the tetragonal phase of BaTiO$_3$ dispersoids, Raman spectra of the nanocomposites were measured. In Fig. 4, Raman spectra taken from the nanocomposites and a monolithic BaTiO$_3$ ceramics are shown. It is well known that the tetragonal BaTiO$_3$ shows a characteristic Raman peak at 305 cm^{-1} [16, 17]. As shown by arrow in Fig. 4, not only the monolithic BaTiO$_3$ ceramics but also MgO/BaTiO$_3$ nanocomposites clearly exhibit the peak at 305 cm^{-1}. Therefore, the BaTiO$_3$ dispersoids must be tetragonal and a ferroelectric phase.

One explanation for the decrease in tetragonality of BaTiO$_3$ dispersoids may be the incorporation of small amounts of MgO. Wada [18] has reported that a small amount of magnesium ion can substitute Ti-site of BaTiO$_3$ and the substitution results in a decrease in tetragonality of BaTiO$_3$. Another explanation may be the effect of particle size on ferroelectricity.

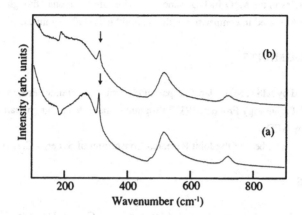

Fig. 4   Raman spectra of (a) monolithic BaTiO$_3$ ceramics and (b) MgO/10vol% BaTiO$_3$ nanocomposite prepared from 300 nm BaTiO$_3$ powder.

Uchino [19] reported that tetragonality of BaTiO$_3$ starts to decrease with decreasing particle sizes below 300 nm, and the tetragonality, c/a, becomes 1 at the size of 120 nm at room temperature. In the MgO/BaTiO$_3$ nanocomposites, some of the BaTiO$_3$ dispersoids are smaller than 300 nm, as shown in Fig. 2. Those smaller particles should have lower tetragonality than bigger ones.

Although the tetragonality of the BaTiO$_3$ dispersoids in MgO/BaTiO$_3$ nanocomposites decreased from the starting material, the BaTiO$_3$ dispersoids remains in ferroelectric tetragonal phase as shown by high-angle X-ray diffraction analysis and Raman spectroscopy. It suggests the strong possibility to utilize the ferroelectric properties in MgO/BaTiO$_3$ nanocomposites, for example for sensing the stress and fracture of the nanocomposites. Further study in dielectric properties of the nanocomposite is in progress.

CONCLUSIONS

MgO matrix composites with BaTiO$_3$ dispersoid were prepared. Microstructure as well as phase stability of the composite were studied using SEM, X-ray diffractometry and Raman spectroscopy. The results are summarized as follows.
(1) MgO is phase compatible with BaTiO$_3$. No reaction phase is produced during the sintering of MgO/BaTiO$_3$ composites.
(2) MgO/BaTiO$_3$ nanocomposites were successfully fabricated by a conventional powder metallurgical technique. In the nanocomposites, submicron sized BaTiO$_3$ particles are dispersed

both within the MgO matrix grains and at the grain boundaries.

(3) $BaTiO_3$ particles in the $MgO/BaTiO_3$ nanocomposites are tetragonal, though the tetragonality of the $BaTiO_3$ is decreased in comparison to the monolithic $BaTiO_3$ ceramics.

ACKNOWLEDGMENTS

Work promoted by AIST, MITI, Japan as part of the Synergy Ceramics Project under the Industrial Science and Technology Frontier (ISTF) Program. Under this program, part of the work has been supported by NEDO.

The authors are members of the Joint Research Consortium of Synergy Ceramics.

REFERENCES

1. K. Niihara and A. Nakahira in Proc. 3rd Int. Symp. on Ceram. Mater. and Component for Engine, edited by Am. Ceram. Soc. (1988) p. 919 - 926.

2. K. Niihara, A. Nakahira, G. Sasaki and M. Hirabayashi in Proc. 1st MRS Int. Meeting on Advanced Mater., (vol. 4, Tokyo, Japan 1989) p. 129 - 134.

3. K. Niihara, J. Ceram. Soc. Jpn. 99, p. 974 (1991).

4. A. Nakahira, T. Sekino, Y. Suzuki and K. Niihara, Ann. Chim. Fr. 18, p. 403 (1993).

5. T. Sekino, T. Nakajima, S. Mihara, S. Ueda and K. Niihara, Ceram. Trans. 44, p. 243 (1994).

6. T. Sekino, T. Kusunose, Y. Hayashi and K. Niihara in Proc. 4th Jpn. Int. SAMPE Symp., (1995) p. 269 - 274.

7. T. Sekino, T. Nakajima and K. Niihara, Mater. Lett. 29, p. 165 (1996).

8. T. Kusunose and K. Niihara, J. Am. Ceram. Soc., [In preparation].

9. G. S. Gong, A. Safari, S. J. Jang and R. E. Newnham, Ferroelectr. Lett. 5, p. 131 (1986).

10. M. H. Lee, A. Halliyal and R. E. Newnham, J. Am. Ceram. Soc. 72, p. 986 (1989).

11. T. Yamamoto, K. Urabe and H. Banno, Jpn. J. Appl. Phys. 32, p. 4272 (1993).

12. H. I. Hsiang and F. S. Yen, Jpn. J. Appl. Phys. 33, p. 3991 (1994).

13. V. F. Janas and A. Safari, J. Am. Ceram. Soc. 78, p. 2945 (1995).

14. A. Herczog, J. Am. Ceram. Soc. 47, p. 107 (1964).

15. A. F. Devonshire, Phil. Mag. 40, p. 1040 (1949).

16. B. D. Begg, K. S. Finnie and E. R. Vance, J. Am. Ceram. Soc. 79, p. 2666 (1996).

17. S. Wada, T. Suzuki and T. Noma, Jpn. J. Appl. Phys. 34, p. 5638 (1995).

18. S. Wada, M. Yano, T. Suzuki and T. Noma in Proc. Annual Meet. Ceramic Society of Japan, (Yokohama, Japan 1996), p. 368.

19. K. Uchino, E. Sadanaga and T. Hirose, J. Am. Ceram. Soc. 72, p. 1555 (1989).

PROCESSING, X-RAY, AND TEM STUDIES
OF QS87 SERIES 56 kΩ/SQUARE THICK FILM RESISTORS

GARY M. CROSBIE, FRANK JOHNSON, WILLIAM T. DONLON
Ford Research Laboratory, Ford Motor Company, MD 3182, SRL Bldg.
P.O. Box 2053, Dearborn, MI 48121-2053, gcrosbie@ford.com

ABSTRACT

Thick film resistors are glass/metal oxide nanocomposites used in hybrid microcircuits. These components have a small temperature coefficient of resistance that is useful in systems that experience a wide range of service temperatures. Test samples were produced by printing, drying, and firing resistor pastes in a laboratory process that simulated production conditions. The process parameters of peak firing temperature, time at peak temperature, and probe current were factors in a 2^3 factorial experiment that measured in-situ resistance (resistance during processing), as-fired resistance, and the temperature coefficients of resistance. As-fired resistance is shown to increase with firing time and temperature. In-situ resistance exhibited a small decrease with increasing firing temperature due to thermally-activated glass conduction at firing temperatures. The temperature coefficient of resistance measurements show that R[T] curve flattens with increasing firing time and temperature. X-ray diffraction revealed Pb-ruthenate, alumina, and Zr-silicate phases to be dispersed in the glass. Transmission electron microscopy in conjunction with energy dispersive x-ray spectroscopy revealed that the conductive phases, Pb- and CuBi-ruthenate particles, increased in size with increasing firing time and temperature. Lattice parameter measurements revealed only a small increase in the ruthenate structure. Resistance changes are attributed to increased separation of the conductive ruthenate particles by coarsening.

INTRODUCTION

For thick film resistors, two electrical properties of concern are resistance (R) and temperature coefficient of resistance (TCR). Controlling resistance values is of obvious importance to the manufacture of thick film resistors. An understanding of how and why processing conditions affect resistance values would be useful for improving resistor performance. Knowing of how microstructure is related to resistance would also be valuable.

This report summarizes the results of a process optimization study of thick film resistors. Readers who are interested in more detailed explanations are referred to an article[1] in press. In these experiments we observed the effects of firing time and temperature on the electrical properties and microstructure of thick film resistors. For these samples, resistance is measured both during (in-situ) and after the firing process. Experimental techniques used include variable-temperature voltage measurement, x-ray diffraction, and transmission electron microscopy.

TCR is a measure of resistor performance. It is calculated by measuring resistance as a function of temperature and calculating the slope of that function between specific temperatures. Two TCR values are calculated, Hot TCR (HTCR) between 25 and 125°C and Cold TCR (CTCR) between -40 and 25°C. These are typical temperature ranges that hybrid microcircuits experience during service. Uncontrolled TCR would seriously impair the function of modules.

The microstructure was observed indirectly with x-ray diffraction (XRD) and directly (5-10 nm scale) by transmission electron microscopy (TEM). Constituent phases, phase composition,

381

Mat. Res. Soc. Symp. Proc. Vol. 457 ° 1997 Materials Research Society

lattice parameters, particle sizes, and particle morphologies were determined with these techniques. These observations were compared with processing conditions and electrical properties to identify trends.

EXPERIMENT

The thick film resistors were DuPont QS87 series pastes blended by Ford at its North Penn Electronics Facility and had a nominal sheet resistivity of 56 kΩ/$\square$. The thick film conductors were Ag/Pd based conductive inks. The components were printed onto 96% alumina substrates by a Panasonic Byoga ink dispensing printer in a variety of geometries. The terminations were arranged in a four-point probe configuration. On some of the samples a 1 cm x 1.5 cm pad was printed for x-ray diffraction experiments. The sample boards were sealed at NPEF and stored in a dry, argon atmosphere to minimize any reactions in the unfired pastes.

A 2^3 factorial experiment was designed with centerpoints and two replicates. The factors are peak temperature, with levels of 825 and 875 °C; peak firing time, with levels of 8 and 12 min; and probe current through the resistor, with levels of 5 and 10 µA. We chose the peak temperature and firing time as factors because earlier experience had shown them to be significant effects for fired resistance. We included current, not because it was expected to be a significant factor, but because it was new parameter to have current flowing during the firing.

The experimental apparatus was designed and built with a temperature-controlled tube furnace used to fire the samples. Each sample was fired separately by manually sliding the sample into and out of the furnace. Voltage and temperature measurements were automatically collected during the experimental runs.

A total of twenty-six experimental runs were performed. Eighteen of these runs were used to provide responses for the experimental matrix. The eight remaining runs were disregarded due to data noise (such as a loose electrical connection) or procedural error.

The TCR measurements were made in an environmental chamber. The boards were mounted in the chamber and the temperature was cycled from -40 to 125°C. Resistance measurements were made at -40, 25, and 125°C. The TCR response variables are expressed in units of ppm/ °C.

X-ray diffraction patterns were collected for one sample from each of the four corners of the time/temperature matrix. The sample set used for x-ray diffraction was also used for electron microscopy. The samples were observed with a JEOL JEM 2000FX TEM. Particle compositions were determined by a Link Systems energy dispersive x-ray spectrometer.

RESULTS

The first response variables analyzed were fired resistance and in-situ resistance. Table I contains the ANOVA table for fired resistance and Table II contains the ANOVA table for in-situ resistance. In Tables I and II, the main and interaction effects are listed for each variable. Paired capital letters, such as AB and AC, represent two-factor interaction effects.

Note that the correlation coefficient (R^2) for the fired resistance is 96.4%. The correlation coefficient for the in-situ resistance is only 42.2%. Thus, the statistical analysis of the fired resistance accounts for a greater fraction of the raw variability of the data than the analysis for the in-situ resistance. Therefore, the in-situ resistance is less well predicted by the experimental factors than the fired resistance.

Fired Resistance ANOVA Table

Effect	Sum of Squares	Df	Mean Sq.	F-Ratio	P-value
A:Peak Temp.	4.27616E+11	1	4.2762E+11	147.11	0.0000
B:Firing Time	1.06572E+11	1	1.0657E+11	36.66	0.0002
C:Current	1.66894E+08	1	1.6689E+08	0.06	0.8185
AB	9.13226E+10	1	9.1323E+10	31.32	0.0003
AC	4.11410E+09	1	4.1141E+09	1.42	0.2646
BC	2.96912E+10	1	2.9691E+10	10.21	0.0109
ABC	4.36387E+10	1	4.3639E+10	15.01	0.0038
block	3.58182E+09	1	3.5818E+09	1.23	0.2958
Total Error	2.61613E+10	9	2.9068E+09		
Total (corr.)	7.32865E+11	17			

$R^2 = 0.964303$ R^2 (Adj. for d.f.) = 0.932572

Table I. ANOVA Table for Fired Resistance

In-Situ Resistance ANOVA Table

Effect	Sum of Squares	Df	Mean Sq.	F-Ratio	P-value
A:Peak Temp.	3.41056E +09	1	3.4106E+09	4.32	0.0675
B:Firing Time	2.44283E +07	1	2.4428E+07	0.03	0.8661
C:Current	8.38682E +08	1	8.3868E+08	1.06	0.3297
AB	1.50001E +08	1	1.5000E+08	0.19	0.6778
AC	1.47623E +08	1	1.4762E+08	0.19	0.6802
BC	3.59700E +07	1	3.5970E+07	0.05	0.838
ABC	2.66424E +08	1	2.6642E+08	0.34	0.5817
block	3.19371E +08	1	3.1937E+08	0.4	0.5473
Total Error	7.10944E +09	9			
Total (corr.)	1.23025E +10	17			

$R^2 = 0.422114$ R^2 (Adj. for d.f.) = 0.0

Table II. ANOVA Table for In-Situ Resistance

The effects analysis and ANOVA tables for Hot TCR and Cold TCR response variables are presented in Tables III and IV. The statistical analysis correlates fairly well with the data (R^2 of 82.2% for Hot TCR and 70.5% for Cold TCR). The significant effects for Hot TCR are peak temperature, firing time, and the interaction effect between the two. The Hot TCR effects are negative. The significant effect for Cold TCR is firing time, which has a positive effect.

X-ray phase analysis identified Pb-ruthenate, alumina, and Zr-silicate (JCPDS card #'s 34-471, 42-1468, 6-266). Alumina is the substrate material, and Zr-silicate modifies the thermal coefficient of expansion of the glass-ruthenate composite. Ten ruthenate peaks were selected for lattice parameter refinement. The ruthenate lattice parameters are about 1.026 nm, which is reasonable for this composition of pyrochlore. The lattice parameter increased only slightly (but significant statistically) from minimum time and temperature to maximum time and temperature.

Hot TCR ANOVA Table

Effect	Sum of Squares	Df	Mean Sq.	F-Ratio	P-value
A:Peak Temp.	1303.75156	1	1303.7516	9.73	0.0123
B:Firing Time	2638.10641	1	2638.1064	19.68	0.0016
C:Current	384.65016	1	384.6502	2.81	0.1245
AB	976.09381	1	976.0938	7.28	0.0245
AC	9.28726	1	9.2873	0.07	0.8010
BC	160.21231	1	160.2123	1.20	0.3027
ABC	94.81891	1	94.8189	0.71	0.4308
block	39.16125	1	39.1613	0.29	0.6076
Total Error	1206.38941	9	134.0433		
Total (corr.)	6812.47105	17			

$R^2 = .822915$ R^2 (Adj. for d.f.) = 0.665505

Table III. ANOVA Table for Hot TCR

Cold TCR ANOVA Table

Effect	Sum of Squares	Df	Mean Sq.	F-Ratio	P-value
A:Peak Temp.	20.40329	1	20.4033	0.16	0.7013
B:Firing Time	1332.57852	1	1332.5785	10.55	0.0100
C:Current	2.73241	1	2.7324	0.02	0.8878
AB	619.31300	1	619.3130	4.90	0.0540
AC	68.83191	1	68.8319	0.55	0.4868
BC	550.46544	1	550.4654	4.36	0.0664
ABC	109.98766	1	109.9877	0.87	0.3847
block	12.86259	1	12.8626	0.10	0.7602
Total Error	1136.43202	9	126.2702		
Total (corr.)	3853.60684	17			

$R^2 = .705099$ R^2 (Adj. for d.f.) = 0.442965

Table IV. ANOVA Table for Cold TCR

Figures 1 and 2 are TEM photographs of the 825°C, 8 min and the 875°C, 12 min samples. EDX analysis identified individual particles as Pb-ruthenate, CuBi-ruthenate, and Zr-silicate. Electron diffraction indicated that the particles were crystalline. The glass matrix was seen to contain Pb and Si. There were no distinguishable features in size, shape, or contrast that identify particles of different phases. The minimum particle size of the 825°c, 8 min sample is about 5 nm, while the minimum particle size of the 875°c, 12 min sample is about 20 nm . This indicates that particle coarsening occurs during firing.

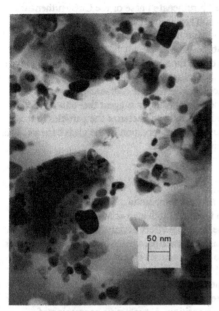 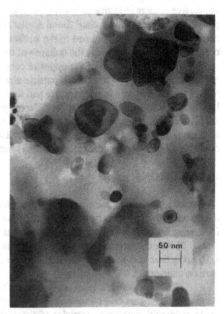

Figure 1. TEM: As fired 825°C 8 min
with particles smaller than 20 nm.

Figure 2. TEM: As fired 875°C 12 min
with most particles larger than 20 nm.

DISCUSSION

The results show that increasing firing time and peak temperature will increase fired
resistance values. The TEM results show a qualitative particle coarsening, with a corresponding
increase in the distance between metal oxide particles. A two-phase, tunneling barrier conduction
model[2,3] has been proposed to explain the conduction of these resistors. An increase in
interparticle distance would decrease the tunneling coefficient and thus increase resistance.

The in-situ resistance values did not correlate well to the experimental factors. A negative
dependence of resistance with peak temperature was observed. This is caused by thermally
activated glass conduction that dominates at high temperatures.

The TCR measurements can be explained in the light of a two-phase conduction model.
The metal oxide particles exhibit a positive linear temperature dependence of resistance, whereas
the glass exhibits a negative exponential temperature of resistance. The result for a thick film
resistor is a shallow concave up R(T) curve with a minimum in the vicinity of room temperature.

Because of the resistance minimum the CTCR, which is a measure of the slope between -
40 and 25°C, is usually negative, while the HTCR, measured between 25 and 125°C, is usually
positive. The results of this study show that firing time and temperature have a negative effect on
HTCR while firing time has a positive effect on CTCR. This can be interpreted as a flattening of
the R(T) curve, which is a desirable effect.

Since Pb and Bi pyrochlores have different properties, a change from CuBi to Pb
pyrochlore might serve as a partial explanation of the effects of time and temperature on R and
TCR. It is possible that Pb from the glass exchanges with Bi in the CuBi-ruthenate during firing.

A shift in lattice parameter would be evidence of such a reaction. However, there was no conclusive change in lattice parameter in either the Pb-ruthenate phase or the CuBi-ruthenate phase. Morten and Prudenziati[4] found similar results. They argue that the Pb-Bi exchange reaction, if it occurs, is confined to the surface regions of the conductive particles, and that the core of the particles prevents the collapse of the lattice.

TEM observations of the opposite corners of the time-temperature matrix seems to reveal, at least qualitatively, an increase in particle size with increasing firing time and temperature. It is logical that an increase in temperature and time should cause particle coarsening. However, the lack of characteristic features for the different particle compositions suggest that either more sophisticated or more indirect methods will be needed to fully characterize the particles in these systems. The absence of particle chains did not allow direct observation of the glass barriers. This could be caused by particle agglomerations obscuring the barriers.

CONCLUSIONS

Fired resistance increases with firing time and peak temperature. In-situ resistance decreases with increasing peak temperature due to thermally activated conduction in the glass. Increasing firing time and peak temperature has the desirable effect of flattening the R(T) curve.

X-Ray diffraction shows only a small increase in lattice parameter of the ruthenate phases with firing time and peak temperature. Particle size measurements do not show evidence of particle coarsening.

TEM observations revealed zircon, Pb ruthenate, and CuBi ruthenate phases dispersed in a Pb glass matrix. Particles of various shapes and sizes (>= 5 nm) were observed, but no differences could be identified between particles of different composition. A qualitative coarsening of particles with increasing firing time and peak temperature was observed.

ACKNOWLEDGMENTS

We would like to acknowledge the advice and support of the following people: R. Haberl, J. Hangas, W. Trela, C. K. Lowe-Ma, D. Smith, R. Soltis, H. K. Juday, J. Healy, D. Leandri, and J. Trublowski (of Ford); J. Chen (Hewlett-Packard); and A. Walker, L. Silverman, and R. Bouchard (DuPont).

REFERENCES

1. F. Johnson, G. M. Crosbie, and W. T. Donlon, "The Effects of Processing Conditions on the Resistivity and Microstructure of Ruthenate-Based Thick Film Resistors," Journal of Materials Science: Materials in Electronics, in press, Vol. 8, No. 2, (1997).

2. G. E. Pike and C. H. Seager, "Electrical Properties and Conduction Mechanisms of Ru-Based Thick Film (Cermet) Resistors," J. Appl. Phys., 48, 5152 (1977).

3. Th. Pfeiffer and R. J. Bouchard, "Modeling of Thick Film Resistors," Ceramic Transactions, 33 405 (1993).

4. B. Morten, A. Masoero, M. Prudenziati, T. Manfredini, J. Phys. D: Appl. Phys., 27, 2227 (1994).

THE GROWTH AND PROPERTIES OF THIN-FILM NANO-COMPOSITES

KEITH L LEWIS, A M PITT and A G CULLIS*
Defence Research Agency, St Andrews Road, Malvern, Worcs WR14 3PS, UK
* now at the University of Sheffield, UK

ABSTRACT

Nano-composites allow a materials engineering approach to be exploited to realise specific characteristics in optical thin film ensembles. For example, films can be produced with refractive indices determined by their average composition on the basis of effective medium approximations, so freeing optical designers from the constraints imposed in pure materials. A summary is presented of the progress made in a fundamental study of films based on diverse materials such as fluorides and sulphides, fabricated using molecular beam deposition techniques. The film properties (eg refractive index, surface morphology, environmental stability) are correlated with microstructure (as determined by cross-sectional TEM techniques). The enhanced properties of the films are discussed in relation to the realisation of periodically modulated graded-index structures of the type required for optical filter applications.

1. INTRODUCTION

Significant interest is being shown in the properties of nano-composite materials in thin film form. Whilst developments in such materials to enhance mechanical performance are becoming well established, rather less work has been reported in the area of optical materials although the gains expected in material response are substantial. Most optical thin films are produced by physical vapour deposition and the necessity to avoid high processing temperatures has driven the exploitation of growth techniques such as electron beam evaporation and sputtering. In the case of electron beam evaporation, insufficient energy is provided to the growing film to ensure effective adatom mobility and the deposited films (especially in the case of refractory materials such as oxides) are rarely fully dense. The relatively porous nature of such films leads to the uptake of moisture from the environment and stability is poor. This is most frequently seen in the shift in the positioning of reflection bands in multilayer stacks, particularly when low index materials are used.

There are three major techniques emerging as a means of controlling the microstructure of thin films. Ion assisted deposition processes use momentum exchange processes to provide the additional energy required to enhance surface mobility. Such techniques have found widespread acceptance in the optical coatings industry, although the resulting levels of lattice damage produced can under certain conditions modify the optical properties of the deposit. Similar effects are also produced as a result of ion impact processes during magnetron and ion beam sputtering. The ion flux may be lower in the latter case, since the primary inert gas beam is directed at the target and the ejected species are usually confined to a well defined low energy regime. Ion plating processes [1-3] rely upon electron beam sources to generate the vapour flux, but immerse the substrate in an intense glow discharge to achieve the desired energy transfer from the inert gas. Unless substrate temperatures are high, films produced by these processes are usually nanocrystalline, with high densities and refractive indices.

This study assesses the ability of two alternative techniques for the control of microstructure. The first involves the realisation of stratified media in which the periodicity is based on a few (5-100) atomic layers of complementary materials. The second technique depends on the exploitation of phase separation processes in mixed materials to form a nanocomposite structure. This is a potentially powerful technique provided that the degree of crystallinity can be controlled and that sufficient phase stability is presented by the choice of materials to avoid processes of Ostwald ripening.

2. EXPERIMENTAL

The films assessed in this study were produced by molecular beam deposition in a load locked UHV system fitted with 4 Knudsen sources capable of operation at temperatures of up to 1300 °C. The configuration of the sources was such that the effusing beams converged at the surface of the substrate. The vapour flux from each cell was measured using individual calibrated quartz crystals fitted with restrictors to limit the acceptance angle to that of the defined source and to avoid any cross-talk. A fibre-coupled broadband optical monitor was used to follow the evolution in reflectance spectra during growth and to determine optical thickness in situ. In-situ surface diagnostics (Auger, XPS) allowed the study of both surface preparation procedures and the examination of transient species produced at film-film interfaces as a result of chemical reaction or diffusion. Substrates (glass, zinc selenide or silicon) were cleaned before film growth using a defocussed raster scanned beam of argon ions (0.5-3keV).

Stratified films were produced by shuttering the molecular beams under computer control according to a defined sequence, which determined the thickness ratios of the component layers. Materials used included ZnS, ZnSe, BaF_2, PbF_2 and BiF_3. In comparison, the phase-separated films were produced by the co-deposition of zinc sulphide and barium fluoride. The composition of the composite film was determined in-situ by X-ray photoelectron spectroscopy. The films were deposited at temperatures of up to 350 °C at deposition rates of approximately 0.8-1Å/sec.

The evolution in film microstructure was assessed by cross-sectional transmission electron microscopy. Specimens were prepared by cleaving, epoxy mounting and abrasive thinning to a 100μm thickness. Final thinning was carried out by ion beam bombardment initially with argon and finally using iodine to avoid milling artefacts.

3. EVOLUTION IN MICROSTRUCTURE OF THIN FILMS DURING GROWTH

The microstructure of pure materials in thin film form is dependent not only on the material being deposited, but also on the choice of growth process and process conditions used. The effect of the major process variables have been explored by a large number of workers, but can be summarised by the generalised behaviour described by Movchan and Demchisin [4] and Thornton [5] in their zone models. These are of fundamental importance in determining the density of the film produced. At low temperatures (denoted by Zone I) there is insufficient adatom mobility to ensure redistribution of adsorbed species across the surface and films produced are either amorphous or of low density due to the formation of fractal or dendritic structures. Porosity is evident in such structures, largely because the growing interfaces become physically separated from the arriving vapour flux. As the substrate temperature is increased, a characteristic columnar morphology is developed where surface diffusion processes are sufficient to produce 2-dimensional redistribution of the coating material (Zone II). The individual columns of such material are usually microcrystals with a degree of atomic ordering sufficient to give relatively sharp X-ray diffraction peaks. The interface between the different layers in a stack and at the film/air interface is usually very smooth on the scale of wavelengths of interest for optical purposes. At the highest growth temperatures (Zone III), surface diffusion rates are sufficient to allow 3-dimensional redistribution of the coating material giving rise to well defined crystal facets and a general roughening of film interfaces.

In materials such as barium fluoride, Zone 1 material is usually obtained when films are deposited at ambient temperatures (ie 25 °C). It is necessary to heat the substrate to temperatures in excess of 300 °C before Zone II material can be obtained. The Zone I material is characterstically dendritic [6] when viewed in cross-section in the TEM and the residual porosity present allows the rapid ingress of water from the atmosphere. The process of water adsorption is reversible and exchange with the environment easily occurs, for example as a result of heating to temperatures in excess of 100 °C.

4. NANOCOMPOSITES BY STRATIFICATION.

The generalised behaviour summarised in the previous section is modified considerably in stratified heterostructures largely as a result of the disruption in the evolution of colunmar morphology in the film. This disruptive effect is illustrated schematically in figure 1. The realisation of a brick-wall effect is dependant not only on the magnitude of the mismatch in lattice constant but also on the adatom mobility in the individual layers. For example, in the case of a heterostructure made from ZnS and ZnSe, both materials share a cubic zinc-blende lattice with a mismatch in lattice constants of only 5%. This is insufficient to prevent column propagation (as shown in figure 2) largely due to the accommodation of micro-strain effects by twinning. In the case of barium and lead fluoride, the difference in lattice constants is even smaller (4%), but the difference in adatom mobility allows a lath-type morphology to be produced (figure 3). Some of the characteristics of the component dendritic BaF_2 are retained, although the high mobility of the PbF_2 produces a basic bulding block some 1000 x 50Å in dimension. When the extent of lattice mismatch is raised to 15% as in the case of ZnS and BaF_2, a brick-wall morphology is produced as shown in figure 4. Here the dimensions of each crystallite is close to 100Å. The well-defined level of ordering in each crystallite is clearly evident, indicating that the material is not fundamentally amorphous, although selected area diffraction patterns cannot resolve the individual phases. There is no evidence of any porosity and such films are highly stable in the environment.

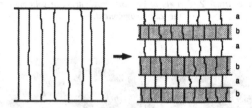

Figure 1   Schematic diagram showing principle of film stratification to control the propagation of columnar morphology

It is notable that multilayer films of exceptional surface quality (from an optical point of view) can be produced using such stratification techniques. The interfaces between the individual layers close to the substrate only vary in elevation by two or three atomic planes on a spatial scale of 160-200Å. The topography of successive interfaces bears no relationship to the previous and the same degree of atomic roughness is present at the top of thick coatings containing a thousand or so individual layers of disparate material. This highlights the ability of the stratification technique to control the propagation of columnar film morphology with resulting benefit in the reduction in optical scatter levels.

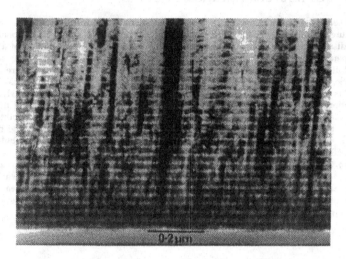

Figure 2   Cross-sectional transmission electron micrograph of stratified a ZnS/ZnSe film highlighting the inability of the lighter contrast ZnS films to prevent propagation of columnar grain morphology of the ZnSe component.

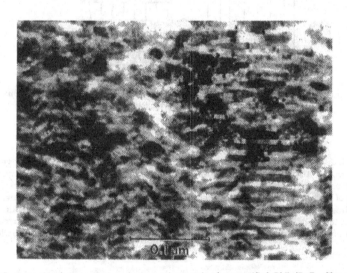

Figure 3   Cross-sectional transmission electron micrograph of a stratified PbF2/BaF2 film showing the development of a lath-type grain morphology.

_____ 50nm

Figure 4   Cross-sectional transmission electron micrograph of a stratified   BaF₂/ZnS film in which the ZnS clearly disrupts the propagation of any conituous grain morphology in the film.

5. CONSEQUENCES OF INTERFACIAL REACTION

Since such stratified coatings contain a large number of interfaces, it is of some concern that interfacial reaction is controlled. Sometimes, such effects can be beneficial to enhance the degree of interfacial bonding. On the other hand, when the optical properties of the product of reaction are likely to lead to extrinsic absorption, steps have to be taken to prevent any reaction occurring. An example would be the formation of PbS at the interface between ZnS and PbF_2. This is undesirable since the optical bandgap or the PbS is considerably smaller than either the PbF_2 or the ZnS. Notably this can be avoided by the incorporation of a thin layer (ca 20Å) of BaF_2 at each PbF_2/ZnS interface [6].

The extent of solid state reaction can be illustrated by considering the case of ZnS and BiF_3 as shown in figure 5. Here the process of interdiffusion has been followed by depositing a monolayer of BiF_3 onto ZnS and following the evolution in chemical composition as a function of time using XPS. It is notable that the reaction to form BiS_x is favourable and that fluorine disappears from the surface atomic layer within a few hours at room temperature. Such studies can only be carried out by carrying out the processes of deposition and chemical analysis in the same equipment without risking contamination from the atmosphere.

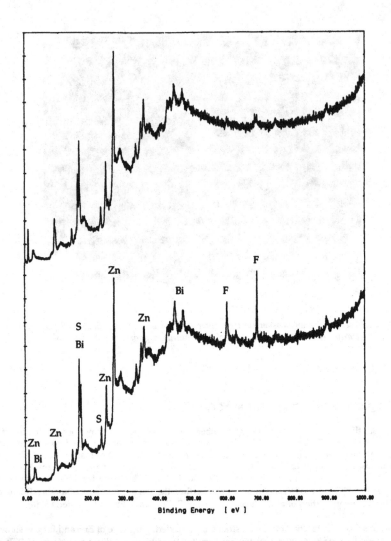

Figure 5a   X-ray photoelectron spectra measured for monolayer of BiF₃ deposited onto ZnS film.   Lower curve as deposited, upper curve measured after 2 days in UHV.   Note lack of oxygen uptake as evidence of cleanliness of system.

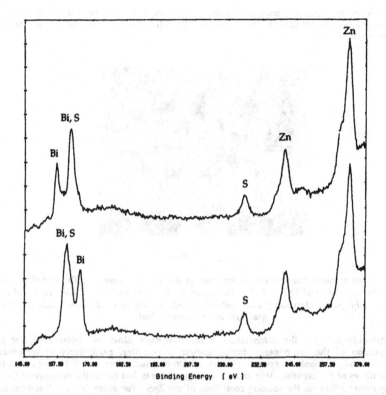

Figure 5b Detail of X-ray photoelectron spectra measured for monolayer of BiF$_3$ deposited onto ZnS film. Lower curve as deposited, upper curve measured after 2 days in UHV. Note shift in positioning of Bi 4f lines near 160eV as chemical environment changes from Bi-F to Bi-S. Differences in peak intensity of these lines are due to underlying sulphur 2p peak.

6. NANOCOMPOSITE FORMATION BY CO-DEPOSITION

Such composites can be produced by co-deposition of two materials known to be immiscible at the substrate temperatures employed. Reports of the production of such inhomogeneous structures abound in the literature [7]. For example, Farabaugh et al [8] have examined the formation of a ZrO$_2$-SiO$_2$ nanocomposite films deposited by electron beam evaporation. TEM studies of pure ZrO$_2$ indicate that the films grow by the formation of tapered polycrystalline columns. As the SiO$_2$ content of the composite is increased, the column diameters decrease and at approximately 25% SiO$_2$ a transition appears to an amorphous microstructure. The porosity of the film is reduced as SiO$_2$ is added with a corresponding increase in refractive index (despite the addition of a low index material to the composite). A maximum in refractive index is obtained at about 20% SiO$_2$.

This behaviour is different to that observed in the present study for the ZnS/BaF$_2$ system, largely because the pure ZnS films are fully dense and have the highest refractive indices. Figure 6 shows a cross-sectional TEM micrograph of the microstructure produced. Over the part of the phase diagram so far examined by transmission electron microscopy, the BaF$_2$ acts as the host, whilst the ZnS component forms the regions of small precipitates. The dimensions of the precipitates are very small, typically only a few tens of nanometres, commensurate with the scale of structures produced by the stratification technique.

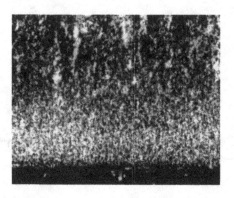

_____ 100nm

Figure 6   Cross sectional transmission electron micrograph of a nano-composite film formed by the co-deposition of BaF₂ and ZnS at 80°C. The composite film is desposited on a BaF₂ layer and is compositionally graded (decreasing ZnS content) towards the top of the diagram where significant grain growth becomes evident.

The relationship between the composition of the composite films and relative vapour phase supersaturation of the components during growth is complex, particularly at temperatures significantly above ambient. In general the composition obtained is not simply in proportion to the molecular fluxes of the separate BaF_2 and the ZnS components, but rather the presence of the BaF_2 has a significant effect on the sticking coefficient of the ZnS. For example, at a Knudsen source temperature of 800°C, a pure ZnS film would deposit at 100°C at a rate of about 0.5μm/hr. However once BaF_2 is added to the vapour flux, the sticking coefficient of the ZnS is reduced, so that the growth rate of the film becomes controlled more by the BaF_2 flux and the resulting film is correspondingly BaF_2 rich. Significantly higher overpressures of ZnS must be used to achieve compositions above the equimolar. This effect is illustrated in Figure 7.

In a similar way, the refractive index of the film (measured in the visible at 530nm) is not simply a linear function of composition as shown in figure 8. However this can explained more on the basis of effective medium approximations (albeit related to film microstructure) rather than simply process dependent factors.

The composite films deposited at 100°C are stable towards the ingress of moisture ingress over a significant part of the composition range. This implies that the films are dense. Significantly under normal circumstances, pure BaF_2 films deposited at the same temperature would only be about 90% theoretical density, a characteristic of zone II material. This suggests that the activity of the ZnS may be enhancing the mobility of the BaF_2 and preventing the propagation of dendritic microstructures. This is arguably akin to a catalytic effect, where the ZnS reduces the energy barrier to surface migration of adatoms. Since the ZnS flux consists largely of Zn atoms and sulphur dimers [9], their recombination at the growing interface is accompanied by the evolution of heat (ca 90kcal per mole). This can be utilised for enhancing the migration of BaF_2 molecules on the surface.

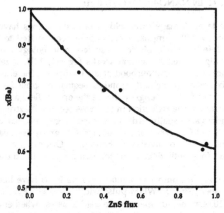

Figure 7 Variation in film composition as a function of ZnS overpressure for a fixed flux of BaF₂ during growth. Flux values are in arbitrary units, related by a single constant to the effective vapour pressure of the ZnS

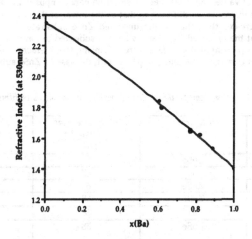

Figure 8 Variation of refractive index of ZnS/BaF₂ nano-composite with mole fraction of film, as deposited at 100 °C.

Clearly there will be a BaF_2 composition above which the film can no longer be dense, since insufficient ZnS is present to provide the requisite level of activation. On the basis of the fact that dense films of BaF_2 deposited at 300 °C have a refractive index of 1.50, the data plotted in figure 8 clearly indicate that porosity may begin to set in at BaF_2 compositions above $x_{Ba} = 0.8$. This appears to have some effect on the laser damage thresholds of the film as discussed in a following section.

7. LASER DAMAGE EFFECTS IN NANO-COMPOSITE FILMS.

The factors controlling the laser damage threshold of optical coatings have been presented many times in previous publications [10]. Emphasis has largely been related to the reduction in linear absorption and control of defects. The evolution in surface morphology around inclusions to form surface hillocks has been highlighted in several works (eg [11, 12]), together with their role in enhancing the optical fluence in the neighbourhood of the inclusion. The effects of laser conditioning are now commonly seen to be a result of the heating and subsequent controlled ejection of such inclusions, leading to defect morphologies which no longer enhance the optical field at the defect site. The other issue of major interest is related to the reduction in the cross-section of extrinsic molecular absorption processes. The major offenders are usually hydroxyl-related species associated with residual porosity in the material. The vibrational frequency of the isolated OH group is usually broadened towards longer wavelengths by the effects of hydrogen bonding. Optical coatings exhibiting residual absorption are not stable and frequently delaminate following a period in the atmosphere.

The laser damage thresholds of a number of nanocomposite films have been measured at $10.6\mu m$ wavelengths using a CO_2 laser (33nsec FWHM pulse width), and at $3.8\mu m$ using a DF laser (55nsec FWHM pulse width). The $10.6\mu m$ measurements were used to assess whether any extrinsic absorption effects were introduced at interfaces within stratified nanocomposites, whereas the $3.8\mu m$ measurements were used to assess the effect of moisture ingress into nanocomposites formed by co-deposition. Values of the laser damage thresholds listed in table 1 are for zero probability of damage in single shot measurements (one shot per site assessed), and are subject to an experimental error of about 5%.

Comparison of the $10.6\mu m$ values indicates that the stratified nanocomposite produced between BaF_2 and ZnSe had a damage threshold no different to that of the lower of the two component materials, and was significantly greater than that of the uncoated ZnSe substrate. This latter effect was attributed to the fact that the uncoated substrates were not subjected to ion beam cleaning, and that residual impurity species present at the surface were enhancing the interaction with the laser beam, leading to a reduction in damage threshold compared with the case of ZnSe substrates coated with ZnSe.

Table 1   Laser damage thresholds of various nanocomposite films

Film	Substrate	$10.6\mu m$ damage threshold J/cm^2	$3.8\mu m$ damage threshold J/cm^2
-	ZnSe	49	-
-	Si	67	50
ZnSe	ZnSe	60	-
BaF_2	ZnSe	68	-
BaF_2	Si	68	-
BaF_2/ZnSe Nanocomposite	ZnSe	58	-
BaF_2/ZnS Nanocomposite (x_{Ba} = 0.89)	Si	-	8.6-10.7
BaF_2/ZnS Nanocomposite (x_{Ba} = 0.61)	Si	-	53-55

The two co-deposited nano-composite ZnS/BaF_2 films were of interest since they had been deliberately chosen on the basis of their composition. There were clear differences in the behaviour observed. For the example chosen with a high BaF$_2$ composition (x_{Ba} = 0.89), the laser induced damage threshold (LIDT) was quite low. In comparison, the film with a BaF$_2$ composition below the critical value of x_{Ba} = 0.8 had a much higher LIDT, and the mode of failure was quite different with comparitively larger areas of film being separated from the substrate during irradiation. The major difference between these two sample was the amount of residual porosity evident in the films as indicated by the differences in the intensities of infra-red absorption due to residual water at 3400cm^{-1}, with the highest absorption found in the x_{Ba} = 0.89 case. The key issue is therefore the correspondence of the DF laser lines with the position of the water absorption band. Whilst the nominal wavelength used was 3.8μm, the laser was not grating tuned. Its pulsed mode of operation, with cascade decay into different energy levels at different times during the pulse meant that the actual wavelength emitted was less well-defined with the inevitable outcome that some of the energy could be coupled directly into the -OH vibrational mode.

8. OPTICAL FILTERS USING NANOCOMPOSITES

The materials engineering approach afforded by nanocomposite materials has been exploited for the fabrication of complex optical filters, particularly those based on inhomogeneous designs. In such structures, reflection bands are produced as a result of interference effects in thin film ensembles with half-wave sinusoidal periodicity in refractive index. The bandwidth of the reflection peak is dependant on the refractive index excursion Δn around the mean value, whilst the peak reflectivity achieved depends on both Δn and on the total number of periods present. Multiline designs can be realised by the superimposition of individual sinusoids [13].

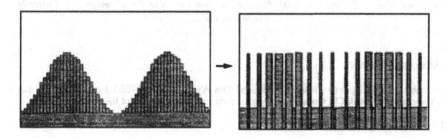

Figure 9   Principle of digital synthesis of refractive index.   Graphs are plots of refractive index (ordinate) against material thickness (abscissa).

Whilst smoothly varying refractive gradients can in principle be produced by analogue control of the Knudsen sources, it is more convenient to adopt a digital approach, in which the analogue slope is synthesised using a staircase. This is then taken a stage further by replacing each step in the staircase by two component films (essentially digital "bits") whose thickness ratio determine the effective refractive index of the step. This process is illustrated in figure 9 and results in a periodicity in the thickness sequence of a stratified nano-composite.

Such structures have been produced under computer control using the ZnS/BaF$_2$ materials system and have proved to be stable and capable of the flexibility that the design procedure allows. A typical filter will be designed around 20 "bits" per period and can require 50 periods to realise the desired optical properties. The ensuing 1000-layer structure places considerable dependance on the integrity of the component material interfaces and on the control system used during growth. Figure 10 shows as an illustration the measured optical density of a digitally synthesised filter constructed on the basis of the superimposition of the characteristic sinusoids for each reflection band using the inhomogeneous ZnS/BaF$_2$ nano-composite.

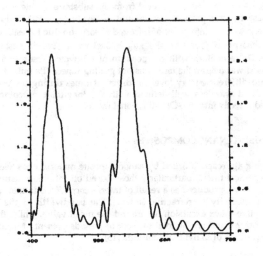

Figure 10   Optical density of a 2-band digitally synthesised filter constructed on the basis of the superimposition of the two individual characteristic sinusoids using the inhomogeneous ZnS/BaF$_2$ nano-composite.

9. ACKOWLEDGMENT

The authors are indebted to Mark Corbett and Tim Wyatt-Davies of DRA Malvern and to Ian T Muirhead formerly of OCLI Optical Coatings for their support in aspects of the research activity.

10. REFERENCES

1. H K Pulker, W Haag, E Moll: Swiss Pat Appl 00928/85-0, Mar 1985

2. K H Guenther: Proc Optical Interference Coatings Topical Meeting, OSA Tech Digest Series **6** 247 (1988)

3. R I Seddon, M D Temple, R E Klinger, T Tuttle-Hart and P M LeFebvre: Proc Optical Interference Coatings Topical Meeting, OSA Tech Digest Series **6** 255 (1988)

4. B A Movchan and A V Demchisin: Phys Met Metallogr. **28** 83 (1969)

5. J A Thornton: J Vac Sci Technol **12** 830 (1975)

6. K L Lewis, A M Pitt, N G Chew, A G Cullis, T J Wyatt-Davies, L Charlwood, O D Dosser and I T Muirhead "Fabrication of fluoride thin films using ultra-high vacuum techniques" *Proc Boulder Damage Symposium* NIST SP 752, 365-386 (1986)

7. R Jacobbsen: Physics of Thin Films **8** 51-98 (1975)

8. E N Farabaugh, Y N Sun, J Sun, A Feldman and H-H Chen "A Study of thin film growth in the ZrO_2-SiO_2 system" *Proc Boulder Damage Symposium* NIST SP 752, 321-331 (1986)

9. K Hall, J Hill and K L Lewis "The vaporisation kinetics of ZnSe and ZnS" *Proc 6th Int Conf on Chem Vap Deposition, Publ Electrochem Soc* V77-5, 36 (1977)

10. A H Guenther: Laser-Induced Damage in Optical Materials - 25 year Index, SPIE V2162 (1994)

11. M R Kozlowski and R Chow "Role of defects in laser damage of multilayer coatings" *Proc Boulder Damage Symposium* SPIE V2114, 640-649 (1993)

12. J R Milward, K L Lewis, K Sheach and R Heinecke "Laser damage studies of silicon oxy-nitride narrow band reflectors" *Proc Boulder Damage Symposium* SPIE V1848, 255-264 (1992)

13. W Gunning, R Hall, F Woodberry and W Southwell: Proc Optical Interference Coatings Topical Meeting, OSA Tech Digest Series **6** 126 (1988)

CRACK DEFLECTION AND INTERFACIAL FRACTURE ENERGIES IN ALUMINA/SiC AND ALUMINA/TiN NANOCOMPOSITES

S. JIAO, M.L. JENKINS
Oxford Centre for Advanced Materials and Composites
Department of Materials, University of Oxford, Parks Road, Oxford OX1 3PH, UK

ABSTRACT

Crack/particle interactions in Al_2O_3/SiC and Al_2O_3/TiN nanocomposites have been observed by TEM on samples containing cracks produced by Vickers indentations. No significant crack deflection by intragranular SiC particles or microcracking around nanoparticles was found. Intergranular cracks were observed to be deflected into the matrix grains by SiC particles on grain boundaries inclined to the direction of crack propagation. TiN particles were not effective in this way. These features are briefly discussed within the framework of the interfacial fracture energies. These were calculated from interfacial energies, which were determined by the measurement of grain boundary-interface dihedral angles.

INTRODUCTION

The deflection of cracks by particles is a potential toughening mechanism in particulate reinforced ceramic matrix composites[1]. There are several possible deflection mechanisms. Intragranular particles may cause crack deflection through the interaction between their stress fields and the crack-tip stress field. The particle stress field might arise from a combination of thermal residual stresses and/or elastic mismatch stresses. It has been suggested that the former mechanism may be particularly effective in alumina/SiC composites because the difference in thermal expansion coefficient between alumina and SiC is large[2]. The stress fields of intergranular particles may also be important, but here interfacial properties may also be significant in affecting the propagation of intergranular cracks.

The present work describes an experimental investigation of crack/particle interactions in alumina/SiC and alumina/TiN nanocomposites by means of transmission electron microscopy (TEM), with the object of determining which if any of the above mechanisms is effective in practice and if any other toughening mechanisms could be identified. The determination of interfacial fracture energies in these systems is also described.

EXPERIMENT

Nanocomposites of alumina containing 5wt%SiC and 5wt%TiN nanoparticles were fabricated by hot-pressing following a conventional powder processing route[3,4]. The nominal mean particle sizes for the TiN and SiC particles were 50nm and 200nm respectively. TEM samples containing cracks induced by Vicker's microindentation (using a 2N load) were prepared by a back-thinning method developed by Hockey[5].

RESULTS

Crack/particle interactions

Scanning electron microscopy (SEM) showed that in the Al_2O_3/TiN nanocomposite fracture is predominantly intergranular, just as in monolithic alumina. Intergranular cracks were clearly seen to propagate along Al_2O_3/TiN interfaces. The addition of nanosized SiC particles to alumina, however, was found to result in a change of fracture mode to predominantly transgranular. A comparison of crack paths in Al_2O_3 and Al_2O_3/SiC composites is given in Fig.1. The large arrows indicate the direction of one of the diagonals of the Vickers indentation, i.e. the average direction of crack propagation (ADOCP for short). The small arrows in Fig.1(b) show sites where the crack changes its nature either from transgranular to intergranular or vice versa. Unlike cracks in monolithic alumina, cracks in the Al_2O_3/SiC composites are very straight, indicating a lack of significant crack deflection by intragranular SiC particles, as demonstrated further in Fig.2 under high magnification. A thorough examination of all cracks showed that stress-induced microcracking around nanosized SiC particles does not occur and cutting of intragranular SiC particles by cracks is also absent.

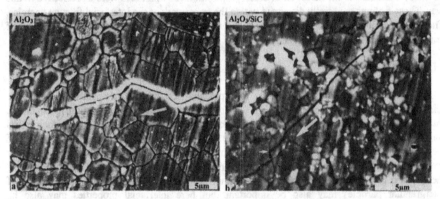

Fig.1 SEM micrographs of crack paths in (a) Al_2O_3 and (b) $Al_2O_3/5wt\%SiC$ nanocomposite. Large arrows indicate the average direction of crack propagation (ADOCP). Small arrows in (b) shows sites where the crack change its nature from intergranular to transgranular or vice versa.

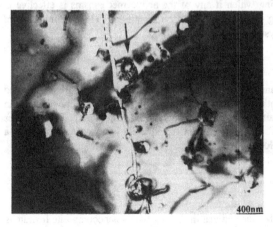

Fig.2 A TEM micrograph under high magnification showing no crack deflection by intragranular SiC particles in $Al_2O_3/5wt\%SiC$ nanocomposite.

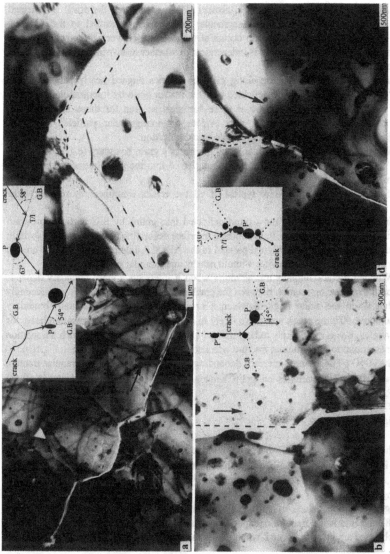

Fig.3 TEM micrographs and corresponding schematic drawings showing crack/particle interactions in Al2O3/5wt%SiC nanocomposites. Dotted lines represent grain boundaries (G.B) and solid lines represent cracks. Angles of deflection are also shown.

Typical interactions between cracks and intergranular particles in Al$_2$O$_3$/SiC composites are shown in the TEM micrographs of Fig.3. The corresponding schematic drawings highlight the salient features. The large arrow in each micrograph indicates the ADOCP. Several features are evident:

(1) When an intergranular crack running along a grain boundary inclined to the ADOCP approaches an intergranular SiC particle (labelled "P" in each case), the crack is deflected strongly by the particle to become transgranular and propagate parallel to the ADOCP (Fig.3(a-c)). Note that (i) the deflection occurs at the interface between the marked particle and the matrix, and (ii) the deflection is independent of whether nearby intragranular particles are present or not.

(2) Intergranular cracks running along grain boundaries aligned nearly parallel to the ADOCP were not observed to be deflected into the matrix grain by intergranular particles. Instead, interface debonding occurred at these particles, as seen for those labelled P' in Fig.3(b) and Fig.3(d). Debonding is also evident for intragranular particles (Fig.3(a)). Debonding does not usually lead to appreciable crack deflection.

(3) When a previously transgranular crack intersects a grain boundary (e.g. at the points marked "T/I" in Fig.3(c) and (d)), the crack is deflected into the grain boundary, becoming intergranular.

The features seen in Fig.3 were typical of all the cracks examined. In summary, the analysis showed:

(i) Intergranular cracks are most often deflected into grains by SiC particles on grain boundaries inclined to the ADOCP. In contrast, interface debonding occurs to particles within grains and on grain boundaries aligned parallel to the ADOCP. Intergranular cracks were not found to become transgranular at sites without nearby intergranular particles.

(ii) Whether a crack remains transgranular when it intersects a grain boundary depends mainly on the relative orientation between the crack and the grain boundary. It was generally found that when the acute angle between a crack and the trace of a grain boundary exceeded about 60°, the crack was more likely to continue propagating transgranularly; for smaller angles, it was more likely to be deflected along the grain boundary. Such a transition from transgranular to intergranular fracture was also observed by SEM, as seen in Fig.1(b).

For the Al$_2$O$_3$/TiN system, interface debonding associated with the intergranular fracture was observed for TiN particles on grain boundaries both parallel to and inclined to the ADOCP.

Interfacial fracture energy

By the measurement of grain boundary-interface dihedral angle, $\theta_d = \theta_1 + \theta_2$, which is defined in Fig.4, from the TEM micrographs of intergranular particles, such as those shown in Fig.5, we can obtain the ratio of interfacial energy/grain boundary energy, γ_i/γ_{gb}, from the following equations [6]

Using appropriate values of surface (free) energies[7] and the grain boundary (free) energy[8], the interfacial fracture energy,G_i, was calculated using the measured γ_i/γ_{gb}. Some values obtained are listed in Table I, where G_{gb} is the fracture energy of the grain boundary.

DISCUSSION

From the observations it is reasonable to conclude that crack deflection by intragranular SiC particles and microcracking do not contribute significantly to the toughening of Al$_2$O$_3$/SiC composites.

We see from Table I that G_i/G_{gb} is much larger than unity in Al_2O_3/SiC composites. The particle morphology, which determines the grain boundary-interface dihedral angle on which this result is based, develops by atomic diffusion at high temperatures when thermal residual stresses are absent. The strong interface in this case therefore arises from strong intrinsic bonding between Al_2O_3 and SiC rather than by a clamping effect due to thermal residual stresses. The low value of G_i/G_{gb} for the Al_2O_3/TiN system indicates a weak interface, which HREM confirms is probably the result of the presence of an amorphous TiO_2 layer[9].

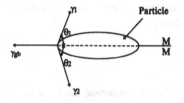

$$\gamma_1/\gamma_{gb} = \sin\theta_2 / \sin\theta_d \qquad (1)$$

$$\gamma_2/\gamma_{gb} = \sin\theta_1 / \sin\theta_d \qquad (2)$$

$$\gamma_i = (\gamma_1 + \gamma_2)/2 \qquad (3)$$

Fig.4 Schematic of a particle on a grain boundary with relevant parameters. 'M' indicates neighbouring matrix grains; grain boundary-interface dihedral angle: $\theta_d = \theta_1 + \theta_2$; interfacial energy: γ_1 , γ_2 or γ_i and grain boundary energy: γ_{gb}.

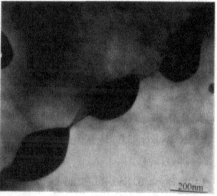

Fig.5 TEM micrographs of (a) a SiC particle and (b) TiN particles on grain boundaries.

Table I Values of some relevant parameters associated with the interface properties.

interface	θ_d	γ_i/γ_{gb}	G_i/G_{gb}
Al_2O_3/SiC	125.6°±17.2°	1.21±0.31	2.97
Al_2O_3/TiN	103.2±17.5	0.80±0.16	0.83

Qualitatively, the presence of intergranular particles is expected to affect the propagation of an intergranular crack, depending on the fracture energies of the matrix/particle interfaces relative to the fracture energy of the grain boundary. In the Al_2O_3/TiN system, an intergranular crack is likely to continue along the weaker interface. However, in the Al_2O_3/SiC system, the stronger interface may have a tendency to hinder debonding. Depending on the inclination of grain boundaries with respect to the direction of crack propagation, either interface debonding or the deflection of an intergranular crack into the matrix grains may occur. These qualitative predictions are in accord with the experimental observations. A detailed quantitative analysis involving an extension of the interface debonding theory developed by He and Hutchinson [10] allows for these features to be understood from the view point of the mechanical-energy-release rate and the interfacial fracture energy, and will be described in another publication [11]

CONCLUSIONS

(i) Microcracking and crack deflection are not significant toughening mechanisms in alumina/SiC nanocomposites.

(ii) Intergranular cracks are frequently deflected into matrix grains by SiC particles on grain boundaries inclined to the direction of crack propagation. On the contrary, intergranular cracks were found to propagate along alumina/TiN interfaces.

(iii) The determination of interfacial fracture energies showed that SiC particles strengthen grain boundaries whereas TiN particles weaken them. The values of the interfacial fracture energies provide a key to the understanding of crack/particle interactions.

REFERENCES

1. K.T. Faber and A. G. Evans, Acta Metall., 31, 565-84 (1983).

2. K. Niihara, The Centennial Memorial Issue of the Ceramic Society of Japan, 99, p. 974-982 (1991).

3. C.N. Walker, C.E. Borsa, R.I.Todd, R.W.Davidge and R.J. Brook, British Ceram. Proc. 53, 249-264 (1994).

4. C.E. Borsa, S. Jiao, R.I. Todd & R.J. Brook, J. Microscopy., 177, 305-312 (1995).

5. B.J. Hockey, J.Am.Ceram.Soc., 54, 331 (1971).

6. C.S. Smith, Trans A I M E , 175, 15 (1948)

7. R.H. Bruce, Science of Ceramics. 2, p. 359-367 (1965)

8. C.A. Handwerker, J.M. Dynys, R.M. Cannon, and R.L. Coble, J.Am.Ceram.Soc., 73, 1371-77 (1990).

9. S. Jiao, M.L. Jenkins and R.W. Davidge, Acta Mater. (1996) in press.

10. M.Y. He, and J.W. Hutchinson, Int. J. Solids Strut., 25, 1053-67 (1989) .

11. S.Jiao and M.L. Jenkins, to be published in Phil. Mag. A.

PERCOLATION THRESHOLD IN SUPERHARD NANOCRYSTALLINE TRANSITION METAL-AMORPHOUS SILICON NITRIDE COMPOSITES: THE CONTROL AND UNDERSTANDING OF THE SUPERHARDNESS

STAN VEPŘEK* S. CHRISTIANSEN**, M. ALBRECHT** and H.P. STRUNK**
* Institute for Chemistry of Inorganic Materials, Technical University Munich, Lichtenbergstr. 4, D-85747 Garching/Munich, Germany, Email: veprek@inf.chemie.tu-muenchen.d400.de
** Institute for Materials Science, University Erlangen-Nürnberg, Cauerstraße 6, D-91058 Erlangen, Germany

ABSTRACT

The hardness of the recently developed novel superhard nanocrystalline composites exceeds 5000 kg/mm^2 (50 GPa) and the elastic modulus 550 GPa. This is due to a special microstructure which is formed when the fraction of the amorphous component reaches the percolation threshold. Experimental data are presented and discussed.

INTRODUCTION

Superhard materials are usually defined as those whose hardness exceeds 4000 kg/mm^2 (about 40 GPa). This is significantly more than the hardness, H, of steels (H $\leq$ 1200 kg/mm^2) and of the conventional hard materials such as transition metal nitrides, carbides, sapphire and others [1]. Superhard materials include diamond (H = 7000-10 000 kg/mm^2), cubic boron nitride (c-BN, H $\leq$ 4 000 kg/mm^2) , transition metal superlattices (H $\leq$ 5 000 kg/mm^2) [2,3], amorphous boron carbide (H = 5 000 kg/mm^2) [4] and turbostratic CN$_x$ (x=0.25-0.35, H $\approx$ 5 000 kg/mm^2) [5,6]. Diamond and c-BN are high pressure modifications whose preparation requires rather extreme conditions. The deposition of transition metal superlattices requires a precise control of the repetive deposition of a large number of about $\leq$ 10 nm thin layers of different materials by periodically changing the sputtering source. All these problems made it so far difficult to introduce these materials into a large scale industrial application. The recently developed superhard composite materials consist of a few nanometer small crystallites of a transition metal nitride, nc-Me$_n$N, such as TiN [7,8], W$_2$N [9] or others [10-13], embedded within a 0.6-1 nm thin matrix of amorphous silicon nitride, a-Si$_3$N$_4$. These materials are thermodynamically stable, relatively easy to prepare by plasma chemical vapor deposition at a temperature of 550°C which is compatible with steel substrates, they are resistant against oxidation at high temperatures and chemically compatible with the majority of industrial substrates including steel. The hardness of such materials exceeds 5 000 kg/mm^2 [8-11] and approaches that of diamond [13]. Therefore the interest in these materials is growing fast. This paper aims to contribute to the understanding of the origin of the superhardness which strongly exceeds that given by the rule-of-mixture. It is shown, that the thermodynamically driven segregation and percolation threshold determines the formation of the microstructure which is decisive for the occurrence of the new phenomena of superhardness.

EXPERIMENTAL

The preparation procedure, control of the composition and characterization of the nc-Me$_n$N/a-Si$_3$N$_4$ films was described in [8-13]. Therefore only a very brief summary is given here. As the formation of the cn-Me$_n$N/a-Si$_3$N$_4$ nanostructure is based on a thermodynamically driven seg-

407

Mat. Res. Soc. Symp. Proc. Vol. 457 ° 1997 Materials Research Society

regation of the both phases [8] a preparation technique which provides a high chemical activity of nitrogen and a fast kinetics (activation energy) at a deposition temperatures below 700°C is needed. For reasons outlined in [7,8] we have chosen the plasma CVD using the respective volatile transition metal halides and silane as the sources of the metals and silicon, respectively. The deposition took place at a substrate temperature of 500-550°C under conditions of a large excess of hydrogen and nitrogen. The use of silane instead of silicon tetrachloride has the advantage of a much lower chlorine content of 0.2-0.3 at.%, whereas the films prepared by our colleagues [15] who used SiCl₄ contain typically 1-2 at.% of Cl which causes a film degradation upon a long term exposure to air.

The films were characterized by means of X-ray diffraction, XRD, scanning electron microscopy, SEM, combined with energy dispersive analysis of X-rays, EDX, X-ray photoelectron spectroscopy, XPS and hardness measurements [8-9]. More recently, high resolution transmission electron microscopy, HR TEM, fully confirmed the conclusions and predictions regarding the nanostructure, composition and crystalline size [16].

RESULTS AND DISCUSSION

The excellent agreement between the crystallite size determined from XRD and HR TEM for nc-TiN/a-Si₃N₄ films is shown in Fig. 1. The somewhat higher average crystallite size obtained from HR TEM is most probably due to a human factor, because one can easily overlook the small crystallites when evaluating the large number of micrographs showing direct lattice image of these crystallites. One notices that the crystallite size reaches a minimum at a silicon content corresponding to about 8 at.% of Si. Similar results were found also for nc-W₂N [9] and nc-VN/a-Si₃N₄ [10].

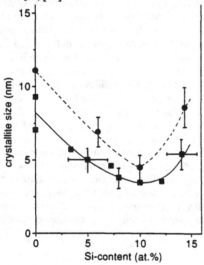

Fig. 1: Dependence of the average crystallite size, as determined from the HRTEM and XRD, on the silicon content in the films (see text).

When combining two materials which form a solid solution, such as TiN and TiC [10], the hardness H follows the rule-of-mixture, i.e. it increases monotonously according to equation (1) (see [10]).

$$H(TiN_{1-x}C_x) = [(1-x)H(TiN) + xH(TiC)] \tag{1}$$

In contrast, the hardness of the novel nanocrystalline composites nc-Me$_n$N/a-Si$_3$N$_4$ (Me being a transition metal which forms stable and hard nitrides) increases with increasing content of a-Si$_3$N$_4$ up to a maximum of about 16 - 20 mol.% of Si$_3$N$_4$, and decreases again when the fraction of silicon nitride further increases. Moreover, the maximum hardness exceeds that of both components by factor of 3 or more [7-13] thus indicating that new phenomena are responsible for this increase. This is illustrated by Fig. 2 for the nc-W$_2$N/a-Si$_3$N$_4$ films. Similar dependence was found also for nc-TiN/a-Si$_3$N$_4$ [8] and nc-VN/a-Si$_3$N$_4$ films [10]. One notices, that the crystallite size reaches a minimum and the hardness a maximum at a silicon content of about 7 at.% which corresponds to about 16 mol.% of Si$_3$N$_4$. It has been verified that the silicon in the films is incorporated as Si$_3$N$_4$ [8]. The other data shown in Fig. 2 will be discussed further below.

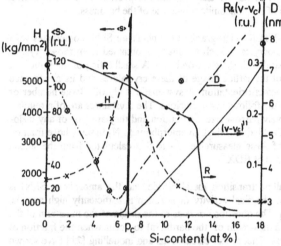

Fig. 2: Measured hardness (H) average crystallite size of W$_2$N (D), reflectivity of the films (R), the theoretically calculated average cluster size <S> and the scaling law for electrical conductivity (V-V$_C$)$^{1.1}$ vs. the silicon content of the nc-W$_2$N/a-Si$_3$N$_4$ films. (See text.)

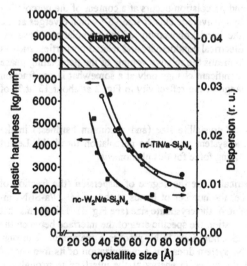

Fig. 3: Correlation between the measured hardness of nc-TiN/a-Si$_3$N$_4$ and nc-W$_2$N/a-Si$_3$N$_4$ and the crystallite size. Even a conservative extrapolation suggests that the hardness of diamond should be reached at a crystallite size of ≤ 2.5 nm. The open symbols show the fraction of atoms at the surfaces of the crystallites (see text).

The correlation of the hardness with the crystallite size D shown in Fig. 3 exhibits a strong increase when the crystallite size decreases below 5 nm. In contrast, the classical mechanism of hardening based on the Hall-Petch relationship $H(D) = H_0 + const.D^{-1/2}$ (for the theory see [17,18]), applies only up to a crystallite size $D \geq 10$ nm. Below that value a softening due to the lack of dislocation multiplication and pile-up, and due to a strong increase of the fraction of the softer material within the grain boundaries is found [19]. Also the epitaxial [2] and polycrystalline [3] heterostructures, suggested in a theoretical work of Koehler [20] show softening when the lattice period decreases below about 6-10 nm [2]. These results clearly show that new phenomena are responsible for the strong increase of the hardness in our nanocrystalline composite materials. In the following part of this paper we shall show that the nanocrystals are indeed free of any dislocations. Afterwards, we shall discuss the possible mechanism of the formation of the microstructure and of the concomitant increase of the hardness.

Figure 4 shows an example of an HR TEM micrograph of the nc-TiN/a-Si$_3$N$_4$ composites with a direct lattice image of several nanocrystals. Such an image is obtained from those crystals whose planes are oriented exactly parallel to the electron beam. A small tilting of the sample in the microscope leads to vanishing of the lattice image of these crystallites and its appearance from others which come into the correct orientation. A systematic study of a large number of such images [16] allows us to draw the following conclusions: The crystallites are of a regular, almost spherical shape with a relatively narrow size distribution and they are free of any dislocations. The TiN-crystallites are embedded into the thin amorphous Si$_3$N$_4$ tissue whose fraction corresponds - within the accuracy of these measurements - to that calculated from the silicon content which was measured by means of EDX.

The segregation of the nanocrystalline transition metal nitride and the amorphous Si$_3$N$_4$ is thermodynamically driven when the chemical activity of nitrogen is sufficiently high and the temperature below about 700°C [8]. The question arises as to the origin of the minimum of the crystallite size (and simultaneously a maximum of the hardness) to occur when the fraction of a-Si$_3$N$_4$ reaches about 16 - 20 mol.%. Theoretical calculations and modelling [21] have shown that in a three dimensional network the bond percolation occurs at a content of the percolating phase of 16 %. This is indicated in Fig. 2 by the average cluster size <S> which diverges at the percolation threshold. This means, that the bond connectivity of the percolating phase (a-Si$_3$N$_4$) occurs through the whole sample. Electrical conductivity and related properties, such as reflectivity, scale differently (see Fig. 2, V$_C$ means the relative volume of the percolating phase at the threshold). Therefore, they show a significant change only at a somewhat larger fraction of the percolating phase as seen by the change of the reflectivity in Fig. 2 at about 13 at.% of Si.

These data clearly show that the minimum crystallite size (and maximum hardness) in the Me$_n$N/a-Si$_3$N$_4$ composites occurs when the system reaches the percolation threshold. Let us now discuss the question of the possible driving force for this phenomena.

For regular crystallites of an almost spherical shape the degree of dispersion (the fraction of atoms at the surface of the crystallites) i.e. the number of bonds at the nc-Me$_n$N/a-Si$_3$N$_4$ increases proportionally to the reciprocal value of the crystallite size (see Fig. 3). This means that at the minimum crystallite size during percolation the specific area of the interface between the nanocrystalline transition metal nitride and amorphous Si$_3$N$_4$ reaches a maximum. The only logical explanation why the nc-Me$_n$/a-Si$_3$N$_4$ system does it is a minimization of its free energy in this case. This however requires that the cohesive energy at the interface is strongly increased due to some yet unknown effect. In [9] we have speculated if electronic phenomena related to quantum confinement could be responsible for such effect. However, a detailed theoretical treatment of problem is still lacking but highly desirable.

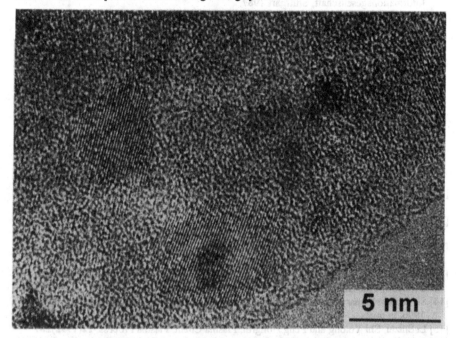

Fig. 4: A typical HRTEM image of a sample with 33 mole % of Si_3N_4 showing a direct image of the (111) and (200) lattice planes of those TiN nanocrystals which have the correct orientation with respect to the electron beam. The defect free nc-TiN crystallites of a fairly regular, almost spherical shape are imbedded in the amorphous a-Si_3N_4 matrix. (See text.)

CONCLUSIONS

The superhardness phenomena found in nc-Me_nN/a-Si_3N_4 composites is associated with the thermodynamically driven segregation accompanied by a percolation threshold. The experimental results and a logical reasoning show that some yet unknown phenomena cause a strong increase of the cohesive energy across the nc-Me_nN/a-Si_3N_4 interface. The data also indicate that the hardness of diamond should be reached or even superpassed if the crystallite size at the percolation would decrease to 2-2.5 nm. This is a great challenge for further more detailed experimental and theoretical studies.

ACKNOWLEDGMENT

We would like to thank Dr. S. Reiprich for the preparation of the samples. This work has been supported in part by the Deutsche Forschungsgemeinschaft.

RFERENCES

[1] H. Fischmeister and H. Jehn, Hartstoffschichten zur Verschleissminderung, DGM-Informationsgesellschaft, Stuttgart 1987.

[2] S.A. Barnett, in: Physics of Thin Films. Vol. 17, Mechanics and Dielectric Properties, eds. M.H. Francombe and J.L. Vossen, Academic Press, Boston 1993.

[3] W.D. Sproul, Int. Conf. on Metallurgical Coatings, San Diego, USA, April 22-26, 1996, in press.

[4] S. Veprek and M. Jurcik-Rajman, Proc. 7th Int. Symp. on Plasma Chem., Eindhoven 1985, p. 90.

[5] H. Sjöström, S. Stafström, M. Boman and J.-E. Sundgren, Phys. Rev. Lett. 75(1995)1336 (see also errata in 76(1996)2205).

[6] H. Sjöström, L. Hultman, J.-E. Sundgren, S.V. Hainsworth, T.F. Page and G.S.A.M. Theunissen, J. Vac. Sci. Technol. A 14(1996)56.

[7] S. Veprek, S. Reiprich and Li Shizhi, Appl. Phys. Lett. 66(1995).

[8] S. Veprek and S. Reiprich, Thin Solid Films 268(1996)64.

[9] S. Veprek, M. Haussmann and S. Reiprich, J. Vac. Sci. Technol. A 14(1996)46.

[10] S. Veprek, M. Haussmann and Li Shizhi, Proc. 13th Int. Conf. on CVD, Los Angeles 1996, Electrochemical Soc. Proc. Vol. 96-5(1996)619.

[11] S. Veprek, M. Haussmann, S. Reiprich, Li Shizhi and J. Dian, Proc. Int. Conf. on Metallurgical Coatings and Thin Films, San Diego, April 1996, in press.

[12] S. Veprek, Proc. of the 5th Int. Conf. on Plasma-Surface-Engineering, Garmisch-Partenkirchen, Germany, September 9-13, 1996, in press.

[13] S. Veprek, Proc. of the 10th Int. Conf. on Thin Films, Salamanca, Spain, September 23-27, 1996, Thin Solid Films, in press.

[14] S. Veprek, Plenary talk at the European Mater. Res. Soc., Strasbourg, France, June 3-7, 1996, in press.

[15] Li Shizhi, Shi Yulong and Peng Hongrui, Plasma Chem. Plasma Process. 12(1992)287

[16] S.Christiansen, M. Albrecht, H.P. Strunk and S. Veprek, J. Vac. Sci. Technol., submitted

[17] R.W. Hertzberg, Deformation and Fracture Mechanics of Engineering Materials, 3rd ed., Wiley, New York 1989.

[18] A. Kelly and N.H. MacMillan, Strong Solids, 3rd. ed., Clarendon, Oxford 1986.

[19] J.E. Carsley, J. Ning, W.W. Milligan, S.A. Hackney and E.C. Aifantis, NanoStructured Materials 5(1995)441.

[20] J.S. Koehler, Phys. Rev. B 2(1970)547.

[21] R. Zallen, The Physics of Amorphous Solids, John Willey, New York 1983.

CONSOLIDATION AND EVOLUTION OF PHYSICAL-MECHANICAL PROPERTIES OF NANOCOMPOSITE MATERIALS BASED ON HIGH-MELTING COMPOUNDS

R.A.ANDRIEVSKI[1], G.V.KALINNIKOV[1], and V.S.URBANOVICH[2].
[1]Institute for New Chemical Problems, Russian Academy of Sciences, Chernogolovka, Moscow Region, 142432, Russia
[2]Institute of Solid State Physics and Semiconductors, Belarus Academy of Sciences, Minsk, 220726, Belarus

ABSTRACT

Nanocomposite materials of TiN/TiB_2 system are prepared both by high pressure sintering of ultrafine powders and the films deposition using magnetron sputtering. Properties and structure of obtained particulate and film materials derived from hardness measurements as well as TEM and SEM analyses are discussed and compared in detail.

INTRODUCTION

In the past few years, the development of nanophase and nanocomposite materials (NM) has emerged as an important step in creating a new generation of materials because of the attractive properties observed in these NM. This is also typical for both particulate materials produced by numerous methods from ultrafine powders and films deposited by many versions of CVD/PVD technique. Electrodeposition and crystallization from amorphous state are also popular for NM preparation. Different consolidation methods of NM have been analyzed elsewhere [1-3]. It seems to be important to note that NM preparation by all methods is essentially accompanied by recrystallization and partial annihilation of nanocrystalline structure. In this connection only high-energy and low-temperature consolidation methods such as high pressure sintering, shock compaction, hot forging, and magnetron sputtering are considered to be the most useful for NM preparation.

In elaboration of our investigations [1, 3-7], this report is concentrated on TiN/TiB_2 nanocomposites prepared by high pressure sintering and magnetron sputtering. The TiN/TiB_2 system has been the focus of attention of many investigations (e.g., [8-20]) but only some of them, mainly as regarding films [10-13, 17] and partly consolidated bulks [15, 16, 18, 20], are devoted to NM. As Holleck has firstly noted this system forms coherent interfaces [10]. However, many features of this nanocomposite and others based on high-melting compounds seem to be not studied in detail.

RESULTS AND DISCUSSION

(a) Consolidated Bulks

The details and features of our experimental technique of high pressure sintering have previously been reported [4]. Table I shows the hardness and density values for different TiN/TiB_2 composites including individual compounds after high pressure sintering at 1200°C and 1400°C (P=4 GPa, t=5 min). Initial values of mean diameter of particles determined from specific surface area values were of ~80 nm (TiN) and ~400 nm (TiB_2). The first powder was prepared by plasmachemical synthesis [4] and the second one was from the H.C.Starck Co. These powders were mixed by ball milling.

Table I. Vickers Microhardness (H_v) and Density (γ) of TiN/TiB$_2$ Composites Consolidated at 1200°C and 1400°C

Composition (wt%)	T_{cons}=1200°C					T_{cons}=1400°C			
	γ (g/cm³)	H_v (GPa) at F=0.5N; T_{test}=20°C		H_v at F=5N		γ (g/cm³)	H_v at F=0.5 N ; T_{test}=20°C		H_v at F=0.5 N T_{test}=20°C
		TiN phase	TiB$_2$ phase	T_{test}=20°C	T_{test}=900°C		TiN phase	TiB$_2$ phase	
TiN ($d \sim 80$ nm)	5.2	25.3±1.1	—	14.5±0.5	7.1±0.2	5.24	27.8±1.2	—	23.1±1.8
75TiN/25TiB$_2$	5.2	22.2±1.0	(19.6)	17.0±1.2	7.2±0.7	5.25	26.1±1.2	28.4±1.6	23.3±1.5
50TiN/50TiB$_2$	4.77	22.3±0.8	27.1±1.0	18.0±1.8	9.0±1.2	4.80	25.2±1.2	28.5±1.4	23.0±0.8
25TiN/75TiB$_2$	4.55	21.7±1.3	28.4±1.2	18.8±1.85	9.6±0.5	4.59	25.1±1.0	32.2±1.3	24.5±0.8
TiB$_2$ ($d \sim 400$ nm)	4.31	—	25.6±2.0	18.2±1.9	9.1±0.85	4.33	—	31.5±3.2	24.3±2.4

Let us discuss these results. The porosity value of specimens was usually smaller than 1-3%. As it is evident the increase of preparation temperature is accompanied by small increase of density, and by more significant one of hardness. The later is revealed to a greater extent in the case of TiB_2 phase because of harder sinterability of these coarser powders. The difference in TiN and TiB_2 hardness can be brought out only at the low load (F) measurements. At $F=(2-5)$ N the influence of composition on hardness is not so high. The availability of hardness size effect is also noteworthy, the difference between measurements at F of 0.5 N and 2.0 N and especially between 0.5 N and 5.0 N is not small and is higher than the standard deviation. The hot hardness measurements have not revealed any anomalies such as superplasticity in the studied temperature interval up to 900°C.

Notice that the general level of hardness is quite high. Our results especially for TiN and TiN rich composites are much higher as compared with data for conventional cast, sintered and hotpressed specimens [8, 9, 14, 15, 19]. The features of hardness of nanocrystalline TiN have been discussed elsewhere [3, 6, 7]. It was found that the values of crystallites size (L) obtained by different methods were not uniform. So the L values were of 50-70 nm by XRD analysis, 70-150 nm by SEM examination (Fig. 1, a) and 20-80 nm by TEM (Fig. 2, b). Although slightly larger crystallites can also be observed in the micrographs [6, 7].

The structure of TiN/TiB_2 composites seems to be similar. XRD analysis has revealed the availability of only two phases; the values of their lattice parameters were essentially the same for all compositions. Optical microanalysis has indicated that the particles of TiB_2 have large distribution of sizes (from ~0.1 μm to 3-5 μm).

Turning to the analysis of Table I it seems to be important to keep in mind that increase of processing temperature has resulted in hardness increase in spite of possible recrystallization effect. It can be connected not only with a small density increase but also with "strengthening" of grain boundaries as we have pointed out [21] and also noted by Watanabe [22]. It seems very likely that a further reduction of TiB_2 powder size and an optimization of high pressure sintering parameters or other high-energy consolidation methods will further increase the hardness level and the foregoing data are not the limiting values.

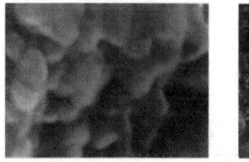 

a   b

Figure 1: FE-SEM micrograph of fracture surface (a) and bright field TEM micrograph (b) of TiN specimen obtained at $T=1200^{\circ}$C and $P=4$ GPa.

(b) Films

Table II shows our and other investigators data concerning to TiN/TiB$_2$ films. The numerous results of TiN hardness determination are not included in Table II but only the most representative ones are listed. It is well known that there are many factors, such as the preparation method, deposition parameters, films thickness, test load, substrate, type of hardness measurement and so on effect the films hardness. The whole discussion of boride, nitride and other interstitial phases films hardness has been published elsewhere [27].

It is difficult to discuss all features of Table II because of the absence of full information on structure (crystallite sizes and stress situation) and composition (phase analysis and admixture content). However, as can be seen from Table II, our results which have been obtained by using rather thin films are in reasonably good agreement with literature data. So the application of the procedure proposed by Jonsson and Hogmark [28] has permitted to estimate the hardness value of boride film in itself. These Vickers values were found to be equal of 50-60 GPa (these values are in round brackets) which are in close agreement with such information for "thick" boride films [12,13]. Our TEM, SEM, AES, and other examination of TiN/TiB$_2$ films obtained by DCMNRS and RFMNRS are now in progress and we are reporting some preliminary results.

Table II. Vickers Microhardness (H_v) of TiN/TiB$_2$ Films

Compound	Method[1]	Thickness (μm)	Load (N)	Substrate	H_v (GPa)	Source
Ti(B$_{0.8}$N$_{0.05}$O$_{0.1}$C$_{0.05}$)$_{2.5}$	DCMNRS	1.8	0.3	(100)TiB$_2$	44.7±4(~51)	[5]
Ti(B$_{0.69}$N$_{.024}$O$_{0.04}$C$_{0.03}$)$_{1.65}$	RFMNRS	2.4	0.3	Si	25±1.1(~37)	
Ti(B$_{0.34}$N$_{0.49}$O$_{0.12}$C$_{0.05}$)$_{1.5}$	RFMNRS	~1.0	0.3	Si	18.6±0.9(~49)	
Ti(B$_{0.23}$N$_{0.49}$O$_{0.16}$C$_{0.12}$)$_{2.1}$	RFMNRS	~0.6	0.3	Si	16.3±0.8(~57)	
TiB$_{\sim2}$	DCMNRS	5	0.5	WC/Co	34	[11]
TiB$_{\sim2.8}$	DCMNRS	4.1	0.15-0.3	WC/Co	51	[12]
TiB$_{\sim2}$	RFMNRS	5	0.25	Cr steel	37.2	[23]
TiB$_{\sim1.9}$	DCMNRS	~8	0.5	HSS	~62.5	[13]
TiN$_{0.96}$	DCMRS	5	0.01-0.03	SUS 304	~30	[24]
TiN$_x$	CAPD	5	0.5	HSS	~43	[25]
TiB$_{1.6}$N$_{0.6}$	DCMRS	4.9	0.2-0.3	WC/Co	57	[12]
TiB$_2$N$_{0.4}$C$_{0.4}$	DCMRS	4.9	0.2	WC/Co	50	[12]
TiN$_{\sim1.0}$(5-16at%B)	EBIP	~3	0.25	HSS	35-45	[26]
TiB$_{2.1}$N$_{0.6}$	DCMRS	5	0.5	HSS	41.5	[13]
TiBN$_{0.4}$	RFMNRS	1-3	(100 nm)[2]	HSS	55	[17]
TiB$_2$N$_{0.6}$	RFMNRS	1-3	(100 nm)	HSS	36	[17]
TiB$_{1.4}$N$_{0.5}$	RFMNRS	1-3	(100 nm)	HSS	48	[17]
TiB$_{0.7}$N$_{0.5}$	RFMNRS	1-3	(100 nm)	HSS	52	[17]
TiB$_{0.6}$N$_{0.5}$	RFMNRS	1-3	(100 nm)	HSS	27	[17]

[1] Methods of preparation: DCMNRS - d.c.magnetron non-reactive sputtering; DCMRS - d.c. magnetron reactive sputtering; RFMNRS - r.f.magnetron non-reactive sputtering; CAPD - cathodic arc plasma deposition; EBIP - electron beam ion plating. [2] A depth of nanoindenter penetration.

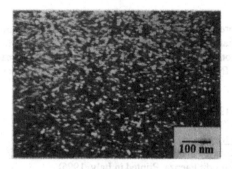

a   b

Figure 2: Dark field TEM micrograph (a) and corresponding selected area electron diffraction pattern (b) of Ti($B_{0.69}N_{0.24}O_{0.04}C_{0.03}$)$_{1.65}$ film obtained by RFMNRS (T_{substr} =150°C, N=0.5 kWt).

The specific difficulty in analysis of Ti(B,N)$_x$ films is in their characterization of chemical composition and phase. Because of highly non-equilibrium processing, the variety of the structures and correspondingly properties versions is significant larger in the case of films in comparison with bulk specimens (see Table I and II). We are indicating the approximate formulas of our boride films on the base of the AES analysis. It is also assumed that the one-phase composi-tion exists and all nonmetallic impurities substitute at the boron sublattice. It should be recorded that both overstoichiometric borides and superstoichiometric ones have been revealed in our experiments. Such situation has also been pointed in literature (e.g., [12, 13, 17, 26, 27]).

The dark field TEM micrograph of boride film obtained by RFMNRS is shown in Fig. 2, a. It can be seen that the L value ranges up to about 5-10 nm. The assignment of the diffraction rings (Fig. 2, b) has revealed the existence of the TiB$_2$ lattice (AlB$_2$ type) which is also confirmed by XRD analysis. According to the XRD and SAED data, the lattice parameters were found to be: a=0.3083±0.0024 nm and c=0.3246±0.0025 nm (XRD); a=0.311 nm and c=0.323 nm (SAED). These values are higher then conventional standard ones for TiB$_2$ (a=0.3028-0.3040 nm and c=0.3228-0.3234 nm) [29] that can likely be connected with films composition.

In the case of Ti($B_{0.36}N_{0.49}O_{0.12}C_{0.05}$)$_{1.5}$ film the L value was found to be of 3-6 nm by TEM examination. However, its phase composition is not so simple and need further refinement. There are some considerations on phase composition of Ti(B,N)$_x$ films [17]. It seems plausible that the further optimization of deposition regimes and composition can led to new interesting data. The special attention must be given to the effect of nature of crystallites boundaries and impurities.

SUMMARY

This study indicates that it is possible to prepare TiN/TiB$_2$ nanocomposites using different consolidation methods both for particulate materials and films. The parameters of high pressure sintering and PVD deposition exert great action on structure and properties of NM. At present the hardness level of particulate NM is lower than that for films. However, the hope to increase this level seems to be optimistic in both cases.

ACKNOWLEDGMENTS

The authors would like express their appreciation to Dr. II.V.Gridneva, Dr.S.I.Chugunova, Dr.I.I.Timofeeva, Dr.A.Bloyce, and Dr.D.Shtansky as well as Mrs. I.V.Goncharova, Mr.K.M.Ma, and Mr.O.A.Babij for their active cooperation. We also grateful to the Russian Basic Research Foundation (Project No 95-02-0318) and the INTAS Program (Project No 94-1291) for supports.

REFERENCES

1. R.A.Andrievski, J. Mater. Sci., **29**, 614 (1994)
2. C.Suryanarayana, Int.Mater.Rev., **40**, 41 (1995)
3. R.A.Andrievski, in Advances in Powder Metallurgy & Particulate Materials, (in press)
4. R.A.Andrievski, V.S.Urbanovich, N.P.Kobelev and V.M.Kuchinski, in FOURTH EURO CERAMICS, ed.A.Bellosi, Vol.4 (Gruppo Edit.Faenza, Printed in Italy, 1995)
5. R.A.Andrievski, G.V.Kalinnikov, A.F.Potafeev, M.A.Ponomarev and S.Yu.Sharivker, Inorg. Mater., **31**, 1395 (1995)
6. R.A.Andrievski, in Processing and Properties of Nanocrystalline Materials, eds. C.Suryanarayana, J.Singh, and F.H.Froes (TMS, Warrendale, 1996)
7. R.A.Andrievski, Nanosrtuct. Mater., 1997, accepted for publication
8. V.D.Chupov, V.I.Unrod, and S.S.Ordanyan, Powder Metall (in Russ.), N1, 62 (1981)
9. Yu.G.Tkachenko, S.S.Ordanyan, D.Z.Yurchenko, V.K.Yulyugin, and D.V.Chupov, ibid, N2, 70 (1983)
10. H.Holleck, J. Vac. Sci. Technol., **A4**, 2661 (1986)
11. H.Holleck and H.Schulz, Surf. Coat. Techn., **36**, 707 (1988)
12. C.Mitterer, M.Rauter, and P.Rodhammer, Surf. Coat.Techn., **41**, 351 (1990)
13. O.Knotek and F.Loffler, J. Hard Mater., **3**, 29(1992)
14 T.Giaziani, C.Melaudri, and A.Bellossi, ibid, **4**, 29 (1993)
15 M.Nowakowski, K.Su, L.Sneddon, and D.Bonnel, in Nanophase and Nanocomposite Materials , eds.S.Komarneni, J.Parker, and G.Thomas, Vol.286 (MRS, Pittsburgh, 1993)
16. V.Szabo, M.Ruhle, K.Su, L.G.Sneddon, and D.Bonnel, ibid, p.443
17. P.Hammer, A.Steiner, R.Villa, M.Baker, P.N.Gibson, J.Haupt, and W.Gissler, Surf. Coat. Techn., **68/69**, 194 (1994)
18. J.Y.Dai, Y.G.Wang, D.X.Li, and H.Q.Ye, Phil. Mag., **70A**, 905 (1994)
19. M.Moriyama, H.Aoki, and K.Kamata, J. Ceram. Soc. Jap., **103**, 844 (1995)
20. G.Zhang, Ceram. Int., **21**, 29 (1995)
21. R.A.Andrievski, G.V.Kalinnikov, A.F.Potafeev, and V.S.Urbanovich, Nanostr. Mater., **6**, 353 (1995)
22. T.Watanabe, in Mechanical Properties and Deformation Behaviour of Materials Having Ultra-Fine Microstructure, eds. M.Nastasi, D.M.Parkin, and H.Gleiter (Kluwer Acad. Publ., Dodrecht, 1993)
23. W.Herr, B.Matthes, E.Broszeit, and K.H.Kloos, ibid, **A140**, 616 (1991)
24. W.C.Oliver, C.J.McHargue, and S.J.Zinkle, Thin Solid Films, **153**, 185 (1987)
25. R.A.Andrievski, I.A.Belokon', and Z.Kh.Fuksman, in High Performance Ceramic Films and Coatings, ed.P.Vincenzini (Elsevier, Amsterdam, 1991)
26. L.S.Wen, X.Z.Chen, Q.Q.Yang, Y.Q.Zheng, Y.Z.Chuang, Surf. Eng., **6**, 41 (1990)
27. R.A.Andrievski, Russ. Chem. Rev., 1997, accepted for publication
28. B.Jonsson and S.Hogmark, Thin Solid Films, **114**, 257 (1984)
29. J.L.Murray, P.K.Liao, and K.E.Spear, Bull. Alloy Phase Diagr., **7**, 550 (1986)

NANOCOMPOSITE THIN FILMS OF TRANSITION METAL CARBIDES FABRICATED USING PULSED LASER DEPOSITION

W. F. BROCK*, J. E. KRZANOWSKI**, AND R. E. LEUCHTNER*
*Physics Department and **Mechanical Engineering Department, University of New Hampshire, Durham, NH 03824

ABSTRACT

A study has been conducted on the pulsed laser deposition (PLD) of transition metal carbides in order to examine alloying and phase formation in binary systems. Alternating layers of TiC/ZrC and ZrC/VC were deposited at 400 C and 5 mTorr Ar with nominal period thicknesses of 0.6 nm, 10.0 nm, and 50.0 nm. ZrC/VC x-ray diffraction analysis showed that the alloys were amorphous and the TiC/ZrC alloys were crystalline. The thicker films showed a higher degree of phase separation of the two compounds. Transmission electron microscopy confirmed the amorphous structure in the 0.6 nm ZrC/VC film, while the 50.0 nm film showed a layered structure and extremely fine grain size.

INTRODUCTION

Transition metal carbide coatings have long been used to improve the operating life of mechanical components due to their relatively low friction coefficients and high hardness [1]. Presently, the most common carbide coating material is titanium carbide. However, despite its attractive properties, the hardness level of TiC (30 GPa) is still well below that of other competitive materials, such as diamond (80 GPa). Improving the hardness levels of carbides can be achieved by engineering nanophase composites based on transition metal carbide mixtures.

Films in this study were deposited using pulsed-laser deposition (PLD), a thin film deposition technique that enables direct replication of target compositions in the deposited films [2]. Therefore, carbide films can be deposited directly from carbide targets without the use of a reacting gas. Furthermore, sub-monolayer coverage is produced on each laser shot and this makes PLD ideal for producing nanocomposite mixed carbide films.

In this study, nano-composite thin films have been fabricated using the TiC/ZrC and ZrC/VC systems. The lattice mismatch for TiC/ZrC is 8%, and is believed to be completely soluble at high temperatures but undergoes spinodal decomposition at lower temperatures [3]. The mismatch for ZrC/VC is 12%, and these carbides are expected to be largely insoluble [4]. However, the films deposited by PLD are condensed far from equilibrium so the microstructures obtained and the nature of the phases present require further investigation. In this paper we report on the microstructures obtained and their relation to film deposition conditions.

EXPERIMENTAL

PLD of carbides was conducted in a high-vacuum system containing a heated substrate holder and computer controlled target carousel. The target-substrate distance was 4.6 cm. A KrF excimer laser beam (248 nm, 30 nsec FWHM, 9 Hz) was focused through a 0.5 m focal length lens onto the target. Laser energy densities ranged from 7.3 - 10.6 J/cm^2. The typical base pressure for the ablation chamber was 5×10^{-6} torr; this pressure was increased to 5×10^{-3} torr with ultra-pure argon gas during ablation. Films were deposited directly onto <111> Si, glass and sapphire (a-plane) substrates at 400 C. The gas flow rate was held at 3.0 sccm and the pressure adjusted by throttling the pump speed. The target carousel held two half-circular target segments composed of high-purity (99.9%), hot-pressed ZrC, VC, or TiC. This system enabled deposition of individual layers with a defined bilayer thickness in each sample. Films were deposited using target combinations of TiC/ZrC and VC/ZrC by oscillating the targets under the laser beam for a specified number of laser shots to produce the desired thickness for each layer. Table I shows a

Table I: Film deposition conditions.

Sample	Substrate Temperature (°C)	Background Ar Pressure (mTorr)	Individual Layer Thickness (nm)
TiC/ZrC	400	5	0.3
TiC/ZrC	400	5	5.0
TiC/ZrC	400	5	25.0
ZrC/VC	400	5	0.3
ZrC/VC	400	5	5.0
ZrC/VC	400	5	25.0

summary of the deposition conditions for the films. Layer thicknesses per shot were calibrated through the use of atomic force microscopy (AFM) on masked deposited films, as well as quartz crystal microbalance measurements. This yielded a deposition rate on the order of 0.1 Å per shot at an energy density of 8.3 J/cm².

Transmission electron microscopy (TEM) was used to study the grain structure of films using both plan-view and cross-section specimens. X-ray diffraction (XRD) using CuKα radiation was used for phase and crystal orientation analysis of the films.

RESULTS
X-Ray Diffraction

X-ray diffraction was carried out on the films listed in Table I. Figure 1a shows an XRD spectrum of a ZrC film grown at 400 C and 5 mTorr Ar. The only film peak observed was the <111> ZrC reflection at 33.09° (2θ) with all other reflections corresponding to the sapphire substrate. When ZrC was combined with TiC in 0.3 nm layers (Figure 1b) there was a strong reflection at 34.09°, falling

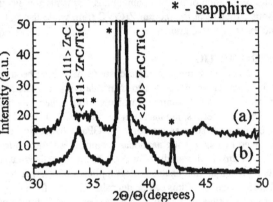

Figure 1: XRD patterns of (a) pure ZrC and (b) mixed ZrC/TiC alloy with 0.3nm layer periods.

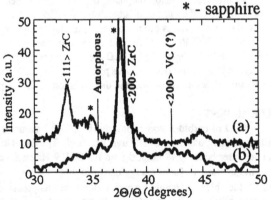

Figure 2: XRD patterns of (a) pure ZrC and (b) mixed ZrC/VC alloy with 0.3nm layer periods.

approximately half way between the expected <111> reflections for ZrC (33.04°-2θ) and TiC

(35.91° -2θ). The presence of a single reflection indicates a TiC/ZrC solid solution alloy formed. Figure 1b also shows a peak at 39.79° (2θ) corresponding to the <200> ZrC-TiC alloy peak. Therefore, a solid solution alloy formed, despite the deposition of alternating 0.3 nm layers, due to atomic mixing or surface diffusion processes.

When ZrC was combined with VC in 0.3 nm layers (Figure 2b), there was an amorphous background reflection surrounding the large sapphire peak, suggesting the formation of an amorphous alloy. A broad peak near the <200> ZrC reflection was also observed, indicating the possible presence of an ultra-fine crystalline precipitate. It is believed the amorphous structure of the ZrC/VC film occurs due to the larger lattice mismatch (12%) compared to the TiC/ZrC system.

XRD spectra for the 5.0 nm (Figure 3a,b) and 25.0 nm (Figure 4a,b) layered films on <111> Si showed a higher degree of crystallinity with phase separation for the TiC/ZrC <111> and <200> reflections appearing in the 5 nm sample and complete phase separa-

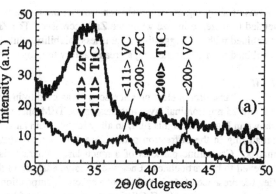

Figure 3: XRD patterns of (a) TiC/ZrC and (b) mixed ZrC/VC alloy with 5.0 nm layer periods.

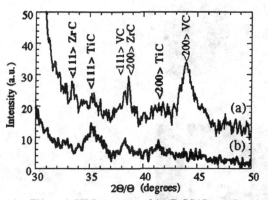

Figure 4: XRD patterns of (a) ZrC/VC and (b) mixed TiC/ZrC alloy with 25.0 nm layer periods.

tion occurring by 25.0 nm. In contrast the amorphous reflection for ZrC/VC disappeared in the 5.0 nm layered film as indicated by the partial phase separation of <111> VC at 37.1° (2θ) and <200> ZrC at 38.1° (2θ). At 25.0 nm layer observed thickness, phase separation was complete and the two peaks at angles of 37.34° (<111> VC) and 38.34° (<200> ZrC) were observed. The thickness of the individual layers directly affects the separation of the crystal phases. The <111> ZrC reflection was present but lower in intensity than the <200> ZrC reflection. The peaks in all of the films appeared to be very broad suggesting a small grain structure. Table II shows the

Table II: Peak shifts due to alloying and calculated lattice constants in 5.0 nm/layer films.

Nominal layer material	Alloying compound:	Lattice Constant Change
ZrC	VC	0.4656nm (0.6% decrease)
ZrC	TiC	0.4569nm (2.5% decrease)
TiC	ZrC	0.4410nm (2.1% increase)
VC	ZrC	0.4240nm (1.4% increase)
TiC/ZrC alloy	-	0.4590nm (1.8% increase)

effects of alloy ing on the lattice constants of the layers in the 5.0 nm/layer films. TiC and VC tended to increase in size while the ZrC decreased. The ZrC had a 0.6% decrease in size when combined with VC suggesting there is less solubility, while ZrC had a 2.5% decrease in size when combined with TiC, reflecting a greater ability to alloy.

TEM

Transmission electron microscopy was used to investigate the ZrC/VC films with the 0.3 nm and 25 nm nominal layer thicknesses. A TEM plan-view image of a ZrC/VC 0.3 nm film is shown in Figure 5. In this film, an alloy formed with an amorphous matrix, as shown by the first broad ring in the diffraction pattern. Small (~ 7 nm) crystalline particles were observed in the matrix. Crystalline particles of a much larger characteristic size were also observed and are believed to be particulates. Energy-dispersive analysis of the film was also conducted and verified the presence of Zr, V and C, but a specific composition analysis was not conducted. A TEM cross-section image of a ZrC/VC multilayer (25 nm) film is shown in Figure 6. This film was also deposited on Si (111) at 400oC/ 5 mTorr Ar, but with an increased dwell time on each target.

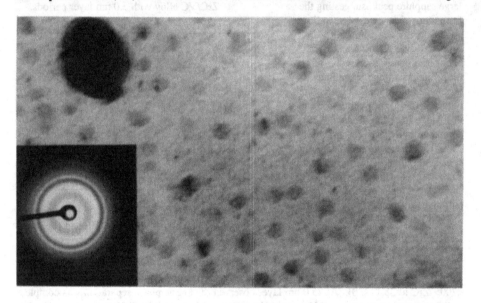

30 nm

Figure 5: A bright-field TEM plan-view image of the ZrC/VC film fabricated by depositing 0.3 nm layers. An amorphous matrix is observed with small (~ 7 nm) crystalline particles in the matrix. The larger crystalline particles are believed to be particulates.

A layered structure can be seen, with a bilayer period of about 75 nm, slightly more than the targeted bilayer thickness of 50 nm. The grain structure shows an ultra-fine grain size of about 6 nm, but a columnar or fibrous structure is clearly absent. The density of the film also appears to be high, with no inter-columnar voids observed. It is believed that the high film density is a result of the highly energetic nature of the ablated species in the PLD process.

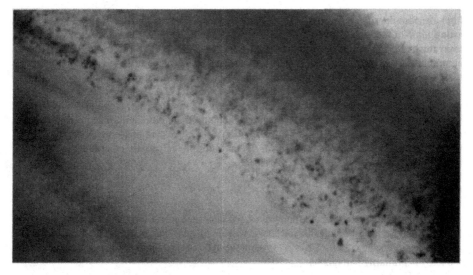

Fig. 6: TEM cross-section image of the ZrC/VC multilayer film deposited with a nominal layer thickness of 25 nm. A layered structure can be seen, with a bilayer period of 75 nm. The grain structure shows an ultra-fine grain size of about 6 nm, but a columnar or fibrous structure is clearly absent.

DISCUSSION

Deposition of films in this study was carried out using separate targets of each carbide. This suggests that distinct carbide phases should result in the absence of any diffusion or atomic mixing processes. The deposition temperature of 400 C represents a homologous temperature of 0.2Tm for TiC, with similar values for ZrC and VC. Consideration of the diffusion of C in Ti (D ~ 10-13 cm²/s [5]) gives a diffusion distance of 144 nm after 1800 seconds. Therefore, the deposition temperature does not allow for significant long-range diffusion. However, the PLD process produces a significant fraction of highly energetic depositing species [6] of energies on the order of 200-400 eV. The high energies of these species can enhance atomic mixing and surface diffusion processes at the depositing film, which contrasts with other deposition processes where most of the depositing species are at thermal energies.

Deposition of the 0.3 nm TiC/ZrC film resulted in a solid solution, indicating that at this layer thickness the diffusion/mixing processes allow for complete intermixing. At larger layer thicknesses, there may be some alloying at the interface, but the x-ray results, for the example at 5 nm layer thickness, suggests that distinct layers form. Analysis of these results are complicated by the fact that the x-ray patterns show broad peaks, and unfortunately this effect could be due to either the fine grain size or some alloying. In the ZrC/VC films, the 0.3 nm layer thickness sample resulted in an amorphous structure. Since the ZrC/VC compounds are insoluble, diffusion must occur in order to allow phase separation into the two crystalline phases. Since diffusion is limited, the film is deposited with an amorphous structure. If this film was heated to transform to the equilibrium state, both crystallization and phase separation need to occur. Low-temperature phase separation of this amorphous structure would necessarily result in an extremely high interface density and energy. The process of crystallization of this film is therefore an interesting topic of study and is currently under investigation. ZrC/VC films deposited with larger layer thickness are crystalline, with ultra-fine grain structures as shown in the TEM results. No amorphous layers were observed, even at the interface, which suggests that the layers are distinct with appropriate compound compositions. However, measured lattice constants for each layer suggest more alloying; for example the shift in lattice constant of 0.6% for the ZrC layer suggests, using a rule of mixtures, a VC content of 13%, clearly well beyond the solubility limit. It is not yet clear whether this effect is due to alloying beyond the solubility limit, or another effect such as film stress. Nonetheless, these results show that an interesting and unique variety of film microstructures can be deposited from mixed transition metal carbide compositions using the PLD technique.

CONCLUSIONS

1. Mixed transition metal carbide films have been deposited in the form of alternating layers by PLD. At the limit of individual layer thicknesses being near that of an atomic layer thickness, alloy formation occurs. In the TiC/ZrC system, the alloy is crystalline, while in the ZrC/VC case the alloy is primarily amorphous.

2. Increasing the individual layer thickness in the ZrC/VC films to 5 or 25 nm results in crystalline film formation which supports the concept of compositionally distinct layers. X-ray diffraction results for the TiC/ZrC films shows some phase separation at 5 nm layer thickness, but broadening of peaks due to the ultrafine grain size limits the analysis.

3. Transmission electron microscopy analysis verified the amorphous nature of the ZrC/VC 0.3 nm film, and revealed the presence of a fine crystalline precipitate within the amorphous matrix. Examination of the ZrC/VC 25 nm film showed an ultrafine grain structure and the absence of a columnar or voided structure.

REFERENCES

1. Toth, Louis E. "Transition Metal Carbides and Nitrides", Academic Press, New York (1971).
2. Pulsed Laser Deposition of Thin Films, D.B. Chrisey and G.H. Hubler, eds. Wiley & Sons, 1994.
3. O. Knotek and A. Barimani, *Thin Solid Films*, v. 174, pp. 51-56 (1989).
4. W .B. Pearson, A Handbook of Lattice Spacings and Structures of Metals and Alloys, Pergamon Press, New York (1958).
5. Smithells Metals Reference Book, Eric A. Brandes, Ed., pp. 13-23, Buttersworth, London (1983).
6. Pulsed Laser Deposition, D. Chrisey & G. Hubler, eds., Wiley 1994.

PREPARATION AND CHARACTERIZATION OF NANOCOMPOSITE COMPOSED OF TiO2 AS ACTIVE MATRIX

T. Sasaki*, R. Rozbicki**, Y. Matsumoto***, N. Koshizaki*, S. Terauchi* and H. Umehara*
*National Institute of Materials and Chemical Research (NIMC), Agency of Industrial Science and Technology, MITI, 1-1 Higashi, Tsukuba, Ibaraki 305, Japan.
**College of Engineering, Boston University, MA 02215
***Department of Applied Chemistry and Biochemistry, Faculty of Engineering, Kumamoto University, 2-39-1 Kurokami, Kumamoto 860, Japan.

ABSTRACT

Pt/TiO2 nanocomposite films were deposited on quartz glass and ITO glass substrates by the co-sputtering method. As-deposited composite films were amorphous and content of Pt in the films could be easily controlled by the amount Pt wire placed on the TiO2 target. Pt/Ti atomic ratio in the nanocomposite increased as the length of Pt wire on the TiO2 target increased. It was determined by XPS that the chemical states of Pt in as-deposited nanocomposites were Pt metal, Pt-O-Ti and PtO2, which were dependent on the Pt/Ti atomic ratio in the nanocomposite. The size of Pt nanoparticles in the composite films increased as the temperature of heat-treatment and Pt/Ti atomic ratio in the composite films increased. Pt nanoparticles in the nanocomposite films inhibited grain growth of TiO2 during heat-treatment.

INTRODUCTION

Nanocomposites doped with semiconductors or metal have been extensively studied because of their unique optical properties. It is well known that Au/SiO_2, $Au/BaTiO_3$ and CdS/SiO_2 have high third-order nonlinear susceptibility $\chi^{(3)}$ [1-3]. Nanocomposites of Si/SiO_2, C/SiO_2 and Ge/SiO_2 have photoluminescence in the visible light wavelength range [4-6]. Recently, new optical functionality of nanocomposites has also been developed, i.e., CoO/SiO_2 nanocomposite changes its optical transmittance by ambient gases such as NO [7]. These unique optical and/or chemical properties result from quantum size effects of the embedded nanoparticles in the matrix, and interface and/or surface effects between nanoparticles and the matrix.

Typically, silicon dioxide (SiO_2) is used as the main matrix material for these optical functional nanocomposites, because it is an inert material which is transparent in the wavelength range of visible light. In these optical nanocomposites, the role of the matrix is solely to support and protect the nanoparticles and the functionality of the matrix is seldom utilized. Titanium dioxide (TiO_2) has photoactivities, which have been used previously for photocatalyst and photoelectrode applications. The use of TiO_2 as the photo-active matrix in the nanocomposite can create a new type of functional nanocomposite.

Pt/TiO2 composite is used as a photocatalyst and/or photoelectrode for the decomposition of water [8,9]. This composite is mainly prepared from TiO_2 powder and a H_2PtCl_6 solution [8]. In this paper we report on the preparation and characterization of Pt/TiO2 nanocomposites using the r.f. magnetron co-sputtering method, and demonstrate the effect of heat treatment on the structure of Pt/TiO2 nanocomposite.

EXPERIMENT

Pt/TiO2 nanocomposite films were deposited on quartz glass and ITO glass (Kinoene Opto. Industry) substrates by the co-sputtering method using an r.f. sputtering apparatus (Shimadzu HSR-521). Platinum wires of 0.5 mm in diameter were placed symmetrically on a hot-pressed rutile

425

titanium dioxide target of 100 mm in diameter. All of the sputter deposition was done under a constant pressure of 0.53 Pa in argon with 100 W in sputtering power at room temperature. The length of Pt wires on the TiO_2 target was changed to obtain nanocomposite films with various Pt content. Sputtering time was 120 minutes for all the samples unless otherwise stated. The thickness of as-deposited nanocomposite films measured from the edge profile by a surface roughness meter (Tokyo Seimitsu, Surfcom-555A) ranged from 350 to 800 nm. All the samples were heated in air at 300 - 700 °C for the purpose of crystallization, since the as-deposited films prepared at room temperature were always amorphous. The structures of the films after the heat-treatment were examined by X-ray diffraction analysis using CuKα radiation (Rigaku, RAD-C). Nano-scale inhomogeneity of the films was analyzed by TEM observation using JEOL, JEM2000-FXII. The chemical states of Ti and Pt, and the Pt/Ti atomic ratio in the nanocomposites were examined by X-ray photoelectron spectroscopy (PHI, 5600ci).

RESULTS AND DISCUSSION

Figure 1 shows the Pt/Ti atomic ratio in the deposited nanocomposite films as a function of Pt wire length placed on the TiO_2 target. The Pt/Ti atomic ratio was calculated from Ti2p and Pt4f peak areas in X-ray photoelectron spectra of nanocomposite films. The Pt/Ti atomic ratio in the deposited nanocomposite films increased with an increase in Pt wire length, suggesting that the content of Pt in the deposited Pt/TiO_2 nanocomposite can be easily controlled by the amount of Pt wire placed on the TiO_2 target. The Pt/Ti atomic ratio of 0.18 corresponds to 30 wt% of Pt in the

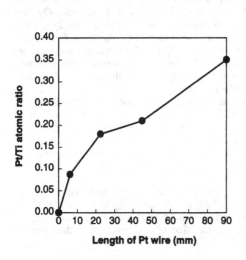

H——H
10 nm

Fig. 1 Pt/Ti atomic ratio in Pt/TiO_2 nanocomposite films as a function of Pt wire length on sputtering target of TiO_2.

Fig. 2 TEM photograph of as-deposited Pt/ TiO_2 nanocomposite film with a Pt/Ti atomic ratio of 0.18.

Pt/TiO$_2$ nanocomposite. This value is much larger than that of Pt/TiO$_2$ composite for photocatalyst, where the percentage of Pt is usually less than 15 wt% [10]. Pt/TiO$_2$ photocatalyst is mainly prepared by the heat-treatment of TiO$_2$ powder immersed in a H$_2$PtCl$_6$ solution or by UV irradiation on TiO$_2$ suspended in a H$_2$PtCl$_6$ solution [8]. In these cases, Pt is deposited only on the surface of TiO$_2$ powder, which can be covered with Pt metal at high Pt content in the composite, resulting in the lowering of photocatalytic activity. Figure 2 shows a typical TEM photograph of as-deposited Pt/TiO$_2$ nanocomposite film with a Pt/Ti atomic ratio of 0.35. As we can see, there are nanoparticles with a diameter from 1 to 2 nm, indicating that as-deposited Pt/TiO$_2$ nanocomposite films have an inhomogeneous nanostructure even at relatively high content of Pt.

Figure 3 shows XPS spectra of the Pt4f$_{5/2}$ and Pt4f$_{7/2}$ levels in as-deposited Pt/TiO$_2$ nanocomposite films. These peak positions in binding energy apparently changed among the nanocomposite films with different Pt/Ti atomic ratios. Table I shows XPS Pt4f$_{7/2}$ binding energies for several Pt species [11]. As shown in Fig. 3, Pt4f$_{7/2}$ binding energies of the nanocomposite

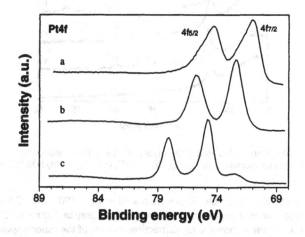

Fig. 3 XPS spectra of Pt4f level in as-deposited Pt/TiO$_2$ nanocomposite films with various Pt/Ti atomic ratios; (a)0.35, (b)0.21, (c) 0.18.

Table I   XPS Pt4f$_{7/2}$ binding energies for various Pt species.

Pt species	Pt4f$_{7/2}$ binding energy / eV
Pt	71.0
Pt-O-Ti	72.0
PtO	74.2
PtO$_2$	74.9

films with Pt/Ti atomic ratios of 0.35 and 0.18 were 71.1 eV and 74.8 eV, respectively. These data indicate that the chemical states of Pt nanoparticles in the nanocomposite films with Pt/Ti atomic ratio of 0.35 and 0.18 were almost always Pt metal and PtO$_2$, respectively. In contrast, Pt4f$_{7/2}$ binding energy of the nanocomposite film with Pt/Ti atomic ratio of 0.21 was 72.4 eV, which is close not to Pt metal, PtO and PtO$_2$ but to Pt-O-Ti, indicating that some complex oxide species (Pt-O-Ti) is present in the nanocomposite film. Ti2p$_{3/2}$ binding energy in as-deposited nanocomposite films was observed in an energy range from 458.1 to 458.6 eV which is very close to the Ti2p$_{3/2}$

binding energy of TiO_2 (458.7 eV) [12], indicating that the chemical state of Ti in as-deposited films was almost always TiO_2 with a slight reduction during the sputter deposition process in argon atmosphere.

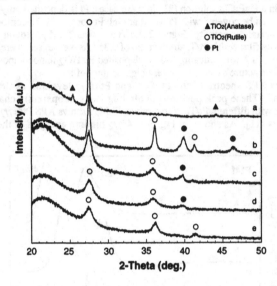

Fig. 4 XRD patterns of heated TiO_2 (a) and Pt/TiO_2 (b-e) nanocomposite films at various temperatures in air; (a), (b) 700 °C, (c) 600 °C, (d) 500 °C, (e) 400 °C.

As-deposited Pt/TiO_2 composite films were heated in air at 300 - 700 °C for the purpose of crystallization, since the as-deposited nanocomposite films prepared at room temperature were always amorphous. Figure 4 shows X-ray diffraction patterns of the nanocomposite films with a Pt/Ti atomic ratio of 0.18 heated at various temperatures. Here, an XRD pattern of heated TiO_2 films at 700 °C is also apparent. The crystallization of TiO_2 in the nanocomposite film was observed after heat-treatment at 400°C and above, while crystallization of Pt metal was observed after heat-treatment at 500°C and above, as shown in Fig.4. Although an anatase phase was observed for heated TiO_2 films, only a rutile phase was observed for heated nanocomposite films, indicating that the Pt nanoparticles in the nanocomposite films may inhibit the formation of anatase type TiO_2. In addition, the crystallite sizes of TiO_2 in heated TiO_2 film and the nanocomposite film with a Pt/Ti atomic ratio of 0.18 at 700 °C were estimated from the peak width to be 30 and 20 nm, respectively. This result indicates that Pt nanoparticles in the nanocomposite films may inhibit not only the formation of anatase type TiO_2, but also grain growth of TiO_2. Thus, nanoparticles can control the crystallization process of matrix phase.

Figure 5 shows TEM photographs of the heated nanocomposite films with a Pt/Ti atomic ratio of 0.18 at 500 and 600 °C. As shown in this figure, nanoparticles with a diameter of 5 to 10 nm were observed in heated nanocomposite film. These nanoparticles increased in size with an increase in temperature. According to the XPS analysis for the heated nanocomposite films with a Pt/Ti atomic ratio of 0.18 at 500 °C, the chemical states of Pt in the composite film were Pt metal and Pt-O-Ti. After heat-treatment at above 600 °C, the chemical state of Pt in these nanocomposite films was changed to Pt metal. In addition, Pt metal nanoparticle size rapidly increased at such

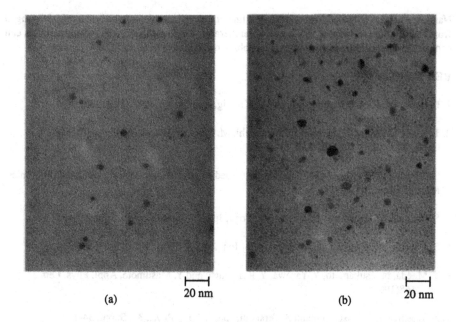

(a) 20 nm   (b) 20 nm

Fig. 5 TEM photographs of heated Pt/TiO$_2$ nanocomposite films with a Pt/Ti atomic ratio of 0.18 at (a) 500°C and (b) 600°C.

temperatures. It is well known that PtO$_2$ decomposes into Pt metal at 585 °C [13]. Thus, the Pt metal particles rapidly grow via decomposition of PtO$_2$ at temperatures higher than 600 °C. In heated nanocomposite films with a Pt/Ti atomic ratio of 0.35 at 600 °C, the nanoparticle size ranged from 10 to 100 nm. The size of these Pt nanoparticles in the heated nanocomposite films increased with an increase in the Pt/Ti atomic ratio of the films.

Pt/TiO$_2$ nanocomposite films can be easily prepared by the r.f. magnetron co-sputtering method. The chemical state of Pt and nanoparticle size in the composite films were strongly dependent upon the Pt/Ti atomic ratio in the nanocomposite films and heat-treatment temperature. A study is currently being made on the capabilities of Pt/TiO$_2$ nanocomposite film prepared by this method as a photoelectrode.

CONCLUSIONS

Pt/TiO$_2$ nanocomposite films were deposited on quartz glass and ITO glass substrates by the co-sputtering method. As-deposited composite films were amorphous and the content of Pt in the films was dependent on the amount of Pt wire placed on the TiO$_2$ target. The Pt/Ti atomic ratio in the nanocomposite increased with an increase in the length of Pt wire on the TiO$_2$ target. It was determined by XPS analysis that the chemical states of Pt in as-deposited nanocomposites were Pt metal, Pt-O-Ti and PtO$_2$, which were dependent on the Pt/Ti atomic ratio in the nanocomposite. These chemical states of Pt changed to Pt metal after heat-treatment at temperatures higher than 600 °C, irrespective of Pt/Ti atomic ratio. The size of Pt nanoparticles in the composite films increased as the temperature of heat-treatment and Pt/Ti atomic ratio in the composite films increased.

Nanoparticle size in heated nanocomposite film with a Pt/Ti atomic ratio of 0.18 at 600 °C ranged from 5 to 10 nm, and rapidly increased at temperatures higher than 600 °C. Nanoparticles can control the crystallization process of matrix phase of TiO_2.

REFERENCES

1. F. Hache, D. Ricard, C. Flytzanis and U. Kreigig, Appl. Phys., **A47**, 347(1988).

2. T. Kineri, M Mori, K. Kadono, T. Sakaguchi, M. Miya, H. Wakabayashi and T. Tsuchiya, J. Ceram. Soc. Jpn., **101**, 1340(1993).

3. S. Ohtsuka, K. Koyama and S. Tanaka, Extended Abstracta Inter. Sympo. Nonlinear Photonics Mater., Tokyo, 207 (1994).

4. S. Yoshida, T. Hanada, S. Tanabe and N. Soga, Jpn. J. Appl. Phys., **35**, 2694(1996).

5. S. Hayashi, M. Kataoka and K. Yamamoto, Jpn. J. Appl. Phys., **32**, L274(1993).

6. Y. Maeda, N. Tsukamoto, Y. Yazawa, Y. Kanemitsu and Y. Masumoto, Appl. Phys. Lett. , **59**, 3168(1991).

7. N. Koshizaki, K. Yasumoto and S. Terauchi, Jpn. J. Appl. Phys., **34**, Suppl. 34-1, 119(1994).

8. S. Tabata, H. Nishida, Y. Masaki and K. Tabata, Catalysis Lett., **34**, 245(1995).

9. L. Avalle, E. Santos, E. Leiva and V. Macagno, Thin Solid Films, **219**, 7(1992).

10. M. Takahashi, K. Mita and H. Toyuki, J. Mater. Sci., **24**, 243(1989).

11. M. Davidson, G. Hofiund, L. Niinisto and H. Laitinen, J. Electroanal. Chem., 228, 471(1987).

12. S. Perrin and C. Bardolle, J. Microsc. Spectrosc. Electron, **1**,175(1976).

13. C. L. McDaniel, J. Solid State Chem., **9**,139(1974).

HETEROGENEOUS NANOCOMPOSITE MATERIALS BASED ON LIQUID CRYSTALS AND POROUS MEDIA

G.P. SINHA and F.M. ALIEV
Department of Physics and Materials Research Center, PO BOX 23343, University of Puerto Rico, San Juan, PR 00931-3343, USA

ABSTRACT

An effective way of preparing a variety of liquid crystal based nanocomposite materials is to disperse LC in porous media with different porous matrix structure, pore size and shape. We present the results of investigations of quasiequilibrium and dynamical properties of nematic and smectic liquid crystals (LC) dispersed in porous matrices with randomly oriented, interconnected pores (porous glasses) and parallel cylindrical pores (Anopore membranes) by light scattering, photon correlation and dielectric spectroscopies. Confining LC to nanoscale level leads to quantitative changes in physical properties and appearance of new behavior which does not exist in either of the components. Relaxation of director fluctuations which is characterized by single relaxation time in the bulk LC are transformed to a process with a spectrum of relaxation times in pores, which includes extremely slow dynamics typical for glass formers. Existence of developed interface in these materials leads to new dielectric properties such as an appearance of a low frequency relaxation of the polarization and modification of dipole rotation.

INTRODUCTION

Heterogeneous systems with one of the component as liquid crystal (LC) form a new class of anisotropic materials that have been intensively studied during the last decade [1-9]. These systems are anisotropic (at least at short scales) and heterogeneous materials characterized by a very developed interface. A variety of new properties and phenomena were discovered and studied [1,2] in liquid crystals based heterogeneous materials. The most commonly known materials of this kind are polymer dispersed liquid crystals [1] in which liquid crystal is confined in organic polymer matrix. This material has become extremely important for both applications and fundamental physics of confined systems. The understanding of new phenomena arising in confined liquid crystals is not only interesting for applications, but also for the fundamental physics of interfaces and finite systems. Although there has been great success in the investigations of the physical properties of confined fluids and liquid crystals, there are still open questions in the understanding of the influence of confinement on the dynamical behavior anisotropic liquids.

We performed dynamic light scattering and dielectric measurements in nematic LC confined in porous matrices. The experiments show significant changes in physical properties of confined LC and suggest that there is some evidence for glass-like dynamical behavior, although bulk liquid crystal does not have glassy properties in both anisotropic and isotropic phases. This paper is a continuation of our previous research on nanocomposite materials based on nonorganic porous matrices and complex organic fluids [3] such as polymers [4] and liquid crystals [3,5-8].

MATERIALS AND EXPERIMENT

We have investigated pentylcyanobiphenyl (5CB) confined in porous glasses with interconnected randomly oriented pores, and in Anopore membranes with parallel cylindrical pores. The sizes of the pores in the porous glasses were 10 nm and 100 nm. The diameters of the pores in the Anopore membranes were 20 nm and 200 nm. These matrices were impregnated with 5CB. The phase transition temperatures of bulk 5CB are T_{CN}=295 K and

T_{NI}=308.27 K. For LC dispersed in porous matrices using static light scattering experiments we obtained that the nematic - isotropic phase transition in 100 nm random pores, 20 nm and 200 nm cylindrical pores is smeared out, the transition is not as sharp as in the bulk LC, it occupies a finite temperature region, and the temperature of this transition is depressed compared to the bulk value. We determine nematic- isotropic phase transition temperatures for 5CB in pores as: 307.5 K (100 nm random pores), 307.6 K (200 nm cylindrical pores) and 307.0 K (20 nm cylindrical pores). In 10 nm pores we did not observe well defined phase transition from nematic to isotropic phase.

We performed photon correlation measurements using a 6328 Å He-Ne laser and the ALV-5000/Fast Digital Multiple Tau Correlator (real time) operating over delay times from 12.5 ns up to 10^3 s with the Thorn EMI 9130/100B03 photomultiplier and the ALV pream-plifier. The depolarized component of scattered light were investigated. In the dynamic light scattering experiment, one measures the intensity-intensity autocorrelation function $g_2(t) = \langle I(t)I(0)\rangle/\langle I(0)\rangle^2$. This function $g_2(t)$ is related to dynamic structure factor f(q,t) of the sample by

$$g_2(t) = 1 + kf(q,t)^2, \tag{1}$$

where k is a contrast factor that determines the signal-to-noise ratio and $q = 4\pi n \sin(\Theta/2)/\lambda$, ($n$ is the refractive index, and Θ - the scattering angle). All dynamic light scattering data we discuss below were obtained at the scattering angle Θ=30°.

Measurements of the real (ϵ') and the imaginary (ϵ'') parts of the complex dielectric permittivity in the frequency range 0.1 Hz - 3 MHz were carried out at different tempera-tures using a computer controlled Schlumberger Technologies 1260 Impedance/Gain-Phase Analyzer. For measurements in the frequency range 1MHz - 1.5GHz we used HP 4291A RF Impedance Analyzer with a calibrated HP 16453A Dielectric Material Test Fixture. The analysis of data from dielectric experiment shows that in our materials dielectric relaxation cannot be described by Debye equation and a suitable description is provided by Cole and Cole formula for a system which has more than one relaxational process:

$$\epsilon^* = \epsilon_\infty + \sum_{j=1}(\epsilon_{js} - \epsilon_\infty)/(1 + i2\pi f\tau_j)^{1-\alpha_j} - i\sigma/2\pi\epsilon_0 f^n, \tag{2}$$

where ϵ_∞ is the high-frequency limit of the permittivity, ϵ_{js} the low-frequency limit, τ_j the mean relaxation time, and j the number of the relaxational process. The term $i\sigma/2\pi\epsilon_0 f^n$ takes into account the contribution of conductivity σ and n is a fitting parameter $(n \simeq 1)$.

DYNAMIC LIGHT SCATTERING

Dynamic light scattering in bulk nematic liquid crystals is very well understood, and the main contribution to the intensity of scattered light is due to the director fluctuations. In the single elastic constant approximation, if we assume that the six Leslie coefficients have the same order of magnitude and are $\sim \eta$ (η is an average viscosity), then the relaxation time is $\tau = \eta/Kq^2$ and is of order of magnitude $\sim 10^{-5}s$. The corresponding decay function is exponential: $f(q,t) = a \cdot \exp(-t/\tau)$. This is illustrated by curve (1) in Fig. 1 which represents intensity/intensity autocorrelation function for bulk nematic multidomain 5CB.

The difference between the dynamic behavior of bulk 5CB in nematic phase and 5CB dispersed in porous media can be seen by comparing curves (1), (2), (3) and (4) in Fig. 1. Slow relaxational process which does not exist in the bulk LC and a broad spectrum of relaxation times $(10^{-8} - 10)s$ appear for 5CB in both random and cylindrical pores if LC is in anisotropic phase. It is clear from Fig. 1 that the relaxation processes in 5CB confined in the both matrices are highly nonexponential. We are not able to find the correlation function (or a superposition of correlation functions) which would satisfactorily and quantitatively describe the whole experimental data from $t = 10^{-4}ms$ up to $t = 10^6ms$. However we found that in the time interval $10^{-3}ms - 10^3ms$ (6 decades on the time scale) the decay function :

$$f(q,t) = a \cdot exp(-t/\tau_1) + (1 - a) \cdot exp(-x^z), \tag{3}$$

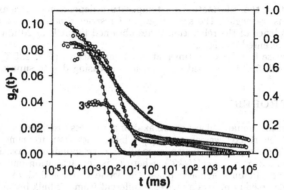

Figure 1: Intensity/intensity autocorrelation functions for 5CB: (1) - bulk 5CB, T=306.2 K (right scale), (2) - in 10 nm random pores, T=295.8 K (left scale), (3) - in 20 nm cylindrical pores, T=306.9 K, (4) - in 200 nm cylindrical pores, T=307.3K (left scale). Opened circles-experimental data, solid lines-fitting.

where $x = ln(t/\tau_0)/ln(\tau_2/\tau_0)$, and in our case $\tau_0 = 10^{-8}s$ provides suitable fitting. For 5CB in 10 nm random pores the second term in relationship (3) dominates, whereas for 20 nm pores the contribution from the first term is much more visible. The fitting parameters corresponding to curves (2), (3) and (4) in Fig. 1 are: (2) - z=2.3, τ_2=0.04 ms (exponential decay is neglected); (3) - τ_1=0.07 ms, z=3.6, τ_2=3s; (4) - τ_1=0.06 ms, z=4, τ_2=23s.

We found that the relaxation time of the slow process for 5CB in 10 nm pores strongly increases when temperature decreases from 300 K up to 270 K varying from 1.7×10^{-4} s to 14 s. The data analysis shows that the temperature dependence of the relaxation times for 5CB in 10 nm random pores, in the temperature interval (270-300) K, follows the Vogel-Fulcher law: $\tau = \tau_0 exp(B/(T - T_0))$ with parameters: $\tau_0 = 1.4 \cdot 10^{-11}s$, $B = 847K$ and $T_0 = 246K$. The dynamical behavior of 5CB confined in cylindrical and large random pores (100 nm)

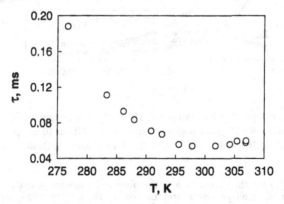

Figure 2: 5CB in 20 nm cylindrical pores. Temperature dependence of relaxation times of director fluctuations.

was closer to the bulk behavior as we expected. In these matrices the decay due to director fluctuations dominates. However a slow decay, origin of which is under question appears in the anisotropic phase of LC (Fig.1).

In order to obtain the information on temperature dependence of relaxation times of director fluctuations we neglect the slow decay and restrict time range by $t < 0.1$ s. The temperature dependence of the relaxation times obtained for 5CB in 20 nm pores by using this procedure is presented in Fig. 2.

Noticeable increase in relaxation times at low temperatures $(T < 295K)$ corresponding to solid phase in bulk 5 CB is caused by viscosity increasing due to supercooling of LC in pores.

DIELECTRIC PROPERTIES

For bulk 5CB in the nematic phase one relaxational process associated with the molecular rotation around it's short axes was observed [9]. The characteristic frequency of this process is in MHz range. The temperature dependence of the corresponding relaxation times obeys the Arrhenius formula: $\tau = \tau_0 exp(U/kT)$, where U is the activation energy and k is the Boltzmann constant. In bulk nematic phase the value of U_b is 0.61 eV. The dielectric behavior of 5CB dispersed in porous matrices is different from it's bulk behavior. In confined 5CB we observe at least three identified relaxational processes. The first low frequency process (1 Hz - 10 KHz) which does not exist in the bulk is illustrated in Fig. 3 for 5CB in 10 nm pores (curves 1 and 1a) and in 100 nm pores (curves 2 and 2a). The second process is very clearly seen in the MHz frequency range and the last one, which is less visible appears in the frequency range $f > 30$MHz. Both processes are shown in Fig. 3, curves 1b, 1c (10 nm pores) and 2b, 2c (100 nm pores). The solid lines in Fig.3 represent the results

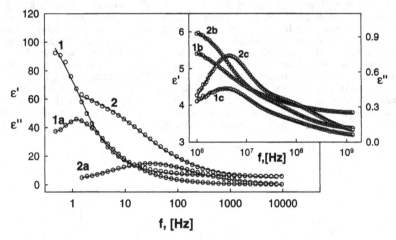

Figure 3: Frequency dependence of real (ϵ' (1, 1b, 2, 2b) and imaginary (ϵ'' (1a,2a, 1c, 2c) of dielectric permittivity of LC - porous glass composition. 5CB in 10 nm pores: 1, 1a at 13°C and 1b, 1c at 23°C. 5CB in 100 nm pores: 2, 2a at 19°C and 2b, 2c at 23°C. Open symbols - experimental data, solid lines - fitting.

of using formula (2) for the description of the observed dielectric spectra. The parameters describing these relaxation processes are: curves (1, 1a) - $\tau_1 = 0.13s$, α_1=0.2; (1b, 1c) - $\tau_2 = 4.7 \times 10^{-8}s$, α_2=0.2, $\tau_3 = 2.2 \times 10^{-9}s$, α_3=0.4; (2, 2a) - $\tau_1 = 0.01s$, α_1=0.3; (2b, 2c) - $\tau_2 = 3.8 \times 10^{-8}s$, α_2=0.1, $\tau_3 = 1.7 \times 10^{-9}s$, α_3=0.3. Note that the low frequency relaxational processes presented in Fig. 3 correspond to T = 13.0 °C and 19.0 °C, which are below the bulk crystallization temperature. We observed that all the dielectrically active modes were not completely frozen even at temperatures about 20 degrees below the bulk crystallization temperature,. This property is very different from the behavior expected in the solid phase. The data analysis shows [6] that the temperature dependencies of these relaxation times in both random pores in the temperature interval (275-295) K for 100 nm pores and in (275-305) K for 10 nm pores follow the Vogel-Fulcher law with parameters: $\tau_0 = 1.7 \cdot 10^{-9}$ s, $B = 1240K$, $T_0 = 212K$ for 5CB in 100 nm pores and $\tau_0 = 1.2 \cdot 10^{-5}$ s, $B = 627K$, $T_0 = 220K$ for 5CB in 10 nm pores. The fact that relaxation times of the first process are strongly temperature dependent and there exists a spectrum of relaxation times suggests that the first relaxational process is probably not related to low frequency dispersion given by the Maxwell-Wagner mechanism. Possibly at low frequencies we observe the relaxation of interfacial polarization not due to the Maxwell-Wagner effect but rather due to the formation of a surface layer with polar ordering on the pore wall. In this case a new cooperative and slow process may arise.

The relaxational process in the MHz range with $\tau \sim 10^{-8}s$ is bulk-like and corresponds to the rotation of the molecule around the short axis. In cylindrical pores at frequencies $f > 1MHz$ the main contribution to observed dielectric relaxation is due to molecular rotation around short axis, and the process with $\tau \sim 10^{-10}s$ was less visible [7] than in random pores. The temperature dependence of relaxation times corresponding to the rotation of molecules around short axis for 5CB in 10 nm random pores is presented in Fig. 4. This dependence for

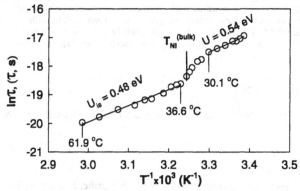

Figure 4: 5CB in 100 nm random pores. Dielectric relaxation times corresponding to molecular rotation around short axis as function of inverse temperature. Symbols - experimental data, solid lines - fitting.

5CB in pores is different from that in the bulk nematic phase. From Fig. 4 we see that there is no indication of a sharp or well identified nematic-isotropic phase transition. Instead we observe a gradual change of relaxation times in a wide temperature range 36.5 °C< T <30.0 °C. The temperature range T < 36.5°C, where the temperature dependence of $ln\tau$ deviates from linear dependence, corresponds to the anisotropic phase and $ln\tau$ is not a linear function of $1/T$ in this range. However if we consider the temperature regions T < 30.0°C and $T > 36.5°C$ separately then $ln\tau = f(1/T)$ in these regions is reasonably well approximated by a linear function and the corresponding activation energies are $U = 0.54eV$ and $U_{is} = 0.48eV$. The first activation energy U is less than the activation energy of bulk nematic phase $U_b = 0.61eV$. This fact could be considered as an evidence for smectic type order formation at pore wall - LC interface in this temperature range. Qualitatively for 5CB in cylindrical pores the temperature dependence of relaxation times corresponding to molecular rotation around short axis is close to that in 10 nm pores. For 5CB in 100 nm random pores we found that the temperature range (T < 34.5°C) corresponds to the anisotropic phase of 5CB and $ln\tau$ is not a linear function of $1/T$. Again if we consider the temperature regions 34.5°C < T < 20°C and 19.5°C < T < 9°C separately then $ln\tau = f(1/T)$ in each of these regions could be reasonably well approximated by a linear function and the corresponding activation energies were found to be $U_1 = 0.74eV$ and $U_2 = 0.53eV$. The first activation energy U_1 is greater than the activation energy of bulk nematic phase but $U_2 < U_b$. We attribute the temperature range $34.5°C < T < 20°C$ to nematic phase. The activation energy in pores in nematic phase is greater because the pore wall imposes additional potential due to pore wall - molecule interaction. This potential is 0.13 eV $(2 \cdot 10^{-13} erg)$, and taking into account that number of molecules per unit area is $(2-3) \cdot 10^{14} cm^{-2}$ we estimate surface potential of molecule-wall interaction $U_{surf} \sim 50 erg/cm^2$. The fact that $U_2 < U_1$ at the temperatures below 19.5°C is due the same reason as in 10 nm pores - the formation of smectic type order in this temperature range. The process with $\tau \sim 10^{-10}s$ could be related to the oscillation of long molecular axis around the director.

CONCLUSION

We have shown that heterogeneous nanocomposite materials based on nanoporous dielectric matrices and liquid crystal have new properties appear. Each of the components of the composition separately does not have these properties. The photon correlation and dielectric experiments show significant changes in the physical properties of liquid crystals confined in porous media. We found that the relaxational processes in confined LC are highly non-exponential and they are not frozen even about $20°C$ below bulk crystallization temperature. The temperature dependence of relaxation times of the slow process is described Vogel-Fulcher law which is characteristic of glass-like behavior. The differences in dynamical behavior of confined LC from that in the bulk mainly are due to finite-size effects and the existence of developed pore wall - liquid crystal interface and the structure of pores is less important.

ACKNOWLEDGEMENTS

This work was supported by US Air Force grant F49620-95-1-0520 and NSF grant OSR-9452893.

REFERENCES

1. P.S. Drzaic, Liquid Crystal Dispersions, (World Scientific, Singapore, 1995).
2. G.P. Crawford and S. Zumer, Liquid crystals in complex geometries, Taylor & Francis, London, 1996).
3. F.M. Aliev, in Access in Nanoporous Materials, edited by T.J. Pinnavaia and M.F. Thorpe, (Plenum Press, New York, 1995), pp. 335-354.
4. F.M. Aliev in: Advances in Porous Materials, edited by S. Komarneni, D.M. Smith, and J.S. Beck (Mater. Res. Soc. Proc. 371, Pittsburgh, PA 1995), p. 471-476.
5. F.M. Aliev and V.V. Nadtotchi in: Disordered Materials and Interfaces, edited by H.Z. Cummins, D.J. Durian, D.L. Johnson, and H.E. Stanley, (Mater. Res. Soc. Proc., 407, Pittsburgh, PA, 1996), p. 125-130.
6. F.M. Aliev and G.P. Sinha in: Electrically based Microstructural Characterization, edited by R.A. Gerhardt, S.R. Taylor, and E.J. Garboczi (Mater. Res. Soc. Proc. 411, Pittsburgh, PA 1996), p. 413-418.
7. F.M. Aliev and G.P. Sinha in: Liquid Crystals for Advanced Technologies, edited by T.J. Bunning, S.H. Chen, W. Hawthorne, T. Kajiyama, N. Kolde (Mater. Res. Soc. Proc. 425, Pittsburgh, PA 1996), p. 305-310.
8. F.M. Aliev and G.P. Sinha in: Microporous and Macroporous Materials, edited by R.F. Lobo, J.S. Beck, S.L. Suib, D.R. Corbin, M.E. Davis, L.E. Iton, and S.I. Zones (Mater. Res. Soc. Proc. 431, Pittsburgh, PA 1996), p. 505-510.
9. P.G. Cummins, D.A. Danmur, and D.A. Laidler, MCLC 30, p. 109 (1975).

Part VI

Organic–Inorganic and Sol–Gel Nanocomposites

CHARACTERIZATION OF NANOSIZED SILICON PREPARED BY MECHANICAL ATTRITION FOR HIGH REFRACTIVE INDEX NANOCOMPOSITES

DORAB E. BHAGWAGAR, PETER WISNIECKI AND FOTIOS PAPADIMITRAKOPOULOS*

Department of Chemistry, Institute of Materials Science, University of Connecticut, Storrs, CT 06279, papadim@mail.ims.uconn.edu

* To whom correspondence should be addressed

ABSTRACT

High pressure nanomilling provides an inexpensive, environmentally conscious method to fabricate large quantities of nanoparticles. The presence of large particle sizes, inherent in mechanical attrition processes pose obstacles in identifying the optical properties of these nanosized particles. The high refractive index and relatively small absorption coefficient of silicon (Si) directed our research efforts towards Si nanoparticles. We presently report simple separation procedure which allows us to utilize a range of tools to characterize and exploit properties in the *nano* size range. Employing these Si nanoparticles, high refractive index nanocomposites in gelatin were fabricated with values as high as 3.2.

INTRODUCTION

The observation of visible luminescence from passivated Si, an indirect band-gap semiconductor, in the nano crystalline size regime of less than 10 nm has generated tremendous attention in the field of optoelectronics.[1] Our interests in applications of nanosized silicon has been more elementary in nature. Composites of high refractive index can find diverse usage in novel photonic applications. The deagglomeration and uniform incorporation of nanoparticle filler in a polymer matrix could yield an effective and rigorous approach for increasing the refractive index of the resultant nanocomposites. With the filler dimensions approximately an order of magnitude less than the wavelength of visible light, losses due to scattering should be limited. Materials suitable as nanosized fillers must themselves posses high refractive index, as well as low absorption coefficients. Past studies[2-4] have successfully demonstrated the increase in refractive index with chemically synthesized PbS nanoparticles embedded in polymer matrices, either by spin coating or pelletization. The refractive index of the composite was found to vary linearly with the volume fraction of the PbS. Additionally for PbS, the particle size limit where the refractive index approaches the bulk material value appears to be ca. 25 nm.[4] Theoretical arguments also point to significant variations of refractive indices in this size range.[5]

Initial attempts have been underway in our laboratory to utilize Si as the high refractive index nanoparticle additive in fabricating high refractive index nanocomposites. Over the visible wavelength, crystalline Si has amongst the highest refractive indices, while it has a lower absorption coefficient than PbS.[6] As a result of its wide application as semiconductor material, the electronic properties and surface chemistries of bulk Si have been extensively investigated. Moreover, the experience and knowledge gained in separation and chemical modification can be applied to other potential applications of Si nanoparticles.

Si nanoparticles can be synthesized by methods such as controlled pyrolysis of silane[7] and etching of single crystal wafers to porous Si,[8] yielding dimensions as small as 2 to 5 nm. Rather than these sophisticated chemical synthesis routes, we have favored high pressure nanomilling to produce nanosized powder in large quantities by a relatively benign procedure.[9] This paper presents some of our initial observations on the separation and characterization of the milled Si particles. The dispersion of these nanoparticles in gelatin results in refractive indices as high as 3.2.

EXPERIMENTAL

Polycrystalline Si powder of nominal purity of 99.5% and 325 mesh size was procured from Alfa Aesar. Milling was performed in a round-ended hardened steel vial with two 1/2" and four 1/4" hardened steel balls provided by Spex Sample Prep. 2.0 g of Si powder were loaded into the vial while inside a glove box to maintain an inert nitrogen atmosphere during milling. A high energy Spex 8000 shaker mill was used for the ball milling.

Transmission Electron Microscopy (TEM) observations on a Phillips EM 300 at 80 keV consisted of placing drops of Si powder sonnicated in ethanol on carbon coated copper grids. Particle size distribution on dilute solutions by Dynamic Light Scattering (DLS) was obtained on a Nicomp 370 Submicron Particle Sizer. A Nicomp distribution analysis was used for fitting the autocorrelation function.[10] X-ray diffraction (XRD) was performed on a Norelco/Phillips diffractometer using CuKα radiation ($\lambda=1.5418$Å). UV-Vis spectroscopy was recorded on a Perkin-Elmer Lambda Array 3840 in quartz cuvettes from ethanol solutions. In all experiments a Fisher Scientific FS9 Ultrasonic Cleaning bath was used for dispersing the solutions.

Refractive indices were measured on a J.A. Woollam Co. Variable Angle Spectroscopic Ellipsometer (VASE). Data was collected at wavelengths between 4000 and 10000 Å with a 100 Å interval, and at angles between 65 and 80°. The Si nanoparticles were mixed with a gelatin polymer (Eastman Kodak) in a water/ethanol solution and spun at 750-1200 rpm on precleaned Si wafers. The films were annealed at 150°C for 4 hrs in a vacuum oven prior to measurement.

RESULTS AND DISCUSSION

I. Characterization of Si Nanoparticles

Due to its technological importance, nanosized Si has been fabricated by a variety of methods. Recently, it has been shown that the crystal to nanocrystalline transformation for semiconductor elements such as Si can be attained by ball milling.[11] Milling was performed for 4 and 5 hours and the black powder samples so obtained were labeled Si(4h) and Si(5h), respectively. Employing mild sonnication by an ultrasonic cleaning bath resulted in stable dispersions in a host of organic solvents such as THF, ethanol, and water. The solutions were dark black in color and remained suspended for long duration of time (days). Both TEM and Dynamic Light Scattering (DLS) showed a wide distribution in particle sizes. Micrographs, such as shown in Figure 1A, depict the as-milled Si containing particles well in excess of 1000 nm. This was also substantiated by DLS results. For most optoelectronic and optical application which rely on the unique properties of nanosized dimensions the need to remove higher size fractions is imperative.

A number of methods such as filtration and centrifugation were attempted in this regard. After repeated exercises in experimental trial and error, the most efficient technique to separate these particles was found to be centrifugation. By optimizing the solvent, speed, and centrifugation time, large quantities of Si powder could be readily separated. A typical procedure was to first suspend the Si powder in ethanol employing an ultrasonic cleaning bath for ca. 12 hours. The suspension was then centrifuged at 3000 rpm for 90 minutes. The supernatant was clearly distinguishable and could be easily decanted off. The most obvious difference between the centrifuged Si solution and the as-milled solution was the color. Whereas the milled Si suspension is black, after centrifuging and removal of the larger particles, the Si solution is deep orange in color. Such changes in the color as a function of particle size have been observed with a number of direct and indirect band-gap semiconductors.[1,12] The TEM micrographs (Figure 1B) clearly show the absence of any large size disparities in the centrifuged Si.

The particle size distribution by DLS shown in Figure 2 provides quantitative assessment of particle dimension. Milling at 5 hrs. followed by centrifuging shows a smaller average diameter at 21 nm than the corresponding 4 hrs milling at 38 nm. All samples examined also showed a small fraction of particles in the 130 to 150 nm size range. This is believed to be due a degree of agglomeration that cannot be avoided even at dilute concentrations. Such agglomeration is probably intrinsically present in nanoparticle dispersions.[7,12] Indeed, passage through a 100 nm microfilter does not eliminate the small fraction of particles at 130 to 150 nm.

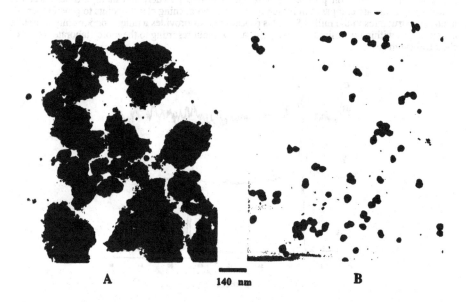

Figure 1: Bright-field Transmission Electron Micrographs of Si(4h) (A) as-milled, and (B) after separation by centrifugation.

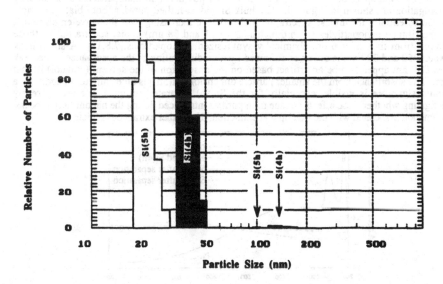

Figure 2: Number average particle size distribution obtained by Dynamic Light Scattering on centrifuged Si(4h) and Si(5h) samples.

The X-ray diffraction profile for bulk Si, as-milled powder, and nanoparticles after the centrifuging cycle are compared in Figure 3. The line broadening clearly points to procurement of nanosized structures in the milled Si. This procedure thus provides a unique opportunity to obtain nanoparticles within the desired size range,[4] and without resorting to the more elaborate routes of chemical synthesis.

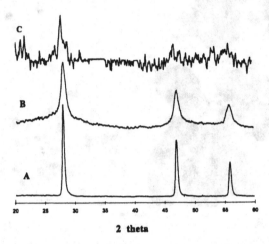

2 theta

Figure 3: Radial X-ray diffraction profiles of (A) bulk Si, (B) as-milled sample Si(5h), and (C) Si(5h) after separation by centrifugation.

The UV-Vis spectrum of the bulk Si, milled Si powder, and nanoparticles separated by centrifugation are shown in Figure 4. The bulk Si and as-milled powder (both black in color) show an essentially featureless spectra, with absorbance increasing over the wavelength range. The centrifuged nanoparticles (with dimensions of 21 and 38 nm) depict spectra unlike those observed from the under 6 nm chemically synthesized nanoparticles.[7,8,13] A sharp peak between 250 and 300 nm is observed followed by a gradual tailing of the absorbance. The peak below 250 nm appears to be an artifact based on the absorption of the solvent (ethanol) in that range. The absorbance from the smaller particle dimension [Si(5h)] is blue shifted. In addition, a vibronic signature is visible, especially for the Si(5h) separated sample. We are currently investigating whether these effects are due to impurities introduced during the nanomilling process, phenomenon associated with the Mie optical extinction,[1] or other extraneous effects.

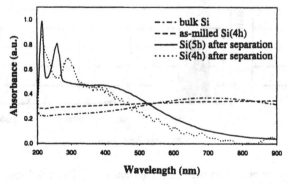

Wavelength (nm)

Figure 4: UV-Vis absorbance spectra of different Si samples.

II. Refractive Index of Si Nanocomposites

Preliminary results on the applicability of the separated Si nanoparticles as inorganic additives in increasing the refractive indices of the matrix polymers were obtained by ellipsometry. A roughly 50/50 (w/w) mixture of centrifuged Si(5h) and gelatin polymer was spin coated on polished Si wafers. The experimental ellipsometry parameters psi (ψ) and delta (Δ) over the entire wavelength range of 4000 to 10000 Å and angular range from 65 to 80° were numerically fit by the refractive index (n) and absorption coefficient (k). The best fit results depicted in Figure 5 show a gradual decrease in refractive index over the wavelength range from about 3.2 to 2.5. The insert shows the same refractive index of the nanocomposite compared to that for bulk Si and the gelatin polymer. The graph clearly shows that the index of refraction of the composite is an additive mixture of the indices of its two components. Past experiments,[2-4] have shown the refractive index of the nanocomposite as the volume fraction average of the constituent phases ($n = n_1 v_1 + n_2 v_2$, where n_i and v_i are the refractive indices and volume fractions, respectively). X-ray Photoelectron Spectroscopy (XPS) and Raman measurements are underway to verify the exact composition of the spin cast films.

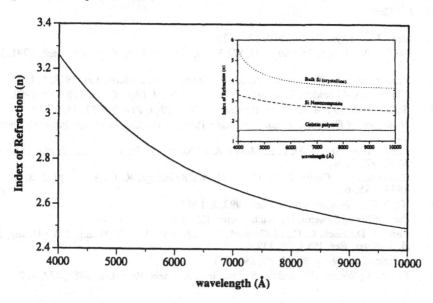

Figure 5: Index of refraction of a roughly 50/50 (w/w) gelatin/Si(5h) nanoparticles film prepared by spin coating. Insert compares the nanocomposite with bulk crystalline Si and gelatin, respectively.

CONCLUDING REMARKS

Si nanoparticles in the 25 to 50 nm range can be fabricated and subsequently isolated by mechanical attrition in large quantities. Upon separation of the large particles, TEM and DLS clearly show a relatively narrow distribution of particle sizes. These particles posses all the characteristics associated with the nanosized range, and provide a useful starting material for producing high refractive index nanocomposites. Initial ellipsometric results indicate an increase in the refractive index of spin coated films on addition of centrifuged Si nanoparticles. The refractive index over all wavelength appears to be an average of the refractive index of the components.

Future work is directed towards gaining a better understanding of the optical and absorbance properties of the milled Si nanoparticles. Considerable effort is also being expended towards the thorough characterization of these and similar type of nanocomposites for high refractive index and other photonic applications.

ACKNOWLEDGMENTS

The authors would like to thank Lamia Khairallah for helping with the TEM micrographs, Tom Fabian for assistance with ellipsometry measurements, and Prof. Faquir Jain for many useful discussions. Financial support from NSF Grant ECS 9528731 and the Critical Technologies Program through the Institute of Materials Science, University of Connecticut are greatly appreciated.

REFERENCES

1. Brus, L. *J. Phys. Chem.* **1994**, *98*, 3575.
2. Weibel, M.; Caseri, W.; Suter, U. W.; Kiess, H.; Wehrli, E. *Poly. Adv. Tech.* **1991**, *2*, 75.
3. Zimmermann, L.; Weibel, M.; Caseri, W.; Suter, U. W. *J. Mater. Res.* **1993**, *8*, 1742.
4. Kyprianidou-Leodidou, T.; Caseri, W.; Suter, U. W. *J. Phys. Chem.* **1994**, *98*, 8992.
5. Schmitt-Rink, S.; Miller, D. A. B.; Chemla, D. S. *Phys. Rev. B* **1987**, *35*, 8113.
6. *Handbook of Optical Constants of Solids*; Palik, E. D., Ed.; Academic Press: Orlando, 1985.
7. Littau, K. A.; Szajowski, P. J.; Muller, A. J.; Kortan, A. R.; Brus, L. E. *J. Phys. Chem.* **1993**, *97*, 1224.
8. Heinrich, J. L.; Curtis, C. L.; Credo, G. M.; Kavanagh, K. L.; Sailor, M. J. *Science* **1992**, *255*, 66.
9. Koch, C. C. *Nanostructured Mat.* **1993**, *2*, 109.
10. Particle Sizing Systems, Inc. Santa Barbra, CA.
11. Shen, T. D.; Koch, C. C.; McCormick, T. L.; Nemanich, R. J.; Huang, J. Y.; Huang, J. B. *J. Mater. Res.* **1995**, *10*, 139.
12. Henglein, A. *Chem. Rev.* **1989**, *89*, 1861.
13. Fojtik, A.; Weller, H.; Fiechter, S.; Henglein, A. *Chem. Phys. Lett.* **1987**, *134*, 477.

GROWTH OF BaTiO₃ IN HYDROTHERMALLY DERIVED (<100° C) BaTiO₃/POLYMER COMPOSITE THIN FILMS

DAVID E. COLLINS AND ELLIOTT B. SLAMOVICH
School of Materials Science and Engineering, Purdue University,
West Lafayette, IN 47907.

ABSTRACT

Composite BaTiO₃/polymer films (<1μm thickness) were processed by the *in-situ* growth of BaTiO₃ particles in a polymer matrix. A solution of a polybutadiene/polystyrene triblock copolymer and titanium diisoproxide bis(ethlyacetoacetate) dissolved in toluene was cast onto a Ag-coated substrate. Subsequent hydrothermal treatment of the films in 1.0 M Ba(OH)₂ solutions at 80° C resulted in the nucleation and growth of BaTiO₃ within the polymer matrix. The volume fraction/connectivity of BaTiO₃ was controlled by varying the relative amounts of titanium precursor and polymer in solution. Growth of BaTiO₃ within the polymer was examined by infrared spectroscopy and electron microscopy. The dielectric constant of the composite films increased with BaTiO₃ content.

INTRODUCTION

Thin film BaTiO₃ capacitors have a large permittivity, but are susceptible to failure by dielectric breakdown and have poor mechanical properties. These limitations can be minimized by additions of a relatively strong dielectric strength polymer. Typical composite fabrication techniques concentrate on mixing BaTiO₃ powder in a polymer/solvent solution or by melt-blending. The mixture is then molded or cast to produce bulk films greater than 100 μm thick [1-3]. One disadvantage of this is inhomogeneous particle dispersion due to agglomerations [2]. Also, the high viscosity associated with high solids loading mixtures complicates thin film processing. Hydrothermal processing is an alternative route in which BaTiO₃ can be coprocessed with a polymer below 100° C [4,5]. This method involves the formation of crystalline materials from metal-organic precursors in an aqueous medium under strongly alkaline conditions to form fine, uniform particles. In this study, thin (< 1 μm), composite BaTiO₃/polymer films were formed by growing BaTiO₃ within a polymer matrix from a titanium alkoxide/polymer precursor solution. The intent of this investigation is to process hydrothermally derived BaTiO₃ within a polymer matrix, and to evaluate the dielectric properties of the composite films as a function of composition and film morphology.

MATERIALS AND METHODS

Relative amounts a polybutadiene/polystyrene triblock copolymer (70 wt% polybutadiene, MW = 30000) (Kraton D1102, Shell Chemical Company) and titanium diisopropoxide bis(ethlyacetoacetate) (TIBE) (Gelest Chemical Co.) were dissolved in toluene and mixed for several days. The volume fraction of BaTiO₃ was varied by preparing precursor solutions containing 90 wt.%, 50 wt.% and 25 wt.% of TIBE relative to the polymer (90/10, 50/50, and 25/75 respectively). Due to density changes associated with BaTiO₃ formation, one would expect the volume fraction of ceramic in the composite to be smaller than the volume fraction of TIBE mixed with the polymer. To maintain consistent spinning conditions between solutions, the viscosity of each solution was adjusted between 3.0 and 5.0 cP using toluene dilutions. The solution

445

was applied to Ag-coated glass substrates with a 0.22 μm filter syringe and spun at two speeds, 200 RPM for 2 s to facilitate a homogenous coverage of the film and 6000 RPM for 20 s. The Ag coating, approximately 500 nm thick, was evaporated onto the glass substrate and served as a bottom electrode for dielectric measurements and a reflective surface for Fourier transform infrared (FTIR) spectroscopy. After drying overnight in a desiccator, the films were placed in an aqueous 1.0 M Ba(OH)$_2$ solution at 80° C for reaction times spanning 6 h to facilitate the nucleation and growth of BaTiO$_3$ in the polymer matrix [6,7]. Ba(OH)$_2$ solutions were made from distilled water, boiled for 15 min to removed dissolved CO$_2$. Once Ba(OH)$_2$ was dissolved in the hot water (> 80° C), the solution was filtered into a polyethylene bottle to remove any BaCO$_3$ formed on the solution surface. The solutions were flushed with Ar, sealed, and stored at 80° C to avoid precipitation of Ba(OH)$_2$. The films were removed from the Ba(OH)$_2$ solution in a nitrogen environment and washed in a hot ammonium hydroxide solution (pH = 11) to reduce barium carbonate formation on the film surface. They were also washed in ethanol to facilitate water removal.

BaTiO$_3$ growth within the polymer was examined by X-ray diffraction (XRD) using Cu Kα radiation, FTIR spectroscopy, scanning electron microscopy (SEM), and transmission electron microscopy (TEM). Samples for XRD analysis were processed without the Ag electrode layer. FTIR spectra of the films were acquired in spectral reflectance. Samples used for TEM were prepared by spin casting the precursor solution onto 200-mesh, Au-gilded Ni grids at 4000 RPM for 25 s. This technique produced a thickness gradient such that the film centers were electron transparent. After drying overnight, the grids were processed at 80°C in 1.0 M Ba(OH)$_2$ for 60 min.

Photolithography was used to apply square 150 x 150 μm^2 electrodes to the film surface. The bottom electrode was exposed by etching a corner of the film with toluene. Capacitance measurements were performed at room temperature using a 1 kHz signal and an applied voltage of 1 V on a Hewlett Packard 427A multi-frequency LCR meter. Film thickness, used to calculate the dielectric constant, was measured using a profilometer on the etched edge of the film.

RESULTS AND DISCUSSION

XRD and FTIR on selected films were used to confirm the presence of polymer, BaCO$_3$, and BaTiO$_3$ after hydrothermal processing. Typical spectra are shown in Figures 1 and 2. The XRD pattern of a 90/10 TIBE/polymer film processed for 60 min confirmed the formation of cubic-BaTiO$_3$, although particle size broadening due to crystallite size may mask some tetragonal peak splitting. The FTIR spectra of the same film composition also indicated the presence of BaTiO$_3$ (440-830 cm^{-1}) in the composite film [7-9]. Bands at 850, 1055 and 1400-1500 cm^{-1} correspond to BaCO$_3$ formed on the film surface during processing. The broad band at 2800-3600 cm^{-1} resulted from various stretching and bending modes of H$_2$0 and -OH groups [8-10]. Peaks at 690, 960 and 2800-3100 cm^{-1} indicate the presence of the polymer.

Figure 3 shows the surface morphology of each composition after exposure to 1.0 M Ba(OH)$_2$ at 80°C for 60 min. Each case produced spherical BaTiO$_3$ particles (0.1-0.2 μm) with a knobby surface. The 25/75 TIBE/polymer film produced isolated BaTiO$_3$ clusters within a polymer matrix. The 50/50 TIBE/polymer film produced a larger number density of BaTiO$_3$ clusters on the surface such that few were isolated by polymer. The 90/10 TIBE/polymer film produced a relatively homogenous surface coverage. Although the presence of polymer was demonstrate by FTIR analysis (Fig. 2), it was not observed in 90/10 TIBE/polymer film structures.

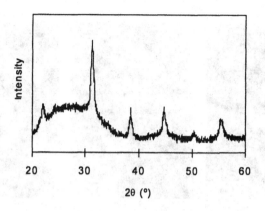

Figure 1   XRD pattern of 90/10 TIBE/polymer composite film processed in 1.0 M Ba(OH)$_2$ at 80° C for 60 min on a glass substrate.

Similar film morphologies were observed using TEM (Fig. 4). Again, the polymer and BaTiO$_3$ could be differentiated in the 50/50 TIBE/polymer film whereas the polymer was not evident in the 90/10 TIBE/polymer film. Carbonate formation was not observed within the films, only on the surface. Figure 5 is a magnified view of a BaTiO$_3$ particle and the corresponding electron diffraction pattern from the film shown in Figure 4b. The bright filed image and electron diffraction pattern suggest that the BaTiO$_3$ clusters are composed of primary particles (5-10 nm) as observed in the SEM.

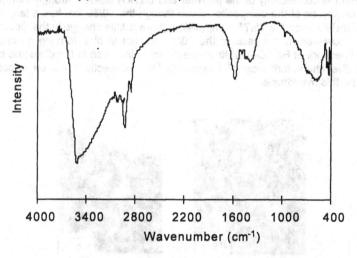

Figure 2   Spectral reflectance FTIR spectra done of 90/10 TIBE/polymer composite films processed in 1.0 M Ba(OH)$_2$ at 80° C for 60 min on a Ag-coated glass substrate.

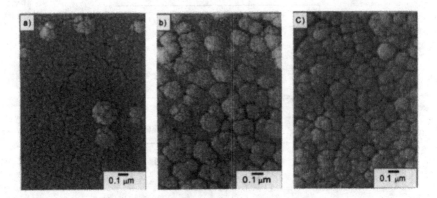

Figure 3   SEM micrographs of a) 25/75 b) 50/50 and c) 90/10 TIBE/polymer composite films 1.0 M Ba(OH)₂ at 80 ºC for 60 min.

Figure 6 shows the dielectric response of the three BaTiO₃/polymer film compositions for processing times up to 6 h in 1.0 M Ba(OH)₂ at 80° C. The error bars are primarily a result of scatter in the film thickness measurements. The dielectric constant of the 90/10 TIBE/polymer film increased with processing time, reaching a value of 16 after 6 h at 80°C, while the dielectric constants of the 50/50 and 25/75 TIBE/polymer films slightly exceeded that of the polymer alone (2.5 [11]). The results suggest that the connectivity of the polymer and BaTiO₃ phases strongly influences the dielectric properties of the composite. Even though the 50/50 film has a larger volume fraction of BaTiO₃ than the 25/75 composition, there is little change in the dielectric constant. However, in the case of the 90/10 composite film, forming a continuously connected network of BaTiO₃ results in a significant increase in the dielectric constant. A quantitative characterization of polymer/BaTiO₃ connectivity will be necessary to substantiate this hypothesis.

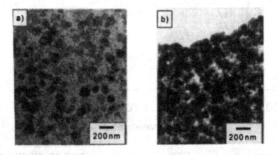

Figure 4   TEM micrographs of 50/50 and 90/10 TIBE/polymer composite film processed in 1.0 M Ba(OH)₂ at 800° C 60 min.

 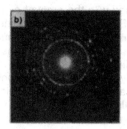

Figure 5   a) Magnification and b) electron diffraction pattern of a BaTiO$_3$ cluster in a 90/10 TIBE/polymer composite film processed in 1.0 M Ba(OH)$_2$ at 80° C for 60 min.

The dielectric constant for other BaTiO$_3$/polymer composite systems range from values of the polymer to approximately 75, depending on the polymer matrix and volume fraction of BaTiO$_3$. For example, Aulagner et al. [12] and Gregorio et al. [13] obtained values of 40 (40 μm thick) and 64 (18 μm thick), respectively, for composites of 40 vol.% BaTiO$_3$ in polyvinylidene. Comparison to other thin BaTiO$_3$ film studies is difficult because values vary greatly depending on particle size and processing. The dielectric constant can range from 10-20 for films prepared by MOCVD and ion assisted deposition [14,15]; 200-500 for films prepared by evaporation and sputtering [16,17]; and 1200 for films prepared by metallo-organic deposition [18]. In one study of BaTiO$_3$ films grown hydrothermally on Ti-coated Si substrates, films produced a dielectric constant of approximately 200 [19].

SUMMARY

Composite BaTiO$_3$/polymer thin films were hydrothermally processed at 80° C from TIBE-polybutadiene/polystyrene block copolymer precursor solutions. The films were composed of BaTiO$_3$ clusters within a polymer. FTIR spectroscopy confirmed the presence of the block copolymer in all of the composite films. The volume fraction of

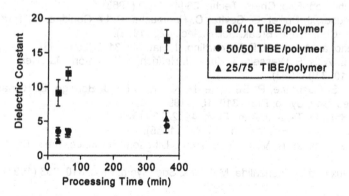

Figure 6   Dielectric constant for 25/75, 50/50, 90/10 TIBE/polymer composite films processed in 1.0 M Ba(OH)$_2$ at 80° C for up to 6 h.

449

BaTiO$_3$ increased with the amount of TIBE used in the precursor solution. The 25/75 and 50/50 TIBE/polymer films contained isolated BaTiO$_3$ clusters whereas the 90/10 TIBE/polymer films contained a continuous network of BaTiO$_3$ clusters. This trend was also evident in the dielectric response. The 25/75 and 50/50 TIBE/polymer films displayed dielectric constants consistent with values listed for polystyrene/polybutadiene block copolymer whereas the 90/10 TIBE/polymer film demonstrated a relatively larger dielectric constant. An increase in the dielectric constant was also shown to increase with processing time for the 90/10 TIBE/polymer film composition.

ACKNOWLEDGEMENTS

We wish to thank Said Mansour for his assistance with TEM. This research was funded by the Purdue Research Foundation and the National Science Foundation (contract #DMR-9623744).

REFERENCES

1. T. Yamamoto, K. Urabe, H. Banno, Jpn. J. Appl. Phys. **32**, 4272 (1993).
2. K. Nagata, S. Kodama, H. Kawasaki, S. Deki, M. Mizuhata, J. Appl. Poly. Sci. **56**, 1313 (1995).
3. A. Balal, M. Amin, H. Hassan, A. Abd El-Mongy, B. Kamal, K. Ibrahm, Phys. Stat. Sol. (a) **144**, (k53) (1994).
4. P. Calvert, R. Broad, Contemporary Topics in Polymer Science Vol. 6, 95-105, ed. W. Culbertson (Plenum, New York, 1989).
5. J. Burdon, P. Calvert, Mat. Res. Soc. Symp. Proc. **255**, 375 (1992).
6. E. Slamovich, I. Aksay, Mat. Res. Soc. Symp. Proc. **346**, 63 (1994).
7. E. Slamovich, I. Aksay, J. Am. Ceram. Soc. **79**, (1) 239 (1996).
8. R. Nyquist and R. Kagel, Infrared of Inorganic Compounds (3800 - 45 cm^{-1}) (Academic Press, New York 1971).
9. D. Hennings, S. Schreinemacher, J. Europ. Ceram. Soc. **9** 41 (1992).
10. K. Nakamoto, Infrared and Raman Spectra of Inorganic and Coordination Compounds, 4th ed. (Wiley, New York 1988).
11. A. McPherson, Rub. Chem. Techn. **36** [4], 1230 (1963).
12. E. Aulagner, J. Buillet, G. Seytre, C. Hantouche, P. Le Gonidee, G. Terzulli, IEEE 5th Inter. Conf. Cond. Break. Sol. Dielec., 423 (1995).
13. R. Gregoria, M. Cestari, F. Bernardino, J. Mat. Sci. **31**, 2925 (1996).
14. H. Lu, L. Wills, B. Wessels, X. Zhan, J. Helfrich, J. Ketterson, Mat. Res. Soc. Symp. Proc. **310**, 319 (1993).
15. W. Liu, S. Cochrane, P. Beckage, D. Knorr, T. Lu, J. Borrego, E. Rymaszewski, Mat. Res. Soc. Symp. Proc. **310**, 157 (1993).
16. Y. Shintani, O. Tada, J. Appl. Phys. **41**, 2376 (1970).
17. C. Feldman, Rev. Sci. Instrum. **26**, 463 (1955).
18. J. Xu, A. Shaikh, R. Vest, IEEE Trans. Ultrason. Ferroelec. Freq. Contr. **36**, 307 (1989).
19. M. Pilleux and V. Fuenzalida, Mat. Res. Soc. Symp. Proc. **310**, 333 (1993).

SOL-GEL ROUTES TOWARDS MAGNETIC NANOCOMPOSITES WITH TAILORED MICROWAVE ABSORPTION

Ph. COLOMBAN*,** and V. VENDANGE*
*ONERA, OM, BP72, 92322 Chatillon, France
**CNRS, LASIR, 2 rue Henri Dunant, 94320 Thiais, France

ABSTRACT

Ceramic microwave absorbents are of considerable interest for the structural applications with the tailored radar cross-section. The main parameters governing the electromagnetic cross-section are discussed in this paper with the specific interest of the dispersion of Co, Fe (and their alloys) particles in a dielectric aluminosilicate matrix (high Curie temperature, significant absorption in the 0.1-15GHz range) and the advantages of the sol-gel routes for the materials preparation. Three methods have been examined : (i) the first one is the mixing of the (sub)micronic metal powder within a liquid matrix precursor ; (ii) the second is the mixing of the alkoxides with an aqueous solution of metal ions ; and (iii) the third is the preparation of a porous host matrix impregnated by a concentrated solution of transition metal nitrates. Nanocomposites with diameter 20-200nm were achieved by heating at temperatures between 600 and 1100°C under a H_2 atmosphere. The samples were characterized by SEM, TEM, magnetic hysteresis, Mössbauer and Raman spectroscopies and microwave absorption.

INTRODUCTION

The impetus for the study of electromagnetic wave-matter interactions in the microwave range is created mainly by the need to improve the stealthiness of aircraft, missiles and ships, against the detection by the energy radar receivers [1,2]. A significant decrease in the radar cross-section -or radar equivalent surface (RES)- may be obtained by the optimization of the shape of mobile devices. However, the variations possible are limited, especially for the high velocity aircrafts and missiles to which the aerodynamic requirements dominate. Consequently, the absorbent materials are required.

Thin or thick films of magnetic materials deposited on metal are efficient for the above purpose. On the other hand, the control of the electromagnetic absorption by the combination of dielectric and conducting materials requires material thickness close to $\lambda/4$ (a few tens of mm) [3]. It is worthy to notice that only ferro and ferrimagnetic materials may be considered, because dia and para-magnetic are poor absorbent. Ferromagnetic materials are metals and they must be dispersed in insulating matrix to absorb the electromagnetic wave. The particle size may be sufficiently small to be homogeneously dispersed without percolation. In addition, the Curie temperature of the materials may be compatible with hot application (≥300°C). The absorption of an electromagnetic wave by the magnetic materials occurs via a relaxation induced by the domain wall motion or alternatively by a natural spin resonance mechanism. The high magnetization density of iron, cobalt and their alloys can lead to a significant absorption and their gyromagnetic resonance are expected between ~0.5GHz (iron) and 16GHz (hexagonal Co). Transition metals are good conductors, therefore the penetration of the electromagnetic wave is limited at the skin, typically a few tens of nanometers in the GHz range, so nanocomposites are thus required.

The phenomenon of the absorption loss is a consequence of the direct electromagnetic wave-matter interaction. This interaction can be described as the complex permittivity (ε^*) and permeability (μ^*). These complex values are written respectively $\varepsilon^*=\varepsilon'-j\varepsilon''$ and $\mu^*=\mu'-j\mu''$. ε', μ' (real parts) are the wave propagation inside the materials and ε'', μ'' (imaginary parts) are energy dissipation. The absorption of an electromagnetic wave can be obtained by the combination of dielectric and conducting materials (Sallisbury screen) or by a magnetic film (Dällenbach screen). Figure 1 shows the resonance criteria for a film from a magnetic materials dispersed in a dielectric matrix [5]. The absorption is maximum when the reflexion ($R_1+R_2+...=0$) of the incident microwave is zero. Absorption is large when $\mu'/\varepsilon' \sim 0.3$ and $\mu''.d.f = 1/2\pi\sqrt{\mu_0\varepsilon_0} = 47.7$ (d in mm, f in Ghz), i.e. when the screen impedance is equal to that of air. Two regimes can be recognized : (i) the thin film regime when the film thickness (d) is small versus the incident wavelength ($\varepsilon'd/\lambda<0.02$). In this case, the absorption depends on μ'' only ; (ii) the thick film regime when $\varepsilon'd/\lambda>0.6$, i.e. when the thickness is close to $\lambda/4$ as for a dielectric absorbent. In the latter case both μ' and μ'' values depend on the permittivity. The thickness ranges between 1 to 10cm and thus this type of absorbent can be used for ships and fixed plants, only.

Sol-gel routes have been widely used for many centuries in earthenware, china and enamel production : the use of a liquid precursor, a slurry of colloïdal particles, clays for instance, which is transferred into gels by the decrease of water content and then dried and fired, allows the preparation of film, bulk, ... For the last two decades, the sol-gel technology has been extensively used to prepare advanced ceramics, fibers and films [6,7]. This method has many advantages : lowering the densification temperature of multicomponent oxides, controlling the chemical purity and crystallinity [6,7] and tailoring the micro and nanostructure. Films, powders and monoliths can be prepared with gels. A compact from gel powder or a gel monolithic can be densified by sintering. Preliminar thermal treatments below 600°C are often used to control their sintering behaviour.

Tailoring of the microstructure of the sol-gel derived composites has been investigated in the past. Metal-ceramic microcomposites have been prepared by Roy *et al.* [8,9], Petrullat *et al.* [10] and Breval *et al.* [9] e.g. the elaboration of Ni particles in a silica matrix. Glass-metal nanocomposites with iron, nickel and copper particles in a silica matrix have been prepared by Chatterjee and Chakvavorty [12]. Static magnetic properties of nanocomposites have been studied by Shull *et al.* [13] and we have also reported preliminar static [14] and frequency dependent [15] magnetic investigations.

This paper compares several sol-gel routes for the preparation of metal-ceramic nanocomposites and reviews the key parameters to control the microwave absorption of the materials. Materials have been studied in form of optically monolithic pieces which are representative of films. Two dielectric matrices, alumina ($\varepsilon'=9$) and mullite-like aluminosilicate ($\varepsilon'\sim5$) and two metals, iron and cobalt, were examined.

EXPERIMENTAL

Synthesis

Three methods have been examined :
i) The first method is the mixing of micronic metal powder (cobalt for instance, diameter $\sim1.8\mu m$) with a liquid matrix precursor before its hydrolytic polycondensation. The precursor is a mixture of water, isopropanol and $(OC_4H_9)_2-Al-O-Si(OC_2H_5)_3$ ester (from Hüls France). The volume ratio of the water to solvent to alkoxide is 1:4:2, one volume

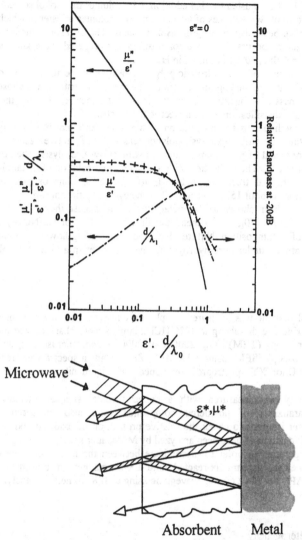

Fig. 1 : Real (μ') and immaginary (μ'') permeability plots for a film of dielectric ($\varepsilon' \neq 0$, $\varepsilon''=0$) magnetic materials as a function of the reduced thickness ($\varepsilon'd/\lambda_0$) (λ_0 wavelength in vacuum) giving a total reflexion equal to zero (R≤-85dB). The corresponding relative band pass at R≤0.1 (-20dB) is given on the right side scale (after P. Hartemann and M. Labeyrie, Revue Technique Thomson-CSF **19(3-4)**, p. 413-430 (1987)). A scheme showing the absorption mechanism for a Dällenbach screen made of absorbent material deposited on a metallic substrate is given. Resonant absorption occurs when the reflection is nil i.e. if the impedance of the screen (absorbent + metal support) is equal to that of air.

of water, pH=2.5 adjusted by HCl addition, two volumes of alcohol solvent were poured into a mixture of two volumes of alcohol and two volumes of ester in which the particles of metal have been dispersed by ultrasonic means. The ultrasonic stirring is maintained until the gelation occurs. The 1:3:6 composition was prepared in the same way. The mass content of cobalt is 30% after firing in H_2.

ii) In the second method, the hydrolytic polycondensation of the alkoxide is obtained by the addition of water containing metal nitrate. The metal concentration is chosen to obtain a final 20% mass content after the thermal treatment under H_2 inducing the reduction of metal salt and densification of the dielectric host matrix.

iii) The third method was the preparation of a porous host matrix in the gel or xerogel (glass) state by the hydrolysis-polycondensation of above ester (1:3:6, 1:4:2 compositions) and of pure aluminium s-butoxide (slow hydrolysis), then drying at 40°C or firing at 600°C. The infiltration with a concentrated solution of metal nitrates is made before the thermal treatment under H_2 to monoliths, films or powder. The metal concentration is about 15% in weight after one cycle of infiltration-firing.

Nitrates have been chosen as the metal precursors because their high metal content in solution (e.g. up to 270g of Co metal per liter of solution of $Co(NO_3)_2$) and their low temperature of decomposition. Nitrates decomposition leading however to nitrogen oxides, a thermal treatment under pure hydrogen is needed to obtain a nearly complete reduction into metal.

Techniques

The metal content was measured by plasma emission after melting the samples with sodium peroxide and dissolving in 10% HCl. Samples were characterized by transmission electron microscopy (TEM) using 200CX or 2000FX Jeol microscopes, and by scanning electron microscopy (SEM) using a DSM960 Zeiss. Raman spectra were recorded using a multichannel Dilor XY spectrograph equipped with liquid nitrogen cooled Wright CCD detector.

Susceptibility was measured with a coiled toroidal sample assembly. Quasi-static magnetic parameters of powdered composites were also measured with a SIIS hysteresismeter in order to check the relative proportion of oxidized and reduced metals. Iron containing materials have been analyzed by Mössbauer spectroscopy at room and liquid nitrogen temperature in order to discriminate between the metal core, the oxide shell of particles as well the shell-matrix reacted zone. Electromagnetic measurements were carried out with an APC7 cell in a coaxial waveguide using a HP8510 network analyzer.

RESULTS

Degree of heterogeneity

Typical scanning electron micrographs for materials prepared using methods I, II and III and thermally treated at 1100°C under H_2 are shown in figure 2. The differences are obvious. Large aggregates ranging from 2 to 40µm are observed in materials prepared using cobalt powder (method I). Agglomeration resulted from the magnetic attraction between metal particles although the strong ultrasonic stirring was used. The poor reproducibility of the dispersion limits a good control of the magnetic properties. However, the size and shape of magnetic particles can modify their permeability.

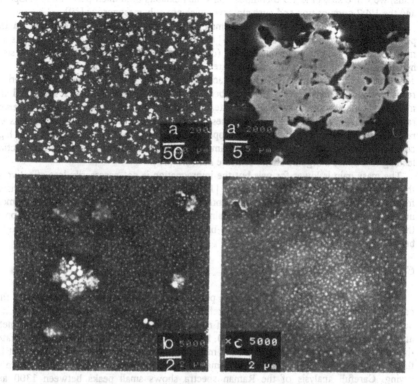

Fig. 2 : Scanning backscattered electron micrographs of cobalt-$Al_2O_3 2SiO_2$ composites prepared by method I (a,a'), II (b) and III (c). Materials have been thermally treated at 1100°C under H_2. 1:4:2 matrix with addition of 35% molar of B_2O_3 is used (bars : scale in micron).

A more homogeneous dispersion is achieved with method II. Dispersion is prepared at the submicronic scale but large agglomerates, up to 2µm in diameter, are still observed. The best homogeneity is achieved with method III.

Infiltration of metal precursor using method III

A gel is a network which is soaked by a liquid, generally water. More or less full dehydration produces the departure of the water logged in the network pores without drastic modification of the solid network. In the previous work [16,17] we have shown that gels and glasses (or stable xerogels) having pores in the meso or microporous range can preserve their porosity up to 800°C and more. The different ways to control the pore distribution for a given Si/Al composition have been studied [18]. Microporous materials can be prepared [19] using a liquid precursor which have a high concentration of alkoxides (e.g. 1:3:8 water-solvent-alkoxide volume ratio) but the time between the mixing of reagents and the gelation is too short to allows the preparation of films and monoliths.

Thus, we will discuss the 1:3:6 composition which exhibits both micropores and mesopores and the 1:4:2 composition and pure alumina which are only mesoporous [19].

Study on the local structure using small angle X-ray scattering [16,17] and on the porosity using gas absorption analysis [14] has given a description of these solids : gels are made of primary, rather globular entities (diameter 0.8-2nm) which are more (using slow hydrolysis of high alkoxide concentrated solution) or less (rapid hydrolysis, low concentrated alkoxide solution) densely packed, leading to micro or mesoporous materials, respectively. The thickness of a "wall" between adjacent pores is always small and the matter in contact with the surface dominates. The high values of hydroxyl surface coverage ratio (~6OH$^-$/nm^2 and 2OH$^-$/nm^2 for mesoporous alumina and aluminosilicate glasses, respectively, about 3OH$^-$/nm^2 for microporous aluminosilicate [20]) explain the high correlation between the hydroxyl group elimination and the nucleation-densification reaction leading to a dense inorganic aluminosilicate network [16].

The mass gain is lower for gels but the impregnation is not necessarily less effective as the metal content after pyrolysis indicates [19], according to the larger mass uptake arising from the filling of empty pore by concentrated nitrate solution. After thermal treatment under H$_2$, monoliths are black. Iron containing gels loose their optically clearness soon at the infiltration steps and exhibit often a brown color. This behaviour will be discussed below.

Cobalt precursor infiltration

The metal concentration inside the pores is always less than or equal to that corresponding to 4 mol/l solution and therefore the use of a highly concentrated solution is not useful to obtain the highest content of metal infiltration (typically 10-15% wt of metal after one cycle of infiltration and H$_2$ firing). IR [19] and EXAFS [21] analysis demonstrated the solid structure of the cobalt inside the mesoporous host lattice and that a large part for the impregnated nitrate in micropores remains liquid, for same time and temperature of drying. Carefull analysis of the Raman spectra shows small peaks between 1300 and 1500cm^{-1} assigned to the fingerprint of uni or bidentate complexes of nitrate ions, characteristic of adsorbed species on pore walls. This phenomenon of adsorption on the pore walls is significant and may explain the slower penetration of the solution in microporous samples : the kinetics of the impregnation is about 500 times (saturation time ratio) slower in microporous than in mesoporous aluminosilicates gels and glasses [19]. Figure 3 shows a scheme of the ions diffusion inside the pore with or without trapping at the pore opening and resulting in the formation of impregnated materials. A schematic view of the formation of nanoprecipitates by thermal treatment is also given.

Iron precursor infiltration

A reaction between infiltrated ions and pore walls has been observed in the materials immersed in iron nitrate solution. Raman scattering demonstrates that iron ions diffuse inside the Al-O-Si framework to form a Fe$_{1-x}$(Al$_y$,Si$_{1-y}$)$_x$OOH phase at the pore opening whereas the nitrate ions diffuse far away in the pore. The reaction is limited using pure alumina gels and glasses. Figure 4 compares the Raman spectrum recorded on brown superficial pieces impregnated by iron nitrate with that recorded on the core of a monolith. The spectrum of the latter displays a strong line at 1042cm^{-1} characteristic of the NO$_3^-$ ion stretch. The spectrum differs however from the 1047-1056cm^{-1} doublet of the salt and from the narrow 1058cm^{-1} peak of the solution. In contrast, the spectra recorded on brown scales shows a weaker peak at 1049cm^{-1}, the weakest signal being observed on impregnated

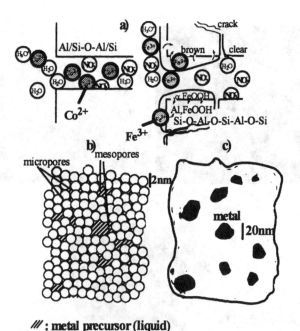

/// : metal precursor (liquid)

Fig. 3 : Scheme of the ions diffusion/adsorption mechanism at the pore opening (a) and of the porous structure of the impregnated host matrix before (b) and after (c) thermal treatment. The size of basic polymeric moities (schematic drawn as circle) is close to 1-3nm.

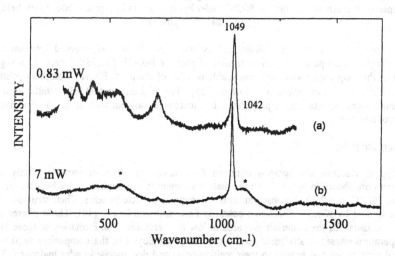

Fig. 4 : Micro-Raman spectra (λ=514.5nm spectra) of iron nitrate impregnated aluminosilicate (1:4:2) glass : (a) brown scales, (b) colourless core of monoliths. The used laser power is given. Stars indicate the contribution of the lenses used for collecting the scattered light.

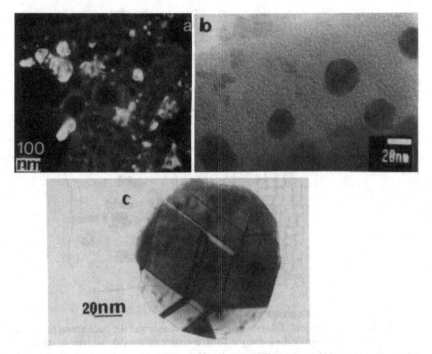

Fig. 5 : Transmission electron photomicrograph of metal-aluminosilicate composites prepared by thermal treatment at 800°C under hydrogen : (a) iron precipitates (dark field), (b,c) cobalt precipitates.

microporous host lattice. In addition, at least five broad bands are observed between 250 and 720cm^{-1}. This pattern is similar to that of goethite FeOOH [22,23], except for a slight shift of the frequencies and the considerable breadth of the lines. EXAFS analysis confirms that γFeOOH-like materials was formed [21]. The highest content of metallic iron is obtained using mullite matrix prepared using acetone as solvent with addition of a drying control additive.

Nanoprecipitates

Typical electron micrographs obtained from nanocomposites of various metals and matrices are shown in figure 5. All the matrices remain still amorphous after heat treatment below 900°C, in which the observation of the metallic particle benefits. Their structure has been studied by X-ray diffraction : cobalt is f.c.c. and iron is c.c. [14]. These phases are rather unusual for the materials prepared at low temperatures. The formation of these high temperature phases is attributed to the conditions employed in the composites synthesis. Metal nitrates are fast heated to their melting point and decomposed under hydrogen. The reaction between NO_x and H_2 can be explosive and the temperature in the pores can be much higher than the measured. Formation of a diamond-like precipitates is consistent with this mechanism [21].

The size of the metallic precipitates may depend on the initial pore structure of the host lattice and on the heating temperature. Using microporous gels and glasses as host matrices, the metal particle size ranges from 5 to 20nm for a 600°C thermal treatment and reaches typically 100nm using mesoporous host matrix. Only the iron precipitates have no particular geometry whereas the cobalt has globular shape with sometimes nice facets (octahedral habit, see figure 5). The irregular form of iron precipitates can be related to its preferential adsorption and reaction with the pore walls.

Thermal stability

Initial susceptibility measurement versus temperature by the induction method allows a determination of the use limit of the materials. The susceptibility is constant up to our measurement limit (550°C in air). DSC and X-ray analysis show that the formation of Co_3O_4 and Fe_3O_4 (magnetite) or γFe_2O_3 (maghemite) takes place above 600°C [14]. Iron oxydation can led to FeO, Fe_2O_3 or Fe_3O_4. The discrimination between maghemite and magnetite spinel from their X-ray powder patterns is not straightforward. Mössbauer analysis discriminates Fe^{3+} magnetic spinel component by its medium isomeric shift (IS ~0.3-0.5mm/s) and its large field parameter (H=40-50T). These components are clearly observed in materials where the metal content is small (e.g. Fe in alumina matrix or $Fe_{0.7}/-Co_{0.3}$ alloys in $2Al_2O_3SiO_2$ matrix). Different spinel components are observed, which indicates a variety of sites and environment. This can be explained by formation of Fe_2CoO_4 or Al substituted magnetite.

In situ alloying

The technique for preparing the nanocomposites of the alloy precipitates inside the dielectric matrix can tailor the absorption range of the materials. Infiltration using nitrate solution of iron and cobalt (Fe/Co=1) has been made using microporous aluminosilicate gel as host lattice. The relative amount of metallic iron versus oxydized iron (Fe^{2+}, Fe^{3+} ions) has been determined by Mössbauer spectroscopy. The lack of intensity increase when temperature is decreased from 300 to 80K indicates that particles prepared using mesoporous host matrix are not superparamagnetic. Different components are observed :
i) A quadrupolar rather symmetrical doublet located at 0.39mm/s versus Fe (isomeric shift : IS) with a quadrupolar splitting close to 0.7mm/s (QS) assigned to Fe^{3+} ion in octahedral site. Only this component is observed before thermal treatment. The larger band width (B) observed for alloys (B~0.4mm/s instead of 0.3mm/s for pure iron derivative) indicates a higher local disorder.
ii) After iron reduction by thermal treatment under H_2, new features are associated to the Fe^{3+} doublet : furthermore small shifts are observed for IS and QS. Band at IS~0.93 (QS=2.15, B=0.5mm/s) and 1.23 (QS=2.25, B=0.45mm/s) are characteristic of Fe^{2+}ions, according to the formation of γFe_2O_3 shell around the particle. This oxydation is coherent with the γFe_2O_3 fingerprint on X-ray powder patterns. The first band (IS~0.9mm/s) may correspond to tetrahedral site and the second (IS~1.2mm/s) to the octahedral site.
iii) Band at IS~0 corresponds to metal particles. The relative proportion of the Fe^{2+}, Fe^{3+} ions and Fe metal can be calculated, roughly, from the peak area. The fraction of metal is rather poor (~10%) but the maximum occurs when infiltration is made in the glass state which decrease the reaction between iron nitrate solution and the pore walls. In some case, another component (IS~0.3-0.5, QS~0, B~0.5-1.5mm/s) characteristic of Fe^{3+} is

observed. The complex shape of this component is coherent with the formation Al substituted spinel.

Electromagnetic properties

The measured values of saturation magnetization are lower than the expected value (reduction close to 20% for Co and to 80% for Fe). As discussed above, this may be explained by the oxydation of metal precipitates according X-ray and Mössbauer analyses. The thickness of the oxydized Co_3O_4 shell and the core size can be calculated from relative intensities to be 6nm and 50nm, for the shell and the core respectively. The thickness of Fe_2O_3 shell can be 60nm for a 40nm Fe core size.

Figure 6 compares the X-band properties (0.1-10GHz) of a Co nanoparticles dispersed in mesoporous aluminosilicate matrix. 35% mole B_2O_3 has been added to the 1:4:2 $Al_2O_3 2SiO_2$ composition to make possible the film deposit onto convenient substrate.

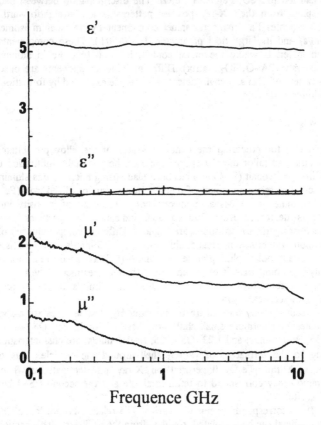

Frequence GHz

Fig. 6 : Evolution with frequency (0.1 to 10GHz) of real (ε') and imaginary (ε'') permittivity and real (μ') and imaginary (μ'') permeability for a thick film (thickness 2mm) made of a dispersion of cobalt nanoparticles (8% in volume) in an aluminoborosilicate glass (1:4:2 with addition of 35% molar B_2O_3).

The real part of permittivity is close to 4 for the pure aluminosilicate glass. The aluminoborosilicate matrix exhibits a low dielectric constant for two reasons : first, silica or mullite-type material have a very open framework, without highly polarizable or mobile ions. This means that relaxations, in the microwave range, do not exist and the resulted real permittivity is low and the immaginary part is zero ; second, as the samples retain some mesoporosity, a proportion of material has an ε' value close to 1. Therefore, the matrix is nearly "transparent" to the microwave energy. The dielectric parameter of the nanocomposite increases to 5, due to the contribution of Co metal dispersion (metal content 8% in volume, 30% in weight). The magnetic loss (μ'') decreases with frequency up to 1GHz, corresponding to the relaxation induced by the wall motion. The peak between 6 and 10GHz corresponds to the gyromagnetic resonance of cobalt. Alloying with iron will decrease the resonance frequency. Iron oxydation will decrease the magnetization at saturation but the magnetization of resulted spinel remains sufficiently large to contribute significantly to the absorption of the electromagnetic wave. Furthermore, the resonance frequency of magnetite is expected to be ~4GHz.

CONCLUSIONS

The magnetic nanocomposites were prepared by the impregnation with metal salt solution to the porous host matrices prepared by the sol-gel route. Homogenous dispersion are achieved at the submicronic scale. About 5% in volume (15% in weight) can be incorporated in one cycle of impregnation/firing. The metal content can be increased by successive impregnation-firing cycles (up to ~50% in volume). This process is a novel route to prepare microwave absorbent materials. Combined infiltrations with different metal and alloy precursors can tailor the absorption versus frequency between 0.1 and 10GHz. These new materials can find "hot" applications as their magnetic properties are maintained up to 550°C.

ACKNOWLEDGMENTS

The authors thanks Dr. E. Tronc for her help with Mössbauer spectra and their interpretation.

REFERENCES

1. C. Deleuze, A. Mathiot, P. Zamora and G. Zerah, Chocs **6**, 14-29 (1992).

2. F. Boust, in Proc. JCMM96, Quatrièmes Journées de Caractérisation Micro-Onde des Matériaux, April 3-5 1996, Chamberry (Fr), TP ONERA 1996-34, Chatillon, France.

3. E. Mouchon and Ph. Colomban, J. Mater. Sci. **31**, 323-334 (1996).

4. Ph. Colomban, in Proc. Third International Conference on Intelligent Materials-Third European Conference on Smart Structures and Materials, 3rd ICIM-ECSSM Lyon 96, June 3-5 1996, edited by P.F. Gobin and J. Tatibouët, (SPIE-The International Society for Optical Engineering, Washington, 1996), **2779** pp. 813-818.

5. E.F. Knott, J.F. Staeffer and M.T. Tuley, Radar Cross Section, Arteen House (1985) ; P. Hartemann and M. Labeyrie, Revue Technique Thomson-CSF **19**(3-4), 413-430 (1987)

6. L.C. Klein (Ed.), Sol-Gel Technology, Noyes Publications, Park Ridge (NJ) 1988

7. Ph. Colomban, Ceram. Int. **15**, 23-50 (1989).

8. R. Roy, S. Komarneni and D.M. Roy, Mater. Res. Soc. Symp. Proc. **32**, 347-359 (1984).

9. R.A. Roy and R. Roy, Mater. Res. Bull. **19**, 169-177 (1984).

10. J. Petrullat, S. Ray, U. Schubert, G. Guldner, Ch. Egger and B. Breitscheidel, J. Non-Cryst. Solids **147 & 148**, 594-600 (1992).

11. E. Breval, Z. Deng, S. Chiou and C.G. Pantano, J. Mater. Sci. **27**, 1464-1471 (1992).

12. A. Chatterjee and D. Chakrarorty, J. Phys. D **23**, 1097-1104 (1990).

13. R.D. Schull, J.J. Ritter, A.J. Shapiro, L.J. Swartzendruber and L.H. Bennett, J. Appl. Phys. **67**, 4490-4497 (1990).

14. V. Vendange and Ph. Colomban, in Proc. EMRS Meeting, Nanophase Materials Symposium, November 3-6 1992, Strasbourg, Mater. Sci. Eng. **A168**, 199-203 (1993).

15. V. Vendange, E. Flavin and Ph. Colomban, J. Mater. Sci. Lett. **15**, 137-141 (1996).

16. Ph. Colomban and V. Vendange, J. Non-Cryst. Solids **147&148**, 245-250 (1992).

17. V. Vendange and Ph. Colomban, J. Sol-Gel Sci. Techn. **2**, 407-411 (1994).

18. V. Vendange and Ph. Colomban, J. Mater. Res. **11**, 518-528 (1996).

19. V. Vendange, Ph. Colomban and F. Larché, Microporous Mat. **5**, 389-400 (1996).

20. V. Vendange and Ph. Colomban, J. Porous Materials **3** (1996) in press.

21. V. Vendange, D.J. Jones and Ph. Colomban, J. Phys. Chem. Solids (1996) in press.

22. R.J. Thibeau C.W. Brow and H. Heidersbach, Appl. Spect. **32**, 532-540 (1978).

23. P. Forbes, J. Dubessy and C. Kosztolangi, Geo-Raman-86, Terra Cognita 7, 16 (1977).

PHYSICAL AND CHEMICAL PROPERTIES OF CaCl$_2$/H$_2$O AND LiBr/H$_2$O SYSTEMS CONFINED TO NANOPORES OF SILICA GELS

Yurii I. Aristov[*], Mikhail M. Tokarev[*], Gaetano Cacciola[**], Giovanni Restuccia[**], Gaetano DiMarco[+++] and Valentin N.Parmon[*]
[*] Boreskov Institute of Catalysis Russian Academy of Sciences
Prospect Lavrentieva, 5, Novosibirsk, 630090, Russia,
ARISTOV@catalysis.nsk.su
[**] CNR/ITAE, 98126 Messina, Italy
[+++]CNR/ITS, Messina, Italy

ABSTRACT

Here we present physical and chemical properties of CaCl$_2$/H$_2$O and LiBr/H$_2$O systems confined to nanopores of silica gels. Sorption isobars, isosters and isotherms were measured at temperature 293 - 423K and vapor partial pressure 8 - 133 mbar. Specific heat of the systems was found as a function of temperature (300 - 400K) and sorbed water content (0 - 53 wt.%). Solidification/melting diagrams were measured over 170 - 320K temperature range at salt concentrations 0 - 50 wt.%.
The results obtained evidenced a significant change in the thermodynamic properties of CaCl$_2$/H$_2$O and LiBr/H$_2$O systems due to confinement to the silica gel micropores.

INTRODUCTION

In Refs. [1-6] a new generation of composite materials called "selective water sorbents" (SWSs) has been presented. An SWS material is a two-phase system which consists of a porous host matrix and a hygroscopic substance (commonly an inorganic salt) impregnated into its pores. The goal of this work is to find out the influence of the silica pore structure on thermodynamic properties of the confined salt-water systems.

EXPERIMENTAL

Two commercial silica gels - mesoporous KSKG (average pore diameter D = 15 nm, S$_{BET}$= 350 m^2/g) and microporous KSM (D= 3.5 nm, S$_{BET}$= 600 m^2/g) were used as the porous matrix for sample synthesis. The matrices were filled with CaCl$_2$ and LiBr aqueous solutions, then dried and saturated with water vapor until the needed water content was reached (more details of synthesis may be found elsewhere [4,5]). A vapor sorption equilibrium was studied by a thermal balance method. The amount of water sorbed by the sample was measured with the CAHN C2000 thermal balance in the temperature range of 20 - 150°C at different partial pressures P$_{H_2O}$ ranging from 8 to 133 mbar. The accuracy of the temperature regulation was about ±0.4°C. The water content was calculated as number N of sorbed water molecules with respect to 1 molecule of the salt.
A DSC-111 "SETARAM" unit was used for measuring solidification/melting diagram over the temperature range of 170 – 320K. The samples (m$_o$ ≈ 30 mg) were placed in standard platinum crucibles, cooled down to 170 K at a constant rate (5 K/min), maintained at this temperature for 10 min and finally heated at the same rate up to 320 K. A Mettler TA4000 DSC was used for specific heat measurements over temperature range 313 - 413 K at heating rate 5-10 K/min.

463

RESULTS AND DISCUSSION

Water - calcium chloride system confined to the mesopores
Water sorption equilibrium.

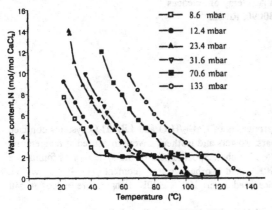

Fig.1. Water sorption isobars for the SWS-1L material.

Water sorption isobars for the system "CaCl$_2$ confined to the mesoporous KSKG silica gel" (SWS-1L) are presented in Fig.1 as N(T) dependence at different partial pressures P_{H_2O}. All the curves were found to be similar but shifted and partially extended along the temperature axis. Each curve has a plateau corresponding to N = 2, which indicates the formation of solid crystalline hydrate CaCl$_2$·2H$_2$O [7,8]. This hydrate is rather stable and possesses no transformation in 10-20°C temperature range showing a monovariant type of the sorption equilibrium typical for gas-solid processes. At high temperatures this hydrate undergoes decomposition to form the lowest possible crystalline hydrate CaCl$_2$·0.33H$_2$O [8,9].

At temperatures lower than that of the left boundary of the plateau, water sorption depends on both temperature and partial vapor pressure, so that the equilibrium becomes divariant, typical for liquid solutions of CaCl$_2$ [7].

Based on isobaric curves water sorption isosters are plotted for N = 1 - 10. They appeared to be straight lines in Ln(P_{H_2O}) vs 1/T presentation Ln(P_{H_2O}) = A(N) + B(N)/T, with a slope depending on the water content N and giving an isosteric water sorption heat $\Delta H_{is}(N) = B(N)$ R, where R is the universal gas constant (Table I). For N > 2 this value (43.9 ± 1.3 kJ/mol) is close to the water evaporation heat from aqueous CaCl$_2$ solutions [10]. A significant increase in ΔH_{is} at

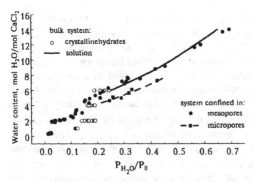

Fig.2. Universal water sorption isotherm for the CaCl$_2$-H$_2$O system confined to micro- and mesopores.

N ≤ 2 is caused by the formation of solid hydrates where water molecules are bound stronger.

If water sorption is presented as a function of the relative vapor pressure η = P_{H_2O}/P_0, all the experimental data satisfactorily follow the same curve that can be considered as a universal isotherm of the water sorption on the SWS-1L material (Fig.2). This isotherm gives a one-to-one correspondence between the sorbed water amount and the relative water vapor pressure. One can observe once more the plateaus corresponding to N=0.4 and N=2 as well as a shoulder at $N \approx 4$. At $P_{H_2O}/P_0 > 0.15\text{-}0.2$ the isotherm is a smooth curve which essentially coincides with the proper curve for the bulk aqueous solutions of calcium chloride [7]. It means that the solution confinement to mesoporous silica gel does not change its water sorption properties as compared to the bulk solution. Our DSC measurements (see below) of the specific heat of $CaCl_2$ aqueous solutions dispersed in the mesoporous matrix also have not established any changes in the C_p value due to the confinement to the pores.

Table I. Isosteric heat of water sorption ΔH_{is} for $CaCl_2$ - H_2O system confined to mesopores.

N, $\frac{mol\ H_2O}{mol\ CaCl_2}$	1	2	3	4	5	6	7	8	9	10
ΔH_{is} (kJ/mol)	63.1	62.3	42.2	43.9	45.6	44.7	44.3	45.6	45.1	43.9

Solidification - melting phase diagrams

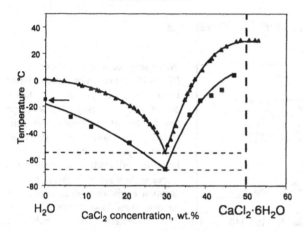

The phase diagram for $CaCl_2$ aqueous solutions confined to the KSKG pores (Fig. 3) appears to belong to the same simple eutectic type as for the bulk solution [7]. The diagram clearly shows the melting temperature depression by 10-30 K over the whole concentration range from pure water (is pointed by an arrow) to the highest stoichiometric crystalline hydrate ($CaCl_2 \cdot 6H_2O$). It is noteworthy that the phase diagrams for both the bulk and dispersed solutions do not follow the Schreder equation

Fig.3. Phase diagram "solidification-melting" for the bulk (▲) and disperse in the KSKG silica gel (■) binary mixtures H_2O - $CaCl_2$.

[11], indicating a strong deviation from the behaviour typical for ideal binary solutions.

Besides the mentioned above melting point depression, a strong supercooling of solution before solidification and a reduction of the phase transition enthalpy were detected. This is an evidence of the significant influence of the pore geometry on the solid-liquid equilibrium in the confined $CaCl_2$-H_2O binary system.

Specific heat

Specific heat of the confined $CaCl_2$-H_2O system ranges from 0.7 to 2.2 J/gK depending on temperature and water content. In order to study the influence of the solution confinement to the silica mesopores a comparison of the Cp-values for the dispersed and bulk solutions has been done. Fig.4 clearly shows no changes in the solution specific heat due to its confinement to the silica mesopores.

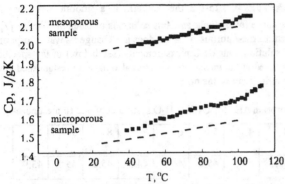

Fig.4. Specific heat of the CaCl₂ aqueous solutions: dispersed in the silica pores(■), bulk (dashed lines).

Water - calcium chloride system confined to the micropores

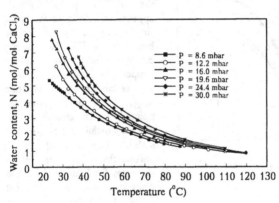

Fig.5. Water sorption isobars for the SWS-1S material.

Water sorption isobars for this system (SWS-1S) appear to be quite different from those measured for CaCl₂ confined to the mesoporous silica gel (Fig.5). No plateaus indicating solid hydrate formation are recorded. Water sorption is found to decrease monotonically with the temperature increase showing a divariant type of sorption equilibrium even at the salt concentration lower than 86 wt.%. This concentration corresponds to 1 molecule of sorbed water per 1 calcium chloride molecule. Contrary to a monovariant gas-solid equilibrium, a divariant equilibrium is typical for liquid salt solutions.

Water sorption isosters on the SWS-1S can be satisfactorily approximated with straight lines. Their slopes give an isosteric heat of water sorption ΔH_{is}. Values of $\Delta H_{is}(N)$ are presented in Table II for N = 2 – 7. This value tends to decrease with the water content increase and at N = 7 it approaches the value of the evaporation heat of bulk CaCl₂ aqueous solutions.

Table II. Isosteric heat of water sorption ΔH_{is} for CaCl₂ - H₂O system confined to micropores

N, $\frac{mol\ H_2O}{mol\ CaCl_2}$	2	3	4	5	6	7
ΔH (kJ/mol)	69.0	62.3	56.0	50.2	46.0	43.9

The water sorption dependence on the relative vapor pressure η (Fig.2) allows further comparison of the confined and bulk solutions. This isotherm lies below the proper curve for the ordinary bulk solutions. It means that the solution confinement to the micropores decreases its ability to bind water. Indeed, at a fixed solution concentration the relative vapor pressure over the solution in micropores is higher than that over an ordinary solution. Of course, this result looks surprising because the vapor pressure of wetting liquids commonly decreases in micropores due to the capillary effect [12]. It is reasonable to suppose that the observed increase in the vapor pressure is caused by a significant change in the thermodynamic properties of the solution due to its confinement to the micropores. Another example of such change is an increase in the solution specific heat in the micropores (Fig.4).

Thermodynamic properties of electrolyte solutions are determined mainly by electrostatic interactions between positive and negative ions of dissociated salt. Debye and Huckel suggested an electrostatic model describing thermodynamic properties of strong electrolyte dilute solutions [13]. They concluded that any ion in solution is surrounded by an envelope of other ions with a characteristic radius R which can be calculated as [13,14]

$$R = A \left[(D\ T)/(\varepsilon^2\ I) \right]^{1/2} \quad (1)$$

Here A is a coefficient, D is the ion diffusion coefficient, T is the solution temperature, I is the solution ionic strength, ε is the solvent dielectric constant. Although the solutions are not dilute under our conditions, equation (1) can be used for a brief estimation of the ion atmosphere diameter. It appears to be about 0.2-0.5 nm. This value is much lower than the average pore diameter of the mesoporous silica gel (d/R = 15 nm/(0.2–0.6 nm) = 25–75 » 1). As a result, the mesopore walls do not influence the ion spatial distribution in the solution and do not change its chemical potential and water vapor pressure. For micropores of the KSM silica gel the ratio d/R = 6–17 still looks low. Nevertheless, the wall effect is expected to be significant if one takes into account that the near-wall layer of 0.4 nm thickness contains 23, 40 or 54% of the solution volume for 3.5 nm pores slit, cylindrical or spherical in shape, respectively. Thus, a large portion of ions are in the wall vicinity and their true spatial distribution may be strongly disturbed by near-wall geometrical restrictions. Indeed, any ion located near an uncharged pore surface has an energy excess with respect to a bulk ion, since it has no neighboring ions in the half-space behind the wall. As a result, thermodynamic properties of solution change radically due to confinement into small pores.

LiBr confined to the silica gel mesopores

Universal water sorption isotherm for the LiBr-H$_2$O system confined to mesopores is presented in Fig. 6. At $\eta > 0.1$ the sorption properties of the disperse and bulk LiBr solutions are close, and the sorption equilibrium is divariant. At lower η, the equilibrium is monovariant that indicates the formation of solid LiBr monohydrate, the water sorption properties of the hydrate in the disperse state being much higher than that at normal (bulk) conditions as it is also a case for CaCl$_2$ (see above). This indicates that despite of the salt nature the salt confinement to the silica gel mesopores does not change the thermodynamic properties of its aqueous solution but may strongly influence the solid hydrate-vapor equilibrium.

At high water content (N>1) the isosteric heat of water sorption (43 kJ/mol) turns out to be close to the water evaporation heat from the bulk LiBr solutions whereas a significant increase of sorption heat (up to 62 kJ/mol) is detected in case of the solid phase formation .

Solidification/melting phase diagram of the disperse LiBr-H₂O system is found to be of a simple eutectic type, and lies by 10-30K below a proper diagram in the bulk state. It is also similar to the low temperature behavior of CaCl₂-H₂O binary system confined to the silica gel mesopores. Strong supercooling of LiBr solution in mesopores and the reduction of its melting enthalpy serve as an extra confirmation of this conclusion.

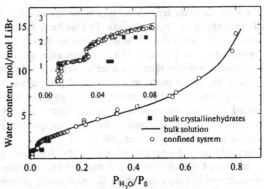

Fig.6. Water sorption isotherm for LiBr-H₂O system confined to the silica gel mesopores.

CONCLUSIONS

Thus, the results obtained allow to formulate of the following properties of the CaCl₂/H₂O and LiBr/H₂O systems confined to the silica nanopores:

in mesopores - the sorption equilibrium is either of mono- or divariant types; the impregnation of the salt solutions into mesopores does not influence their specific heat or vapor pressure, whereas the vapor pressure over the solid crystalline hydrates turns out to be lower than that for the bulk salt; the melting point of the confined solutions is depressed by 10-30 K in comparison with ordinary bulk systems.

in micropores - the sorption equilibrium is divariant, and no solid crystalline hydrates formation is detected; the vapor pressure over the confined solution is higher than that for the bulk one; the specific heat of the CaCl₂/H₂O solutions increases due to their confinement to the micropores.

REFERENCES

1. E.A.Levitskii, Yu.I.Aristov, V.N.Parmon, M.M.Tokarev: *Sol.Enegy Mater. Sol.Cells*, vol.31 (1996).
2. Yu.I. Aristov, M.M. Tokarev, G. Di Marco, G.Cacciola, D.Restuccia, V.N. Parmon: *Russian J. Phys. Chem.*, 1997, **71**, N 2, pp.253-258.
3. Yu.I. Aristov, M.M. Tokarev, G.Cacciola, D.Restuccia: *Russian J. Phys. Chem.*, 1997, **71**, N2, pp.391-394.
4. Yu.I.Aristov, M.M.Tokarev, G.Cacciola, G.Restuccia: *React. Kinet. Cat. Lett.*, **59**, N 2,1996, p.325
5. Yu.I.Aristov, G.Restuccia, M.M.Tokarev, G.Cacciola: *React. Kinet. Cat. Lett.*, **59**, N 2,1996, p.335.
6. Yu.I. Aristov, M.M. Tokarev, G. Di Marco, V.N. Parmon: *React. Kinet. Cat. Lett.*, **61**, N1, 1997.
7. *Gmelins Handbuch der Anorganischen Chemie, Calcium Teil B - Lieferung 2.* /Hauptredakteur E.H.Erich Pietsch. Verlag Chemie GmbH, 1957.
8. *Kirk-Othmer Encyclopedia of Chemical Engineering, 4th Ed.*, v.4, Wiley, New York, 1992.
9. G.S.Sinke, E.H.Mossner, J.L.Curnutt: *J.Chem.Thermodynam.*, **17**, 893 (1985).
10. B.M.Gurvich, R.R.Karimov, S.M.Mezheritskii: *Rus.J.Appl.Chem.*, **59**, 2692 (1986).
11. *Physical Chemistry*, Ed. S.N.Kondratiev, p.196, Vishaya shkola, Moscow 1978 (in Russian).
12. S.J.Greg, K.S.W.Sing: *Adsorption, Surface area and Porosity*, Academic Press, 1982.
13. P.Debye, E.Huckel, *Phys.Z.*, **24**, 185 (1923).
14. D.K.Chattoraj, K.S.Birdi: *Adsorption and the Gibbs Surface Excess*, Plenum, N.Y., 1984, p.99.

PREPARATION AND CHARACTERIZATION OF Ag-CLUSTER IN POLY(METHYLMETHACRYLATE)

NAOHISA YANAGIHARA,* YOSHITAKA ISHII, TAKANORI KAWASE, TOSHIMARE KANEKO, HISASHI HORIE, TORU HARA
Department of Materials Science and Engineering, Teikyo University, 1-1 Toyosatodai Utsunomiya 320, Japan

ABSTRACT

Solid sols of silver in poly(methylmethacrylate), Ag/PMMA, were prepared by bulk polymerization of methyl methacrylate (MMA) with benzoyl peroxide (BPO) as an initiator in the presence of silver(I) trifluoroacetate. Ag/PMMAs were characterized by visible spectroscopy. Effects of the concentration of initiator, the concentration of silver (I) complex and the heat-treatment time on the formation of silver cluster were studied in detail.

INTRODUCTION

Colloidal dispersions of metal nanoclusters are of great attention in various fields of science.[1] For example, in chemistry well characterized and stable transition metal particles of a narrow size distribution (ideally single sized or monodispersed), are of significant interest in catalysis,[2] meanwhile in the fields of physics[3] the metal clusters are very important to create entirely new materials with made-to-order electronic, magnetic and optical properties.[4-6]

In order to use such metal nanoclusters for the materials in optical devices, chemical and optical stabilities are required for the dispersed particles. Moreover, a simple fabrication method for processing at relatively low temperatures is recommended. On the other hand, inorganic glasses have been mainly used for the base matrixes in the aforementioned materials,[7-9] and there have been a few reports on the preparative methods using organic polymer matrixes.[10-12] Among these preparative methods reported, that proposed by Nakao[12] might be one of the most promising methods. According to his procedure, solid sols of both thermoplastic and thermosetting resins[12,13] containing well dispersed various noble particles such as Au, Ag, Pt and Pd can be easily prepared, upon simply polymerizing a monomer-dissolved corresponding metal compound and followed by heat-treatment of the resultant sample at no more than 140 °C. However, in those reports, no consideration has been taken into the effect of concentration of initiator on the preparation of noble metal solid sols. In order to control the particle size, it is important to clarify the growth process in the polymer matrix. In this paper, we report on the preparation and characterization of Ag /PMMAs, which have been prepared by changing various concentrations of BPO and Ag complex, and also heat-treatment times.

EXPERIMENT

Benzoyl peroxide (BPO) was purified by precipitation from chloroform into methanol and then crystallized in methanol at 0 °C. Commercial grade methyl methacrylate (MMA) was distilled over anhydrous sodium sulfate prior to use. Silver(I) trifluoroacetate ($AgCF_3CO_2$) was prepared based on literature and recrystallized from hot benzene.[14]

Preparative procedure of solid sols of Ag in PMMA was almost the same as that reported previously[12], but we made a little modification. A typical preparation was as follows. $AgCF_3CO_2$ (5×10^{-2} M) and BPO (2.0×10^{-2} M) were dissolved in MMA (20 mL). This solution was first heated at 60 °C for 30-60 min. The resulting viscous syrup was then transferred into a mold that consists of two 5 x 10 cm glass plates and a polyethylene flame (1 mm thickness), and heated again at 50 °C for 20 h to complete the polymerization. After removal from the mold, a clear PMMA plate containing Ag cluster was obtained. The PMMA plates thus obtained were then post-heated for various times at 120 °C.

The variation of the Ag cluster formation in PMMA with varying initial concentrations of $AgCF_3CO_2$ and BPO (hereafter they are abbreviated as $[Ag]_0$ and $[BPO]_0$, respectively) and heat treatment (post-heating) time was investigated.

The absorption spectrum of Ag colloids was used mainly to characterize the formation of Ag clusters. Optical absorption spectra of the plates were measured in the range from 300 to 700 nm by a Shimadzu UV-3100 spectrometer at room temperature. The diameter of the Ag particles was observed directly by transmission electron microscopy (JEOL, JEM- 2000FXII).

RESULTS AND DISCUSSION

Figure 1 shows the absorption spectra of Ag/PMMA that were prepared by various concentrations of $[Ag]_0$ with $[BPO]_0 = 2.0 \times 10^{-2}$ M, heat-treated at 120 °C for 24 h. As expected, an absorption peak was observed around 420 nm. The absorption band grows depending on the $[Ag]_0$ and shifts slightly to shorter-wavelength side.

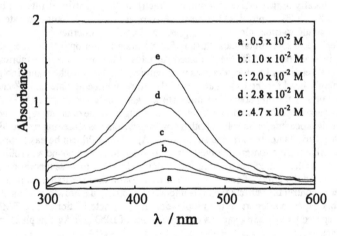

Fig. 1. Absorption spectra of Ag/PMMA prepared by various $[Ag]_0$ concentrations. ($[BPO]_0 = 2.0 \times 10^{-2}$ M, heat-treated at 120 °C for 24 h.)

Shown in Fig. 2 are the absorption spectra of Ag/PMMA that were post-heated at 120 °C for 7-42 h, while $[Ag]_0$ and $[BPO]_0$ were kept constant in the polymerizations. The behavior of peak growth is similar to that observed in Fig. 1, i.e., the longer the duration of heat-treatment, the larger the absorbance. However, when a sample was post-heated for more than 42 h, it was observed that metallic silver, like a silver mirror, begins to separate on the surface of Ag/PMMA plate. The absorption band of such sample tends to be rather broad as shown in Fig. 2(e).

It is known that the reduction of silver ions generally leads to a yellow sol of colloidal particles

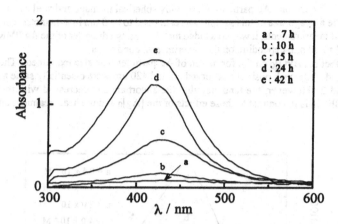

Fig. 2. Absorption spectra of Ag/PMMA prepared by various heat-treatment durations at 120 °C. ($[Ag]_0 = 2.8 \times 10^{-2}$ M, $[BPO]_0 = 2.0 \times 10^{-2}$ M)

with several nanometers in a diameter.[15] The absorption spectrum of such a colloid contains a rather narrow band peaking around 380-430 nm.[16] This band is caused by surface plasmon absorption of the electron gas in the particles, and once this band is observed one may conclude that one is dealing with particles having metallic properties.[15,16] Moreover, it has been understood both theoretically and experimentally that the absorbance is increased and the full width at half maximum of the absorption band is decreased with increasing the diameter of Ag particles.[9,11,17,18] Thus, the results of UV measurements obtained in the present study are consistent with the evidences that the Ag/PMMA samples contain metallic Ag particles, and that the growth of Ag particles depends on the $[Ag]_0$ and the heat-treatment time.

The formation and growth of the Ag particles were reconfirmed by TEM measurements. Figure 3 is a typical TEM photograph of the yellow colored Ag/PMMA sample, which was made by the following condition; $[Ag]_0 = 2.8 \times 10^{-2}$ M, $[BPO]_0 = 2.0 \times 10^{-2}$ M, and heat-treated at 120

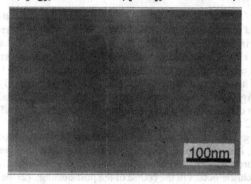

Fig. 3. Transmission electron micrograph of a typical Ag/PMMA sample. ($[Ag]_0 = 2.8 \times 10^{-2}$ M, $[BPO]_0 = 2.0 \times 10^{-2}$ M, heat-treated at 120 °C for 24 h.)

471

°C for 24 h. It is clear that Ag particles are roughly spherical in shape and well but not uniformly dispersed. The average size of this sample was estimated to be 5 nm in a diameter. On the basis of the other TEM measurements, it was concluded that the average diameter of the Ag/PMMA samples varies from 3 to 10 nm, depending on the experimental conditions.

The effect of $[BPO]_0$ on the formation of Ag particles was also examined. The results are shown in Fig. 4. The peak positions observed around 420 nm were essentially same as those seen in Figs. 1 and 2. However, the tendency that the absorbance is decreased with increasing the amount of $[BPO]_0$ is in contrast to those effects of the $[Ag]_0$ and the heat-treatment duration.

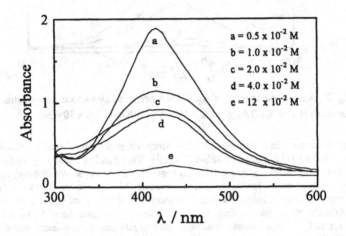

Fig. 4. Absorption spectra of Ag/PMMA prepared by various $[BPO]_0$ concentrations. ($[Ag]_0 = 2.8 \times 10^{-2}$ M, heat-treated at 120 °C for 24 h.)

Since the role of BPO in the formation of Ag particles seems to be complicated, we attempted to carry out additional experiments. Two cast films of the Ag/PMMA were prepared by a distinct manner, keeping the initial concentrations of $[Ag]_0$ and $[BPO]_0$ to be the same. The preparative procedure to make one film (procedure A) was almost identical to the procedure aforementioned in this paper: i.e., upon obtaining a Ag/PMMA plate, a cast film was made by dissolving the plate sample in toluene. The other cast film was made as follows (procedure B); firstly a PMMA was obtained, and then $AgCF_3CO_2$ was dissolved into toluene solution containing PMMA and this polymer solution was cast to prepare a film of Ag/PMMA. Both of the Ag/PMMA cast films were heat-treated at 120 °C for 24 h.

As shown in Fig. 5, the absorbance of the Ag/PMMA prepared by procedure A is very strong and clear. However, it is noteworthy that no growth of the absorption peak is observed for the sample prepared by procedure B. These results reveal that: (1) neither the reduction of Ag ion nor the formation of the cluster occurs in the course of the post-heating; (2) the reduction of Ag ion to metallic Ag must be done before the heat-treatment in order to form the Ag cluster. Therefore, it is reasonable to consider that the nucleation (formation of very small Ag particles) followed by the reduction of Ag ion and the growth (aggregation of small particles and formation of lager Ag particles) proceed independently.

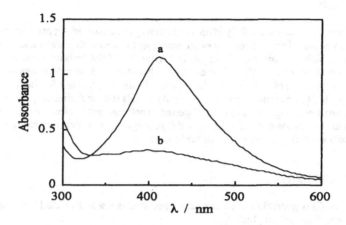

Fig. 5. Absorption spectra of cast film of Ag/PMMA: (a) prepared from
Ag/PMMA sample; (b) prepared by dissolving AgCF₃CO₂ in PMMA.
([Ag]$_0$ = 2.8 x 10^{-2} M, [BPO]$_0$ = 2.0 x 10^{-2} M, heat-treated at 120 °C for 24 h.)

If so, a question which chemical species governs the reduction of Ag ion is raised. Without any kinetics data on the polymerization of MMA, it is ambiguous to clarify the reducing species. It is, however, true that BPO itself does not participate in the reduction of Ag ions, since the absorbance is decreased with increasing the concentration of BPO (see Fig. 4.). Thus, it is probable that Ag ions are reduced by either monomer radicals or growing polymer radicals to form a number of small metallic during the polymerization. Once they formed, these small Ag particles are dispersed in the amorphous polymer matrix after the polymerization.

As already shown in Fig. 2, the peak intensity due to the surface plasmon resonance of Ag particles is increased with increasing the duration of post-heating of the Ag/PMMA. It is pointed out that heat-treatment is a useful technique to control the size of metal particles or semiconductor microcrystallites in the glass.[19, 20] It is also reported that the particles grow by diffusion in the coalescence process when the concentration of the particles in the supersaturated glass solution approaches the solubility limit.[21,22]

In our case, the temperature applied for the post-heating of Ag/PMMA samples is just above the glass transition temperature and near the melting (or heat deflection) point of PMMA. In this temperature region, PMMA is no longer in the glassy state, but liquid-like motion of polymer segments is allowed. Thus, not only the polymer segments but also Ag particles dispersed in the polymer matrix can move easily even at the relatively low temperature of 120 °C. However, the ease of the diffusion of Ag particles is probably not the same, and depends on the length of the polymer segments. It is well known that the higher the initiator concentration, the lower the number-average molecular weight (i.e., the shorter the average polymer length). This implies that the polymer having longer segments trends to form coiled polymer chains. Consequently, the small Ag particles dispersed in the PMMA matrix prepared with the low concentration of BPO are caged in the long coiled chains. The Ag particles under such circumstances do not move far away. Therefore, they easily aggregate to form large clusters during the heat-treatment. On the other hand, it might be somewhat difficult to form large clusters for the Ag particles in the polymer matrix that consist of shorter segments, since they are too dispersed to aggregate.

CONCLUSIONS

We have prepared successfully PMMA containing the dispersed Ag particles of nanometer size by simply polymerizing the corresponding monomer together with initiator and Ag compounds and by subsequent heat-treatment. It was proved that the size of the Ag is increased with increasing the initial concentrations of $[Ag]_0$ and the duration of heat-treatment, and decreases with increasing the initial concentration of $[BPO]_0$. A following mechanism is proposed for the formation of Ag cluster in PMMA: (1) the reduction of Ag ions to Ag metals is done during the polymerization; (2) these small particles of Ag that have been dispersed in polymer matrix after the polymerization act as seeds for the nucleation of small clusters; and (3) the aggregation of clusters to grow up to larger clusters take place finally in the course of heat-treatment.

ACKNOWLEDGMENTS

The author (N.Y.) is grateful for the financial support of this research by the Foundation of Sinsei Shigen Kyokai, Fujikura Co., Ltd.

REFERENCES

1. K. Rademann, Ber. Bunsenges. Phys. Chem. **93**, 653(1989).
2. G. Schmid, Aspects Homogeneous Catal. **7**, 1(1990).
3. M.A. Duncan and D. H. Rouvray, Scientific American 1989, 110.
4. E.J. Heolweil and R.M. Hochestrasser, J. Chem. Phy. **82**, 4762(1985).
5. D. Ricard, P. Roussignol and C. Flytzanis, Opt. Lett. **10**, 511(1985).
6. F. Hache, D. Ricard, C. Flytzanis and U. Kreibig, Appl. Phys. **A47**, 347(1988).
7. U. Kreibig, Appl. Phys. **10**, 255(1976)
8. T. Akai, K. Kadono, H. Yamanaka, T. Sakaguchi, M. Miya, and H. Wakabayashi, J. Ceram. Soc. Jpn. **101**, 105(1993).
9. I. Takahashi, M. Yoshida, Y. Manabe, and T. Yohda, J. Mater. Res. **10**, 362(1995).
10. A.K.St. Clain and L.T. Taylor, J. Appl. Polym. Sci. **28**, 2393(1983).
11. S. Ogawa, Y. Hayashi, N. Kobayashi, T. Tokizaki, and A. Nakamura, J. Appl. Phys. **33**, L331(1994).
12. Y. Nakao, J. Chem. Soc., Chem. Commun. 1993, 826.
13. Y. Nakao, Kobunshi **43**, 852(1994); Zairyou Kagaku **31**, 28(1994).
14. A.N. Soto, N. Yanagihara, and T. Ogura, J. Coord. Chem. **38**, 65(1996).
15. T. Linnert, P. Mulvaney, A. Henglein, and H. Weller, J. Am. Chem. Soc. **112**, 4657(1990).
16. U. Kreibig and L. Genzel, Surface Science **156**, 678(1985).
17. D. Fornasiero and F. Grieser, J. Colloid Interface Sci. **141**, 168(1991).
18. S.M. Heard, F. Grieser, and C.G. Barraclough, J. Colloid Interface Sci. **93**, 545(1983).
19. L.C. Liu and S.H. Risbud, J. Appl. Phys. **68**, 28(1990).
20. A.I. Ekimov, A. L. Efros, and A.A. Onushchenko, Solid State Commun. **56**, 921(1985).
21. J. Fu, A. Osaka, T. Nanba, and Y. Miura, J. Mater. Res. **9**, 493(1994).
22. H. Nasu, S. Kaneko, K. Tsunetomo, and K. Kamiya, J. Ceram. Soc. Jpn. **99**, 266(1991).

CARBON BLACK-FILLED POLYMER BLENDS:
A SCANNING PROBE MICROSCOPY CHARACTERIZATION

Ph. LECLÈRE [1], R. LAZZARONI [1], F. GUBBELS [2], C. CALBERG [2], Ph. DUBOIS [2], R. JÉRÔME [2], and J.L. BRÉDAS [1]

[1] Service de Chimie des Matériaux Nouveaux, Centre de Recherche en Electronique et Photonique Moléculaires, Université de Mons-Hainaut, B-7000 Mons, Belgium
[2] Centre d'Etude et de Recherche sur les Macromolécules (CERM) Université de Liège, Institut de Chimie, B-4000 Sart-Tilman, Belgium

ABSTRACT

Conducting polymer composites, that consist of a conducting filler randomly distributed throughout an insulating polymer or polymer blend, attract interest in several application fields such as sensors or electromagnetic radiation shielding. The macroscopic electrical resistivity of the filled polyblend strongly depends on the localization of the filler. Here, we investigate the morphology of Carbon Black (CB)-filled polymer blends in order to determine the parameters governing the selective localization of CB in one phase of the blend components or at the interface between the components. The dispersion of the CB particles in the polymer blend is observed by means of Lateral Force Microscopy (LFM) as a function of the blend composition and the load in CB. The selective localization of CB at the interface enables the reduction of the percolation threshold down to 0.5 wt%; as a result, the mechanical properties of the polymer blend can be fully retained. Different techniques can be used to locate the CB at the interface; we compare their efficiency experimentally.

INTRODUCTION

Polymers made electrically conductive by loading with a conductive filler have been known and used for decades [1, 2]. For instance, we can cite their use as antistatic or electromagnetic shielding materials as well as piezoresistive materials [3] (pressure sensors, switches electrical safety devices, and self-regulated heaters [4]).

A better knowledge is still required of the actual structure of the clustered particles, the structure formation during material processing, and its relationship with the macroscopic properties. Tools such as electronic microscopes and scanning probe microscopes (Scanning Tunneling Microscopy and Atomic Force Microscopy) can therefore be of major help. In this context, scanning probe potentiometers [5, 6] have been used to examine electrostatic forces on the surface, however with a low lateral resolution. Recently, Electric Force Microscopy (EFM) has been proposed [7] as a new type of scanning probe microscopy that is able to measure electric field gradients near the surface of a sample when using a sharp conductive tip.

Note that a major problem in the production of such composites is the filler content. This must be kept as low as possible since otherwise processing becomes difficult, the mechanical properties of the composites are poor, and the final cost is high (high-grade conductive fillers are indeed expensive). In this context, our aim is to set up a strategy to decrease the filler content by *combining the advantages of composites* (polymer/filler combinations) and *polymerblends* (polymer/polymer combinations). Polymer blends with a co-continuous two-phase structure, *i.e.*, a morphology with a dual phase continuity, have been extensively discussed in relation to percolation theory [8-10]; percolation of the filler particles in one of the continuous phases or at the interface of a co-continuous binary polyblend is a complex but very attractive situation, since a double percolation (one for the polymer phase and one for the filler in this phase or at the interface) phenomenon results in significant electrical conductivity at a very low filler content.

EXPERIMENTS

High-density polyethylene (PE) (Solvay Eltex B3925: M_n = 8,500, M_w = 265,000, density 0.96, melt index < 0.1), polystyrene (PS) (BASF Polystyrol 158K: M_n = 100,000, M_w = 280,000, density 1.05, melt index 0.39) and carbon black particles (CB) (Degussa Printex XE-2 (XE) or Cabbot Black Pearls BP-1000 (BP)) are introduced in an internal mixer (Brabender Rheomixer) at 200°C. Electrical measurements are performed with the four-probe technique (to prevent resistance from the sample/electrode contacts). Atomic Force Microscopy (height and friction) images are recorded with a Digital Instruments Nanoscope III microscope, operated in contact mode at room temperature in air, using a 100 μm triangular cantilever (spring constant of 0.58 Nm^{-1}).

RESULTS

Electrical and morphological characterization of CB-filled homopolymers

For low CB concentrations, the resistivity is close to that of the polymer matrix, on the order of 10^{11} to 10^{16} Ω.cm [11]. When the CB concentration increases, the resistivity undergoes a fast decrease by several decades over a narrow concentration range corresponding to the percolation threshold; it then decreases more slowly towards the limiting resistivity of the compressed filler powder of order 10^4 to 1 Ω.cm. The resistivity [12] of the composites obeys a power law of the form $\rho = (p - p_C)^{-t}$ near the transition, where ρ is the bulk resistivity of the composite, p is the concentration of the conductive component, p_C is the percolation threshold concentration, and t is a universal exponent. Prediction of the exact percolation threshold remains difficult, as the critical volume concentration value can be observed in the 5-30 % range [8-13]; understanding such a broad range of critical concentrations is not easy since the main results of percolation theory is that the threshold should be close to 20% [13]. For XE, the values of p_C are 5% and 8% in PE and PS, respectively; for BP, the corresponding values are 12% and 25% [11]. From these results and for a given type of CB, it is seen that substitution of a monophasic polystyrene by a two-phase semicrystalline polyethylene favors a decrease in the percolation threshold (*i.e.* from 8% and 25% down to 5% and 12%, for PS and PE respectively). This is consistent with the selective localization of CB particles in the amorphous phase of PE; increasing the degree of crystallinity would thus be a potential way of decreasing further the percolation threshold.

Characterization of the CB-filled polymer blends

These systems can be characterized in direct space at the nanometer scale by using Lateral Force Microscopy (or Friction Force Microscopy) [11]. Since their friction coefficients μ are different, it is possible to distinguish PE, PS, and CB particles (Figure 1). It appears from the collected images that PE is characterized by a higher value of μ, which translates on the image to a lighter color on the gray-scale; CB appears in black, since there is almost no interaction between the tip and the CB particles; the μ value for PS is intermediate and PS therefore appears in dark gray on the image. The black spots attributed to CB particles are observed to be dispersed only in the PE phase of a PE/PS polymer blend. This indicates that in these experimental conditions (45% PE, 55% PS, and 1% XE), CB prefers to localize in the PE phase. For this system, the percolation threshold is about 2.9%. In the case of BP-type CB, the CB particles then localize exclusively in the PS phase and the percolation threshold is about 10.9%.

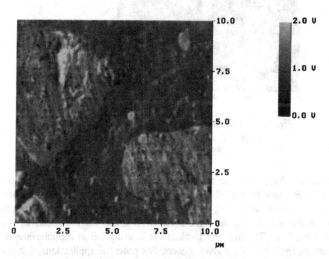

Figure 1. Detailed LFM image of a PE/PS/XE (45/55/1) sample.
PE is brighter and PS darker; black spots in PE phase correspond to CB aggregates.

The percolation threshold can be decreased by selective localization of CB in the smaller phase of a co-continuous PE/PS blend [14, 15]. To further decrease the value of p_C, one can exploit topology arguments that indicate that the interface of a co-continuous morphology is continuous through the volume: localization of the CB particles at the PE/PS interface should thus drastically reduce the p_C value. We describe below two ways to achieve such a localization via either a kinetic or thermodynamic process.

Kinetic localization of CB at the PE/PS interface. The idea, previously proposed by Gubbels (16), is to mix first the CB particles with the less preferred phase (for instance XE with PS or BP with PE) and thereafter to add the second polymer. The CB particles then tend to migrate slowly from the first polymer to the more "attractive" one. By stopping the mixing at regular time intervals and recording an LFM image, we can determine the optimal time when the CB particles are mainly located at the PE/PS interface. Figure 2 gives the corresponding LFM image for the optimized mixing time for the system (45% PE, 55% PS and 1% of BP-CB): there clearly appears to be a CB layer between PE and PS. The thickness of this layer is about 100 nm. For this system, the percolation threshold value is as low as 0.60%.

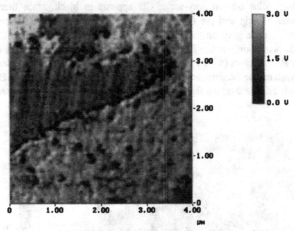

Figure 2. LFM characterization of the localization of CB
at the PE/PS interface by a kinetic process
(PE/PS/CB-BP = 45/55/1).

With mixing time, electrical resistivity decreases to reach a value of 2.3×10^4 Ω.cm at the optimized mixing time t_C for which CB is located at the PE/PS interface. For mixing times longer than t_C, resistivity starts going up again with time, due to dilution of CB in the PE phase. When using BP instead of XE, the same behavior is found for the resistivity evolution as a function of mixing time. The optimized values for locating CB at the interface is in a narrow range and must be optimized for a given system. For potential applications, it is thus preferable to rely on another process that avoids this problem.

Thermodynamic localization of CB at the PE/PS interface. We have indicated above that BP prefers to be located in the PS phase and XE in the PE phase. Since XE is more graphitic and the BP particles present a larger number of irregularities (such as holes and steps) and a higher oxidation rate due to the presence of oxygen-rich functional groups at the surface, the particle pH appears to be a good parameter to characterize the behavior of the CB particles. The idea proposed by Gubbels [16] is to slowly increase the CB particle pH. We then examine the corresponding resistivity evolution and the LFM images [11]. We observe the migration of CB particles from the PS phase (at low pH) to PE phase (at high pH). For intermediate values of pH, we can expect to locate the CB particles at the PE/PS interface independently from the mixing time. Figure 3 corresponds to a CB particle characterized by an intermediate pH. In this case, a CB layer about 120 nm thick (appearing in black on the LFM image) is clearly seen at the PE/PS interface. The percolation threshold in this system is 0.46%, a remarkably low value.

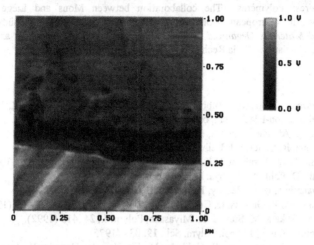

Figure 3. LFM image of a PE/PS/CB polymer blend
for which the CB particle pH is intermediate.

CONCLUSIONS

In an amorphous homopolymer matrix, such as PS, the percolation threshold p_C is close to 8%. Substitution of a monophasic PS by a two-phase semicrystalline PE favors a decrease in p_c down to 5%, consistent with the selective localization of CB particles in the amorphous phase of PE. The experimental results reported here emphasize that co-continuous polymer alloys of insulating immiscible polymers (PE and PS) can be endowed with electrical conductivity at even smaller concentrations of conductive CB particles. The key tools in the design of such conducting polymer composites are: (i) polymer blend co-continuity and (ii) selective localization of CB at the interface. Since the components are characterized by different friction coefficients, LFM constitutes a useful technique for the morphology characterization of such systems at the nanometer scale. A double percolation is the basic requirement for electrical conductivity. Provided that CB is selectively localized at the polymer alloy interface, the CB percolation threshold p_C can be as low as 0.4 wt %, *i.e.*, a striking 0.002 volume fraction. This strategy is not restricted to the loading of polymer blends with carbon black fillers; particles of intrinsically conducting polymers could be used as well.

ACKNOWLEDGEMENTS

The authors are grateful to R. Deltour and M. De Vos for some of the electrical measurements. The research in Mons is supported by the "Ministère de la Région Wallonne (DGTRE: Programme mobilisateur ALCOPO)", the Belgian Federal Government Office of Science Policy (SSTC) "Pôles d'Attraction Interuniversitaires en Chimie Supramoléculaire et

Catalyse", the Belgian National Fund for Scientific Research FNRS/FRFC, and an IBM Academic Joint Study. The research in Liège is supported by the SSTC "Pôles d'Attraction Interuniversitaires: Polymères". The collaboration between Mons and Liège is partially supported by the European Commission (Human Capital and Mobility Network: *Functionalized Materials Organized at Supramolecular Level*). RL and PhD are chercheurs qualifiés du Fonds National de la Recherche Scientifique (FNRS - Belgium).

REFERENCES

(1) R.M. Norman, Conductive Rubbers and Plastics; Elsevier: New-York, NY, 1970.

(2) E. Sichel, Carbon-Black Composites, Eds.; Dekker:New-York, NY, 1982.

(3) F. Carmona, Ann de Chim. Fr., **13**, 395 (1988).

(4) F. Carmona, R. Canet, P.J. Delhaes, Appl. Phys., **61**, 2550 (1987).

(5) Y. Martin, D.W. Abraham, H.K. Wickramasinghe, Appl. Phys. Lett., **52**, 1103 (1988).

(6) P. Muralt, D. Pohl, Appl. Phys. Lett., **48**, 514 (1986).

(7) R. Viswanathan, M.B. Heaney, Phys. Rev. B, **75**, 4433 (1995).

(8) G. Geuskens, J.L. Gielens, D. Geshef, R. Deltour, Eur. Polym. J., **23**, 993 (1987).

(9) S. Asai, K. Sakata, M. Sumita, K. Miyasaka, Polym. J., **24**, 415 (1992).

(10) C. Klason, J. Kubát, J. Appl. Polym. Sci., **19**, 831 (1975).

(11) Ph. Leclère, R. Lazzaroni, F. Gubbels, M. De Vos, R. Deltour, R. Jérôme, and J.L. Brédas, to be published in ACS Symposium Series "Scanning Probe Microscopy in Polymers", edited by V. Tsukruk and B. Ratner (1997).

(12) S. Kirkpatrick, Rev. Mod. Phys., **45**, 574 (1973).

(13) J.P. Clerc, G. Giraud, J.M. Laugier, J.M. Luck, Adv. Phys., **39**, 190 (1990).

(14) F. Gubbels, R. Jérôme, Ph. Theyssié, E. Vanlathem, R. Deltour, A. Calderone, V. Parente, J.L. Brédas, Macromolecules, **27**, 1972 (1994).

(15) F. Gubbels, S. Blacher, E. Vanlathem, R. Jérôme, R. Deltour, F. Brouers, Ph. Teyssié, Macromolecules, **28**, 1559 (1995).

(16) F. Gubbels, Ph. D. thesis, University of Liège, 1995.

Synthesis and Characterization of
Aerogel-derived Cation-substituted Hexaaluminates

Lin-chiuan Yan and Levi T. Thompson

Department of Chemical Engineering, The University of Michigan,
Ann Arbor, MI 48109-2136 USA

ABSTRACT

New methods have been developed for the synthesis of high surface area cation-substituted hexaaluminates. These materials were prepared by calcining high temperature (ethanol extraction) or low temperature (CO_2 extraction) aerogels at temperatures up to 1600°C. Cation-substituted hexaaluminates have emerged as promising catalysts for use in high temperature catalytic combustion. In comparing unsubstituted and cation-substituted hexaaluminates, we found that the phase transformations were much cleaner for the cation-substituted materials. $BaCO_3$ and $BaAl_2O_4$ were intermediates during transformation of the unsubstituted materials, while the cation-substituted materials transformed directly from an amorphous phase to crystalline hexaaluminate. Moreover, the presence of substitution cations caused the transformation to occur at lower temperatures. Mn seems to be a better substitution cation than Co since the Mn-substituted materials exhibited higher surface areas and better heat resistances than the Co-substituted materials. The low temperature aerogel-derived materials possessed quite different characteristics from the high temperature aerogel-derived materials. For example, phase transformation pathways were different.

INTRODUCTION

The development of low emission, high efficiency combustors is encouraged by strict environmental regulations and global energy shortage issues. High temperature catalytic combustion has received a great deal of attention as an approach to simultaneously eliminate NO_x and CO emissions (1-3). The use of a catalyst results in an enhanced combustion efficiency as well as a decreased NO_x emission due to lower operating temperatures.

Conventional combustion catalysts such as noble metals and perovskites are not suitable in high temperature applications due to severe sintering problems which result in the loss of active surface area (1). Furthermore, it is believed that, at high conversion levels, complete combustion of residual hydrocarbons is diffusion controlled rather than reaction controlled, so that the reaction rate is not dependent on the number of active sites but the available surface area (2). Therefore, high heat resistant catalysts with high surface areas are very desirable.

Cation-substituted hexaaluminates have received a great deal of attention for use as high temperature combustion catalysts because they maintain high surface areas and display good catalytic activities at high temperatures (3,4). Hexaaluminates are stabilized alumina; their crystalline structures are shown in Figure 1. Large cations such as Ba, La and Sr are introduced into alumina to form these layered hexaaluminate structures. Smaller cations like Mn and Co are added to enhance catalytic activity. These cations occupy some of the aluminum lattice sites and the resulting materials are called cation-substituted hexaaluminates. Hexaaluminates maintain high surface areas at high temperatures because their anisotropic layered structures suppress crystal growth (5,6).

In an attempt to fabricate high surface area hexaaluminates, aerogel precursors were employed. The calcined aerogels were expected to maintain higher surface areas because of their more open microstructures. A second objective of our work was to enhance their catalytic properties by substituting active species like Mn and Co into the hexaaluminate structure.

Mat. Res. Soc. Symp. Proc. Vol. 457 © 1997 Materials Research Society

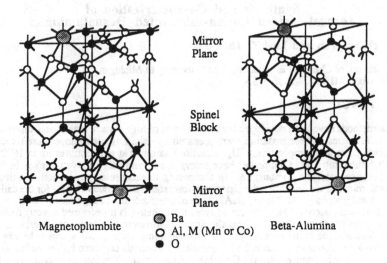

Figure 1. Crystalline structures of hexaaluminates.

EXPERIMENTAL

Preparation and drying of wet gel

Cation-substituted and doubly cation-substituted barium hexaaluminates, $BaMAl_{11}O_{19-\alpha}$ (M = Co and/or Mn)) were synthesized using sol-gel processes with composite metal alkoxides and nitrates as starting materials. First, manganese nitrate $(Mn(NO_3)_2)$ and/or cobalt nitrate $(Co(NO_3)_3)$ along with ethyl acetoacetate were dissolved in ethanol. The solution was mixed with the appropriate amounts of barium ethoxide $(Ba(OC_2H_5)_2)$ and aluminum sec-butoxide $Al(OC_3H_7)_3$. The resulting mixture was kept at 80°C for 30 minutes then hydrolyzed with water diluted in ethanol. A small amount of ammonia hydroxide was added to promote gelation. The resulting wet gels were aged at 60°C for at least 3 days before solvent extraction.

The supercritical solvent extraction was accomplished using two methods. In the first method, the wet gels were dried under supercritical conditions for ethanol (270°C and 1350 psi). The resulting materials were referred to as high temperature aerogels. The second method involved exchanging ethanol with liquid CO_2 then extracting the CO_2 under its supercritical condition (55°C and 1350 psi). The resulting materials from this method were referred to as low temperature aerogels since CO_2 has a much lower critical temperature than ethanol. For comparison, alumina aerogels were prepared in a similar fashion with $Al(OC_3H_7)_3$ as the starting material. Barium hexaaluminate xerogels were synthesized by drying the wet gels under ambient temperature and pressure.

Characterization of samples

The specific surface areas and pore size distributions were measured using a Micromeritics ASAP 2000. The crystalline phases for the dried gels and calcined samples were determined by X-ray diffraction (Rigaku DMAX-B) using CuKα radiation (λ=1.542Å).

RESULTS AND DISCUSSION

Specific surface area and pore size distribution

Surface areas of the aerogel-derived alumina and cation-substituted hexaaluminates, and the xerogel-derived hexaaluminates are compared in figure 2. The aerogel-derived hexaaluminate had higher surface areas than the alumina aerogels at high temperatures. This result demonstrates the superior heat resistance of the hexaaluminates. Furthermore, the xerogel-derived hexaaluminates exhibited significantly lower surface areas than their counterpart aerogel-derived materials at all temperatures. Similar results were reported by Mizushima and Hori (7) for unsubstituted aerogel-derived hexaaluminates. The higher surface areas for the aerogel-derived materials compared to the xerogel-derived materials may be a consequence of different pore size distributions. Figure 3 compares the pore size distributions of these materials. The aerogel-derived materials possessed much larger pores than the xerogel-derived samples.

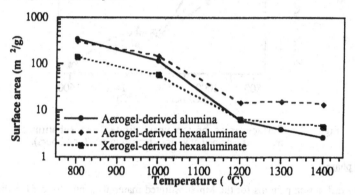

Figure 2. Surface areas of the aerogel- and xerogel-derived materials at various calcination temperatures (for 5 hours).

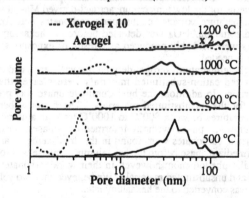

Figure 3. Pore size distributions of aerogel-derived and xerogel-derived materials at various calcination temperatures (for 5 hours).

Figure 4 illustrates the effect of calcination temperature on the surface areas of the high temperature aerogels. A dramatic decrease in surface area occurred for calcination temperatures higher than 1000°C and 800°C for the unsubstituted and substituted materials, respectively. This surface area loss appears to be a consequence of phase transformations which will be discussed in the next section. The Mn-substituted hexaaluminates displayed better heat resistance and higher surface areas than the Co-substituted materials when the calcination temperatures were higher than 1000°C. This result suggests that Mn is a better substitution cation than Co.

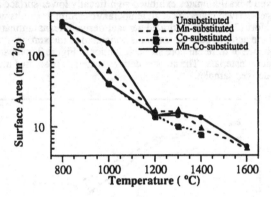

Figure 4.   Surface areas of the unsubstituted and substituted materials calcined at various temperatures (for 5 hours).

Crystal phases

X-ray diffraction patterns for the Mn-substituted materials synthesized from the high temperature aerogel precursors are shown in figure 5. The hexaaluminate phase was formed directly from an amorphous material. In contrast, the low temperature extracted samples demonstrated quite different phase transformation pathways. Figure 6 illustrates x-ray diffraction patterns for the low temperature aerogel-derived Mn-substituted materials. Unlike their high temperature extracted counterparts, carbonates were found immediately after extraction and some $BaAl_2O_4$ was detected before the hexaaluminate phase was formed. Carbonates were probably produced as a result of exposure to CO_2, however, further investigation is needed.

Phases for the high temperature aerogel-derived materials are listed in Table I. Phase transformations for the cation-substituted materials were cleaner than those for the unsubstituted materials. In addition, the hexaaluminate phase was produced at lower temperatures than for the unsubstituted materials. The cation-substituted materials transformed at temperatures between 800°C to 1000°C from an amorphous phase to the crystalline hexaaluminate phase without intermediate phases. However, for the unsubstituted materials, carbonates were found in the dried aerogels and $BaAl_2O_4$ was observed as an intermediate phase before the hexaaluminate phase was formed. Groppi et al.(8) observed similar phenomenon; however, in their work carbonates were present in both the substituted and unsubstituted materials. We believed that two solid state reactions occurred as $BaCO_3$ was converted to the hexaaluminate.

$$BaCO_3 + Al_2O_3 \quad \text{--->} \quad BaAl_2O_4 + CO_2$$

$$BaAl_2O_4 + 5\,Al_2O_3 \quad \text{---->} \quad BaAl_{12}O_{19} \text{ (hexaaluminate)}$$

The second reaction is very slow and requires long calcination times at temperatures above 1200°C (9). The results indicate that the dramatic loss in surface area at temperatures between 800 and 1200°C was partly due to phase transformations.

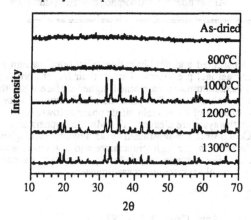

Figure 5. X-ray diffraction patterns of the Mn-substituted high temperature aerogel-derived materials.

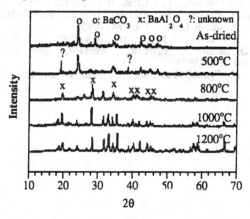

Figure 6. X-ray diffraction patterns of the Mn-substituted low temperature aerogel-derived materials.

485

Table I Phases present in the high temperature aerogel-derived materials following calcination at various temperatures for 5 hours.

Temperature	as dried	800°C	1000°C	1200°C
Unsubstituted	$BaCO_3$	$BaCO_3$	$BaCO_3$ $+ BaAl_2O_4$	$BaAl_2O_4$ $+$ Hexaaluminate
Substituted	Amorphous	Amorphous	Hexaaluminate	Hexaaluminate

SUMMARY

New cation-substituted and doubly cation-substituted barium hexaaluminates were synthesized using aerogel precursors. Due to their larger pore sizes and volumes, the aerogel-derived hexaaluminates maintained higher surface areas than their counterpart xerogel-derived hexaaluminates. The presence of substitution cations such as Mn and Co also promoted formation of the hexaaluminate phase at lower temperatures, and the phase transformation pathways were cleaner. The Mn ion appears to be a better substitution cation than the Co ion since Mn-substituted hexaaluminates had higher surface areas at high temperatures. In addition, the high temperature and low temperature aerogels had quite different characters. For example, they followed different phase transformation pathways in going from the aerogel to crystalline hexaaluminate.

REFERENCE

1. Trimm, D.L., Appl. Catal., 7, 249-282 (1983).
2. Zwinkels, M.F.M., Jaras, S.G., and Menon, P.G., Catal. Rev. Sci. Eng., 35(3),319-358 (1993).
3. Arai, M. and Machida, M., Catal. Today, 10, 82-94 (1991).
4. Arai, M. and Machida, M., J. Catal., 103, 385-393 (1987)
5. Machida, M., Eguchi, K and Arai, H., J. Catal., 120, 377-386 (1989).
6. Busca, G., Catalysis Letter, 31, 65-74 (1995).
7. Mizushima, Y. and Hori, M., J. Mater. Res., 9(9), 2272-76 (1994).
8. Groppi, G., Bellotto, M., Cristiani, C., Forzatti, P. and Villa, P.L., Appl. Catal. A, 104, 101-108 (1993)
9. Machida, M., Eguchi, K and Arai, H., J. Am. Ceram. Soc., 71(12), 1142-47 (1988).

Part VII

Nanocomposites of Layered and Mesoporous Materials

ENHANCEMENT OF ION MOBILITY IN ALUMINOSILICATE-POLYPHOSPHAZENE NANOCOMPOSITES

J.C. HUTCHISON, R. BISSESSUR, D.F. SHRIVER
Department of Chemistry and Materials Research Center, Northwestern University, Evanston, IL, 60208-3113

ABSTRACT

Nanocomposites of poly(bis-(2(2-methoxyethoxy)ethoxy)phosphazene) (MEEP) or cryptand[2.2.2] with the aluminosilicate Na-montmorillonite (NaMont) were studied to develop new solid electrolytes with high conductivity and a unity cation transport number. An aluminosilicate was chosen because the low basicity of the Si-O-Al framework should minimize ion pairing. To further reduce ion pairing, solvating molecules or polymers such as cryptand[2.2.2] or MEEP were introduced into the aluminosilicate. When compared to pristine Na-montmorillonite, impedance spectroscopy indicates an increase in conductivity of up to 100 for MEEP·NaMont intercalates, and of 50 for cryptand[2.2.2]·NaMont intercalates. The MEEP·NaMont intercalate exhibits high ionic conductivity anisotropy with respect to the montmorillonite layers ($\sigma_{para}/\sigma_{perp.} = 100$), which is consistent with increased tortuosity of the cation diffusion path perpendicular to the structure layers. The temperature dependance of the conductivity suggests that cation transport is coupled to segmental motion of the intercalated polymer, as observed previously for simple polymer-salt complexes. Nanocomposites of solvating polymers or molecules with aluminosilicates provide a promising new direction in solid-state electrolytes.

INTRODUCTION

Solvent-free electrolytes are of great interest because of fundamental questions over their charge transport mechanisms and the possible applications of these materials in electrochemical devices.[1-3] Clay minerals such as NaMont have appreciable ionic conductivities when swollen by water,[4-6] and work has shown that polar polymers also mobilize Na^+ in NaMont.[7-16] NaMont, a naturally occurring mineral from the clay group *smectites*, has a structure consisting of extended anionic layers balanced by mobile interlayer cations and a unit formula of $Na_{0.6}[(Mg_{0.6}Al_{3.4})Si_8O_{20}(OH)_4]$. Smectites are interesting subjects for studies of cation mobility for a number of reasons, most notably because they are polyelectrolytes with fixed anions. The identity of current carriers is less ambiguous in this type of electrolyte than in salt solutions where both anions and cations are mobile. The mobility of a single type of ion also can be advantageous in electrochemical devices. Additionally, the charged sites in montmorillonite are well separated so ion pairing with the mobile cation is attenuated. To further reduce attractive forces between the cation and the aluminosilicate sheets, a variety of solvating species have been intercalated into clays such as poly(ethylene oxide) (PEO),[9-15] MEEP,[7,8] poly(oxymethylene oligo(ethylene oxide)),[11] cryptands,[7,16] and crown ethers.[16] The cation mobility of these composite electrolytes is highly anisotropic, and it is greatly enhanced in comparison with the parent clay. The role of the intercalated compound and the mechanism of ion conductivity is difficult to fully characterize experimentally. For PEO·montmorillonite composites, Ruiz-Hitzky and co-workers have ascribed the conductivity enhancement to increased layer separation and factors associated with relaxations of the polymer chain.[12,13] Lerner and co-workers suggested that the polymer decreases interactions between the cation and the negatively charged clay surface and thereby increases conductivity.[11] Giannelis, Zax and coworkers have probed polymer dynamics and lithium ion transport in PEO•lithium-montmorillonite

489

composites with a variety of solid-state NMR techniques.[17] In the present work on MEEP·NaMont and cryptand[2.2.2]·NaMont composites, we examine the role of the intercalated species in long range ion transport by an analysis of the temperature-dependant conductivity and we describe the methodology and results of conductivity anisotropy measurements.

EXPERIMENTAL

Where appropriate in the synthesis and characterization, inert atmosphere techniques were employed to prevent adventitious H_2O from affecting the impedance measurements.

Materials

Montmorillonite (SWy-1) obtained from Source Clays and converted to the Na^+ form by cation exchange with 1M NaCl (Aldrich) followed by rinsing with DI H_2O until the [Na^+] of the rinse was less than 0.1 ppm as measured by a Na^+ ion selective electrode (ISE). Following the cation exchange, the NaMont was dried on a high-vacuum line (ca. 3×10^{-5} torr at 100°C). MEEP (molecular weight = 10,300 by gel phase chromatography) was prepared by the literature method[18] and dried on a high-vacuum line (ca. 3×10^5 torr at 100°C). Cryptand[2.2.2] (Aldrich) was dried on a high-vacuum line (ca. 3×10^{-5} torr at 40°C). CH_3CN (Aldrich) was freshly distilled from CaH_2 (Aldrich).

Synthesis and Characterization

Inside a N_2 atmosphere dry-box, measured amounts of finely ground NaMont and either MEEP or cryptand[2.2.2] were added to a Schlenk flask and sealed. On a schlenk line CH_3CN was added to the flask by syringe, and the resulting slurry was stirred magnetically. The progress of the intercalation was monitored by powder XRD on films cast from alequots of the reaction mixture. After the mixture had a homogeneous appearance (ca. 3 days) and there was no evidence of unintercalated NaMont by XRD, the reaction was considered complete. Samples for conductivity measurements parallel to the clay layers were prepared by letting the mixture settle on a medium porosity glass frit (nominal pore size 10-20 μm) followed by slow evaporation of the solvent resulting in a pellet (thickness ca. 2 mm). Sample for conductivity measurements perpendicular to the clay layers were dried under dynamic vacuum and pressed into pellets. Solvent removal was confirmed in all cases by FTIR spectroscopy of the dried products. ^{23}Na MAS NMR was performed on powdered samples.

Impedance Spectroscopy

Impedance spectra were obtained over a frequency range of 1-60,000 Hz using a Solartron 1286/1250 potentiostat/frequency response analyzer combination interfaced to a PC. The temperature of the sample was varied over a range of 30-100°C with a Sun Oven environmental chamber. Impedance data were fit to a parallel resistor-capacitor equivalent circuit (Figure 1) using a non-linear least squares fit routine[19] The conductivity of the material was calculated from the resistance R. The impedance cell consisted of spring loaded stainless steel electrodes inside an O-ring sealed Kel-F housing. Sample faces were Au-sputtered to enhance the electrode-sample contact.

RESULTS

A MEEP to NaMont ratio of one polymer repeat unit, $(CH_3OC_2H_4OC_2H_4O)_2PN$, to two clay Si_8O_{20} units (abbreviated 1:2 MEEP·NaMont) yields a material with a d-spacings of 19 Å as measured by powder XRD. Cryptand[2.2.2] intercalated NaMont at loadings of 0.6 cryptand[2.2.2] to 1 Si_8O_{20} repeat unit and 1.2 cryptand[2.2.2] to 1 Si_8O_{20} repeat unit (*i.e.* 1 cryptand[2.2.2] to 1 interlayer Na$^+$ and 2 cryptand[2.2.2] to 1 interlayer Na$^+$ respectively) both have a d-spacing of 17.9 Å. An x-ray pole analysis of samples indicates preferential orientation of the clay layers parallel to the surface upon which they were cast.

The ^{23}Na MAS NMR spectrum of pristine NaMont consists of a diffuse feature centered at -17 ppm with reference to aqueous 1M NaCl. The MEEP·NaMont intercalate has a broad feature at -20 ppm, and the cryptand[2.2.2] intercalate has a sharper resonance at -16 ppm. This is similar to the solid state spectra of Na$^+$-cryptates and Na$^+$ salt-polyether complexes.[20]

The conductivity of NaMont shows single exponential dependance as a function of 1/T with an apparent activation energy of 0.64 eV.

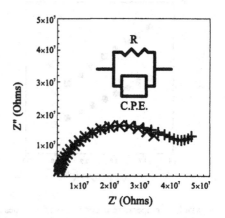

Figure 1. Superposition of a typical impedance spectrum of MEEP·NaMont measured parallel to the clay layers at 67°C (✛), and a non-linear least-squares fit of the high frequency region (✖). The equivalent circuit is represented above.

1:1 cryptand[2.2.2]·NaMont shows similar behavior, but with a higher activation energy of 0.92 eV and enhanced conductivity in the experimental temperature range. Both of these apparent activation energies are higher than those for proton conduction in NaMont where activation energies are typically around 0.2 eV.[6] (Figure 2) The conductivity of the 1:2 MEEP·NaMont and cryptand[2.2.2]·NaMont is substantially enhanced over pristine NaMont, and the conductivity anisotropy ($\sigma_{para.}$ / $\sigma_{perp.}$) of 1:2 MEEP·NaMont is about 100. (Figure 3) The conductivities of the 1:2 MEEP·NaMont composite, both the parallel and perpendicular to the clay layers, were fit to the Vogel-Tammann-Fulcher (VTF) equation[21] (Table 1) which is commonly used to fit the conductivities of polymer electrolytes.:[1,22-28]

$$\sigma = \sigma_0 e^{-\beta/R(T-T_0)}$$

Table 1. VTF Parameters for 1:2 MEEP·NaMont

Sample	σ_0 ($\Omega^{-1}cm^{-1}K^{1/2}$)	β (J)	T_0(K)
$\sigma_{perp.}$	2.2×10^{-3}	1.2×10^4	218
$\sigma_{para.}$	2.1×10^{-1}	1.3×10^4	204

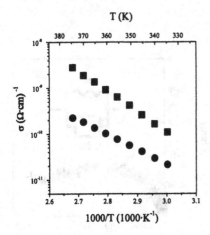

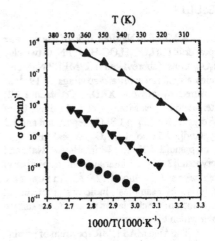

Figure 2. Temperature dependant conductivities (σ) of 1:1 cryptand[2.2.2]•NaMont measured perpendicular to the composite layers (■), E_a = 0.92 eV; Pristine NaMont (●) measured perpendicular to the layers, E_a = 0.64 eV.

Figure 3. Temperature dependant conductivities (σ) of: MEEP•NaMont measured parallel to the composite layers (▲), VTF fit (—); MEEP-NaMont perpendicular to the composite layers (▼), VTF fit (– –); Pristine NaMont (●) measured perpendicular to the layers.

CONCLUSIONS

Expansion in the d-layer spacing as measured by powder XRD indicates that cryptand[2.2.2] and MEEP both form nanocomposites with NaMont. The changes in ^{23}Na MAS NMR spectra of these nanocomposites indicates that there is some interaction between the intercalated species and the Na$^+$, and this interaction is probably solvation of the Na$^+$ by the etheric oxygens. Solvation of the interlayer cations by the etheric oxygens affects the conductivity of both the cryptand[2.2.2] and the MEEP nanocomposites.

The conductivity of the cryptand[2.2.2] nanocomposite displays Arrhenius behavior like the pristine NaMont, but both a higher activation energy and enhanced conductivity. This is a puzzling result, and more work is needed to elucidate the reason for this. In contrast, the conductivity of MEEP·NaMont nanocomposites appears to be non-Arrhenius. The VTF dependance of conductivity in the MEEP·NaMont composites is strongly suggestive of coupling between polymer high amplitude segmental motion and long range cation transport. Furthermore, from the similar temperature dependencies of σ_{para} and σ_{perp} in the 1:2 MEEP·NaMont composite, it appears that the coupling between Na$^+$ transport and polymer motion is similar perpendicular and parallel to the composite layers. The anisotropy arises as a consequence of the σ_o term, which reflects the tortuosity of Na$^+$ motion. Apparently, the path of a Na$^+$ moving perpendicular to the smectite layers is more convoluted than that of a Na$^+$ moving parallel to the smectite layers.

ACKNOWLEDGMENTS

The author would like to thank The Electrochemical Society for the Colin Garfield Fink Summer Fellowship, the Army Research Office (DAAH-04-94-60066), the National Science Foundation (Award No.s DMR-9120521 and CHE-9256486), and Northwestern University. The author also gratefully acknowledges Dr. Gary Beall and Dr. Semeon Tsipursky of American Colloid Company for providing smectites and advice.

REFERENCES

(1)   Ratner, M. A.; Shriver, D. F. *Chem. Rev.* **1988**, *88*, 109.

(2)   MacCallum, J. R.; Vincent, C. A. *Polymer Electrolyte Reviews*; Elsevier: London, 1987, 1989; Vols. 1, 2.

(3)   Tonge, J. S.; Shriver, D. F. In *Polymers for Electronic Applications*; Lai, J. H., Ed.; CRC Press: Boca Raton, FL, 1989; pp.194- 200.

(4)   Wang, W. L.; Lin, F. L. *Solid St. Ionics*, **1990**, *40/41*, 125.

(5)   Fan, Y.Q. *Solid St. Ionics*, **1988**, *28-30*, 1596.

(6)   Slade, R.; Barker, J.; Hirst, P. *Solid St. Ionics* **1987**, *24*, 289.

(7)   Hutchison, J. C.; Bissessur, R.; Shriver, D. F. In *Molecularly Designed Nanostructured Materials*, Gan-Moog Chow, ed., *ACS Symposium Series,622*, 263-272.

(8)   Hutchison, J. C.; Bissessur, R.; Shriver, D. F.; *Chem. Mater.,* **1996**, *8*, 1597-1599.

(9)   Ruiz-Hitzky, E.; Aranda, P.; Casal, B.; Galván, J. *Adv. Mater.* **1995**, *7*, 180.

(10)   Vaia, R.; Vasudevan, S.; Wlodzimierz, K.; Scanlon, L.; Giannelis, E. *Adv. Mater.* **1995**, *7*, 154.

(11)   Wu, J.; Lerner, M. *Chem. Mater.* **1993**, *5*, 835.

(12)   Aranda, P. *Adv. Mater.* **1993**, *5*, 334.

(13)   Aranda, P.; Ruiz-Hitzky, E. *Chem. Mater.* **1992**, *4*, 1395-1403.

(14)   Aranda, P.; Galvan, J.; Casal, B.; Ruiz-Hitzky, E. *Electrochem. Acta* **1992**, *37*, 1573.

(15)   Ruiz-Hitzky, E.; Aranda, P. *Adv. Mater.* **1990**, *2*, 545-547.

(16)   Ruiz-Hitzky, E.; Casal, B. *Nature*, **1978**, *276*, 596-597

(17)   Wong, S.; Vasudevan, S.; Vaia, R.; Giannelis, E.; Zax, D. *J. Am. Chem. Soc.*, **1995**, *117*, 7568-7569.

(18)   Allcock, H. R.; Austin, P. E.; Neenan, T. X.; Sisko, J. T.; Blonsky, P. M.; Shriver, D. F. *Macromolecules* **1986,** *19*, 1508.

(19)   Boukamp, B. A., Equivalent Circuit vs. 3.6. University of Twente, P.O. Box 217, 7500 A.E. Entshede, Netherlands.

(20)   Rawsky, G.; Shriver, D. F. unpublished results.

(21)   Vogel, H. *Phys. Z.* **1921**, *22*, 645. Tamman, G.; Hesse, W. *Z. Anorg. Allg. Chem.* 1926, *165*, 254. Fulcher, G. S. *J. Am. Ceram. Soc.* **1925,** *8*, 339.

(22)   Cheradame, H. In *IUPAC Macromolecules*; Benoit, H.; Rempp, P. Eds.; Pergamon Press: New York, 1982; p. 251.

(23)   Killis, A.; LeNest, J. F.; Cheradame, H. J. *J. Polym. Sci., Makromol. Chem., Rapid Commun.* **1980,** *1*, 595.

(24)   Killis, A.; LeNest, J. F.; Gandini, A.; Cheradame, H. J. *J. Polym. Sci. Polym. Phys. Ed.* **1981,** *19*, 1073.

(25)   Killis, A.; LeNest, J. F.; Gandini, A.; Cheradame, H. J.; Cohen-Addad, J. P. *Solid State Ionics* **1984,** *14*, 231.

(26)   Druger, S. D.; Ratner, M. A.; Nitzan, A. *Phys Rev. B.* **1985,** *31*, 3939.

(27)   Tipton, A. L.; Lonergan, M. C.; Shriver, D. F. *J. Phys. Chem.* **1994,** *98*, 4148.

(28)   Lonergan, M. C.; Ratner, M. A.; Shriver, D. F. *J Am. Chem. Soc.* **1995,** *117*, 2344.

DIRECT OBSERVATION OF FRACTURE MECHANISMS IN POLYMER-LAYERED SILICATE NANOCOMPOSITES

Evangelos Manias, Wook Jin Han[§], Klaus D. Jandt[†], Edward J. Kramer
Emmanuel P. Giannelis

Department of Materials Science and Engineering,
Cornell University, Bard Hall, Ithaca NY 14853.
[§] *Motorola Korea, Seoul, Korea.*
[†] *Department of Oral and Dental Sciences, University of Bristol, Bristol, UK.*

Abstract

Conventional three point bending and TEM techniques are employed to determine the fracture toughness and identify the failure mechanisms in *model* layered-silicate polymer nanocomposites.

Introduction

Layered-silicate based polymer nanocomposites have become an active area of scientific research due to their possible technological applications. In their pristine form most of these layered mica-type silicates contain a hydrated layer of cations between the silicate planes and only certain polar polymers can be intercalated. On the other hand, one can modify these inorganic host lattices by tethering cationic surfactant molecules on the silicate surfaces and in this way a very broad range of polymers –from non-polar polystyrene (PS) to strongly polar nylon– can be intercalated in them [1].

Due to the diversity of the available layered hosts, as well as the variety of surfactants that can be used to organically modify them, one can synthesize a variety of polymer-silicate hybrids. The role of the surfactant is to lower the surface energy of the inorganic host and improve the wetting characteristics with the polymer [2]. In general, with decreasing affinity of the polymer to the silicate, three types of hybrids can be formed (figure 1):

• **delaminated** or **exfoliated** in which the 10Å silicate layers are dispersed throughout the macromolecular matrix and are seperated by tenths or hundreds of nanometers of polymer
• **intercalated** where the galleries between the inorganic planes expand to accomodate an ultrathin 10-20Å polymer slab, while they remain in registry, stacked parallel to each other
• finally, the surfactant can be chosen in such a way that the polymer-silicate interaction is not favourable enough to allow polymer intercalation and this results in an **immiscible** state, where organo-silicate tactoids and can be viewed as conventional fillers inside the polymer matrix.

Although composite materials and polymer-filler systems are already widely used in very diverse areas of structural materials and consumer products, polymer-layered silicate nanocomposites (PLS) is a relatively new class of materials. Due to their nanometer scale dispersion and the high aspect ratio of the silicates, these hybrids offer markedly improved properties compared to the respective pure polymer or the conventionally filled counterparts. For instance, these nanocomposites exhibit higher modulus and thermal stability [3, 4] dramatically improved diffusional barrier properties and solvent resistance characteristics [6, 7], increased fire retandancy [8] and in some cases markedly improved strength and mechanical properties [4, 5]. Moreover are far lighter than conventionally filled polymers[1].

495

Mat. Res. Soc. Symp. Proc. Vol. 457 ℮ 1997 Materials Research Society

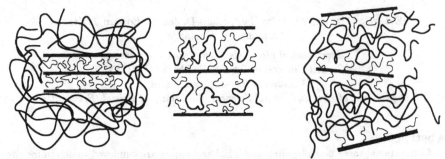

Figure 1 *Schematic representation of the possible hybrid formations with organically modified silicates and polymers. From left to right immiscible, intercalated and delaminated or exfoliated hybrids are shown.*

In this paper we study the fracture mechanisms and mechanical properties of montmorillonite based nanocomposites with polystyrene. Although these are by far not examples of systems with improved mechanical properties they provide very good *model systems* that enable us to test the methods we use, as well as to compare against the failure behaviour of the pure polymer and its conventional filled systems.

Experimental

Organically modified layered silicates were prepared from Na montmorillonite (charge exhange capacity: cec ~0.9 meq/gr) by ion-exchange reactions with protonated quaternary amine surfactants, such as dimethyl-dioctadecyl-ammonium and dimethyl-benzyloctadecyl-ammonium. Commercially available polystyrene with M_w=200000 and a polydispersity of 1.06 was used in most of the studies. Nanocomposites were created by direct melt intercalation or via extrusion. Tensile bars were prepared by using a hydraulic press and loads 5-7 tons at 150°C [9].

The elastic moduli of the nanocomposites were determined through ultrasonic techniques by measuring the time of flight of longitudinal and shear waves through several directions of polished speciments. The experimental details are reported elsewhere [9]. Fracture toughness experiments were performed on an Instron Model 1125 mechanical tester using a variable displacement transducer, a strain gauge load cell and chevron-notched three point bend speciments 2 × 2 × 8mm). The critical stress intensity factor K_{IC} was determined from load-displacement curves. The work of fracture was also calculated and the fractured surfaces were characterized by Scanning Electron Microscopy (SEM).

Direct observation of failure mechanisms was realized by Transmission Electron Microscopy (TEM) studies of nanocomposite films (0.5-1.5 μm thick) under strain in a JEOL-1200EX operating at 120KV. The samples were either microtomed from the tensile bars used in the 3 point bend technique or spin casted. The copper grid method was previously developed by E. J. Kramer et al.[10] and was employed to study the failure mechanisms at the polymer-polymer interface [11], in oriented polystyrene [12] and other systems.

Results & Discussion

All the samples for mechanical characterization and fracture toughness measurements were made by "compression molding" inside a hydraulic press under high loads and temperatures above the softening temperature of the polymer. Due to the high aspect ratio of the silicate layers (with $1nm$ thickness and μm lateral dimensions) a preferential orientation may be induced normal to the direction of compression. For this reason fracture toughness was determined both parallel and normal to this preferential orientation. Nanocomposites with 10wt% loading of organically modified montmorillonite were studied for the case of melt-intercalated, extruded-intercalated and immiscible systems. In addition, the pure polystyrene as well as a 20wt% intercalated nanocomposite were also studied. The elastic moduli for these systems were determined by time-of-flight measurements using 10 MHz frequency for the longitudinal waves and 5 MHz for the shear, and are given in Table I.

Table I *The longitudinal and shear Young moduli for selected nanocomposites are shown.*

	E_1 (GPa)	E_3 (GPa)
Polystyrene		3.92
10 wt% intercalated	4.42	4.17
20 wt% intercalated	5.43	4.52
10 wt% extr. interc.	4.35	4.31
10 wt% immiscible	4.59	4.41

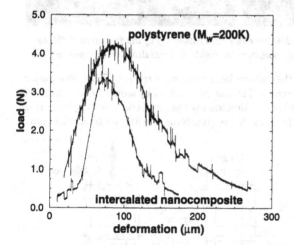

Figure 2 *The normalized load-deformation curve for the polystyrene and the intercalated nanocomposite from the 3-point bend measurement. Interestingly, the nanocomposite fails in a manner similar to that of the pure polymer.*

The fracture toughness of the same samples was measured by Chevron-notched samples in a 3-point bending geometry. Interestingly, the load-deformation curve (figure 2) shows a failure qualitatively similar to that of the pure polymer, although the K_{IC} decreased (Table II). At this point we should mention that there are ways to increase the fracture toughness of the nanocomposites [4] but our aim in the present study is to explore methods for mechanical characterization of the nanocomposites and identify their failure mechanisms, as it will become obvious by the TEM results. Moreover, by observing the fractured surfaces with an SEM (figure 3) it is revealed that the fractured surface of the nanocomposite is much rougher than the one of the pure polystyrene, in contrast with the usual behavior of materials with higher K_{IC} that exhibit a more *tortuous fracture path*.

Table II *The fracture toughnesses for the same systems as in table I are shown with respect to the silicate preferential orientation. K_{IC} was determined from the load-deformation curves in a 3-point bend geometry.*

	K_{IC} (MPa $\sqrt{m}$)	
	horizontal	vertical
Polystyrene	1.89	
10 wt% intercalated	0.86	0.76
20 wt% intercalated	0.97	0.65
10 wt% extr. interc.	0.93	0.71
10 wt% immiscible	0.76	0.68

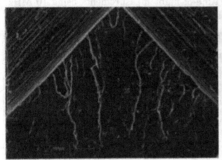

Figure 3 *Scanning Electron Microscopy (SEM) imaging of the fractured surfaces of the samples in figure 2. Left: polystyrene, right: intercalated nanocomposite.*

In order to get more insight on the fracture behaviour, we carried out TEM observations of thin nanocomposite films under strain. This can be achieved by mounting μm-thin slices of the nanocomposite –obtained either by microtoming the tensile bars or by spin casting– on a ductile copper grid with a 1x1mm mesh size [10]. Subsequently, we strain the system

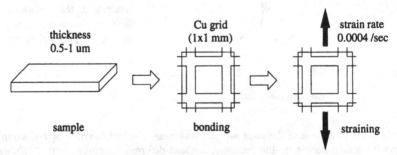

Figure 4 *Schematic of the TEM sample preperation.*

with a very small strain rate to a 5-10% deformation (figure 4). The ductile copper remains deformed and can keep the composite material under strain inside the TEM. In order to minimize as much as possible failure mechanisms that originate from surface properties of ultra-thin films, we used samples thicker than 0.5 μm and up to 1.5 μm, beyond which it becomes difficult to observe them with our TEM.

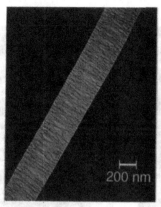

Figure 5 *TEM studies of a strained polystyrene film. Crazes propagate in a straight line parallel to each-- other. The average size of the craze and the structure of the fibrils depend on the strain applied and are in good agreement with the typical craze morphology in glassy polymers [13].*

20 μm

200 nm

The typical crazing behaviour of pure polystyrene can be seen in figure 5. Crazes are formed and they run almost parallel to each other mainly normal to the straining direction. One should notice that the craze sides are quite smooth and this is reflected to a smooth fractured surface in the 3-point bend experiment. On the other hand, in figure 6 the craze runs much less smoothly and straight and becomes very deformed in regions rich in silicate layers. For the montmorillonite based intercalated hybrids some of the dominant fracture mechanisms include (figure 6 from top left clockwise):

(i) the failure of the craze in the vicinity of a silicate layer parallel to the craze propagation direction, (ii) the craze splits up in two parts and moves around a tactoid (group of parallel silicate layers), (iii) the craze propagation stops at silicate layers normal to its propagation direction and (iv) a small tactoid "opens-up" for the craze to go through it and the craze fails between the silicate layers (the i and iv behaviour is highlighted in figure 7).

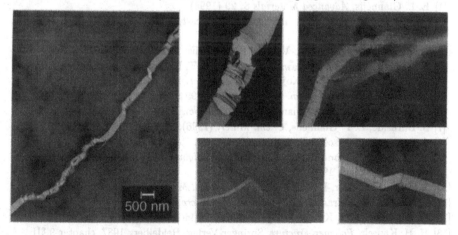

500 nm

Figure 6 *Typical failure mechanisms for intercalated polystyrene nanocomposites*

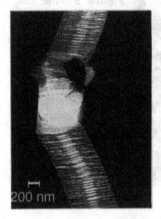

Figure 7 *Two failure mechanisms often observed for crazes running parallel to silicate tactoids are shown here. Left: the craze fails by creating a void adjacent to silicate layers. Right: a silicate tactoid found in the propagation way of the craze "opens-up" and a void is created.*

Summarizing, ultrasonic and 3-point bend techniques were successfully employed for the mechanical characterisation of nanocomposite materials. Furthermore, TEM studies of strained composite films revealed the failure mechanisms located mainly at the polymer-silicate interface. A way to improve the toughness of these materials is through the strengthening of the polymer-surface binding [4, 6]

Acknowledgments This work was supported by generous gifts from DuPont, Exxon, Hoerst Celanese, Hercules, Monsanto, Nanocor, Southern Clay Products and Xerox.

References

[1] E. P. Giannelis, *Advanced Materials* **8**, 29 (1995).

[2] R. A. Vaia, *Polymer melt intercalation in mica-type silicates*, Ph.D. thesis, Cornell University, 1995, chapter 5

[3] R. K. Krishnamoorti, R. A. Vaia, E. P. Giannelis, *Chem. Mat.* **8**, 1728 (1996).

[4] Y. Kojima, A. Usuki, M. Kawasumi, A. Okada, T. Kurauchi, O. Kamigaito, *J. Polym. Sci. A* **31**, 983 (1993). Y. Kojima et al, *J. Mater. Res.* **8**, 1185 (1993).

[5] T. Lan, T. Pinnavaia, *Chem. Mater.* **2**, 2216 (1994). *ibid* **7**, 2144 (1995).

[6] P. B. Messersmith, E. P. Giannelis, *J. Polym. Sci. A* **33**, 1047 (1995).

[7] S. Burnside, E. P. Giannelis, *Chem. Mater.* (1996).

[8] J. D. Lee, T. Takekoshi, E. P. Giannelis, Mater. Res. Soc. Proc. *current issue.*

[9] W. J. Han, *Mechanical characterization of polymer nanocomposites*, Master thesis, Cornell University, 1996

[10] B. D. Lauterwasser, E. J. Kramer, *Philosophical Magazine A* **39**, 469 (1979).

[11] C. Creton, E. J. Kramer, G. Hadziioannou, *Macromolecules* **24**, 1846 (1991).

[12] C. Maestrini, E. J. Kramer, *Polymer* **32**, 609 (1991).

[13] H. H. Kausch, *Polymer Fracture*, Springer-Verlag, Heidelberg 1987, chapter 9 §II

FLUOROPHLOGOPITE AND TAENIOLITE: SYNTHESIS AND NANOCOMPOSITE FORMATION

GREGORY J. MOORE, PETER Y. ZAVALIJ and M. STANLEY WHITTINGHAM*
Chemistry Department and Materials Research Center, State University of New York at Binghamton, Binghamton, NY 13902-6000, USA

ABSTRACT

Sodium fluorophlogopite and lithium taeniolite have been synthesized by new routes for application in lithium batteries. The fluorophlogopite synthesized by a high temperature solid state reaction, was found to be non-water-swellable and unreactive towards several mono- and divalent ions. However it was found to readily undergo ion-exchange with both copper and iron ions, with concomitant swelling to a bilayer water state. This swelled material reacted readily with long chain amines and other molecules and ions behaving like a regular swellable silicate. A taeniolite precursor was synthesized by mild hydrothermal reactions, and annealed into a well crystalline layer solid, that reacted readily with organics to form ordered composites that have potential use as battery electrolytes and cathodes.

INTRODUCTION

There has been much interest in clay-like materials for electrolytes in batteries, because of their ready availability, electronically insulating behavior and high ionic conductivity. However, the high ionic conductivity is only found for the bilayer hydrated forms where the alkali ions diffuse through a water medium . Even removing just one water layer, drops the ionic conductivity by two orders of magnitude from 3 x 10^{-4} S/cm to around 10^{-6} S/cm for sodium in vermiculite [1]. The presence of water makes these materials undesirable for battery application involving alkali metals. It should be possible to form related materials with the water replaced by simple organic solvents or by polymeric species such as polyethylene oxide. The latter is itself under extensive study as an electrolyte in lithium batteries; however, the predominant ionic current carriers are anions rather than lithium. Our thinking is to intercalate organic solvating species into clay-like materials where the negatively charged clay layers act as the anion so that only the cations will be mobile.

Initially this study is looking at two types of layered silicates, sodium fluorophlogopite and lithium taeniolite. The former contains fluoride groups in place of the normally present hydroxyl groups giving it greater thermal stability and anticipated greater stability to lithium. Taeniolite, a much less studied material but where the preferred cation is in the structure as made, has a high surface area and should therefore be readily swellable.

Fluorophlogopite was studied as the main component in a new electroluminescent material, where the luminescently active ion was placed in the silicate layer and a conducting polymer would be intercalated in the interlayer region. A field could then be applied between the conducting polymer layers, exciting the electroluminescent ion [2]. The fluorine allowed the use of high temperatures, up to 1300°C in its synthesis. Fluorophlogpite is a non-swellable 2:1 trioctahedral mica with the structure shown schematically in figure 1, but in it's hydroxy form is naturally occurring, swellable and has the ideal formula $K(Mg_3)(AlSi_3O_{10})(OH)_2$ [3, 4]. Taeniolite has a similar structure but the silicate tetrahedral sheets contain only silicon, so unlike phlogopite where 25%

501

of the silicate sites are occupied there is no possibility of forming stable Alkali-O-Al positions so lithium diffusion through the lattice should be enhanced. It's formula can be represented by $Li_x(Mg_{3-x}Li_x)(Si_4O_{10})F_2 \cdot nH_2O$.

Layered silicates have been of interest recently for use as electroluminescent displays, polyelectrolytes [5-8], and organic/inorganic composites for creating stress relieving points in engineered materials [9-11]. In order to incorporate polymers within the layers it is necessary to swell the mica-like layers greater than that found for the alkali metal alone. Several types of silicates, such as montmorilonite and fluorohectorite which have low charge densities, will expand almost continuously when placed in aqueous solutions. This generally leads to a controlled amount of polymer adsorbing onto the layers by controlling the component ratios [12]. However, there are no reports of the swelling of fluorophlogopite so a means must be found to incorporate polymers between the sheets.

This paper discusses recent progress toward forming sheet silicate structures that might be used as components of electrochemical devices such as sensors and batteries.

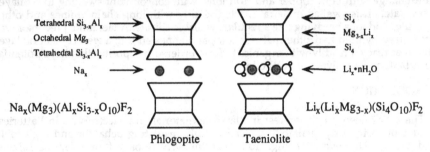

Fig. 1. Schematic structure of layered trioctahedral clays

EXPERIMENTAL

The fluorophlogopite was synthesized by a solid state reaction which consisted of firing the oxides or fluorides at 1300°C. The final compound had the chemical formula of $NaMg_3(Al,Si_3)O_{10}F_2$, which was confirmed by microprobe analysis. The analysis of the crystal structure was done using a Scintag powder diffractometer. The water content and thermal stability were analyzed using a Perkin Elmer TGA7 thermal analyzer. Chemical analysis was done using a Jeol electron microprobe.

First the fluorophlogopite was reacted with a 1M acetic acid solution to exchange the sodium for protons and to swell the lattice as found to be successful for vermiculite [13, 14]. This would form a swelled acid that could be readily reacted with bases such as aniline. A partial substitution of the hydrogen ions by copper would then provide the oxidation catalyst for the in-situ polymerization of the aniline. No swelling of the lattice was observed with acetic acid, and reactions with stronger acids destroyed the crystallinity of the lattice.

Since the phlogopite did not swell with acids, reaction with a simple amine, octyl amine, was attempted at room temperature and at 70°C, but again no swelling resulted.

As it was desired to use copper to initiate the polymerization of monomers such as aniline, the sodium fluorophlogopite was reacted with 2M $Cu(NO_3)_2$; the layers swelled and the first reflection went from 9.8Å to 14.4Å. The reaction time was initially 10 days at room temperature but was decreased to 4 days by increasing the temperature to 70°C. Other reactions were tried in order to swell the layers but efforts with $AgNO_3$ and

$Ni(NO_3)_2$ under similar conditions were fruitless. The sodium fluorophlogopite was also successfully reacted with ferric nitrate solution.

Once the fluorophlogopite was swelled it became much more reactive. It was reacted with a range of straight chain amines, hexyl, heptyl, nonyl, dodecyl, tetradecyl, and hexadecyl, for 3 days at 70°C.

The lithium taeniolite was synthesized by mild hydrothermal reaction of Li_2CO_3, MgO, $(NH_4)_2SiF_6$ and SiO_2 in water at 200°C for 3 days. This precursor was then heated to 1000°C in oxygen when highly crystalline material was formed.

RESULTS AND DISCUSSION

Sodium Fluorophlogopite

The initial Na-fluorophlogopite was indexed and found to have a triclinic unit cell with dimensions of a=5.308(1), b=9.179(2), c=9.788(2), alpha=99.223(5), beta=91.473(5), and gamma=90.025(8). It's powder x-ray pattern is shown in figure 2. A WDS elemental analysis on the electron microprobe led to the formula of $NaMg_3(AlSi_3O_{10})F_2$ by using an average over five different crystals which had a mesh size of 200-300. A thermogravimetric analysis carried out in nitrogen showed negligible weight loss, less than 0.5%, up to 1000°C therefore demonstrating no interlayer water or structural hydroxyl groups.

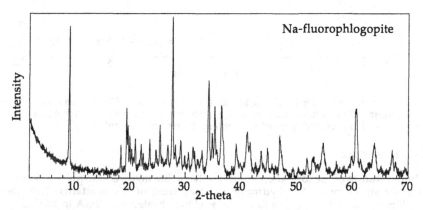

Fig. 2 X-ray diffraction pattern of sodium fluorophlogopite

The Cu-fluorophlogopite formed by ion-exchange was determined to have the composition $Cu_{0.4}Na_{0.2}Mg_3(Al_{1.7}Si_{2.3}O_{10})F_2 \bullet H_2O$ with the lattice expanding to 14.34Å, with unit cell a= 5.322, b= 9.239, c=28.89Å and β= 96.98°.

Ferric nitrate was also capable of swelling the layers of the Na-fluorophlogopite by reaction with an 0.5M solution of $Fe(NO_3)_3$ for 2 days at 60°. This expanded the layers similarly to Cu(II), with the repeat being 14.23Å. Silver and nickel nitrate solutions did not cause expansion of the lattice. One reason for the effectiveness of copper and iron aqueous solutions may be their greater acidity which leads to the presence of lower charge species in the interlayer region, such as $[Cu(OH)(H_2O)_5]^{1+}$ as well as hydronium ions which may be ion exchanged for additional copper [15].

Reaction of the copper fluorophlogopite with alkylamines caused a substantial expansion of the silicate lattice as shown in figure 3. The lower slope line corresponds to a single layer of amine with the nitrogen adjacent to the oxygen of the silicate and the hydrocarbon chains interleaved. The higher expansions are typical of that for a bilayer of amine, as found for vermiculite and other readily swellable clays. The slopes correspond to an angle of the chains of around 60° to the silicate layer. Unlike the vermiculite where an ammonium salt is formed, in this material the amine is acting as a solvent for the cations. TGA showed that around 0.3 or 0.7 amine groups were incorporated per $(Si,Al)_4O_{10}$ group for the two configurations shown in the figure, again consistent with single and bilayer amine configurations.

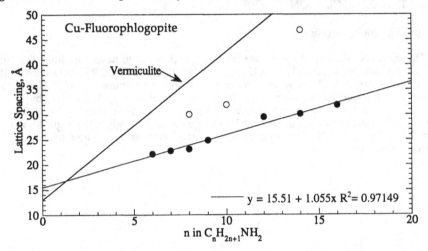

Fig. 3. Lattice spacing of copper fluorophlogopite on intercalation of long chain amines. Closed circles correspond to an interleaved monolayer, whereas open circles are probably associated with a bilayer of amines, and approach the spacing observed in vermiculite [14].

Lithium Taeniolite

Initial studies of the formation of lithium taeniolite involved the preparation of the precursor structure by the hydrothermal treatment of the reactants. The poorly crystalline material formed, see figure 4a, was then heated on a TGA to 1000°C when the structure became highly crystalline as shown in figure 4b. Analysis of this x-ray data indicated a unit cell with a= 24.305 (2 x 12.152), b= 8.786, c= 5.917, β= 92.61 and space group C 2/m. The lattice repeat distance of 12.1Å indicates a single layer of water between the silicate sheets.

Lithium taeniolite was found to be much more reactive than the phlogopite phase. Thus, addition of a drop of water to the powder caused an immediate expansion of the lattice and broadening of all the x-ray diffraction lines.

Ionic Mobility Measurements

The ion mobility was measured by complex impedance methods using a Solartron Bridge, but the resistance of both sodium fluorophlogopite and lithium taeniolite were found to be well in excess of 10^6 ohms as shown in table 1. As expected these are higher

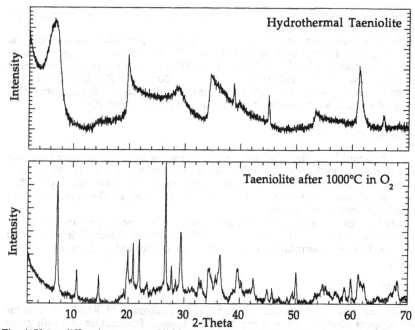

Fig. 4. X-ray diffraction pattern of lithium taeniolite (top) after hydrothermal treatment and (bottom) after heating to 1000°C.

Table 1. Conductivity of Silicates

Compound	Lattice Spacing, Å	Conductivity, S/cm
Sodium vermiculite [1]	14.82	3×10^{-4}
Sodium fluorophlogopite	9.18	7×10^{-8}
Copper vermiculite [5]	14.32	6×10^{-6}
Copper fluorophlogopite	14.34	2×10^{-9}
Lithium vermiculite [5]	14.5	3×10^{-5}
Lithium taeniolite	12.1	2×10^{-7}
Lithium taeniolite (no anneal)		1.3×10^{-7}
Lithium PEO taeniolite	14.3	2×10^{-8}

than expected for the bilayer vermiculite compounds. The drop of almost four orders of magnitude in the conductivity between vermiculite and phlogopite is consistent, as going from the bilayer to the monolayer causes a 100 fold decrease and when no water is present the ions are locked into the trigonal sites. Copper is a poor ionic conductor even in vermiculite, but the much lower value in phlogopite is not understood, but might be related to the grain size. The difference between the lithium ion conductivity in vermiculite and taeniolite is expected from the lattice spacing reflecting the additional water layer in the former.

Poly(ethylene oxide) was incorporated into the taeniolite lattice by reacting the taeniolite with a 1M acetonitrile solution of PEO at 60°C for one day. This resulted in an expansion of the lattice to 14.30Å, and the resulting TGA was very different from that of taeniolite itself showing an overall 10% weight loss by 350°C, with 7% above 150°C. The conductivity was even less than that observed for the lithium taeniolite suggesting that a monolayer of water is a better solvent for lithium ion mobility than PEO.

CONCLUSIONS:

Inert Na-fluorophlogopite, synthesized at elevated temperatures, may be made reactive by partial ion-exchange with aqueous copper or iron species. It is thought that this is a result of reduction of the effective charge density between the layers. The copper compound was found capable of intercalating alkylamines like vermiculite and other oxide and sulfide layer materials.

Taeniolite was formed as a highly crystalline powder by firing a precursor powder, formed hydrothermally, at 1000°C in oxygen. This easily expanded silicate exhibited ionic mobility, but the intercalation of poly(ethylene oxide) reduced the mobility of lithium ions.

ACKNOWLEDGEMENTS

We thank the Advanced Research Projects Agency, through OSRAM-Sylvania for the initial support of this work and the National Science Foundation through grant DMR-9422667 for partial support of this work. We also thank Mr. Bill Blackburn for the electron microprobe studies.

REFERENCES

1. M. S. Whittingham, Solid State Ionics, 25 (1987) 295.
2. R. Karam, R. ButchiReddy, and M. S. Whittingham, U.S. Patent Application, (1992)
3. B. K. G. Theng, *The Chemistry of Clay-Organic Reactions*. 1974, London: Adam Hilger Ltd.
4. B. K. G. Theng, *Formation and Properties of Clay-Polymer Complexes*. Developments in Soil Science, 1979, Amsterdam: Elsevier.
5. H. Maraqah, J. Li, and M. S. Whittingham. *Ion Transport in Single Crystals of the Clay-like Aluminosilicate, Vermiculite*. in *Solid State Ionics II*. 210 (1991) 351. Boston, MA: Materials Research Society.
6. J. Li and M. S. Whittingham, Materials Res. Bull., 26 (1991) 849.
7. J. C. Hutchison, R. Bissessur, and D. F. Shriver, Chem. Mater., 8 (1996) 1597.
8. C. O. Oriaki and M. M. Lerner, Chem. Mater., 8 (1996) 2016.
9. E. P. Giannelis, Advanced Materials, 8 (1996) 29.
10. D. C. Lee and L. W. Jang, J Appl Polymer Science, 61 (1996) 1117.
11. R. Krishnamoorti, R. A. Vaia, and E. P. Giannelis, Chem. Mater., 8 (1996) 1728.
12. Y.-J. Liu, J. L. Schindler, D. C. DeGroot, C. R. Kannewurf, W. Hirpo, and M. G. Kanatzidis, Chem. Mater., 8 (1996) 525.
13. H. Maraqah, J. Li, and M. S. Whittingham, J. Electrochem. Soc., 138 (1991) L61.
14. H. Maraqah, J. Li, and M. S. Whittingham, Solid State Ionics, 51 (1992) 139.
15. M. Suzuki, M. Yeh, C. R. Burr, M. S. Whittingham, K. Koga, and H. Nishina, Phys. Rev., B40 (1989) 11229.

SYNTHESIS AND LCST BEHAVIOR OF THERMALLY RESPONSIVE POLY(N-ISOPROPYLACRYLAMIDE)/LAYERED SILICATE NANOCOMPOSITES

Phillip B. Messersmith and F. Znidarsich
Departments of Restorative Dentistry and Bioengineering, University of Illinois at Chicago, 801 S. Paulina St., Chicago, IL 60612.

ABSTRACT

Stimuli responsive polymeric hydrogel composites were synthesized by room temperature copolymerization of N-isopropyl acrylamide and methylene bisacrylamide (crosslinking monomer) in an aqueous suspension of Na-montmorillonite. Hydrogels containing 3.5 weight% of montmorillonite exhibited a lower critical solution temperature (LCST) similar to unmodified PNIPAM hydrogel (approximately 32°C), and underwent a reversible 60-70% volume shrinkage when heated from ambient temperature to above the LCST. However, hydrogels containing 10 weight% montmorillonite did not exhibit a measurable LCST, and underwent considerably less shrinkage when heated. A solvent exchange reaction was used to replace the water with an acrylic monomer, which was polymerized in-situ to create a delaminated montmorillonite/polymer nanocomposite.

INTRODUCTION

Nanostructured composites consisting of mica-type silicate layers embedded within various polymer matrices have been intensively studied in recent years.[1] Technological interest in these types of nanocomposites is spurred by their enhanced mechanical and barrier properties compared to conventionally prepared filled composites.[2-7] We are interested in developing nanostructured composites which are capable of altering their structure and properties under the influence of an applied stimulus. Stimuli-responsive nanocomposites could provide a spectrum of optical, mechanical, and barrier properties depending on ambient conditions of temperature, humidity, etc. As a first step towards this goal, we have chosen to explore the synthesis of nanocomposites of mica-type silicates (MTS) and poly(N-isopropyl acrylamide) (PNIPAM), a thermally responsive polymer which exhibits a lower critical solution temperature (LCST) of 32°C in the presence of water.[8] PNIPAM is a hydrophilic polymer below the LCST, but is strongly hydrophobic above the LCST. Crosslinked hydrogels of PNIPAM are readily prepared by polymerization of an aqueous solution of N-isopropyl acrylamide monomer in the presence of a bifunctional monomer such as methylene bisacrylamide, and exhibit considerable volume shrinkage in association with the LCST.

Lightly crosslinked PNIPAM hydrogels have been extensively investigated as potential matrices for drug release and as chemomechanical systems,[9-11] with the LCST being exploited to alter properties in response to a temperature change. For example, in controlled release applications the diffusion of a therapeutic agent from the PNIPAM matrix is significantly altered at the LCST,[9,10] whereas for chemomechanical systems the volume shrinkage associated with the LCST can be used to generate a force to do work.[11] Given the well established impact of delaminated MTS on the diffusional and mechanical behavior of polymer matrices, we surmise that MTS/PNIPAM nanocomposites may possess novel properties. Here, we report the synthesis and LCST behavior of PNIPAM hydrogels containing delaminated layered silicate.

Mat. Res. Soc. Symp. Proc. Vol. 457 © 1997 Materials Research Society

MATERIALS AND METHODS

N-isopropyl acrylamide (NIPAM), methylene bisacrylamide (MBA), ammonium persulfate (APS), and tetramethylethylenediamine (TEMED) were obtained from Aldrich and used as received. Na-montmorillonite was obtained from Southern Clay Products as a thixotropic 3.8 weight% suspension in water and was used as received or was dried at 60°C and used as a powder. MTS-free hydrogels were synthesized by first dissolving 1.57g NIPAM, .026g MBA and .08g APS in 20g distilled water. Polymerization was initiated by addition of 48μl TEMED to the monomer solution. Hydrogels were polymerized between glass plates for 24 hours at room temperature, and stored in distilled water until use. Hydrogels containing 3.5 and 10.7 weight% MTS were synthesized by substituting Na-montmorillonite for water in the formulation described above. Delamination of the montmorillonite was aided by brief probe sonication, during which gelation of the sample occurred.

Volume shrinkage of the hydrogels was estimated using a previously reported method.[2] Briefly, 3mm x 3mm x 3mm cubes of hydrogel were aged in distilled water for 30 minutes at various temperatures then weighed to determine the amount of water displaced by shrinkage. LCST's were detected using dynamic scanning calorimetry (DSC) of hydrogel samples (10-20mg). X-ray diffraction (XRD) was performed on a Siemens D5000 diffractometer using a Cu K_α source.

Nanocomposites of MTS, PNIPAM, and poly(methylmethacrylate) (PMMA) were synthesized by a solvent exchange procedure as follows. MTS/PNIPAM hydrogel was immersed in excess methylmethacrylate (MMA) monomer containing 0.5 weight% benzoyl peroxide (BPO). Diffusion of water out of and MMA into the swollen MTS/PNIPAM gel was aided by replacement of the excess solvent with fresh MMA/BPO every few hours for 1-2 days. During this period, slight volume expansion of the MTS/PNIPAM gel occurred (this was not quantified). The MMA-swollen MTS/PNIPAM gel was then polymerized by heating at 80°C for 2 hours. Thin sections of the MTS/PNIPAM/PMMA composites were then obtained by glass knife ultramicrotomy, and imaged unstained in a JEOL 100CX TEM.

RESULTS AND DISCUSSION

In contrast to most layered silicate/polymer nanocomposites, exchange of the inorganic gallery cations (e.g. Na+) of the pristine silicate with organic cations (e.g. alkylammonium cations) is not necessary to delaminate Na-montmorillonite in water. The relative ease of MTS delamination in water facilitates the synthesis of composite hydrogels; water soluble monomers can be dissolved in a delaminated MTS gel and polymerized to yield a crosslinked polymeric hydrogel in which delaminated MTS layers are embedded. Although not quantified, a significant increase in hydrogel modulus was noted for the MTS-containing samples, the magnitude of which was roughly proportional to the weight% of MTS in the sample. XRD analysis of the 3.5 and 10.7 weight% MTS/PNIPAM hydrogels (Figure 1) revealed no major basal plane (00l) reflections, suggesting delamination of the montmorillonite layers within the polymeric hydrogel.

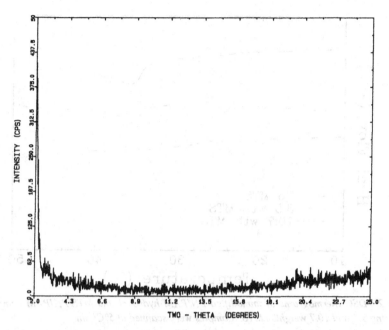

Figure 1. Powder XRD pattern of MTS/PNIPAM composite hydrogel containing 10.7 weight% MTS. The composite containing 3.5 weight% MTS was similar in appearance (data not shown).

The LCST behavior of the MTS/PNIPAM composite hydrogels was investigated using dynamic scanning calorimetry in combination with volume shrinkage experiments. In the DSC thermogram of the unmodified PNIPAM gel the LCST is detected as a broad endotherm of low intensity at approximately 35°C (Figure 2), in agreement with the LCST of 32°C observed for linear and lightly crosslinked PNIPAM systems.[8] While a nearly identical LCST was observed for the 3.5 weight% MTS/PNIPAM composite, the LCST was noticeably absent in the 10.7 weight% composite; no major thermal transitions were observed between 10 and 50°C (Fig. 2).

Three distinct stages of thermally induced contraction can be seen for unmodified PNIPAM hydrogel (Figure 3). Between 0 and 25°C the unmodified PNIPAM hydrogel exhibits only slight contraction, on the order of 15%. Between 28 and 32°C considerable shrinkage occurs in conjunction with the LCST; above 35°C further shrinkage was minimal. For the nanocomposite containing 3.5 weight% MTS/PNIPAM composite, the shrinkage behavior was qualitatively similar to the unmodified PNIPAM hydrogel, except that the MTS-containing hydrogel exhibited a slight increase in shrinkage temperature (approx. 1-2°C) compared to the unmodified hydrogel. Total volume shrinkage of both materials was approximately 70% at 60°C. In contrast, the composite containing 10.7% MTS showed uniform shrinkage throughout the experimental temperature range (0-70°C), but no major volume reductions indicative of a well-defined LCST. Furthermore, total volume shrinkage of the 10.7 weight% MTS sample was less than half (approx. 23% shrinkage at 60°C) of the unmodified and 3.5 weight% PNIPAM hydrogels. The data suggests that the LCST behavior of PNIPAM gels is little affected by MTS at low content, but is suppressed or even eliminated at higher MTS content.

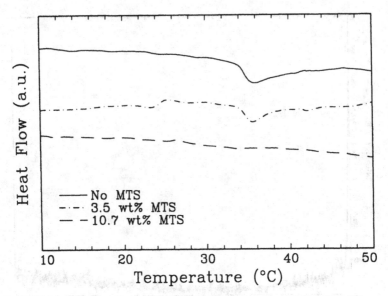

Figure 2. DSC thermograms of unmodified PNIPAM hydrogel and MTS/PNIPAM composites containing 3.5 and 10.7 weight% MTS. Samples were scanned at 5°C/min.

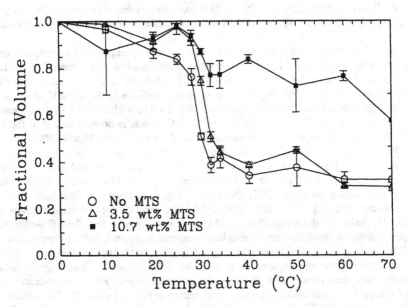

Figure 3. Relative shrinkage data for PNIPAM and MTS/PNIPAM composites containing 3.5 and 10.7 weight% MTS. Fractional volume was calculated using the volume of the samples in the fully expanded state (0°C).

We also demonstrated a new route for the synthesis of bulk polymer nanocomposites from MTS-containing hydrogels. To accomplish this, we replaced the water of the MTS/PNIPAM hydrogel with a polymerizable monomer (MMA) by a solvent exchange process. We then polymerized the MMA monomer to yield an MTS/PNIPAM/PMMA composite consisting of approximately 89% PMMA. As shown in Figure 4, TEM micrographs of thin sections of this composite revealed the presence individual delaminated montmorillonite layers embedded within the polymer matrix. Given the method of preparation, it is likely that delamination of the MTS layers was initially achieved in water, locked in place by formation of the crosslinked three-dimensional PNIPAM network, and ultimately preserved during MMA exchange and subsequent polymerization. This general route to the synthesis of delaminated MTS/polymer composites has the potential advantage of utilizing unmodified Na-montmorillonite as opposed to organic cation-exchanged forms of the mineral which are expected to be more costly as raw materials.

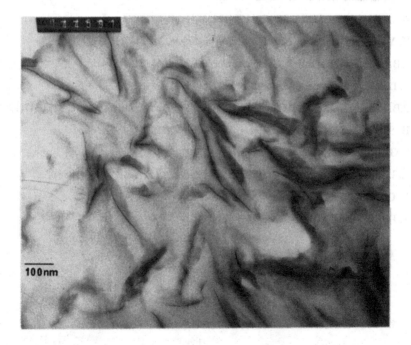

Figure 4.   Transmission electron micrograph of a thin section of a MTS/PNIPAM/PMMA composite synthesized from a MTS/PNIPAM hydrogel containing 3.5 weight% MTS.

CONCLUSIONS

Thermoresponsive nanocomposites consisting of mica-type silicate layers dispersed within a PNIPAM hydrogel were synthesized by polymerization of suspensions containing water, monomer, and delaminated silicate. The LCST transition of PNIPAM, and the volume shrinkage associated with this transition, is only slightly affected at small montmorillonite loadings. However, the LCST was essentially absent in composite hydrogel containing 10.7 weight% silicate. Replacement of the hydrogel water with organic monomer, and polymerization of the infiltrated monomer, was demonstrated as a new route to the synthesis of delaminated MTS/polymer nanocomposites.

ACKNOWLEDGMENTS

The authors would like to thank Professor Steve Guggenheim for assistance with the x-ray diffraction experiments.

REFERENCES

1. E.P. Giannelis, *Adv. Mater.*, **8**, 229(1996).

2. A. Usuki, et al., *J. Mater. Res.* **8**, 1179(1993).

3. K. Yano, et al., *J. Polym. Sci. Part A: Polym. Chem.*, **31**, 2493(1993).

4. P.B. Messersmith and E. P. Giannelis, *Chem. Mater.*, **6**, 1719(1994).

5. T. Lan, T.J. Pinnavaia, *Chem. Mater.* **6**, 2216(1994).

6. P.B. Messersmith and E.P Giannelis, *J. Polym. Sci. Part A: Polym. Chem.*, **33**, 1047(1995).

7. H. Shi, T. Lan, T.J. Pinnavaia, *Chem. Mater.* **8**, 1584(1996).

8. H.G. Schild, *Prog. Polym. Sci.*, **17**, 163(1992).

9. A.S. Hoffman, A. Afrassiabi, L.C. Dong, *J. Controlled Release*, **4**, 213(1986).

10. Y. Okuyama, et al., *J. Biomater. Sci. Polymer Edn.*, **4**, 545(1993).

11. Z. Hu, X. Zhang, Y. Li, *Science*, **269**, 525(1995).

FIRE RETARDANT POLYETHERIMIDE NANOCOMPOSITES

Jongdoo Lee, Tohru Takekoshi and Emmanuel P. Giannelis
Department of Materials Science and Engineering, Cornell University, Ithaca, NY 14853

ABSTRACT

Polyetherimide-layered silicates nanocomposites with increased char yield and fire retardancy are described. The use of nanocomposites is a new, environmentally-benign approach to improve fire resistance of polymers.

INTRODUCTION

As use of synthetic polymers has grown dramatically over the last three decades so have efforts to control polymer flammability. Developments to that end have included intrinsically thermally stable polymers, fire retardant fillers and intumescent fire retardant systems [1]. An effective way to improve fire resistance has relied on the introduction of highly aromatic rings into the polymer structure. An increase in the aromaticity yields high char residues that normally correlate with higher oxygen index and lower flammability. The often high cost of these materials and the specialized processing techniques required, however, have limited the use of these polymers to certain specialized applications. The effectiveness of fire retardant fillers is also limited since the large amounts required make processing difficult and might inadvertently affect mechanical properties.

Polymer nanocomposites, especially polymer-layered silicate, PLS, nanocomposites, represent a radical alternative to conventionally filled polymers [2]. Because of their nanometer-size dispersion the nanocomposites exhibit markedly improved properties when compared to their pure polymer constituents or their macrocomposite counterparts. These include increased modulus and strength, decreased gas permeability, increased solvent resistance and increased thermal stability. For example, a doubling of the tensile modulus and strength is achieved for nylon-layered silicate nanocomposites containing as little as 2 vol.% inorganic. In addition, the heat distortion temperature of the nanocomposites increases by up to 100 °C extending the use of the composite to higher temperature environments, such as automotive under-the-hood parts. Furthermore, the relative permeability and solvent uptake of the nanocomposites decreases by almost an order of magnitude.

Polymer-layered silicate, PLS, nanocomposites exhibit many advantages including: (a) they are lighter in weight compared to conventionally filled polymers because high degrees of stiffness and strength are realized with far less high density inorganic material; (b) they exhibit outstanding diffusional barrier properties without requiring a multipolymer layered design, allowing for recycling; and (c) their mechanical properties are potentially superior to unidirectional fiber reinforced polymers because reinforcement from the inorganic layers will occur in two rather than in one dimension.

Mat. Res. Soc. Symp. Proc. Vol. 457 © 1997 Materials Research Society

Melt intercalation of high polymers is a powerful new approach to synthesize polymer-layered silicate, PLS, nanocomposites. This method is quite general and is broadly applicable to a range of commodity polymers from essentially non-polar polystyrene, to weakly polar poly(ethylene terephthalate) to strongly polar nylon. PLS nanocomposites are, thus, processable using current technologies and easily scaled to manufacturing quantities. In general, two types of hybrids are possible: **intercalated**, in which a single, extended polymer chain is intercalated between the host layers resulting in a well ordered multilayer with alternating polymer/inorganic layers, and **delaminated**, in which the silicate layers (1 nm thick) are exfoliated and dispersed in a continuous polymer matrix.

In this paper we report the synthesis, thermal properties and flame resistance of polyether imide nanocomposites. Both the decomposition temperature and the char yield of the nanocomposites is much higher than that of the polymer or a conventionally filled system at similar loadings to the nanocomposites.

Experimental

Polyetherimides were synthesized according to the published methods [3,4] from 4,4'-(3,4-dicarboxyphenoxy)diphenylsulfide dianhydride and a series of aliphatic diamines using m-cresol as solvent. Organically modified layered silicates were prepared as previously outlined, by a cation-exchange reaction between Li fluorohectorite (cation exchange capacity, CEC of 1.5 meq/g, particle size 10 μ) or Na montmorillonite (CEC = 0.9 meq/g, particle size 1 μ) and the corresponding protonated primary amine. Nanocomposites were synthesized either statically or in a microextruder. In the former the silicate and the polymer were mechanically mixed and formed into a pellet (25 mm^2 x 5 mm) using a hydraulic press and a pressure of 70 MPa. The pellets were subsequently annealed in vacuum at 170 °C for several hours. The microextruded samples were processed at 195 °C for 30 min. and a rate of 30 rpm.

X-ray diffraction analysis was performed using a Scintag θ-θ diffractometer and Cu Kα radiation. Thermal analysis was performed on a DuPont Instruments 9900 thermal analyzer at a heating rate of 10°C/min under flowing air or nitrogen.

Results and Discussion

The family of aliphatic polyether imides used and their characteristics are summarized in Table I. In contrast to their aromatic counterparts, the aliphatic PEI are amorphous with no evidence for any melting transitions. As the length of the aliphatic chain, m, increases the glass transition of the polymer decreases. The narrow MW range was accomplished by carefully controlling the reactant stoichiometry during polymerization.

Pristine mica-type layered silicates usually contain hydrated Na$^+$ or K$^+$ ions. Ion exchange reactions with cationic surfactants including primary, tertiary and quaternary ammonium ions render the normally hydrophilic silicate surface organophilic, which makes intercalation of many engineering polymers possible. The role of the alkyl ammonium cations in the organosilicates is to lower the surface energy of the inorganic host and improve the wetting characteristics and, therefore, miscibility with the polymer.

TABLE I

Properties of Synthesized Polyetherimides (PEI)

Polymer	Length of Aliphatic Amine, m	$T_g/°C$	M_w	M_n	M_w/M_n
PEI-6	6	123	45,000	38,000	1.18
PEI-7	7	114	36,000	31,000	1.16
PEI-8	8	110	43,000	35,000	1.23
PEI-9	9	85	43,000	35,500	1.21
PEI-10	10	83	42,000	33,000	1.27

X-ray diffraction measurements provide a quick measure for nanocomposite formation. Generally intense reflections in the range $2\theta = 3 - 9°$ indicate either an ordered intercalated hybrid or an immiscible system. The former, however, shows an increase in the d-spacing corresponding to the intercalation of the polymer chains in the host galleries while in the latter the d-spacing remains unchanged. In delaminated hybrids, on the other hand, XRD patterns with no distinct features in the low 2θ range are anticipated due to the introduced disorder and the loss of structural registry in the silicate layers.

Table II summarizes the x-ray diffraction analysis of montmorillonite and fluorohectorite nanocomposites with PEI-10. Na-montmorillonite being hydrophilic leads to an immiscible system. In contrast, nanocomposites are formed with the organically modified silicates. As the length of the organic cation (from C12 to C18) or the charge density of the host increases a transition from exfoliated to intercalated nanocomposites is observed. This behavior is in accord with the predictions of the mean-field thermodynamic model for hybrids developed in our group [5].

Figure 1 shows the TGA in air of pristine PEI and three hybrids containing 10 wt.% silicate. The corresponding TGA in nitrogen is shown in Figure 2. Both the intercalated and delaminated nanocomposites show a delayed decomposition temperature compared to the unfilled polymer. Interestingly, the immiscible hybrid containing the same amount of silicate shows no improvement suggesting that formation of the nanostructure is responsible for the increases in thermal stability.

Figures 3 and 4 show the isothermal TGAs for the above systems in air at 450 and 500 °C, respectively. The intercalated nanocomposites show a much higher char yield than any of the other systems. For example, for the 450 °C isothermal the intercalated nanocomposite retains about 90 and 45 % of its weight after 20 and 120 min., respectively. The corresponding numbers for the pure polymer are 45 and 15 %. In the 500 °C isothermal the polymer is completely lost after 40 min. while the char yield of the intercalated nanocomposite is ~55 %.

TABLE II
X-ray Diffraction Analysis of PEI-10 Nanocomposites

Organosilicate	Nanostructure
Dodecylammonium montmorillonite, MC12	Delaminated
Tetradecylammonium montmorillonite, MC14	Delaminated
Hexadecylammonium montmorillonite, MC16	Delaminated
Octadecylammonium montmorillonite, MC18	Intercalated
Dioctadecyldimethylammonium montmorillonite, M2C18	Intercalated
Na montmorillonite	Immiscible
Dodecylammonium fluorohectorite, FC12	Intercalated
Tetradecylammonium fluorohectorite, FC14	Intercalated
Hexadecylammonium fluorohectorite, FC16	Intercalated
Octadecylammonium fluorohectorite, FC18	Intercalated

While there appears to be a difference between the intercalated and delaminated nanocomposites we have found no difference between the fluorohectorite and montmorillonite based nanocomposites as long as they exhibit the same nanostructure suggesting that the particle size of the silicates is not an important factor. Additionally, the thermal stability was independent of the cation in the organosilicate with the nanostructure again been the predominant variable.

Even though the mechanism is unknown at present the nanocomposites show significant fire retardancy when compared to the pure polymer (Figure 5). In both cases the specimens were exposed to open flame for about 10 sec. The pure polymer persisted burning after the flame was removed until it

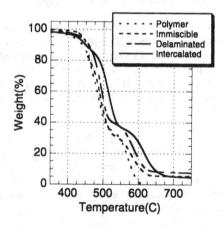

Figure 1. TGA of PEI-10 and PEI-10 hybrids in air.

was externally extinguished. In contrast, the nanocomposite became highly charred but maintained its original dimensions and ceased burning after the flame was removed.

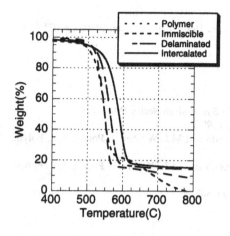

Figure 2. TGA of PEI-10 and PEI-10 hybrids in nitrogen.

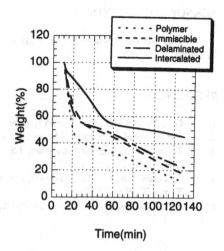

Figure 3. Isothermal at 450 °C in air.

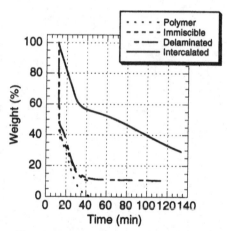

Figure 4. Isothermal at 500 °C in air.

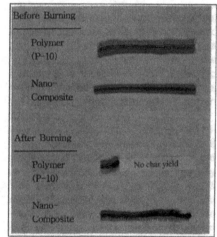

Figure 5. Fire test of PEI and PEI nanocomposite.

ACKNOWLEDGEMENTS

This work was supported by a grant from the Department of Transportation (FAA, Dr. R. Lyon).

REFERENCES

1.   G.L. Nelson, Ed., Fire and Polymers, ACS Symposium Series 599, 1995.
2.   E.P. Giannelis, Advanced Materials, $\underline{8}$, 29 (1996).
3.   T. Takekoshi, J.E. Kochanowski, J.S. Manello and M.J. Webber, J. Polym. Sci., Polym. Chem. Ed. $\underline{23}$, 1759 (1985).
4.   T. Takekoshi, J.E. Kochanowski, J.S. Manello and M.J. Webber, J. Polym. Sci., Polym. Symp. $\underline{74}$, 93 (1986).
5.   R.A. Vaia and E.P. Giannelis, submitted for publication.

SELF-ASSEMBLY OF LAYERED ALUMINUM SILSESQUIOXANES: CLAY-LIKE ORGANIC-INORGANIC NANOCOMPOSITES

L. UKRAINCZYK*, R. A. BELLMAN**, K. A. SMITH*** AND J. E. BOYD***
*Dept. of Agronomy, Iowa State Univ., Ames, IA 50011
**Dept. of Materials Science and Engineering, Iowa State Univ., Ames, IA 50011
***Dept. of Chemistry, Univ. of New Mexico, Albuquerque, NM 87131

ABSTRACT

A series of layered silicate-like structures with a wide range of Si/Al ratios that have an organic functionality directly bonded to the structural Si atom by Si-C bond were prepared by template sol-gel synthesis at room temperature and pressure. XRD patterns indicate that organic functionalities in the interlayers are in paraffin-like arrangement and do not interpenetrate. Structural ordering is primarily governed by the assembly of the organic functionalities into lamellar micelles. Nanocomposites were studied by solid state ^{29}Si and ^{27}Al NMR to determine the degree of condensation of inorganic framework. The results indicate that Si-O-Al linkages do not form in gels precipitated at low pH. Stable Si-O-Al linkages form when pH of the precipitates is raised. The highest degree of Si-O-Al bonding is obtained when Al solutions are prehydrolyzed prior to the addition of silane.

INTRODUCTION

In the recent years it has been recognized that layered silicate structures, as well as many other ordered inorganic materials, can be prepared under non-hydrothermal conditions by biomimetic template synthesis using self-organized assemblies of organic molecules.[1] In particular, synthesis routes using liquid crystal and surfactant micelles as templates have been subject of intense research.[1,2] The surfactants commonly used in the template synthesis have polar ends that do not become part of the inorganic framework. A possibility of the polar surfactant headgroup becoming a part of inorganic framework has been explored previously for synthesis of layered organic/inorganic structures[2,3] or thin films grown on substrates by self-assembly of multilayers.[4]

We have synthesized smectite-like organic/inorganic nanocomposites (layered Al- and Mg-silsesquioxanes) using trialkoxysilanes where polar ends are silanol groups that polymerize following the hydrolysis of alkoxy groups, and become integral part of the inorganic framework (Figure. 1).[5] The materials were precipitated at room temperature by addition of base to an alcohol solution containing a mixture of $AlCl_3$ or $MgCl_2$ and a trialkoxysilane with an alkyl or a phenyl functionality. The Si/Al and Si/Mg ratios of the reaction mixtures were 2:1 and 4:3, respectively, and were chosen to match the composition of clay minerals pyrophyllite and talc. The materials have good structural integrity and disperse in solvents of low polarity. The most ordered products were formed from long chain n-alkylsilanes with Al. The dependence of ordering on chain length suggests that the formation of the layered structure is due to self-assembly of the hydrolyzed trialkoxysilanes into lamellar micelles. The micelles act as a template for the formation of a clay-like inorganic framework by condensation between silanols and aqueous metal species attracted by the negatively charged surfactant layers. These clay-like organic-inorganic nanocomposites may find use as fillers for polymers and coatings,

519

Mat. Res. Soc. Symp. Proc. Vol. 457 ⁰ 1997 Materials Research Society and as sorbents and barriers for environmental applications.

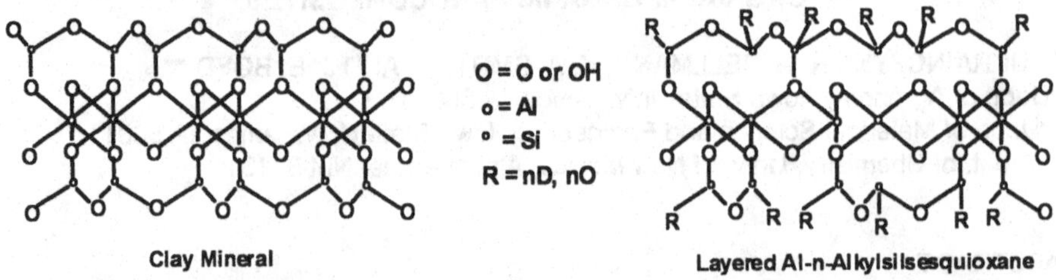

O = O or OH
o = Al
° = Si
R = nD, nO

Clay Mineral   **Layered Al-n-Alkylsilsesquioxane**

Figure. 1   Side-view of a 2:1 aluminosilicate clay mineral layer and schematic presentation of inorganic
framework of layered Al-n-alkylsilsesquioxane.

The structural integrity of the nanocomposites will be dependent on the extent of formation of Si-O-Al bonds. The objective of this study was to investigate conditions favoring formation of Si-O-Al linkages during self-assembly of Al-n-dodecyltriethoxysilane (NDS) and Al-n-octyltriethoxysilane (NOS).

EXPERIMENTAL

NDS was obtained from Pfaltz & Bauer, and NOS was obtained from Gelest. **AlnDS** and **AlnOS** were prepared from a silane/AlCl$_3$ solution with Si/Al molar ratio of 2:1. A partially hydrolyzed Al solution, prepared by adjusting pH of a volume of 0.2 M AlCl$_3$ x 6 H$_2$O in ethanol to 3.7 with a 0.5 M NaOH, was used. A volume of freshly prepared 0.4 M trialkoxysilane solution in ethanol was added to the Al solution, and the mixture was then titrated with 0.5 M aqueous NaOH while stirring until pH was 5.5. The precipitate was aged for 24 hours at room temperature, and then washed repeatedly with distilled deionized water until no Cl$^-$ could be detected in the filtrate. The white solid was air-dried at room temperature. **AlnDS21a** (reaction mixture Si/Al = 2) was prepared by adjusting pH of 0.2 M AlCl$_3$ in ethanol to 3 using 0.05 M NaOH prior to the addition of 0.4 M NDS solution in ethanol. The initially cloudy mixture was stirred until it was clear, indicating silane has hydrolyzed, and then titrated with 0.05 M NaOH until it turned into a thick gel. The pH of the gel was 3. The gel was aged for 48 hours, washed and dried as described above. **AlnDS21b, AlnDS41a, AlnDS31b, AlnDS31c,** and **AlnOS31** were prepared in the similar way as **AlnDS21a,** except that: for **AlnDS41a** (reaction mixture Si/Al = 4) 0.8 M NDS in ethanol was used; for **AlnDS31c** (reaction mixture Si/Al = 3) 0.6 M NDS in ethanol was used, gel was allowed to dry and was not washed with water; for **AlnDS21b** (reaction mixture Si/Al = 2) base addition was continued until final pH was 6; for **AlnDS31b** (reaction mixture Si/Al = 3) 0.6 M NDS in ethanol was used and base addition was continued until final pH was 6; for **AlnOS31b** (reaction mixture Si/Al = 3:1) 0.6 M NOS in ethanol was used and base addition was continued until final pH was 3. **nDS** was prepared by slowly titrating 0.2 M NDS in ethanol with 0.05 M aqueous NaOH until white gel-like precipitate was formed (final pH=11) and the product was washed and air-dried as described above to give a white greasy solid. All other air-dried products were white, xerogel-like solids that were gently ground and vacuum-desiccated for further analysis.

Elemental analysis, XRD, FTIR, TEM and SEM were performed as described[5] previously. ^{29}Si NMR spectra were acquired on Bruker ASX 300 spectrometer at the University of New Mexico NMR Facility using CP/MAS, spinning rates of 3.5 kHz and 5 s recycle time. ^{27}Al MAS NMR spectra were acquired with a single pulse, 1 s recycle time and spinning rates of 10 kHz.

RESULTS AND DISCUSSION

Composition data in Table 1 show that the Si/Al ratios of Al-silsesquioxanes for which the final pH of the synthesis mixture was high (5.5-6) have Si/Al ratios close to those of the initial reaction mixture. On the contrary, the products that were prepared by gel formation at low pH are highly Al deficient indicating that Al was leached out during washing of the products. Layered structures with Si/Al≠2 could only be precipitated through gelation process; attempts to make them from solutions containing prehydrolyzed Al resulted in greasy, oily products.

A representative XRD pattern, showing layered structure of one of the products, is presented in Figure 2. In layered Al-silsesquioxanes basal spacings is consistent with a

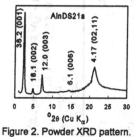

Figure 2. Powder XRD pattern.

~10 Å inorganic layer and a bilayer arrangement of R groups.[5] All the products made from NDS had similar patterns, with slight variations in d_{001} (Table 1). With exception of **nDS** (which retained water), the d_{001} spacing increased with the increasing amount of Al in the product, which can be attributed to the decrease in the thickness of the inorganic layer at low Al contents. The XRD patterns of the products made from NOS were similar to those of NDS, except that only d_{002} and d_{003} reflections are observed and their intensities are lower than those of NDS products. This decrease in intensity is likely due to a lower degree of ordering of NOS products.[5]

The layered morphology of the products is also evident in the SEM and TEM micrographs (Figure 3). Unlike Al containing products, **nDS** layers exhibited curling. The fringing observed in the TEM micrographs, confirms that products are crystalline.

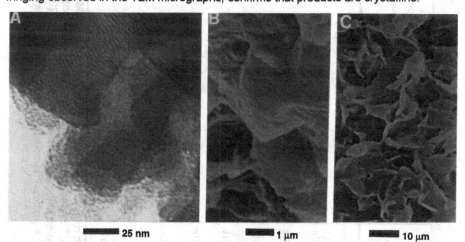

Figure 3. (a) TEM micrograph of **AlnDS**, and (b-c) SEM micrograph of **(b) AlnDS31b**, and (c) of **nDS**.

Solid state ^{29}Si and ^{27}Al NMR spectra were highly sensitive to the synthesis method

Table 1. Summary of composition data, solid state ^{29}Si and ^{27}Al NMR and XRD basal spacings for layered Al-n-alkyl silsesquioxanes.

Sample	Si/Al Molar Ratio		C	H	Peak positions[a] (areas[b]) obtained from ^{29}Si CP/MAS NMR					Deg. of cond. (%)[c]	Peak positions[a] (areas[b]) obtained from ^{27}Al MAS NMR			$^dT^3/Al^{IV}$	d_{001} (Å)
	Initial	Product	(wt. %)		T^3	T^2	T^1	T^0	T_m		Al^{IV}	Al^V	Al^{VI}		
AlnDS	2	2.1	54.3	9.9	-64.0 (24)	-56.3 (66)	-47.8 (10)			71	55 (41)	40 (29)	0 (30)	0.9	38.5
AlnDS21a	2	12.9	63.2	10.9	-67.4 (25)	-57.8 (53)	-48.7 (11)	-39.8 (4)	-52.9 (7)	64	55 (49)		0 (51)	6.6	36.2
AlnDS21b	2	2.0	53.8	9.7	-65.2 (20)	-57.4 (55)	-49.8 (25)			65	55 (30)		0 (70)	1.3	38.9
AlnDS41a	4	11.5	51.5	9.6	-67.1 (32)	-57.8 (57)	-49.0 (11)			74	55 (54)		0 (46)	6.8	36.2
AlnDS31b	3	3.0	56.8	9.8	-65.1 (22)	-57.6 (54)	-49.0 (17)	-39.8 (1)	-52.9 (5)	64	54 (38)		0 (62)	1.7	37.5
AlnDS31c	3	3.0	nd	nd	-67.6 (28)	-57.7 (57)	-48.6 (8)	-39.8 (2)	-52.8 (5)	69			0 (100)	∞	36.9
nDS	∞	∞	64.8	10.1	-67.6 (64)	-58.0 (35)	-48.9 (<1)			88				na	37.2
AlnOS	2	2.1	45.7	8.1	-61.2 (31)	-56.8 (60)	-45.1 (9)			74	54 (58)		0 (42)	1.1	25.8
AlnOS31	3	3.0	49.8	8.5	-65.2 (36)	-57.8 (56)	-46.3 (8)			76	54 (49)		0 (51)	2.2	25.5

[a] In ppm relative to TMS (^{29}Si; ± 1 ppm) and 0.1 M Al(NO$_3$)$_3$ (^{27}Al; ± 3 ppm).
[b] ±10%. ^{29}Si peak areas were corrected for cross-polarization efficiency.[6]
[c] Calculated using:[7] degree of condensation (%) = [(0.5 x area T^1) + (area T^2) + (1.5 x area T^3)]/1.5
[d] T^3/Al^{IV} = (moles T^3)/(moles Al^{IV}) where 0.9 = (moles T^3)/(moles Al^{IV} + 0.5 x moles Al^V).

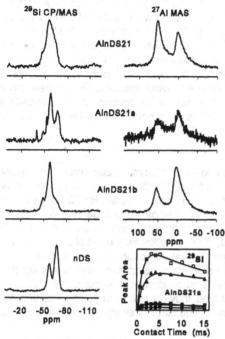

Figure 4. ^{29}Si CP/MAS and ^{27}Al MAS NMR spectra of layered silsesquioxanes and a plot of cross-polarization efficiency for ^{29}Si. All ^{29}Si spectra shown obtained with contact time of 5 ms.

(Figure 4, Table 1). All the spectra have three peaks in the chemical shift range of the trifunctional Si (T) and no resonance in the silicate region, confirming that Si-C bond remained intact during synthesis.[8] In the T^n notation used to describe Si sites for silsesquioxanes superscript refers to the number of bonds to other Si atom through an oxygen bridge: T^0, T^1, T^2 and T^3 stand for $^*Si(R)(OR')_{3-n}(OSi)_n$ (where R=organic moiety bonded to Si by Si-C bond; R'=H, or alkyl group depending on the degree of hydrolysis). The small FWHM of the peak at ~52 ppm suggests that this peak is due to an oligomeric T^2_m species (where m is most likely 3).[8] Because all of the products except **AlnDS31c** were extensively washed with distilled water, it is unlikely that any alkoxy groups remained in the solid. Thus, the T^n Si's in all prodcts except **AlnDS31c** should be bonded to either an OH, or to Si or Al through an oxygen bridge. The T^0 and T^2_m are only observed in Al-deficient structures formed by gelling at low pH.

The ^{29}Si chemical shifts of **nDS** are similar to those found for grafted trifunctional Si.[6] As expected, the degree of condensation is highest for **nDS** which was precipitated at high pH. In **AlnDS**, **AlnDS21b**, and **AlnDS31b** the ^{29}Si peaks are shifted downfield relative to **nDS** which is consistent the presence of Si-O-Al linkages.[5,10] Because T^2 and T^1 peaks in Al products have very small shifts relative to the corresponding peaks in **nDS**, Al substitution in the "tetrahedral" sheet of AlnDS can be ruled out.[5] **AlnDS** exhibits the largest shifts of all three high-Al products and the broadest ^{29}Si peaks which suggests that the degree of Si-O-Al bonding is greatest in this product. This product also has the lowest proportion of Al in 6-coordinate sites.

The T^3 peak in the spectra of the products prepared by gelling (**AlnDS21a**, **AlnDS41a**, **AlnDS31c**) is shifted <1 ppm relative to **nDS**. In **AlnDS21a** and **AlnDS41a** this can be attributed to low Al content, while in **AlnDS31c**, a gel which was not washed, this is likely due to lack of formation of Si-O-Al bonds. Thus, it can be concluded that the Si-O-Al bonds do not form in the gels prepared at low pH where Al is not hydrolyzing, and pH dependent negative charge of silanol groups is low. If the base addition to the gels is continued and pH is increased, as was the case for **AlnDS21b** and **AlnDS31b**, then the Si-O-Al linkages are established but to a lesser extent then in **AlnDS**. Similar trend is observed for **AlnOS** and **AlnOS31b**: the first one, made from prehydrolyzed Al solution, has T^3 peak shifted downfield by 4 ppm relative to **AlnOS31b** made by raising the pH of the gel.

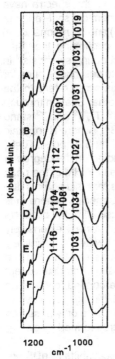

^{27}Al NMR spectra of all the products except **AlnDS31c** exhibit broad peaks indicative of wide variety of bond distances and angles in the inorganic layer consistent with the structure shown in Figure 1.[5] **AlnDS31c**, which likely contains no Si-O-Al linkages, exhibits a sharp Al resonance at 0 ppm.

In the IR spectra of the products (Figure 5) there is a shift in the silicate bands towards lower frequency with increasing Al substitution. This is consistent with increasing amount of Si-O-Al bonds which should shift silicate band towards lower frequency.

CONCLUSIONS

Long-chain *n*-alkylsilsesquioxanes can order into lamellar micelles and form surfactant templates with and without Al. Formation of Si-O-Al bonds occurs only at pH>3.5 where silanol groups have pH-dependent negative charge and Al starts to hydrolyze. If silanes are hydrolyzed and gelled at pH<3 in Al-containing solutions, Si-O-Al linkages are not likely to form and Al is lost by washing of the gel. Although Si-O-Al linkages can be produced by raising the pH of the gel, their number is smaller then when initial hydrolysis and condensation occur at pH>3.5. Once Si-O-Al bonds are formed, they are not destroyed by washing with water and exhibit hydrolytic stability similar to that found in clay minerals.

Figure 5. FTIR spectra of (a) **AlnDS**, (b) **AlnDS21b**, (c) **AlnDS31b**, (d) **AlnDS31c**, (e) **AlnDS41a**, and (f) **nDS**.

ACKNOWLEDGMENTS

This research was supported in part by Laboratory Directed Research and Development grant from Ames Lab (U.S. DOE), and in part by Iowa Agricultural Experiment Station.

REFERENCES

[1] A. Monnier, F. Schuth, Q. Huo, D. Kumar, D. I. Margolese, R. S. Maxwell, G. D. Stucky, M. Krishnamurty, P. M. Petroff, A. Firouzi, M. Janicke, B. F. Chmelka, Science **261**, 1299 (1993).

[2] Q. Huo, D. I. Margolese, U. Ciesla, P. Feng, T. E. Gier, P. Sieger, R. Leon, P. M. Petroff, F. Schuth, G. D. Stucky, Nature **368**, 317, (1994).

[3] Y. Fukushima, M. Tani, J. Chem. Soc. Chem. Commun. 241 (1995).

[4] J. Wood, R. Sharma, Langmuir **10**, 2307 (1994); J. Snover, M. E. Thompson, J. Am. Chem. Soc. **116**, 765 (1994).

[5] L. Ukrainczyk, R. A. Bellman, A. B. Anderson, J. Phys. Chem. (*in press*).

[6] M. Mehring, High Resolution NMR Spectroscopy of Solids; 2nd ed.; Springer-Verlag, New York, 1983.

[7] K. J. Shea, D. A. Loy, O. Webster, J. Am. Chem. Soc. **114**, 6700 (1992).

[8] D. W. Sindorf, G. E. Maciel, J. Am. Chem. Soc. **105**, 3767 (1983).

[9] R. Murugavel, A. Voigt, M. Mrinalini, G. Walawalkar, H. W. Roesky, Chem. Rev. **96**, 2205 (1996).

Synthesis of self-assembled functional molecules in mesoporous materials

*Itaru HONMA, #H.SASABE and #H-S.ZHOU
*Electrotechnical Laboratory, AIST,Tsukuba,Ibaraki,305 Japan, e9513@etl.go.jp
#FRP, The Institute of Physical and Chemical Research (RIKEN), Saitama, Japan

ABSTRACT

Self-assembled functional molecules in mesoporous materials are synthesized directly either by co-assembly of dye-bound surfactant of ferrocenyl TMA with silicate or Pc (phthalocyanine) molecules doped in the $C_{16}TMA$ micelles with oxides framework such as V_2O_5, MoO_3, WO_3 and SiO_2. The process provides well-organized molecular dopoed mesoporous structure by direct and simple procedure.

INTRODUCTION

Mesoporous materials invented by scientists of the Mobil Corporation[1-3] have attracted considerable interest since the first announcement in 1992[4,5]. The formation mechanism of mesoporous materials designated as MCM products have been studied by many groups[6,7] and phase change behavior among lamellae, hexagonal and cubic are understood in terms of charge matching at the interface as well as a free energy argument in the organized organic/inorganic system[8]. The MCM products are widely synthesized in silica (SiO_2) framework and there are interests for extending other oxides such as TiO_2[9], V_2O_5[10,13], WO_3[11,12,13], MoO_3[13], because their potential applicability to adsorbents and catalytic processes. Hexagonal tungsten, vanadium and titanium oxides have been successfully produced respectively, and other oxides mesophase were examined in Pb, Fe, Mn, Zn, Al, Co, Ni to form lamellae structure. Besides extending framework oxides to other class of transition oxides, it's quite attractive to investigate a novel process of mesoporous film formation. Attempts have been made for silica mesoporous film either on the substrate[14] or air-water[15] interface. Those films might be used for device's application as optical, electric and chemical sensors as well as membranes.

In this work, we have investigated to synthesize photo-sensitive mesoporous materials for optical device applications where the photo-absorbing dyes are doped in the mesochannels by direct self-organizing process of the surfactants, not by external doping after the calcination of the channel. If the synthetic path is found to dope functional molecules to mesochannels by self-organized co-assembly process, it will open a wide controllable design to produce functional mesoporous materials for optical application such as sensors or luminescent materials.

Mat. Res. Soc. Symp. Proc. Vol. 457 ° 1997 Materials Research Society

EXPERIMENT&RESULTS

In order to make self-assembled functional molecules in mesoporous materials, we have investigated two different methods for dye-doped MCM products. The first one is by using dye-bound surfactant ; Fe-TMA(11-ferrocenylundecylammonium bromide where the trimethylammonium surfactant has a ferrocenyl dye at the end of the tail. The other is by using C_{16}TMA (hexadecyltrimethylammonium chloride) and the Pc (phthalocyanine ($C_{32}H_{18}N_8$), phthalocyanine blue ($C_{32}H_{16}N_8Cu$)) molecules are doped in a co-assembly within a C_{16}TMA micelles. In the latter case, Pc molecules are, supposedly, self-assembled between the C_{16}TMA's hydrophobic tails and organized with a periodic array of the lipid micelle structures.

In the first case, we have synthesized Fe-TMA-MCM in a similar way as an acidic synthesis of Silica MCM while C_{16}TMA was replaced by Fe-TMA. The molar ratio of the synthetic precursors were as follows; 1.0 TEOS : 9.0 HCl : 0.12 Fe-TMA : 130 H_2O . In spite of the lower concentration of the surfactants, hexagonal phase were successfully produced and the products were colored (light green) by the surfactant's ferrocenyl ligands. Fig. 1(a) shows a low angle XRD scan for Fe-TMA-MCM and the reflection pattern of the hexagonal phase was clearly observed. The d-spacing of the products was 37 A which is as twice long as the Fe-TMA surfactant and those are basically same products as silica MCM where lipid micelle forms the mesochannel structure. Figure 1(b) shows a transmission electron microscopy (TEM) photograph of the Fe-TMA-MCM and the hexagonal array of the mesoporous channels are clearly observed. As far as an eye-observation on the whole area of the low magnification view, only hexagonal phase was produced and other phase such as cubic or lamellae was not observed.

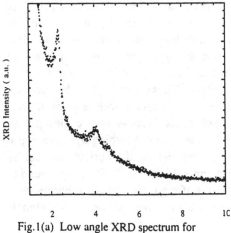

Fig.1(a) Low angle XRD spectrum for
Fe-TMA-MCM

Fig.1(b) Transmission electron micrograph
for Fe-TMA-MCM

The result shows that the Fe-TMA surfactants can synthesize hexagonal mesophase in a similar way as silica MCM and this is a first complete synthesis of the mesoporous products through ferrocenyl surfactants. As a Fe-TMA surfactant carries ferrocenyl dye at the end of the lipid tail, the initial products possess an optical absorption band in a visible light wavelength region and those are colored in light green or yellow green. Figure 2 shows an absorption spectrum of Fe-TMA-MCM powders and two absorption band at 440 nm and 640 nm were clearly seen. The absorption at 440nm was ascribed to a reduced state of the ferrocenyl ligand of Fe-TMA and the one at 640 nm was an oxidized state. Because of the ferrocenyl dyes are incorporated in the channel by self-assembly process, the absorption of the products are coming from concentrated dyes at the channel's center. The absorption band of 440 nm is a reduced state and identical to the absorption of isolated Fe-TMA molecules in the solution, while the one at 640 nm is an oxidized state and the peak appeared by the reaction with air (oxygen) after the synthesis. As the products are exposed to air at ambient temperature, the absorption intensity of 640 nm increases with time. In this new process, it is not necessary to dope molecules into mesoporous channels after the pore become open by calcination. The process provides economic and more efficient way to dope heavily in the mesochannels.

The second method to make self-assembled molecules in a mesoporous materials is by mixing dye molecules among the surfactant's hydrophobic tails where the dye molecules are inserted in a periodic manner within a mesoporous channels, which are basically similar system to chloroplast in the natural plants. We have chosen Pc (phthalocyanine ($C_{32}H_{18}N_8$), phthalocyanine blue ($C_{32}H_{16}N_8Cu$)) molecules as dopants and metal oxides of V_2O_5, WO_3, MoO_3, SiO_2 as mesoporous framework. The synthetic procedure of Pc-doped V_2O_5 MCM was as follows; the ammoniumvanadate (NH_4VO_3) ; 1.0 g, was first dissolved in a aqueous water (18g) with an addition of amount of NaOH (1.2g) to be solved completely. And the $C_{16}TMA$ (0.54g) are dissolved in the above solution with no precipitation at this high pH condition. And, the various amount of Pc (0.043g - 0.34g) were added to the above solution and stirred for 1 hour. Surprisingly, the Pc molecules were solvated quite well in the surfactant solution,which becomes completely blue because of the Pc molecules are isolated within the surfactant micelles. In the present experiment, the molar ratio of Pc

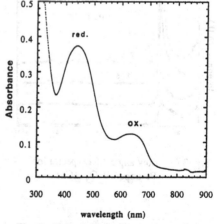

Fig. 2 Optical absorption spectrum for Fe-TMA-MCM

to C_{16}TMA was changed from 0.0 to 1/2.5, so that, in heavy doping case, Pc are inserted in every several surfactant molecules. After stirring Pc/C_{16}TMA solution, HCl (5N) are added drop wise to gradually decrease a solution pH to around 1.0. As HCl adding, vanadium oxides start precipitating at the interface of the micelle and mesoporous vanadium oxides with Pc molecules doped in the mesochannel were made. Similarly, Pc-doped WO_3-MCM as well as Pc-doped MoO_3-MCM were produced; the ammonium tungstate para pentahydrate (($NH_4)_{16}W_{12}O_{41}5H_2O$) were dissolved in water with C_{16}TMA at high pH and Pc were well solved in the above surfactant solution with stirring. The WO_3-MCM was formed by drop wise addition of HCl and green-blue colored powder was precipitated. In the case of Pc-MoO_3-MCM, ammonium phosphomolybdate trihydrate (($NH_4)_3PO_4$ $12MoO_3$ $3H_2O$) was used as a precursor chemical for the MCM products. In silica (SiO_2) case, Pc was first dissolved in the C_{16}TMA surfactant solution at low pH (acidic) condition and TEOS (tetraethoxysilane) was added to precipitate Pc-doped MCM powders. The molar ratio of Pc to C_{16}TMA was changed from 0 to 1/2.5 and the products were characterized by TEM, low angle XRD and optical absorption measurements.

Figure 3 (a) shows a low angle XRD data for Pc-doped V_2O_5 MCM with a variation of Pc-doping ratios. The pure V_2O_5 MCM was also shown in the figure. The as-prepared V_2O_5 MCM products are amorphous structure which is confirmed by a broad fundamental reflection at about 2.5 - 3.0 degree[10,13]. Supposedly, the

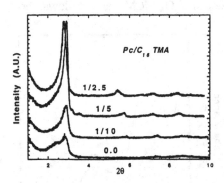

Fig. 3 (a) Low angle XRD spectra for
Pc-doped V_2O_5 MCM

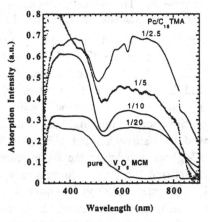

Fig.3 (b) Optical absorption spectra for
Pc-doped V_2O_5 MCM

V_2O_5 primarily building unit is a crystal structure so that it is not possible to form directly hexagonal or cubic phase which have curved interfaces at the mesochannel walls, although, in SiO_2 case, easy by the flexibility of the random amorphous network of the silica. As a Pc-doping ratio increases, the phase becomes unstable, especially at the $Pc/C_{16}TMA$ ratio of around 1/20 where the initial amorphous structure changes and the pattern shifts by minor experimental conditions. And further increase of the Pc-doping makes a MCM structure stable and a crystalline mesophase appeared that is different from the initial amorphous phase. In other words, the insertion of Pc molecules in the mesochannel stabilizes a crystalline phase possibly because of the modification of the channel wall shape. It is also observed that the peak position of the fundamental reflection of the mesophase shifts toward low angle with the Pc doping, i.e., the d-spacing of the V_2O_5 MCM becomes larger slightly by the expansion of the mesochannel. The d-spacing is 30.55A at Pc/C16TMA ratio of 1/5, shifts to 31.87A at the ratio of 1/2.5. Although there is a small expansion from the pure V_2O_5 MCM to Pc-doped MCM, the main peak position does not change significantly, which indicates that the Pc molecules are doped, perhaps, in between the surfactants hydrophobic tails of the lipid micelle, not in the center position of the micelle as in the expander molecules of trimethylbenzene (TMB)[7]. Pc molecules inserted between the surfactant expand slightly the channel size, at the same time, stabilize the vanadium oxides framework. Figure 3 (b) shows an optical absorption data for Pc-doped V_2O_5 MCM and the absorption from the doped Pc was clearly seen. The powder is green and as the doping ratio increases, the absorption intensity increases with the doping. The absorption at short wavelength region is coming from the V_2O_5 framework.

We have investigated structures of Pc-doped MoO_3-MCM. The doped MoO_3 mesoporous structure was made in a similar way as a V_2O_5-MCM. Figure 4(a) shows a low angle XRD scan of Pc-doped MoO_3 -MCM with a variation of the Pc-doping molar ratio ($Pc/C_{16}TMA$) from 0.0 (pure MoO_3 MCM) to 1/2.5. In this materials, the mesostructure was amorphous throughout the whole Pc-doping ratio with a broad scattering between 2 and 9 degree, which is quite different from the V_2O_5-MCM. In this case, the incorporated Pc molecules does not act as stabilizer molecules for the crystalline hexagonal phase. In the amorphous phase, the average pore spacing derived by LAXRD was not changed so much by Pc-doping and ,at the same time, the randomness structure was not changed.Figure 4(b) shows absorption spectra of the doped MoO_3-MCM and the absorption of Pc was observed

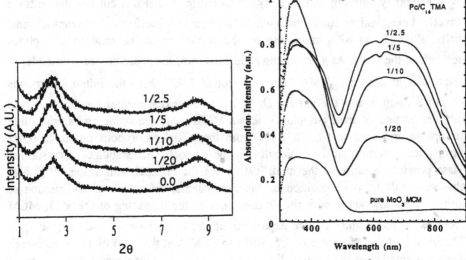

Fig. 4 (a) Low angle XRD spectra for
Pc-doped MoO_3 MCM

Fig. 4 (b) Optical absorption spectra for
Pc-doped MoO_3 MCM

through a transparent MoO_3 framework from 500nm to 900nm wavelength region. As the doping ratio increases, the absorption intensity increases without a shift of the spectra's shape. Additionally, Pc-doped SiO_2-MCM was successfully synthesized and the hexagonal mesophase was retained for high Pc doping condition and optical absorption spectra of the doped phase was similar to those of other oxides.

CONCLUSIONS

We have successfully synthesized Pc-doped metal oxides mesoporous materials in a V_2O_5, MoO_3, WO_3 and SiO_2 framework materials as well as a direct synthesis of dye-doped MCM through a co-assembly of ferrocenyl TMA surfactants . There is a different role of the doping Pc to the host mesophase; in V_2O_5 case, Pc plays a stabilizing agent for crystalline mesoporous phase, i.e., amorphous phase of the pure V_2O_5 MCM transforms to crystalline phase with the doping, although, in the MoO_3-MCM case, the amorphous phase remains by the doping. However, Pc-doped SiO_2-MCM was successfully made to be a hexagonal phase.

REFERENCES

1. J.S.Beck, US Pat., **5** 507 296, (1991)
2. C.T.Kresge,M.E.Leonowicz,W.J.Roth and J.C.Vartuli, US Pat.,**5** 098 684,(1992)
3. J.S. Beck et al., US Pat., **5** 108 725, (1992)
4. C.T.Kresge, M.E.Leonowicz, J.C.Vartuli and J.S. Beck, Nature, **359**, 710(1992)
5. J. S. Beck et al., J. Am. Chem. Soc., **114**, 10834 (1992)
6.A. Monnier et al., Science, **261**, 1299 (1993)
7. A. Firouzi et al., Science, **267**, 1138 (1995)
8. Q. Huo et al., Chem. Mater., **6**, 1176 (1994)
9. D.M.Antonelli and J.Y.Ying, Angew. Chem. Int. Ed. Engl., **34**, 2014 (1995)
10. V. Luca et al., Chem. mater. , **7**, 2220 (1995)
11. A.Stein et al., Chem. Mater., **7**, 304 (1995)
12. U. Ciesla et al., J. Chem. Soc. Commun., 1387 (1994)
13. G.G.Janauer et al., Chem. Mater., **8**, 2096 (1996)
14. I.A.Aksay et al., Science, **273**, 892 (1996)
15. H. Yang et al., Nature, **381**, 589 (1996)

LOW TEMPERATURE SYNTHESIS OF LAMELLAR TRANSITION METAL OXIDES CONTAINING SURFACTANT IONS

GERALD G. JANAUER, RONGJI CHEN, ARTHUR D. DOBLEY, PETER Y. ZAVALIJ, and M. STANLEY WHITTINGHAM
Chemistry Department and Materials Research Center, State University of New York at Binghamton, Binghamton, NY 13902-6000, USA

ABSTRACT

Recently there has been much interest in reacting vanadium oxides hydrothermally with cationic surfactants to form novel layered compounds. A series of new transition metal oxides, however, has also been formed at or near room temperature in open containers. Synthesis, characterization, and proposed mechanisms of formation are the focus of this work. Low temperature reactions of vanadium pentoxide and ammonium (DTA) transition metal oxides with long chain amine surfactants, such as dodecyltrimethylammonium bromide yielded interesting new products many of which are layered phases. $DTA_4H_2V_{10}O_{28} \cdot 8H_2O$, a layered highly crystalline phase, is the first such phase for which a single crystal X-ray structure has been determined. The unit cell for this material was found to be triclinic with space group $P\bar{1}$ and dimensions a=9.895(1)Å, b=11.596(1)Å, c=21.924(1)Å, α=95.153(2)°, β=93.778(1)°, and γ=101.360(1)°. Additionally, we synthesized a dichromate phase and a manganese chloride layered phase. with interlayer spacings of 26.8Å, and 28.7Å respectively. The structure, composition, and synthesis of the vanadium compound are described, as well as the synthesis and preliminary characterization of the new chromium and manganese materials.

INTRODUCTION

Soft chemistry approaches have been utilized in many syntheses resulting in formation of transition metal oxides [1]. One of the most common and successful approaches, hydrothermal synthesis [2], has been employed in studies of tungsten [3, 4], molybdenum [5, 6], and vanadium oxides [7-9]. The discovery of MCM-41 [10], a new class of aluminum silicates with a mesoporous structure sparked a plethora of additional studies of whether transition metal oxides could also be formed via these methods [11-13]. A general mechanism for the formation of both the observed lamellar and hexagonal phase products of transition metal oxides and surfactant templates was proposed by Huo et al [11]. Stein et al, however, in their extensive study [12], had strong evidence that a tungsten oxide reacted with hexadecyltrimethylammonium formed Keggin clusters rather than continuous layers of tungsten oxide. A study on vanadium oxides recently discussed a lamellar vanadium phase formed on reaction of V_2O_5 with dodecyltrimethylammonium bromide [14]. In the first single crystal study on such a lamellar phase it has since been determined that this material also contains clusters rather than continuous layers of vanadium oxide. The discrete decavanadate clusters are joined via hydrogen bonding, and surrounded by the surfactant cations with their head groups stacked in opposite directions. Although this vanadium oxide - surfactant product was originally synthesized via a hydrothermal approach, new synthesis methods have allowed formation of this material via heating on a hot plate [15]. There have been several recent studies where other lamellar transition metal phases have been made via room temperature or near room temperature synthesis methods. In a study by Luca et al [16] several lamellar vanadium oxide phases were formed at room temperature. We have synthesized the first chloride [2], a lamellar manganese chloride, via templating with dodecyltrimethylammonium bromide (DTABr), as well as a new oxide with dichromate. Ayyappan et al [17] also showed the formation of a lamellar chromium oxide was possible at room temperature.

EXPERIMENTAL

The initial formation of the new lamellar phase $DTA_4 H_2V_{10}O_{28} \cdot 8H_2O$ was via a hydrothermal approach documented earlier [14]. It has since been found that combining the reactant materials at room temperature and heating on a hot plate at near boiling conditions, actually produced large crystals in less than 24 hours, rather than weeks for the hydrothermal method. As in the hydrothermal method, DTABr and V_2O_5 are mixed in a 2:1 molar ratio. The 5 M NaOH which was previously added for pH adjustment, however, is now added prior to V_2O_5 addition [15]. After heating for about 1 hour at 65°C, a clump of amorphous material is removed from the reaction mixture via gravity filtration. The product crystals are formed from the supernatant liquid over the following 24 hour period. Crystals proved large enough and of sufficient quality for the single crystal study described here.

The new lamellar dichromate material was synthesized by combining a solution of DTABr with a solution of $(NH_4) Cr_2O_7$ in a 2:1 molar ratio. A precipitate formed immediately, and was filtered and washed thoroughly with deionized water. The product had the appearance of a yellow slurry, but dried to a gel after several days. Over a period of days, this gel became a waxy material which later hardened to a state in which it could actually be ground for X-ray powder diffraction studies. Powder diffraction patterns of all forms of this material yielded virtually identical patterns.

To form this material, solutions of manganese chloride and DTABr were combined to give a ratio of 2:1 DTABr to $MnCl_2$. This solution was then treated hydrothermally at 160°C for 4 days. The product solution (no solid was formed in this reaction), was then evaporated to give the lamellar phase material. This product, like the lamellar chromium material was clear and waxy in nature.

Samples were characterized via X-ray powder diffraction on a Scintag XDS 2000 powder diffractometer using CuK_α radiation, TGA analyses were performed on a Perkin Elmer TGA 7, FTIR analyses on a Perkin Elmer 2000 FTIR, and electron microprobe analysis were done on a JEOL 2000 Electron Microprobe. Single crystal measurements were first done at Ames Lab, Iowa State University on a Siemens P4 single crystal diffractometer, and the final structure determination was made on a Siemens Smart CCD single crystal X-Ray diffractometer using MoK_α radiation at Syracuse University. Pattern refinements and single crystal structure data reduction were done with CSD software [18].

RESULTS AND DISCUSSION

The crystals formed of $DTA_4H_2V_{10}O_{28} \cdot 8H_2O$ are a translucent orange-yellow. The largest crystals are approximately 9 millimeters across and their thickness is of the order of 0.1 millimeters. Analysis via electron microprobe showed that there was no Na left from the pH adjustment, nor was any Br left from the original cationic surfactant material. An X-ray powder diffraction, shown in figure 1, shows the high order and crystallinity of this material. An initial attempt to obtain single crystal data failed due to decomposition over time. The crystals of $DTA_4 H_2V_{10}O_{28} \cdot 8H_2O$ are easily dehydrated, losing 6 water molecules per unit cell. To determine how quickly this material actually lost the waters of hydration, an X-ray study was conducted to collect the powder pattern of a sample of this lamellar phase as it dries. The diffractometer chamber was purged with dry nitrogen, and a fully hydrated sample of $DTA_4H_2V_{10}O_{28} \cdot 8H_2O$ scanned over a 2 hour period of time. As the sample dried, the lattice spacing collapsed by about 3Å, as can be seen in the 3D plot of the X-ray data in this study, figure 2. Nearly 50% of the original peak, due to the hydrated material, has already disappeared after about 20 minutes. This dehydration experiment clearly showed why problems were encountered in the first single crystal study. This study should have run for a period of four days, but by this time the sample had decomposed to form mainly dehydrated material. The study on the Siemens Smart CCD diffractometer at Syracuse University was more successful, as the sample was scanned for 7 hours and encased in glue to enhance

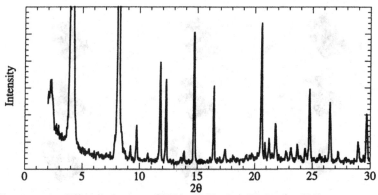

Fig. 1. X-ray powder diffraction pattern of $DTA_4 H_2V_{10}O_{28} \cdot 8H_2O$, after [14].

stability. This study, then, provided adequate data for structure determination. The structure of $DTA_4H_2V_{10}O_{28} \cdot 8H_2O$ is shown in figure 3. In this material, decavanadate clusters are surrounded by water molecules, 2 of which are probably in the form of H_3O^+. These hydronium ions and 4 DTA chains provide charge balance for the $V_{10}O_{28}^{6-}$ clusters. The DTA chains are stacked such that the head groups of neighboring chains are in opposite directions. The additional 6 water molecules are hydrogen bonded to the decavanadate clusters, as can be seen in figure 4. The single crystal data gave a triclinic cell with a space group of P $\bar{1}$, and dimensions a=9.895(1)Å, b=11.596(1)Å, c=21.924(1)Å, α=95.15(1)°, β=93.78(1)°, and γ=101.36(1)°.

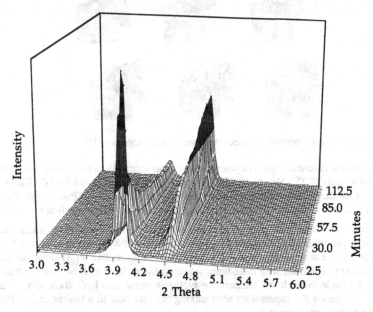

Fig. 2. X-ray diffraction 3D plot of the dehydration of $DTA_4H_2V_{10}O_{28} \cdot 8H_2O$, showing the growth in the higher angle peak as the lower one decreases.

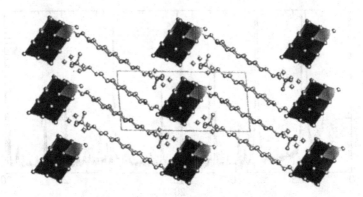

Fig. 3. Structure of $DTA_4H_2V_{10}O_{28}\cdot 8H_2O$ showing the arrangement of the organic chains [15].

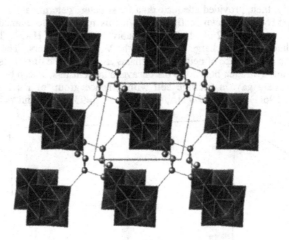

Fig. 4. Water hydrogens bonding the decavanadate clusters together [15].

The lamellar chromium material powder pattern was indexed to a monoclinic cell, space group P2/m with a=26.757(5)Å, b=10.458(2)Å, c=14.829(3)Å, and β =98.01(1)°. This material appears to be similar to that recently studied by Ayyappan et al [17]. A diffraction pattern for this material can be seen in figure 5. Further studies are underway to characterize it more fully.

We found that chlorides, such as manganese dichloride, also form these lamellar structures [2]. The x-ray pattern for this compound is given in figure 6, and the layer repeat was found to be 28.7Å. The exact nature of this material is unknown, as clusters like $V_{10}O_{28}$ are not known for chlorides. Recent experience with room temperature syntheses leads us to believe that this material might also be made through low temperature synthesis rather than hydrothermally. It might be possible to evaporate the supernatant after mixing the reactants in a beaker, rather than going through the hydrothermal process.

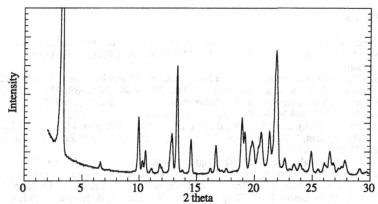

Fig. 5. X-ray diffraction pattern of surfactant chromium oxide material.

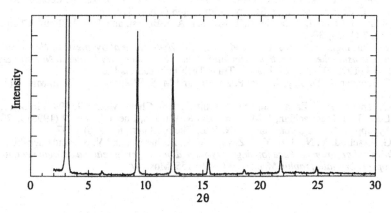

Fig. 6. X-ray diffraction pattern of the manganese chloride surfactant material [2] showing the layer-like nature of the compound.

CONCLUSIONS

A lamellar vanadium phase was shown to contain clusters, and not continuous layers, confirming the conclusion drawn by Stein et al in earlier work [12]. Evidence suggests that many metal oxides and metal chlorides may form this type of compound. Future work with similar materials is likely to aid in determination of mechanisms involved in the formation of such compounds.

ACKNOWLEDGMENTS

We would like to thank NSF-DMR for financial assistance, Dr. Jon Zubieta for use of the NSF regional single crystal facility at Syracuse University, and Robert A. Jacobsen and Vatalij K. Pecharsky at Ames Laboratory, Iowa State University for their single crystal work performed on our vanadium samples.

REFERENCES

1. J. Rouxel, M. Tournoux, and R. Brec, ed. *Soft Chemistry Routes to New Materials - Chimie Douce-*. Materials Science Forum, Vol. 152-153. 1994, Trans Tech Publications: Switzerland.
2. M. S. Whittingham, Current Opinion in Solid State and Materials Science, 1 (1996) 227.
3. J. R. Günter, M. Amberg, and H. Schmalle, Mat. Res. Bull., 24 (1989) 289.
4. K. P. Reis, A. Ramanan, and M. S. Whittingham, Chem. Mater., 2 (1990) 219.
5. J. Guo, P. Zavalij, and M. S. Whittingham, Chem. Mater., 6 (1994) 357.
6. J.-D. Guo, P. Zavalij, and M. S. Whittingham, J. Solid State Chem, 117 (1995) 323.
7. M. S. Whittingham, J. Guo, R. Chen, T. Chirayil, G. Janauer, and P. Zavalij, Solid State Ionics, 75 (1995) 257.
8. T. Chirayil, P. Zavalij, and M. S. Whittingham, Solid State Ionics, 84 (1996) 163.
9. T. A. Chirayil, P. Y. Zavalij, and M. S. Whittingham, Chem. Commun., (1996) in press.
10. J. S. Beck, J. C. Vartuli, W. J. Roth, M. E. Leonowicz, C. T. Kresge, K. D. Schmitt, C. T.-W. Chu, D. H. Olson, E. W. Sheppard, S. B. McCullen, J. B. Higgins, and J. L. Schlenker, J. Amer. Chem. Soc., 114 (1992) 10834.
11. Q. S. Huo, D. I. Margolese, U. Ciesla, P. Feng, T. E. Gier, P. Sieger, R. Leon, P. M. Petroff, F. Schuff, and G. D. Stucky, Nature, 368 (1994) 317.
12. A. Stein, M. Fendorf, T. Jarvie, K. T. Mueller, A. J. Benesi, and T. E. Mallouk, Chem. Mater., 7 (1995) 304.
13. M. S. Whittingham, J. Li, J. Guo, and P. Zavalij. *Hydrothermal Synthesis of New Oxide Materials using the Tetramethyl Ammonium Ion*. in *Soft Chemistry Routes to New Materials*. 152-153 (1993) 99. Nantes, France: Trans Tech Publications Ltd.
14. G. G. Janauer, A. Dobley, J. Guo, P. Zavalij, and M. S. Whittingham, Chem Mat, 8 (1996) 2096.
15. G. G. Janauer, P. Y. Zavalij, and M. S. Whittingham, Chem. Mater., 9 (1997) in press.
16. V. Luca, D. J. MacLachlan, J. M. Hook, and R. Withers, Chem. Mater., 7 (1995) 2220.
17. S. Ayyappan, N. Ulagappan, and C. N. Rao, Chem. Mater., 6 (1996) 1737.
18. L. G. Akselrud, Y. N. Grin, Y. P. Zavalij, V. K. Pecharsky, and V. S. Fundamenskii. *CSD - Universal program package for single crystal and/or powder structure data treatment*. in *12th European Crystallographic Meeting*. 3 (1989) 155. Moscow:

A NEW ROUTE TO THE PREPARATION OF NANOPHASE COMPOSITES VIA LAYERED DOUBLE HYDROXIDES

V.P.ISUPOV, K.A.TARASOV, R.P.MITROFANOVA, L.E.CHUPAKHINA.
Institute of Solid State Chemistry, Siberian Division of the Russian Academy of Sciences, Kutateladze-18, Novosibirsk, 630128, Russia, isupov@solid.nsk.su

ABSTRACT

A promising route to the preparation of nanophase composites with fine particles of transition metals via layered double hydroxides has been shown on the derivatives of Li-Al double hydroxide, LADH-X, where interlayer X anions are complexes of transition metals. Thermal decomposition of such materials in vacuum or an inert gas leads to dehydration and dehydroxylation of the hydroxide matrix and to the collapse red/ox process of the complex anion. The latter results in the carbonisation of the samples and in the appearance of nanoscale (20-500Å) metal particles.

INTRODUCTION

Intercalation compounds of layered double hydroxides are of interest as precursors for the preparation of catalysts, mixed oxides, ceramics etc. Their structure consists of positively charged layers, $[LiAl_2(OH)_6]^+$, separated by intercalated, charge balancing anions and water molecules [1,2]. These compounds can be used for the preparation of nanoscale composites. The type of a material formed depends on composition of an anion, on a chemical process, and on conditions of preparation. During the vacuum thermal decomposition of LADH-X intercalates containing anions of dicarboxylic acids, nanoscale particles of lithium aluminate and carbon are produced [3]. The particle size depends on a type of the organic anion. Fine particles of aluminum oxide and lithium salts are formed upon the vacuum thermolysis of intercalates with inorganic ions (Cl⁻, Br⁻ and SO_4^{2-}) [4]. Intercalates with a common formula $[LiAl_2(OH)_6]_2[Medta]\cdot4H_2O$, (LADH-[Medta]) were used in order to investigate the possibility of production of nanophase composites containing fine metal particles. In their structure the interlayer anions are complexes of copper, cobalt and nickel with ethylenediaminetetraacetic acid (edta). The complexes with these metals under thermolysis in vacuum or an inert gas form organic products and metal phase. In our case thermal decomposition of an aluminum hydroxide intercalate containing complex anions involving transition metal cations and some carboxylate anions could lead to the reduction of the cations to metal. Considering that the nearest interlayer cations must be 8-10Å apart, we expected that the process of formation of metal particles from atoms whould be significally hindered by diffusion. To realize this approach we studied thermolysis of intercalation compounds of lithium-aluminum double hydroxide containing $[Cu(edta)]^{2-}$, $[Co(edta)]^{2-}$ and $[Ni(edta)]^{2-}$ anions. The result to be expected is the formation of sufficiently small particles of metal.

EXPERIMENTS

In order to synthesize the intercalates by the ion-exchange procedure lithium-aluminum double hydroxide with chloride, LADH-Cl, was used [5]. LADH-Cl was obtained by treatment of crystal gibbsite, $Al(OH)_3$, with a concentrated LiCl solution [6]. Chemical composition of the resulting compound was close to $[LiAl_2(OH)_6]Cl\cdot1.5H_2O$ (table I). The synthesized compounds and the products obtained were studied by means of chemical and X-ray phase analysis, optical and IR-spectroscopy. X-ray diffraction patterns were obtained using a DRON-3 diffractometer equipped with Ni filter, CuK_a irradiation. High resolution electron microscopic photographs were obtained with a JEM-4000 microscope at the accelerating voltage of 400 kV. The composition of gas phase products was studied in a vacuum flow set-up with a MX-7303 mass spectrometer. Thermal analysis was carried out by means of a 1500 Q Paulic-Paulic-Erdey derivatograph.

RESULTS AND DISCUSSION.

Thermolysis of LADH-[Medta] in vacuum or an inert gas.

The data obtained by chemical analysis (table I), IR and optical spectroscopy, XRD give the following formulas for the synthesized compounds: $[LiAl_2(OH)_6]_2[Cuedta]\cdot4H_2O$, $[LiAl_2(OH)_6]_2[Coedta]\cdot4H_2O$ and $[LiAl_2(OH)_6]_2[Niedta]\cdot4H_2O$. Edta is an anion of ethylenediaminetetraacetic acid forming stable complexes with transition metals. Thermal decomposition of LADH-X materials leads to dehydration and dehydroxylation of the hydroxide matrix and to the collapse red/ox process of the complex anion with the appearence of transition metal phase. DTA and mass spectral data show that the process of decomposition is a sequence of the following stages. Up to 200°C elimination of adsorbed and interlayer H_2O from the hydroxide prevails. The process of dehydration of interstitially located water molecules proceeds in two stages approximately at 60-90°C (2 molecules) and 140-180°C (2 molecules) for all Medta-intercalates. Heating above 200°C results in the destruction of the hydroxide matrix. The process occurs before the decomposition of organic anion excepting LADH-[Cuedta], where both processes start near-concurrently. On a further increase in temperature, the process of thermolysis proceeds differently. For the nickel-containing intercalate at temperatures up to 300°C, the decomposition process goes with the predominent release of water molecules into the gas phase, which is caused by the destruction of hydroxide layeres. In the case of the copper and cobalt containing intercalates, water removal is accompanied by the simultaneous formation of gaseous products with 28 m/e and 44 m/e, which were referred to CO and CO_2 (fig.1). At temperatures above 300°C, for all compounds, one can observe not only the release of CO, CO_2 and H_2O (18 m/e), but also the appearance of NH_3 and H_2 (16 and 2 m/e), which indicates a deep destruction of edta. The considerable content of carbon is determined in the products of vacuum decomposition, which points to the carbonization of the samples in the course of thermolysis (table I). The C/M and N/M ratios show that approximately one half of the carbon and nitrogen content is removed from the initial intercalates upon thermolysis.

Table I. Elemental composition of the compounds and the products of their thermolysis at 400°C.

Sample	determined calculated , mass %						determined calculated , atom.ratio				
	Li	Al	M	C	N	H	Li/Al	M/Al	C/N	C/M	N/M
LADH-Cl	3.06	24.0	15.5	-	-	-	0.49	-	-	-	-
	3.10	23.9	15.7	-	-	-	0.50	-	-	-	-
LADH-[CoEdta]	1.80	14.1	8.49	15.2	2.99	4.07	0.49	0.28	5.9	8.8	1.5
	1.88	14.5	7.91	16.1	3.76	4.33	0.50	0.25	5.0	10.0	2.0
LADH-[Niedta]	1.75	14.3	7.98	15.2	3.25	4.14	0.47	0.26	5.5	9.3	1.7
	1.88	14.5	7.88	16.1	3.76	4.33	0.50	0.25	5.0	10.0	2.0
LADH-[CuEdta]	1.80	14.4	8.52	14.6	3.12	4.33	0.48	0.25	5.5	9.3	1.7
	1.87	14.4	8.47	16.0	3.73	4.30	0.50	0.25	5.0	10.0	2.0
products of thermolysis :											
LADH-[CoEdta]	2.83	22.3	12.4	19.5	4.74	1.79	0.49	0.25	4.8	7.7	1.6
LADH-[Niedta]	3.06	25.3	13.8	17.6	3.77	1.07	0.47	0.25	5.4	6.3	1.1
LADH-[CuEdta]	2.99	24.3	13.1	17.4	3.77	1.45	0.47	0.23	5.4	7.0	1.2

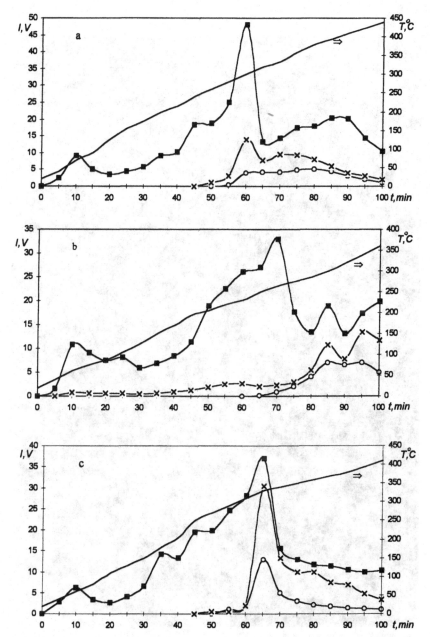

Fig.1. Mass-spectral data of gas release during heating: a - LADH-[Coedta], b - LADH-[Cuedta], c - LADH-[Niedta]; ■ - H_2O, × - CO_2, O - NH_3.

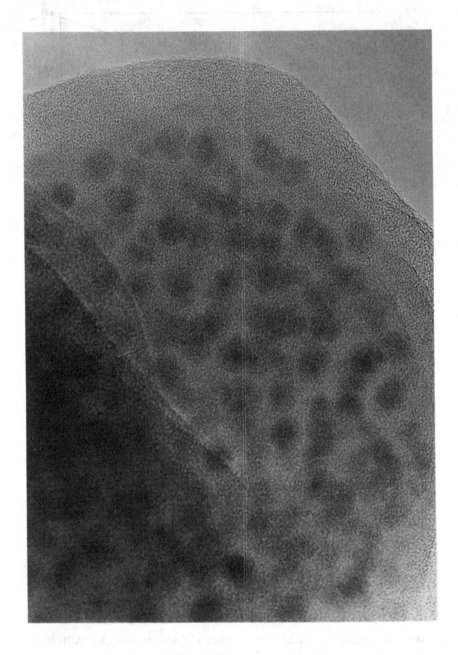

Fig.2. HREM microphotograph of LADH-[Niedta] decomposed in vacuum at 400°C.

Decomposition in vacuum or an inert gas during more than 1 hour results in the appearance of nanoscale metal particles (20-500Å). The data of high resolution transmission electron microscopy (HREM) confirm the formation of the round nickel particles with the narrow size distribution spread uniformly in the matrix upon the thermolysis (400°C) of LADH-[Niedta] (fig.2). On the nickel particles well regulated lattice lines are seen that testifies the particles are well-crystallized and do not possess inner defects. In the case of the [Cuedta]-containing intercalate (400°C), along with the formation of small particles (20-50Å), much larger ones (up to 500 Å) comprising disoriented fragments form. They are located disorderly in the amorphous matrix.

Decomposition of LADH-[Cuedta] in air.

Heating in air for several hours gives rise to formation of metal oxide phase and to subsequent production of spinel forms at high temperatures (above 500°C). Since the destruction of the organic anion for the [Cuedta]-containing intercalate occurs near-concurrently with dehydroxilation of matrix, it was interesing to obtain nanophase particles in the layered hydroxide system in the case of suitable temperature and period of heating. For 20 minutes at 250°C the result was retention of the layered hydroxide and formation not of the CuO composite, but of the metal copper one (fig.3). It is attributable to the slow process of dehydroxilation of the matrix which protects the generated copper phase against O_2.

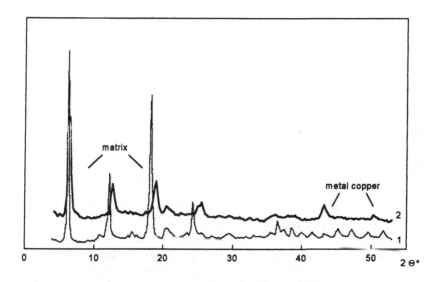

Fig.3. Powder X-ray diffraction patterns of LADH-[Cuedta] before (1) and after heating (2) in air at 250°C during 20 min.

CONCLUSIONS

Thus, thermal decomposition of LADH-intercalates with complexes of transition metals and Edta enables one to obtain a composite containing nanoscale metal particles distributed in a dielectric matrix.

ACKNOWLEDGEMENTS

This research was supported by the Russian Science Foundation (RFFI) under Grant № 95-03-09010.

REFERENCES

1). C.J.Serna, J.L.Rendon, J.E. Iglesias, Clays and Clay Min. **30**, p. 180 (1982).

2). J.P.Thiel, C.K.Chiang, K.R.Poeppelmeier, Chem.Mater. **5**, p. 297 (1993).

3). V. P. Isupov, R. P. Mitrofanova, V. A. Poluboyarov, L. E. Chupakhina, Doklady Academii Nauk (in Russian) **324**, p. 1217 (1992).

4). N.F.Uvarov, B.B.Bokhonov, V.P.Isupov, E. F.Hairetdinov, Solid State Ionics **74**, p.15 (1994).

5). V. P. Isupov, L.E.Chupakhina, Doklady Akademii Nauk (in Russian) **332**, p. 330 (1993).

6). A. P. Nemudry, V. P. Isupov, N. P. Kotsupalo, V. V. Boldyrev, Reactivity of Solids **1**, p. 221 (1986).

Part VIII

Late Paper Accepted

COMPUTER SIMULATION OF THE STRUCTURE OF NANOCRYSTALS

Y. Sasajima, Shigeo Okuda
Ibaraki University, Nakanarusawa 4- 12 - 1, Hitachi Ibaraki 316, JAPAN

ABSTRACT

Structural characteristics of nanocrystals were investigated by molecular dynamics simulation. The structural models were constructed in two dimension, assuming the Lennard-Jones type potential as interaction between atoms. The nanocrystal model consisted of ultra-fine particles as a unit structure, like atoms in normal crystals. In the present study, the diameter of the ultra-fine particles is assumed to be constant and the unit particles are assembled in densely packed structure to form a nanocrystal. The orientations of the unit particles were varied randomly to form various kinds of interfaces between their nearest-neighbor particles. This initial structure was relaxed by the molecular dynamics method with controlling the system temperature by rescaling the velocities of the atoms. The system temperature was set to be so high as to accelerate the relaxation process in the nanocrystal structure. If the diameter of the fine particle is less than 6 (the unit length is nearest-neighbor distance between atoms), the assembled particles were relaxed to form single crystal structure. When the diameter increased larger than 10, the nanocrystal structure was stable and the grain boundaries, vacancies and edge dislocations were remained in the system.

INTRODUCTION

Nanocrystal is aggregation of ultra-fine particles of which radii are in order of nano meters[1]. Nanocrystal can be produced by various methods such as spray deposition, gas condensation followed by consolidation, high energy ball-milling, sol-gel method, CVD and sputtering. The exotic features of nanocrystals cause specific properties such as in solute solubility, specific heat, thermal expansion, optical, magnetic, electric and mechanical properties. Especially, mechanical properties of nanocrystal are outstanding[2, 3]; (1) Young's modulus decreases while hardness increases abnormally, (2) elastic limit is very large (about 2%), (3) Strength obeys the inverse Hall-Petch relationship[4]. It is essential to construct structural mode of nanocrystal to clarify the correlation between structure and properties of nanocrystal. In the present study, a molecular dynamics calculation was performed to obtain the relaxed structural model of nanocrystal and to clarify thermal stability of the nanocrystal model.

METHODS OF CALCULATION

Initial structures of nanocrystal were constructed for molecular dynamics calculation. Nanocrystal can be considered as aggregation of randomly oriented ultra-fine particles. The structural model of nanocrystal proposed here is assembled 12 particles in a hexagonal arrangement with free boundary condition. In this model, the form of a unit grain is circle and the grain size was assumed to be constant for simplicity. The grain size was varied to examine the stability of nanocrystal as a function of the grain size. Using the above structures as the

Mat. Res. Soc. Symp. Proc. Vol. 457 ° 1997 Materials Research Society

input data, structural relaxation was performed by the molecular dynamics method. Lennard-Jones potential,

$$\phi(r) = 4\varepsilon \left[(\sigma/r)^{12} - (\sigma/r)^{6} \right] \ , \tag{1}$$

was assumed as the atomic interaction. The cut off length of the potential was 2.5 times of the nearest neighbor distance. In this paper the distance is measured by the nearest neighbor distance. Temperature is expressed in dimensionless unit as

$$T^{*} = k_{B}T \ / \ \varepsilon \tag{2}$$

The melting temperature of the densely packed lattice of the L-J system is about 0.48 in this unit[5]. Varying the grain size and the system temperature, the two types of the model structure of nanocrystal were relaxed by the molecular dynamics method and thermal stability of nanocrystal was examined. The total MD time steps were 1,000,000, nearly equal to 10,000 periods of atomic vibration.

RESULTS AND DISCUSSION

Figure 1 shows the initial and final structure of the nanocrystal with the diameter 6 at T^{*} = 0.05. The 12 fine-particles were relaxed into a single crystal with some vacancies. The model with the diameter 10 also changed into a single crystal although the relaxation time becomes longer.

(a)   (b)

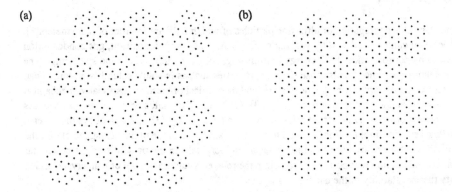

Figure 1. (a) The initial and (b) final structures of the nanocrystal with the diameter 6 at T^{*} = 0.05.

If the diameter of the grain of the nanocrystal becomes 14, the individual particles are stable as shown in figure 2 and the model structure of nanocrystal with the diameter 14 was obtained. During the course of the simulation, each particle was relaxed to produce stable grain boundary structure between the particles. The nanocrystal model with the diameter 16 was also stable at T^{*}

= 0.05 as shown in figure 3(a). Then the thermal stability of this sample was examined by the similar calculation at T^* = 0.10, 0.15 and 0.20 and the obtained structures are shown in figures 3(b), (c) and (d). At T^* = 0.10, the nanocrystal structure was stable but the orientation of one of the particles changed to make more stable interface. The rearrangement of the atom in the particle occurred at the grain boundary but systematic rotation of the particle could not be observed. At T^* = 0.15 and 0.20, the nanocrystal structure was not stable any more. The structure changed into single crystal structure by the same mechanism as the case of T^* = 0.10.

(a)   (b)

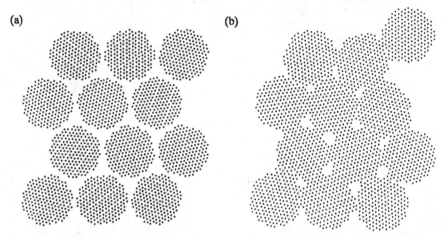

Figure 2. (a) The initial and (b) final structures of the nanocrystal with the diameter 14 at T^* = 0.05.

In summary, these simulations clarified the conditions to obtain stable nanocrystal structure; (1) the grain size is larger than 14 nearest neighbor distance (about 4 nm) and (2) the temperature is less than 0.10 (about 20% of the melting point of the bulk). From the simulation of the coarsening of grains at higher temperature, the hypotheses that large entropy stabilize the nanocrystal structure could not be confirmed.

CONCLUSIONS

Structural characteristics of nanocrystals were investigated by molecular dynamics simulation. The conditions to obtain stable nanocrystal structure were clarified. Many experiments were performed on the mechanical properties of nanocrystal and confirmed the abnormal response to the applied stress. Therefore the structural relaxation of nanocrystal under applied stress is crucial to understand the mechanical properties of nanocrystal. Such investigation is now on progress.

ACKNOWLEDGMENTS

The authors would like to express their thanks to the financial support by Proposal - Based Advanced Industrial Technology R & D program, NEDO.

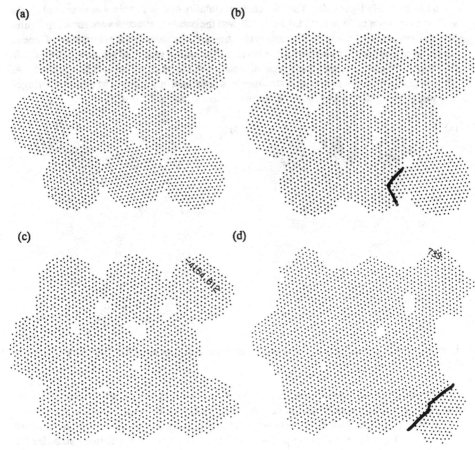

Figure 3. The final structures of the nanocrystal with the diameter 16 at (a)T^* = 0.05, (b)T^* = 0.10, (c)T^* = 0.15 and (d)T^* = 0.20.

REFERENCES

1. H. Gleiter, Progress in Materials Science, **33**, 223 (1989).

2. S. Okuda and F. Tang, Nanostructured Materials, **6**, 585 (1995).

3. F. Tang and S. Okuda, submitted to Mater. Trans. JIM

4. J. Lian and B. Baudelet, Nanostructured Materials, **2**, 415 (1993).

5. B.W.Dodson and P.A.Taylor; Phys. Rev. **B34**, p.2112 (1986).

AUTHOR INDEX

SUBJECT INDEX

nanosized
  fine droplets, 89
  particles in glass, 363
nanostructured iron oxide, 75
nanostructures, 161, 303
Ni-20%Fe, 199
nitride precursors, 213
noble metal, 161
nonequilibrium grain boundaries, 113
nonlinear properties, 143

O-implanted Al, 249
optical
  absorption, 363
    spectra, 363
  filter, 387
  spectra, 105
  thin film, 387
optically transmissive, 237
ordered composites, 501
organic-inorganic nanocomposites, 519
oxidation, 261
oxygen nonstoichiometry, 99

$PbMg_{1/3}Nb_{2/3}O_3$, 39
$PbSc_{1/2}Ta_{1/2}O_3$, 39
percolation threshold, 475
perovskite, 375
phase
  stability, 375
  transformations, 335
phosphors, 105
photo(-)
  active matrix, 425
  catalytic materials, 323
photocatalysts for decomposition
  of water, 323
phthalocyanine, 525
plasma absorption band, 363
plastic, 193
poly(bis-(2(2-methoxyethoxy)ethoxy)
  phosphazene), 489
polycrystalline, 173
polyetherimide, 513
polymerizable complex (PC) technique, 323
polystyrene, 495
porous, 173
  media, 431
  silicon, 161
powders, 131
precipitation-hardened Al(O) alloy layers,
  249
preferential interface reduction, 63
preparation, 33
processing, 381
Pt nanoparticles, 425
Pt/TiO_2 nanocomposite, 425
pulsed laser deposition, 249, 419

Raman scattering, 39

rapid
  consolidation, 347
  heating rate, 347
  physical vapor deposition, 125
recycle, 255
reductive deposition, 161
relaxor ferroelectrics, 39
reorientation, 179
resistance heating, 347
rf plasma torch, 219
ribbons, 279
room temperature ductility, 291
ruthenate, 381

saturation magnetization, 81, 231
scattering, 89
self(-)
  assembled
    functional molecules, 525
    monolayers, 137
    organized growth, 155
  assembly, 519
SEM, 161
short range order, 285
silica, 33
  gels, 463
silicon, 439
  nitride, 187
    composites, 407
silver nanocrystals, 137
sintering, 33, 179
Si-O-Al bonding, 519
sliding wear, 303
small(-)
  angle neutron scattering, 231
  grain sized, 173
soft chemistry, 533
sol-gel, 51, 341, 451
solid
  electrolytes, 489
  sols of silver, 469
  solutions, 297
solidification/melting diagrams, 463
sorption isobars, isosters, and isotherms,
  463
space charge, 369
specific heat, 463
spherule size, 167
sputtering, 149
strength, 279
structural transitions, 309
structure, 205
sulfating, 51
sulphides, 387
superhard, 407
superlattice packing, 137
surface melting, 167
surfactants, 533
suspension, 33
synchrotron, 161
synthesis, 173

Printed in the United States
By Bookmasters